Applied Mathematical Sciences
Volume 139

Springer
New York
Berlin
Heidelberg
Barcelona
Hong Kong
London
Milan
Paris
Singapore
Tokyo

Applied Mathematical Sciences

(continued following index)

Catherine Sulem   Pierre-Louis Sulem

The Nonlinear Schrödinger Equation

Self-Focusing and Wave Collapse

Springer

Catherine Sulem
Department of Mathematics
University of Toronto
Toronto, Ontario, M5S 3G3
Canada
sulem@math.toronto.edu

Pierre-Louis Sulem
CNRS UMR 6529
Observatoire de la Côte d'Azur
Bd. de l'Observatoire, BP 4229
06304 Nice Cedex 4
France
sulem@obs-nice.fr

Editors

J.E. Marsden
Control and Dynamical Systems, 107-81
California Institute of Technology
Pasadena, CA 91125
USA

L. Sirovich
Division of Applied Mathematics
Brown University
Providence, RI 02912
USA

Mathematics Subject Classification (1991): 35Q55, 76B15, 76D33, 78A60, 82D10

With 9 figures.

Library of Congress Cataloging-in-Publication Data
Sulem, C. (Catherine), 1955–
    The nonlinear Schrödinger equation: self-focusing and wave collapse/
Catherine Sulem, Pierre-Louis Sulem.
        p.   cm. — (Applied mathematical sciences; 139)
    Includes bibliographical references and index.
    ISBN 0-387-98611-1 (alk. paper)
    1. Schrödinger equation.   2. Nonlinear theories.   I. Sulem, P.L.
II. Title.   III. Series: Applied mathematical sciences (Springer-
Verlag New York Inc.); v. 139.
QC174.26.W28S85   1999
530.12′4—dc21   98-53840

Printed on acid-free paper.

Production managed by Frank McGuckin; manufacturing supervised by Nancy Wu.
Photocomposed copy prepared from the authors' TEX files.
Printed and bound by Maple-Vail Book Manufacturing Group, York, PA.
Printed in the United States of America.

9 8 7 6 5 4 3 2 1

ISBN 0-387-98611-1 Springer-Verlag New York Berlin Heidelberg   SPIN 10689424

Preface

The nonlinear Schrödinger (NLS) equation provides a canonical description for the envelope dynamics of a quasi-monochromatic plane wave (the carrying wave) propagating in a weakly nonlinear dispersive medium when dissipative processes are negligible. On short times and small propagation distances, the dynamics are linear, but cumulative nonlinear interactions result in a significant modulation of the wave amplitude on large spatial and temporal scales. The NLS equation expresses how the linear dispersion relation is affected by the thickening of the spectral lines associated to the modulation and the resonant nonlinear interactions. In optics, it can also be viewed as the extension to nonlinear media of the paraxial approximation, extensively used for linear waves propagating in random media.

The NLS equation assumes weak nonlinearities but a finite dispersion at the scale of the carrying wave, while in situations where both dispersion and nonlinearities are equally weak, a "reductive perturbative expansion" leads to long-wavelength equations, like the Korteweg–de Vries, the Benjamin–Ono or, in several dimensions, the Kadomtsev–Petviashvili equations (Segur 1978, Ablowitz and Segur 1981). This class of equations also includes the so-called derivative nonlinear Schrödinger (DNLS) equation obeyed by dispersive Alfvén waves propagating along an ambient magnetic field in a quasi-neutral plasma, because of a phase-velocity degeneracy in the dispersionless limit (see Mjølhus and Hada 1997 for a recent review).

The name "NLS equation" originates from a formal analogy with the Schrödinger equation of quantum mechanics. In this context a nonlinear potential (that is nonlocal in the Hartree description) arises in the "mean field" description of interacting particles. When the NLS equation is considered in the wave context, the second-order linear operator, which describes the dispersion and diffraction of the wave-packet, is however not necessarily elliptic, and the nonlinearity arises from the sensitivity of the refractive index to the medium on the wave amplitude.

Special attention is paid in this monograph to the "elliptic" NLS equation, which when written in a frame moving at the group velocity of the carrying wave takes the simple form

$$i\frac{\partial \psi}{\partial t} + \Delta \psi + g|\psi|^2 \psi = 0,$$

with an attractive ($g = 1$) or repulsive ($g = -1$) nonlinearity. This equation together with its generalization to arbitrary power-law nonlinearities $g|\psi|^{2\sigma}\psi$ is addressed in the three first parts of the monograph. The two other parts deal respectively with situations involving a coupling to a mean field that adiabatically follows the carrying wave modulation, and to low-frequency acoustic waves driven by the modulation.

A first question related to the NLS equation concerns the linear stability of a solution that is uniform in space and oscillatory in time. At the level of the original problem, this corresponds to the effect of a slow modulation on a monochromatic wave whose frequency is slightly shifted by the nonlinearity. It is straightforward to show that the elliptic NLS equation is "modulationally" unstable (Benjamin–Feir instability) when $g = +1$ and stable when $g = -1$. Note that some authors restrict the definition of the "modulational instability" to the case of perturbations in the direction of propagation and refer to the "filamentation instability" when the perturbations are in the transverse directions. We shall not follow this restriction here.

The filamentation instability is easily interpreted in the context of nonlinear optics, where in the usual situations, the wave modulation is essentially time-independent. In the NLS equation, the variable t then refers to the coordinate along the beam, and the Laplacian is taken relatively to the directions transverse to the propagation. For $g = +1$, the refractive index increases with the wave intensity, leading to a convergence of light rays from neighborhood sites towards the region of higher amplitude. This in turn further increases the refractive index and leads to more light convergence, resulting in a *self-focusing* of the wave at this location.

The phenomenon of wave-packet contraction can also occur in a one-dimensional setting. The nonlinear development of the modulational instability depends, however, strongly on the space dimension. When the modulation is purely one-dimensional, it leads to the formation of solitonic structures resulting from an exact balance between the dispersive and nonlinear effects. In higher dimensions, in contrast, the nonlinearity dominates when the initial conditions are large enough in a suitable norm, resulting in a *blowup* of the wave amplitude, if additional physical effects like dissipation do not intervene to arrest the process. Since a spatial contraction of the wave packet takes place together with the amplitude blowup, the phenomenon is often called *wave collapse* in the physical literature. This is a basic mechanism to produce a transfer of energy from large to small scales, thus permitting dissipative processes to act and to heat the medium, with possible degradation of the material in the case of a dielectric. In plasmas, the collapsing structures, often called "collapsons," will act as sinks for the wave energy. This phenomenon competes with the more gradual energy transfer to small scales resulting from resonant wave interactions (wave turbulence).

This book mainly reviews the phenomenon of wave collapse, described by the NLS equation or by its generalizations involving coupling to other fields. Several approaches, ranging from rigorous mathematical analysis to formal asymptotic expansions and numerical simulations, are presented, mostly in the case of localized solutions vanishing at infinity. The important issue of the elliptic NLS equation with repulsive nonlinearity for solutions keeping a finite amplitude at infinity is only briefly mentioned. This equation also appears in the description of a Bose condensate, a context where it is often called the Gross–Pitaevskii equation. It admits solutions in the form of coherent structures like vortices that define states that can be excited in superfluid helium.

The validity of the NLS equation breaks down near collapse where the underlying assumptions of small amplitude and large-scale modulation (compared to the frequency and the wavelength of the carrier) no longer hold. The attention paid to the nature of the singularity is, however, not motivated only by interest in the mathematical properties of a fundamental equation of nonlinear physics. The features of the singularity strongly affect the physical processes, even when the collapse is eventually arrested by additional effects that become relevant after sufficiently small scales have been formed.

The book includes several parts that cover various aspects of the problem, in an attempt to put in perspective the rigorous theory of the NLS equation and the physical understanding of the wave-collapse phenomenon.

Part I is an introduction to the physics of quasi-monochromatic waves. Various equivalent derivations of the NLS equation used in the literature are reviewed. An illustration is given in the context of optical waves in a Kerr medium. In usual situations, the duration of a laser pulse is long compared to the carrying wave period, and the modulation can be viewed as stationary. As already mentioned, the amplitude dynamics are then governed by the two-dimensional NLS equation. In contrast, for the ultrashort pulses emitted by power lasers, "normal" or "anomalous" time dispersion effects come into play. The Benjamin–Feir, or modulational, instability is also reviewed, together with the existence of solitons in one space dimension and their instability relative to transverse perturbations. The formal analogy between the NLS equation and the equations of hydrodynamics, which is useful in the context of superfluids, is mentioned. The variational formulation (Lagrangian and Hamiltonian) of the problem, which through the Noether theorem leads to important conservation laws, is presented together with the "variance identity," which is the basic estimate for proving finite-time blowup.

Part II is devoted to a survey of rigorous results on the NLS equation mostly in the case of localized solutions vanishing at infinity. Existence properties are stated briefly, with reference to the original papers or to more mathematically oriented reviews for the details of the proofs. Standing-wave solutions (also called solitary waves) and their stability/instability are also discussed. Special attention is then paid to blowup solutions. Important estimates leading to a quantitative characterization of the collapse are established in detail, in particular at the "critical dimension" ($d_{\mathrm{cr}} = 2$ for the usual cubic NLS equation) where the phenomenon of L^2-norm (or mass) concentration near collapse gives a rigorous basis to the concept of "strong collapse" used in the physical literature.

Part III discusses numerical simulations of blowing up solutions and presents an asymptotic construction of collapsing solutions. In supercritical dimensions, the blowup is self-similar, and a detailed analysis of the profile of the solution is given. At critical dimension, such solutions are not possible, and self-similarity appears to be weakly broken. A delicate asymptotic analysis is presented to characterize the critical blowup, a problem that remained a challenge for more than twenty years but is now well understood. The effects of perturbations that can modify or even arrest the collapse are also considered. They include dissipation, normal dispersion, and saturated nonlinearities.

Part IV considers situations where as in the water-wave problem, a mean field is driven by the amplitude modulation, as a result of quadratic nonlinearities in the primitive equations. The multiple-scale formalism is first presented for a general nonlinear scalar wave equation and then illustrated on a few model equations. An example of a degenerate situation is also discussed. The water-wave problem, which presents some technical difficulties due to the integro-differential character of the primitive equations, is then reviewed and the accuracy of the modulation analysis rigorously estimated. The resulting Davey–Stewartson system governing the coupling of the wave amplitude with the mean field are analyzed in the various regimes which establish according to the parameters of the medium.

Part V addresses situations where the wave modulation drives low-frequency acoustic waves, which can strongly affect the collapse dynamics. Several examples are presented. We first consider Langmuir oscillations, which play a central role in the dynamics of a quasi-neutral plasma. Self-focusing leads to the formation of collapsing density cavities where the oscillating electric field is trapped and where dissipative processes produce a local heating of the plasma. The Zakharov equations governing these dynamics are derived and their properties analyzed. Recent mathematical results for the "scalar model" are reviewed. They include conditions for blowup in a finite or infinite time and also properties of exact self-similar solutions in two dimensions. Examples of progressive waves are also considered for laser beams in plasmas and in the context of the filamentation (transverse collapse) of Alfvén waves propagating along an ambient magnetic field. It turns out that this phenomenon is very sensitive to the coupling to the magnetosonic waves that develop sharp fronts, creating conditions for the onset of strong magnetohydrodynamic turbulence.

The present survey is by no means comprehensive, and we list here a few reviews published in the field. A basic reference on nonlinear waves is the book of Whitham (1974). Asymptotic methods for nonlinear waves in various physical contexts are described in Infeld and Rowlands (1990). A detailed discussion of the physical aspects of the self-focusing phenomenon is found in Vlasov and Talanov (1997). Basic properties of the NLS equation are discussed by Rasmussen and Rypdal (1986). A detailed review of the mathematical theory of the NLS equation is given by Cazenave (1989, 1994), Strauss (1989), and Ginibre (1998) who concentrate on the well-posedness of the Cauchy problem and on the long-time behavior of global solutions. The asymptotic analysis of the collapse is reviewed in Landman, LeMesurier, Papanicolaou, Sulem, and Sulem (1989) and more recently in Sulem and Sulem (1997). A unified presentation of the effect of perturbations on critical collapse is found in Fibich and Papanicolaou (1997b). We refer to Newell and Moloney (1992) for an introduction to nonlinear optics, and to Dyachenko, Newell, Pushkarev, and Zakharov (1992) for a discussion of the two-dimensional cubic NLS equations, including wave turbulence. A recent review of the phenomenon of self-trapping of optical beams in self-focusing media is presented in Segev and Stegeman (1998), while transverse instabilities of solitary waves are discussed in Kivshar and Pelinovsky (1999). The physics of Langmuir collapse is extensively discussed in Zakharov (1984). Other reviews on related topics include Thornhill and ter Haar (1978), Rudakov and Tsytovich (1978), ter Haar and Tsytovich (1981), Goldman (1984), Zakharov, Musher, and Rubenchik (1985), Robinson (1997), Bergé (1998). Various aspects of the nonlinear dynamics of dispersive waves are discussed in the conference proceedings edited by Balabane, Lochak, and Sulem (1989).

We conclude this description by stressing the dynamic relevance of singularity formation. Julia's sentence (quoted by L. Garding, T. Kotake and J. Leray, Bull. Soc. Math. France, **92**, 263, 1964) that we put in epigraph, expresses in a somewhat provocative way the importance of an effect which is sometimes considered as "unphysical." Singularity formation reveals drastic changes in the magnitude and the typical scales of the solution, although in realistic situations other couplings intervene near collapse and act as small-scale mollifiers. From a physical point of view, various situations are, in fact, possible. In some instances, illustrated by the self-focusing of a laser beam, the blowup of NLS solutions reflects the occurrence of violent (although possibly non strictly singular) events at the level of the primitive equations, giving a valuable practical interest to the envelope formalism. In other cases, the NLS singularities only reflect the transition from a weakly to a moderately nonlinear regime. An interesting question then concerns the necessity to return to the primitive equations or the possible existence of an intermediate asymptotics valid beyond the NLS blowup.

The present monograph grew from a suggestion by George Papanicolaou to write a survey on the phenomenon of wave collapse as described by the NLS equation. We acknowledge with gratitude the enlightening collaboration on the nature of the blowup singularities we have had with George and his collaborators B. LeMesurier, M. Landman, and X.P. Wang, during our visits at the Courant Institute. Our work with W. Craig laid the basis of our description of the water-wave problem. Our discussion of the coupling to low-frequency acoustic type waves is closely linked to joint work with T. Passot, S. Champeaux, and A. Gazol on the propagation of Alfvén waves in a magnetized plasma. We thank all of them for these fruitful collaborations that provided the backbone of this book. We are very grateful to L. Bergé, V.S. Buslaev, J. Coleman, V. Dougalis, G. Fibich, N. Koppel, E. Kuznetsov, V. Malkin, D. McLaughlin, H. Nawa, A. Newell, D. Pelinovsky, G. Ponce, J.J. Rasmussen, J.C. Saut, J. Shatah, I.M. Sigal, A. Soffer, W. Strauss, M. Weinstein, J. Xin, and V.E. Zakharov for very useful discussions or correspondence from which we have benefited throughout the years, or for detailed comments and suggestions on earlier drafts of the manuscript. We thank all our colleagues who gave us copies of their papers before publication. We are also grateful to Achi Dosanjh, Frank McGuckin, and the editorial and production staff at Springer-Verlag (New-York) for their amiable and efficient help in preparing the book for publication. We shall maintain a list of errata and corrections that will be available through the world wide web address http://www.obs-nice.fr/sulem/. This work benefited from partial support from NSERC operating grant OGP0046179.

Toronto   *C. Sulem*
Nice   *P.L. Sulem*

March 1999

Contents

V   Coupling to Acoustic Waves   243

Part I

Basic Framework

1
The Physical Context

1.1 Weakly Nonlinear Dispersive Waves

1.1.1 A Weakly Nonlinear Dispersion Relation

The nonlinear Schrödinger (NLS) equation arises in various physical contexts in the description of nonlinear waves (Whitham 1974) such as propagation of a laser beam in a medium whose index of refraction is sensitive to the wave amplitude, water waves at the free surface of an ideal fluid, and plasma waves. It provides a canonical description of the envelope dynamics of a scalar dispersive wave train $\epsilon\,\psi\,e^{i(\mathbf{k}\cdot\mathbf{x}-\omega t)}$ with a small ($\epsilon \ll 1$) but finite amplitude, slowly modulated in space and time, propagating in a conservative system (Benney and Newell 1967, Newell 1974).

Let us consider a scalar nonlinear wave equation written symbolically

$$L(\partial_t, \nabla)u + G(u) = 0, \tag{1.1.1}$$

where L is a linear operator with constant coefficients and G a nonlinear function of u and of its derivatives. For a small-amplitude solution of magnitude $\epsilon \ll 1$, the nonlinear effects can first be neglected, and the equation admits approximate monochromatic wave solutions

$$u = \epsilon\psi e^{i(\mathbf{k}\cdot\mathbf{x}-\omega t)} \tag{1.1.2}$$

with a constant amplitude $\epsilon\psi$. The frequency ω and the wave vector $\mathbf{k}$ are real quantities related by the dispersion relation

$$L(-i\omega, i\mathbf{k}) = 0. \tag{1.1.3}$$

This (algebraic) equation may admit several solutions. We concentrate on one of them:

$$\omega = \omega(\mathbf{k}). \qquad (1.1.4)$$

Although we assume a small-amplitude wave, cumulated nonlinear effects become significant on long time scales and large propagation distances. A regular perturbative calculation of the solution of (1.1.1) about the monochromatic solution (1.1.2) leads indeed to resonant terms in the hierarchy of equations arising at the different orders, which results in secular terms in the perturbative expansion of the solution. A classical approach to such a situation is provided by multiple-scale expansion methods where the complex amplitude of the "carrying wave" (also called carrier) is no longer constant. It depends on the slow variables $T = \epsilon t$ and $\mathbf{X} = \epsilon \mathbf{x}$, and its evolution is prescribed by solvability conditions that eliminate the resonances. This approach is illustrated on a simple example in Section 1.1.3 and then in the physically important context of laser beam propagation in a nonlinear medium where the index of refraction varies linearly with the wave intensity (Kerr medium) in Section 1.2.3. A more general framework involving the driving of a mean field by the modulation is considered in Chapter 10. Situations involving the coupling to low-frequency acoustic waves are addressed in Part V.

A simple heuristic argument can explain the canonical character of the NLS equation, viewed as the expansion at the lowest nontrivial order of a generalized weakly nonlinear dispersion relation (Kadomtsev and Karpman 1971, Kadomtsev 1979, Zakharov, Muster, and Rubenchik 1985, Craik 1985). For this purpose, it is convenient to reinterpret the linear dispersion relation (1.1.4) in the form

$$(i\partial_t - \omega(-i\partial_{\mathbf{X}}))\psi e^{i(\mathbf{k}\cdot\mathbf{x} - \omega t)} = 0, \qquad (1.1.5)$$

where $\partial_{\mathbf{X}}$ is the gradient with respect to $\mathbf{x}$ and $\omega(-i\partial_{\mathbf{X}})$ the pseudo-differential operator obtained by replacing $\mathbf{k}$ by $-i\partial_{\mathbf{X}}$ in $\omega(\mathbf{k})$.

In a weakly nonlinear medium responding adiabatically (i.e., instantaneously) to a finite wave amplitude, the nonlinearity is expected to affect the dispersion relation of the carrying wave (in addition to the generation of harmonics of smaller amplitude). The frequency of the wave becomes dependent of the intensity,[1] which leads to replace the frequency $\omega(\mathbf{k})$ by a function $\Omega(\mathbf{k}, \epsilon^2|\psi|^2)$ with $\Omega(\mathbf{k}, 0) = \omega(\mathbf{k})$. Furthermore, the (complex) wave amplitude ψ is no longer a constant but is modulated in space and time, thus depending on the slow variables $\mathbf{X} = \epsilon \mathbf{x}$ and $T = \epsilon t$. In (1.1.5), the derivatives ∂_t and $\partial_{\mathbf{X}}$ are thus to be replaced by $\partial_t + \epsilon \partial_T$ and $\partial_{\mathbf{X}} + \epsilon \nabla$,

---

[1] For the sake of simplicity, we do not consider here the onset of mean fields or low-frequency waves that in the presence of quadratic nonlinearities, may result from the amplitude modulation (see Parts IV and V).

where ∇ now denotes the gradient with respect to the slow spatial variable $\mathbf{X}$. As a consequence, (1.1.5) is replaced by

$$[i\partial_t + i\epsilon\partial_T - \Omega(-i\partial_{\mathbf{x}} - i\epsilon\nabla, \epsilon^2|\psi|^2)]\psi e^{i(\mathbf{k}\cdot\mathbf{x}-\omega t)} = 0. \qquad (1.1.6)$$

Equivalently, one has the weakly nonlinear dispersion relation

$$[\omega + i\epsilon\partial_T - \Omega(\mathbf{k} - i\epsilon\nabla, \epsilon^2|\psi|^2)]\psi = 0. \qquad (1.1.7)$$

The parameter ϵ being small, this equation is expanded to second order in powers of ϵ about the linear dispersion relation obeyed by the carrying wave, in the form

$$i(\partial_T + \mathbf{v}_g \cdot \nabla)\psi + \epsilon\{\nabla \cdot (\mathcal{D}\nabla\psi) + \gamma|\psi|^2)\psi\} = 0, \qquad (1.1.8)$$

where $\mathbf{v}_g = \nabla_{\mathbf{k}}\omega$ is the group velocity and $\mathcal{D} = (\frac{1}{2}\frac{\partial\omega}{\partial k_j \partial k_\ell})$, with $j, \ell = 1, \cdots d$, is defined as half the Hessian matrix of the frequency, both evaluated at the carrier wavevector $\mathbf{k} \in \mathbf{R}^d$, in the absence of nonlinearities. The coupling coefficient γ is associated to the expansion in powers of the wave intensity and is given by $\frac{\partial\Omega}{\partial(|\psi|^2)}$ evaluated at $|\psi|^2 = 0$ and at the carrier wave vector $\mathbf{k}$.

When (1.1.8) is viewed as an initial value problem in time[2] (absolute dynamics), this equation is conveniently written in a reference frame moving at the group velocity by defining $\boldsymbol{\xi} = \mathbf{X} - T\mathbf{v}_g$. Rescaling time in the form $\tau = \epsilon T$, we get the NLS equation

$$i\frac{\partial\psi}{\partial\tau} + \nabla \cdot (\mathcal{D}\nabla\psi) + \gamma|\psi|^2\psi = 0, \qquad (1.1.9)$$

where the spatial derivatives are now taken with respect to the $\boldsymbol{\xi}$ variable.

It is easily checked that the operator $D = \nabla \cdot (\mathcal{D}\nabla)$ simplifies in isotropic media where the frequency depends only on the modulus of the wave vector. Assuming the carrier wave vector in the x-direction, the operator D can be rewritten $D = \frac{\omega''}{2}\partial_{XX} + \frac{\omega'}{2k}\Delta_\perp$, where ω' and ω'' denote the first and second derivatives of the frequency ω with respect to the wave number k, and $\Delta_\perp$ the Laplacian with respect to the directions transverse to the propagation. In this form, the first term refers to the group velocity dispersion and the second one to the wave diffraction.

The NLS equation is often called *elliptic* when the (real symmetric) operator $D = \nabla \cdot (\mathcal{D}\nabla)$ is so. The elliptic NLS equation arises also in other contexts. In quantum mechanics, it is obtained in localizing the potential of the Hartree equation (Pitaevskii 1961). In chemistry, it appears as a continuous-limit model for mesoscopic molecular structures (Gaididei, Rasmussen and Christiansen 1995).

---

[2]In optics, where the dispersion relation is viewed as prescribing the wavenumber in terms of the frequency, (1.1.8) is usually seen as an initial value problem along the direction defined by the group velocity (convective dynamics). This case is considered in Section 1.2.

1.1.2 Derivation in Terms of Fourier-Mode Coupling

A related derivation of the NLS equation often given in the plasma physics literature (see, e.g., Zakharov, L'vov, and Falkovich 1992) is based on Fourier-mode coupling formalism where the modulation corresponds to a broadening of spectra in frequencies and wave vectors, about the carrier values ω and $\mathbf{k}$. Considering a wave vector $\mathbf{K}$ close to $\mathbf{k}$ and defining $\boldsymbol{\kappa} = \mathbf{K} - \mathbf{k}$ with $|\boldsymbol{\kappa}| \ll k = |\mathbf{k}|$, the corresponding frequency for linear waves is approximated by

$$\omega(\mathbf{K}) \approx \omega(\mathbf{k}) + \mathbf{v}_g \cdot \boldsymbol{\kappa} + \frac{1}{2}\omega_{jl}''\kappa_j\kappa_l. \tag{1.1.10}$$

In the linear theory, the associated Fourier mode $c(\mathbf{K}, t)$ is proportional to $e^{-i\omega(\mathbf{K})t}$ and thus obeys the equation

$$\partial_t c(\mathbf{K}, t) + i\omega(\mathbf{K})c(\mathbf{K}, t) = 0 \tag{1.1.11}$$

where $\omega(\mathbf{K})$ is given by (1.1.10). The nonlinearity contributes dominantly through four-wave interactions where approximately equal wave vectors are involved. This leads to replace in the nonlinear regime (1.1.11) by

$$\partial_t c(\mathbf{K}, t) + i\omega(\mathbf{K})c(\mathbf{K}, t)$$

$$= i\int \mathcal{G}_{\mathbf{K},\mathbf{K}_1,\mathbf{K}_2,\mathbf{K}_3} c^*(\mathbf{K}_1)c(\mathbf{K}_2)c(\mathbf{K}_3)\delta(\mathbf{K} + \mathbf{K}_1 - \mathbf{K}_2 - \mathbf{K}_3)d\mathbf{K}_{123}.$$

$$\tag{1.1.12}$$

where we used the notation $d\mathbf{K}_{123} = d\mathbf{K}_1\, d\mathbf{K}_2\, d\mathbf{K}_3$. For a narrow wave packet centered at the wave vector $\mathbf{k}$, the kernel $\mathcal{G}$ (assumed to be continuous when all the arguments tend to $\mathbf{k}$) is approximated by

$$\mathcal{G}_{\mathbf{K},\mathbf{K}_1,\mathbf{K}_2,\mathbf{K}_3} \approx \mathcal{G}_{\mathbf{k},\mathbf{k},\mathbf{k},\mathbf{k}} \equiv \frac{\gamma}{(2\pi)^3}. \tag{1.1.13}$$

This yields

$$\partial_t c(\mathbf{K}) + i\left(\omega(\mathbf{k}) + \mathbf{v}_g \cdot \boldsymbol{\kappa} + \frac{1}{2}\omega_{jl}''\kappa_j\kappa_l\right)c(\mathbf{K})$$

$$- \frac{i\gamma}{(2\pi)^3}\int c^*(\mathbf{K}_1)c(\mathbf{K}_2)c(\mathbf{K}_3)\delta(\mathbf{K} + \mathbf{K}_1 - \mathbf{K}_2 - \mathbf{K}_3)d\mathbf{K}_{123} = 0.$$

$$\tag{1.1.14}$$

For fixed $\mathbf{k}$, one then defines $\widehat{u}(\boldsymbol{\kappa}, t) = e^{i\omega(\mathbf{k})t}c(\mathbf{k} + \boldsymbol{\kappa}, t)$, which is slowly varying in the t-variable and obeys

$$\partial_t \widehat{u}(\boldsymbol{\kappa}) + i(\mathbf{v}_g \cdot \boldsymbol{\kappa})\widehat{u}(\boldsymbol{\kappa}) + \frac{i}{2}\omega_{jl}''\kappa_j\kappa_l\widehat{u}(\boldsymbol{\kappa})$$

$$- \frac{i\gamma}{(2\pi)^3}\int \widehat{u}^*(\boldsymbol{\kappa}_1)\widehat{u}(\boldsymbol{\kappa}_2)\widehat{u}(\boldsymbol{\kappa}_3)\delta(\boldsymbol{\kappa} + \boldsymbol{\kappa}_1 - \boldsymbol{\kappa}_2 - \boldsymbol{\kappa}_3)d\boldsymbol{\kappa}_{123} = 0.$$

$$\tag{1.1.15}$$

As a consequence, the complex field

$$u(\mathbf{x}, t) = (2\pi)^{-3/2} \int \widehat{u}(\boldsymbol{\kappa}, t) e^{i\boldsymbol{\kappa} \cdot \mathbf{x}} d\boldsymbol{\kappa}$$

$$= (2\pi)^{-3/2} e^{i\omega(\mathbf{k})t} \int c(\mathbf{k} + \boldsymbol{\kappa}, t) e^{i\boldsymbol{\kappa} \cdot \mathbf{x}} d\boldsymbol{\kappa}$$

$$= (2\pi)^{-3/2} e^{-i(\mathbf{k} \cdot \mathbf{x} - \omega(\mathbf{k})t)} \int c(\mathbf{K}, t) e^{i\mathbf{K} \cdot \mathbf{x}} d\mathbf{K}, \qquad (1.1.16)$$

which appears as the envelope of the wave packet $(2\pi)^{-3/2} \int c(\mathbf{K}, t) e^{i\mathbf{K} \cdot \mathbf{x}} d\mathbf{K}$, obeys the NLS equation

$$\partial_t u + \mathbf{v}_g \cdot \nabla u - \frac{i}{2}\omega_{ij}'' \partial_i \partial_j u - i\gamma |u|^2 u = 0. \qquad (1.1.17)$$

written in the primitive physical variables. Since the wave amplitude is small, the function u varies slowly in space and time. The simple rescaling $u(\mathbf{x}, t) = \epsilon\psi(\mathbf{X}, T)$ with $\mathbf{X} = \epsilon\mathbf{x}$ and $T = \epsilon t$ then reproduces (1.1.8). An equivalent approach based on the Hamiltonian character of the problem is given in Section 15.1.2 in a more general case involving the coupling to low-frequency acoustic waves.

1.1.3   Introduction to Multiple-Scale Analysis

As a simple illustration of the multiple-scale method (Dodd, Eilbeck, Gibbon, and Morris 1982, Newell 1985), which provides a systematic derivation of envelope equations, we consider the example of the Klein–Gordon equation in three space dimensions

$$u_{tt} - \Delta u + \delta u = \lambda u^3, \qquad (1.1.18)$$

where δ is a positive parameter associated to the wave dispersion and λ denotes a real coupling constant. A general discussion of this formalism is given in Section 10.1 in the case where the wave equations include quadratic nonlinearities, which induce a coupling to a mean field driven by the modulations.

In the linear regime, (1.1.18) admits a solution in the form of a monochromatic wave

$$u_0(x, t) = \psi e^{i(kx - \omega t)} + \text{c.c.} \qquad (1.1.19)$$

propagating in a direction taken as the x-axis. Here c.c. stands for the complex conjugate, and the frequency ω is given by the dispersion relation

$$\omega^2 = k^2 + \delta. \qquad (1.1.20)$$

In order to analyze the influence of weak nonlinear effects, we expand the solution in powers of a small parameter ϵ measuring the magnitude of the wave amplitude

$$u = \epsilon(u_0 + \epsilon u_1 + \epsilon^2 u_2 + \cdots). \qquad (1.1.21)$$

As is well known, a regular perturbation expansion would lead to the onset of resonant terms resulting from the cumulative effects of weak nonlinearities on long time or large distances. As in the classical example of the anharmonic oscillator (see, e.g., Bender and Orszag 1978), a multiple-scale analysis is thus needed. For this purpose, we introduce the large-scale variables $X = \epsilon x$, $T = \epsilon t$, and in the case of multidimensional modulation, $Y = \epsilon y$ and $Z = \epsilon z$. Looking at the solution as a function of both the fast and slow variables, the linear operator arising in the Klein–Gordon equation can be rewritten as

$$L = L^{(0)} + \epsilon L^{(1)} + \epsilon^2 L^{(2)} + \cdots, \tag{1.1.22}$$

where the two first contributions are given by

$$L^{(0)} = \frac{\partial^2}{\partial t^2} - \frac{\partial^2}{\partial x^2} + \delta, \tag{1.1.23}$$

$$L^{(1)} = 2\left(\frac{\partial^2}{\partial t \partial T} - \frac{\partial^2}{\partial x \partial X}\right). \tag{1.1.24}$$

At order ϵ, we have

$$L^{(0)} u_0 = 0, \tag{1.1.25}$$

and we recover the solution

$$u_0(x,t) = \psi(X,Y,Z,T)e^{i(kx-\omega t)} + \text{c.c.} \tag{1.1.26}$$

of the linear problem, except that the amplitude now depends on the slow variables.

At order ϵ^2, we have

$$
\begin{aligned}
L^{(0)} u_1 &= -L^{(1)} u_0 \\
&= 2i\left(\omega\frac{\partial}{\partial \tau} + k\frac{\partial}{\partial X}\right)\psi e^{i(kx-\omega t)} + \text{c.c.},
\end{aligned}
\tag{1.1.27}
$$

which requires the solvability condition

$$\frac{\partial \psi}{\partial \tau} + \omega'\frac{\partial \psi}{\partial X} = 0, \tag{1.1.28}$$

where, using the dispersion relation (1.1.20), we noted that $\frac{k}{\omega}$ can be identified with the group velocity $v_g = \omega'$ of the wave packet. On the time scale T, the wave packet is just transported at the group velocity and thus depends only on the variable $\xi = X - v_g T$. We then solve

$$u_1 = 0. \tag{1.1.29}$$

Two equivalent points of view are then possible to describe the effect of the next-order contributions. They can be viewed as corrections to the dynamics taking place on the scales associated to the variables $\mathbf{X}$ and T. This approach used in Chapter 10 may be convenient in problems involving coupling between waves propagating at different velocities. Alternatively, we consider here that, as the result of cumulative effects, the next-order contribution affects the leading-order dynamics on a time scale (or in other instances, as discussed in Remark (i) below, a propagation distance) of order ϵ^{-2}. Defining $\tau = \epsilon^2 t$, we thus write $\psi = \psi(\xi, \tau)$ to indicate that in the frame moving at the group velocity, the amplitude evolves on the time scale associated to the variable $\tau = \epsilon^2 t$.

When this second time-scale is introduced, the operator $L^{(2)}$ in (1.1.22) reads

$$L^{(2)} = \frac{\partial^2}{\partial T^2} + 2\frac{\partial^2}{\partial t \partial \tau} - \frac{\partial^2}{\partial X^2} - \Delta_\perp, \tag{1.1.30}$$

where $\Delta_\perp$ denotes the Laplacian with respect to the (slow) transverse coordinates in the case where the wave is also modulated transversally.

At order ϵ^3, we then have

$$\begin{aligned}
L^{(0)} u_2 &= -L^{(2)} u_0 + \lambda u_0^3 \\
&= -\left(\frac{\partial^2}{\partial T^2} - 2i\omega \frac{\partial}{\partial \tau} - \frac{\partial^2}{\partial X^2} - \Delta_\perp \right) u_0 + \lambda u_0^3 \\
&= \left(2i\omega \frac{\partial}{\partial \tau} + \omega \omega'' \frac{\partial^2}{\partial \xi^2} + \Delta_\perp \right) u_0 + \lambda u_0^3, \tag{1.1.31}
\end{aligned}$$

where again from the dispersion relation, the coefficient $1 - \omega'^2$ can be rewritten as $\omega \omega''$. The solvability condition that eliminates the terms proportional to $e^{i(kx - \omega t)}$ (and its complex conjugate) then reduces to the cubic Schrödinger equation

$$2i\frac{\partial \psi}{\partial \tau} + \omega'' \frac{\partial^2 \psi}{\partial \xi^2} + \alpha \Delta_\perp \psi + \beta |\psi|^2 \psi = 0, \tag{1.1.32}$$

where $\beta = \frac{3\lambda}{\omega}$. The coefficient α, here equal to $\frac{1}{\omega}$, holds for the second-order derivative $\frac{\partial^2 \Omega}{\partial K_y^2} = \frac{\partial^2 \Omega}{\partial K_z^2}$ evaluated at $K_x = k$ and $K_y = K_z = 0$, where Ω solves the three-dimensional dispersion relation $-\Omega^2 + K_x^2 + K_y^2 + K_z^2 + \delta = 0$. Due to the isotropy of the dispersion relation, one has $\alpha = v_g/k$. The corrective term

$$u_2 = \psi_2 e^{3i(kx - \omega t)} + \text{c.c.} \tag{1.1.33}$$

is associated to the third harmonics and its amplitude is computed in terms of that of the carrier as

$$\psi_2 = \frac{\psi^3}{-9\omega^2 + 9k^2 + \delta} = -\frac{\psi^3}{8\delta}. \tag{1.1.34}$$

Note that the amplitude equation (1.1.32), identifies with that associated to the Sine–Gordon equation

$$u_{tt} - \Delta u + q \sin ru = 0, \tag{1.1.35}$$

when $qr = \delta$ and $\frac{qr^3}{6} = \lambda$, since at the order of the expansion, only the cubic nonlinearity is relevant. A rigorous discussion of the validity of the modulation equation in this context for the one-dimensional problem is presented by Kirrmann, Schneider and Mielke (1995).

Remarks

(i) Equation (1.1.32) describes the time evolution of the wave amplitude from an initial modulation $\psi_0(X, Y, Z)$, a situation referred to as *absolute dynamics*. In other instances, as in nonlinear optics (see Section 1.2), one often deals with wave packets for which the modulation can be viewed as stationary (or close to this state), compared to the wave period. In such cases, it is convenient to introduce instead of the variables ξ and τ, the delayed time $\theta = T - \frac{X}{v_g}$ and the rescaled propagation coordinate $\zeta = \epsilon X = \epsilon^2 x$. One obtains the NLS equation for the *convective dynamics*, which in an isotropic medium has the form

$$2i\frac{\partial \psi}{\partial \zeta} - k''\frac{\partial^2 \psi}{\partial \theta^2} + \tilde{\alpha}\Delta_\perp \psi + \tilde{\beta}|\psi|^2\psi = 0, \tag{1.1.36}$$

with the time dispersion, $k'' = \frac{d^2 k}{d\omega^2}$ and the diffraction coefficient $\tilde{\alpha}$ given by $\frac{\partial^2 K_x}{\partial K_y^2}$ or $\frac{\partial^2 K_x}{\partial K_z^2}$ evaluated for $\Omega = \omega$, $K_y = K_z = 0$. In the present example, these coefficients are equal to $-1/k^3$ and $-1/k$, respectively. In this context, the NLS equation is to be viewed as an initial problem in terms of the propagation coordinate. If the θ dependence is slow and the time dispersion thus neglected, the amplitude is governed by the elliptic two-dimensional NLS equation.

(ii) It is important to note that the zero dispersion limit $\delta \to 0$ (leading to $\omega'' = 0$ or $k'' = 0$) cannot be taken in (1.1.32) or (1.1.36), since the amplitude ψ_2 of the third harmonics then becomes infinite, indicating that it can no longer be viewed as "slaved" to the carrying wave. A divergence of the coupling coefficient in the NLS equation can even arise when the primitive wave equation includes quadratic nonlinearities because in this case, a contribution comes from the interaction of the carrying wave with the second harmonics u_1 that are nonzero and diverge in the zero-dispersion limit (see Section 10.2).

(iii) We finally mention that an interpretation of the envelope equation was recently given in terms of the renormalization group method (Matsuba and Nozaki 1997).

1.2   The Example of Optical Waves

This section is devoted to the derivation of the NLS equation in the context of a quasi-monochromatic wave train, usually an intense laser beam, propagating in a weakly nonlinear dielectric. It can be viewed as an extension to weakly nonlinear waves of the paraxial approximation for a linear medium with small fluctuations of the refractive index.

1.2.1 Waves in a Dielectric

The starting point is provided by the Maxwell equations in a dielectric medium, which in the absence of free electric charges and currents read

$$\nabla \cdot \mathbf{D} = 0, \tag{1.2.1}$$

$$\nabla \cdot \mathbf{B} = 0, \tag{1.2.2}$$

$$\nabla \times \mathbf{E} = -\frac{\partial \mathbf{B}}{\partial t}, \tag{1.2.3}$$

$$\nabla \times \mathbf{H} = \frac{\partial \mathbf{D}}{\partial t}. \tag{1.2.4}$$

Here the magnetic and electric inductions $\mathbf{B}$ and $\mathbf{D}$ are given in terms of the magnetic and electric fields $\mathbf{H}$ and $\mathbf{E}$, by the constitutive relations

$$\mathbf{B} = \mu\mathbf{H}, \tag{1.2.5}$$

$$\mathbf{D} = \epsilon_0\mathbf{E} + \mathbf{P}. \tag{1.2.6}$$

The magnetic permeability μ and the vacuum permitivity ϵ_0 are related by $\epsilon_0\mu = c^{-2}$, with c denoting the velocity of light in vacuum, where $\mathbf{P} = 0$. In a dielectric medium, in contrast, the polarization $\mathbf{P}$ describes the macroscopic effect resulting from the displacement of the bound electrons when an electric field is applied.

It follows from (1.2.1)–(1.2.4) that

$$\nabla \times (\nabla \times \mathbf{E}) + \frac{1}{c^2}\partial_{tt}\mathbf{E} = -\frac{1}{c^2\epsilon_0}\partial_{tt}\mathbf{P}, \tag{1.2.7}$$

where one can rewrite

$$\nabla \times (\nabla \times \mathbf{E}) = -\Delta\mathbf{E} + \nabla(\nabla \cdot \mathbf{E}). \tag{1.2.8}$$

The problem will be closed by expressing the polarization $\mathbf{P}$ in terms of the electric field $\mathbf{E}$.

1.2.2 The Paraxial Approximation in a Linear Medium

Assume the medium linear and steady, such that $(\varepsilon \ll 1)$

$$\frac{1}{\epsilon_0}\mathbf{P} = (\chi_0 + \varepsilon^2\chi_1(\mathbf{x}))\mathbf{E}, \tag{1.2.9}$$

where χ_0 is a constant and $\varepsilon^2\chi_1(\mathbf{x})$ describes small spatial fluctuations of the electric susceptibility.

In the absence of fluctuations, the wave equation admits solutions in the form of monochromatic linearly polarized plane waves

$$\mathbf{E}_0(\mathbf{x}, t) = (\mathcal{E}e^{i(kz-\omega t)} + \text{c.c.})\,\mathbf{e}, \tag{1.2.10}$$

where $\mathbf{e}$ is a unit vector perpendicular to the wave vector $\mathbf{k} = (0, 0, k)$ with $k^2 = \frac{\omega^2}{c^2}(1 + \chi_0)$. Small fluctuations $\varepsilon^2 \chi_1$ of the electric susceptibility taking place on scales large compared to the wavelength $2\pi/k$ will only be significant on a propagation distance z of order ε^{-2} in wavelength units. The transverse Laplacian will be relevant for transverse scales of order ϵ^{-1}. This leads one to introduce the slow variables $Z = \varepsilon^2 z$, $X = \varepsilon x$, $Y = \varepsilon y$. One also expands

$$\mathbf{E} = \mathbf{E}_0 + \varepsilon^2 \mathbf{E}_1 + \cdots. \tag{1.2.11}$$

The leading order reproduces the unperturbed problem, except that the complex amplitude $\mathcal{E}$ now depends on the slow variables. At the next order, one gets an inhomogeneous linear equation that, for the existence of periodic solutions with respect to the fast variables requires the solvability condition

$$2ik\partial_Z \mathcal{E} + \Delta_\perp \mathcal{E} + \frac{k^2}{1 + \chi_0}\chi_1 \mathcal{E} = 0, \tag{1.2.12}$$

where $\Delta_\perp$ is the Laplacian transverse to the propagation. This is the parabolic equation, also called the paraxial or forward scattering approximation. Karpman (1975) and Litvak, Petrova, Sergeev, and Yunakovskii (1988) mention that this approximation was first introduced by Leontovich (1944) in analyzing the propagation of an electromagnetic wave along the surface of the earth. It is extensively used in the context of wave propagation in random media (Klyatskin and Tatarskii 1970, Klyatskin 1971, Bailly, Clouet, and Fouque 1996 and references therein). Note that in the linear problem where the refractive index is slowly varying, no harmonics of the carrying wave are generated, and dispersive effects are not required.

1.2.3   Static Modulation in a Kerr Medium

The above analysis extends to a nonlinear medium with a cubic nonlinearity (Kerr medium). Assuming formally that the refractive index fluctuations are given by $\chi_1 = \lambda |\mathbf{E}|^2$, when the third-order harmonics are neglected (Kelley 1965, Talanov 1966), one directly gets (1.2.12) with χ_1 replaced by $\lambda |\mathcal{E}|^2$. This approach is, in fact, questionable since the linear problem being nondispersive, third harmonics also lead to secular terms and cannot be a priori ignored.

Dispersive effects actually exist in the physical problem and originate from the fact that matter does not respond instantaneously to stimulation by light (Ablowitz and Segur 1981, Newell and Moloney 1992). In an isotropic medium (we refer to the latter reference for a discussion of more general situations), we have

$$\mathbf{P}(\mathbf{x}, t) = \epsilon_0 \int_{-\infty}^{+\infty} \chi(t - t')\mathbf{E}(\mathbf{x}, t')dt', \tag{1.2.13}$$

where due to causality, the electric susceptibility $\chi(t - t')$ is equal to 0 for $t - t' < 0$. It also rapidly decreases as $|t - t'| \to \infty$.

For an infinitesimal monochromatic wave $\mathbf{E} = (\mathcal{E}e^{i(kz-\omega t)} + \text{c.c.})\mathbf{e}$, the kernel χ is a given function of t, and one easily gets $\mathbf{P} = \epsilon_0 \widehat{\chi}(\omega)\mathcal{E}e^{i(kz-\omega t)} +$ c.c.) $\mathbf{e}$, where $\widehat{\chi}(\omega)$ stands for the Fourier transform of $\chi(t)$. The wave is then dispersive with a dispersion relation $k^2 = \frac{\omega^2}{c^2}(1 + \widehat{\chi}(\omega))$. The dependence of the electric susceptibility $\widehat{\chi}(\omega)$ in terms of the wave frequency is, in general, weak, except when the latter matches a natural frequency of matter. In the following, we assume that we are far from this regime, and the absorption can then be viewed as negligible (real susceptibility).

When the wave amplitude is not infinitesimal, the polarization depends nonlinearly on $\mathbf{E}$. For a centrosymmetric medium (i.e., a medium where when $\mathbf{E}$ is replaced by $-\mathbf{E}$, $\mathbf{P}$ is replaced by $-\mathbf{P}$), the leading nonlinearity (in terms of the wave amplitude), is cubic,[3] and we write

$$\frac{1}{\epsilon_0}\mathbf{P}(t) = \int_{-\infty}^{+\infty} \chi_1(t - t_1)\mathbf{E}(t_1)dt_1$$
$$+ \int_{-\infty}^{+\infty} \chi_3(t - t_1, t - t_2, t - t_3)(\mathbf{E}(t_1) \cdot \mathbf{E}(t_2))\mathbf{E}(t_3)dt_1 dt_2 dt_3,$$

$$(1.2.14)$$

where χ_3 is symmetric in its three arguments.

As noted in Luther, Newell, and Moloney (1994), in the usual conditions, the duration and transverse width of the light pulses are such that the modulation can be viewed as stationary. In expanding $\mathbf{E} = \varepsilon \mathbf{E}_0 + \varepsilon^2 \mathbf{E}_1 + \cdots$, the leading order $\mathbf{E}_0 = (\mathcal{E}(X, Y, Z)e^{i(kx-\omega t)} + \text{c.c.})\mathbf{e}$ contributes to the polarization $\mathbf{P}$ as

$$\epsilon_0(\varepsilon\widehat{\chi}_1(\omega)\mathcal{E}e^{i(kx-\omega t)} + \varepsilon^3\widehat{\chi}_3(\omega, -\omega, \omega)|\mathcal{E}|^2\mathcal{E}e^{i(kx-\omega t)}$$
$$+ \varepsilon^3\widehat{\chi}_3(\omega, \omega, \omega)\mathcal{E}^3 e^{3i(kx-\omega t)} + \text{c.c.})\,\mathbf{e}.$$

One then obtains the envelope equation (Ablowitz and Segur 1981)

$$2ik\partial_z\mathcal{E} + \Delta_\perp \mathcal{E} + 3k^2 \frac{\widehat{\chi}_3(\omega, -\omega, \omega)}{1 + \widehat{\chi}_1(\omega)}|\mathcal{E}|^2\mathcal{E} = 0. \qquad (1.2.15)$$

Note that if the electric susceptibility and thus the refractive index increase with the wave intensity ($\widehat{\chi}_3(\omega, -\omega, \omega) > 0$), then in a region where the wave amplitude is slightly amplified, the refractive index is also enlarged. From geometrical optics, the light rays bend towards this region, which still increases the refractive index and thus bends the light rays even more. This modulational instability of the wave (see Section 1.3.1) leads

---

[3]When the above symmetry is lacking, a quadratic nonlinear term is present, which, as discussed in Chapter 10, induces a mean field driven by the amplitude modulation (see Newell and Moloney 1992 for a discussion of this effect in the context of nonlinear optics).

to the self-focusing (also called filamentation) of the beam, a phenomenon extensively studied in the context of laser beams since the early sixties (see, e.g., Chiao, Garmire, and Townes 1964, Garmire, Chiao, and Townes 1966, Talanov 1964, 1965, Kelley 1965, Askar'yan 1966, Shen 1976, and Vlasov and Talanov 1997 for an extended bibliography). In contrast, if $\widehat{\chi}_3(\omega, -\omega, \omega)$ is negative, the system is stable, the change in the refracting index counteracting the effect of the initial perturbation.

An interesting question is the influence of the inhomogeneities present in the medium on the wave focusing. The effect of a random refractive index that is a white noise with respect to the propagation coordinate, is addressed in Gaididei and Christiansen (1998) (see also references therein). They conclude that random fluctuations can postpone the collapse or even in some instances prevent self-focusing.

1.2.4   Time Dispersion of Ultrashort Wave Trains

When the laser beam is emitted during a very short time, the modulation can no longer be viewed as stationary, and a (slow) time-dependence of the amplitude should be retained. The analysis is then more delicate. We review it here, following Newell and Moloney (1992). Special attention has been recently devoted to such pulses because for the same total energy, they correspond to higher peak intensity with specific consequences such as increased ocular damage and also possible application to eye surgery (Powell, Moloney, and Albanese 1993, Fibich 1995, 1996a and references therein).

Equation (1.2.7) can be rewritten in the form

$$LE = \mathcal{N}\{\mathbf{E}\}, \tag{1.2.16}$$

where both

$$LE \equiv \nabla^2 \mathbf{E} - \nabla(\nabla.\mathbf{E}) - \frac{1}{c^2}\partial_{tt}\mathbf{E} - \frac{1}{c^2}\frac{\partial^2}{\partial t^2}\int_{-\infty}^{+\infty}\chi_1(t-t_1)\mathbf{E}(t_1)dt_1 \tag{1.2.17}$$

and

$$\mathcal{N}\{\mathbf{E}\} \equiv \frac{1}{\epsilon_0 c^2}\frac{\partial^2}{\partial^2 t}\mathbf{P}_{NL}$$

$$\equiv \frac{1}{c^2}\frac{\partial^2}{\partial t^2}\int_{-\infty}^{+\infty}\chi_3(t-t_1, t-t_2, t-t_3)(\mathbf{E}(t_1)\cdot\mathbf{E}(t_2))\mathbf{E}(t_3)dt_1 dt_2 dt_3$$

$$\tag{1.2.18}$$

include a nonlocal dependence with respect to the time variable.

We look for solutions of the form

$$\mathbf{E} = \varepsilon\mathbf{E}_0 + \varepsilon^2\mathbf{E}_1 + \varepsilon^3\mathbf{E}_2 + \ldots, \tag{1.2.19}$$

where ε is a small positive parameter measuring the field amplitude. The leading-order contribution

$$\mathbf{E}_0 = (\mathcal{E}_0(X,Y,Z,T)e^{i(kz-\omega t)} + \text{c.c.})\,\mathbf{e} \qquad (1.2.20)$$

has the form of a quasi-monochromatic wave train, polarized in a direction $\mathbf{e}$, perpendicular to the direction $\mathbf{z}$ of the propagation. The (complex) amplitude $\mathcal{E}$ depends on the slow variables $X = \varepsilon x$, $Y = \varepsilon y$, $Z = \varepsilon z$, $T = \varepsilon t$.

It is convenient at this step to estimate the leading order of the polarization

$$\mathbf{P}_0 = (P_0 + \text{c.c.})\mathbf{e}, \qquad (1.2.21)$$

where

$$P_0(t) = \epsilon_0 \int_{-\infty}^{+\infty} \chi_1(t-t')\mathcal{E}_0(\varepsilon t')e^{i(kz-\omega t')}\,dt'. \qquad (1.2.22)$$

By Fourier transform in time, we have

$$\frac{1}{\epsilon_0}\widehat{P_0}(\Omega) = \widehat{\chi_1}(\Omega)\widehat{E_0}(\Omega) = \frac{1}{\varepsilon}e^{ikz}\widehat{\chi_1}(\Omega)\widehat{\mathcal{E}}(\frac{\Omega-\omega}{\varepsilon}). \qquad (1.2.23)$$

Coming back to the physical space,

$$\frac{1}{\epsilon_0}P_0(t) = \frac{1}{2\pi}\frac{1}{\varepsilon}e^{ikx}\int e^{-i\Omega t}\widehat{\chi_1}(\Omega)\widehat{\mathcal{E}}(\frac{\Omega-\omega}{\varepsilon})\,d\Omega$$

$$= \frac{1}{2\pi}e^{i(kx-\omega t)}\int e^{-i\varepsilon\nu t}\widehat{\chi_1}(\omega+\varepsilon\nu)\widehat{\mathcal{E}}(\nu)\,d\nu \qquad (1.2.24)$$

where we expand $\widehat{\chi_1}(\omega+\varepsilon\nu) = \widehat{\chi_1}(\omega)+\widehat{\chi_1}'(\omega)\varepsilon\nu+\frac{1}{2}\widehat{\chi_1}''(\omega)\epsilon^2\nu^2+\cdots$, where the primes hold for derivatives with respect to the frequency. This yields

$$\frac{1}{\epsilon_0}P_0(t) = e^{i(kx-\omega t)}[\widehat{\chi_1}(\omega) + i\widehat{\chi_1}'(\omega)\frac{\partial}{\partial t} - \frac{1}{2}\widehat{\chi_1}''(\omega)\frac{\partial^2}{\partial t^2} + \cdots]\mathcal{E}_0(\varepsilon t)$$

$$= e^{i(kx-\omega t)}[(\widehat{\chi_1}(\omega) + i\widehat{\chi_1}'(\omega)\epsilon\frac{\partial}{\partial T} - \frac{1}{2}\widehat{\chi_1}''(\omega)\epsilon^2\frac{\partial^2}{\partial T^2} + \cdots]\mathcal{E}_0(T), \qquad (1.2.25)$$

Looking at (1.2.25) as a formal Taylor expansion, we are led to write symbolically

$$\frac{1}{\epsilon_0}P_0(t) = e^{i(kz-\omega t)}\widehat{\chi_1}(\omega + \varepsilon\frac{\partial}{\partial T})\mathcal{E}_0(T)$$

$$= e^{i(kz-\omega t)}\widehat{\chi_1}(i\partial_t + \varepsilon\partial_T)E_0. \qquad (1.2.26)$$

Expanding $L = L_0 + \varepsilon L_1 + \varepsilon^2 L_2 + \cdots$, we obtain at the successive orders

$$L_0\mathbf{E}_0 = 0, \qquad (1.2.27)$$

$$L_0\mathbf{E}_1 = -L_1\mathbf{E}_0, \qquad (1.2.28)$$

$$L_0\mathbf{E}_2 = -L_2\mathbf{E}_0 - L_1\mathbf{E}_1 + \mathcal{N}_3(\mathbf{E}_0), \qquad (1.2.29)$$

where

$$
L_0 \mathbf{E}_0 = \begin{pmatrix} -k^2 + \dfrac{\omega^2 n^2(\omega)}{c^2} & 0 & 0 \\[2ex] 0 & -k^2 + \dfrac{\omega^2 n^2(\omega)}{c^2} & 0 \\[2ex] 0 & 0 & \dfrac{\omega^2 n^2(\omega)}{c^2} \end{pmatrix} \mathbf{E}_0
$$

$$(1.2.30)$$

and $n^2(\omega) = 1 + \widehat{\chi_1}'(\omega)$. The associated dispersion relation reduces to $k = \frac{\omega^2 n^2(\omega)}{c^2}$. Furthermore,

$$
-\frac{1}{c^2}\partial_{tt}\mathbf{E}_0 - \frac{1}{c^2}\frac{\partial^2}{\partial t^2}\int \chi_1(t-t')\mathbf{E}_0(t')dt'
$$

$$
= \frac{(\omega + \epsilon i\partial_T)^2}{c^2}(1 + \widehat{\chi_1}(\omega + \epsilon i\partial_T))\mathbf{E}_0 = k^2(\omega + \epsilon i\partial_T)\mathbf{E}_0
$$

$$
= (k^2(\omega) + 2i\epsilon kk'\partial_T - \epsilon^2(kk'' + k'^2)\partial_{TT})\mathbf{E}_0 + \cdots, \quad (1.2.31)
$$

where in the expansion of the operator $k^2(\omega + \epsilon i\partial_T)$, the prime denotes derivatives and the corresponding functions are evaluated for the frequency ω. It follows that

$$
L_1 \mathbf{E}_0 = \begin{pmatrix} 2ik\partial_Z + 2ikk'\partial_T & 0 & -ik\partial_X \\[1ex] 0 & 2ik\partial_Z + 2ikk'\partial_T & -ik\partial_Y \\[1ex] -ik\partial_X & -ik\partial_Y & 2ikk'\partial_T \end{pmatrix} \mathbf{E}_0,
$$

$$(1.2.32)$$

and

$$
L_2 \mathbf{E}_0 =
$$

$$
\begin{pmatrix} \Delta_{YZ} - \alpha(\omega)\partial_{TT} & -\partial_{XY} & -\partial_{XZ} \\[1ex] -\partial_{XY} & \Delta_{XZ} - \alpha(\omega)\partial_{TT} & -\partial_{XZ} \\[1ex] -\partial_{XZ} & -\partial_{YZ} & \Delta_{XY} - \alpha(\omega)\partial_{TT} \end{pmatrix} \mathbf{E}_0,
$$

$$(1.2.33)$$

where to simplify the notation, we defined $\Delta_{YZ} = \partial_{YY} + \partial_{ZZ}$, $\Delta_{XZ} = \partial_{XX} + \partial_{ZZ}$, $\Delta_{XY} = \partial_{XX} + \partial_{YY}$ and $\alpha(\omega) = (kk' + k'^2)$.

After computing $L_1 \mathbf{E}_1$, (1.2.28) becomes

$$
L_0 \mathbf{E}_1 = -2ik(\partial_Z + k'\partial_T)\mathcal{E}e^{i(kz-\omega t)}\mathbf{e} + ik\nabla \cdot (\mathcal{E}\mathbf{e})e^{i(kz-\omega t)}\mathbf{z}. \quad (1.2.34)
$$

The solvability condition reads

$$
(\partial_Z + k'\partial_T)\mathcal{E}_0 = 0. \quad (1.2.35)
$$

It follows that

$$
\mathbf{E}_1 = \left(\frac{i}{k}\nabla \cdot (\mathcal{E}_0\mathbf{e})\mathbf{z} + \mathcal{E}_1\mathbf{e}\right)e^{i(kz-\omega t)} + \text{c.c.}, \quad (1.2.36)
$$

where we included an element of the null space in the direction **e**. Equation (1.2.29) becomes

$$L_0\mathbf{E}_2 = -\begin{pmatrix} (\partial_{YY} + \partial_{ZZ} - \alpha(k))\partial_{TT})\mathcal{E}_0 e_x - \partial_{XY}\mathcal{E}_0 e_y \\ -\partial_{XY}\mathcal{E}_0 e_x + (\partial_{XX} + \partial_{ZZ} - \alpha(k)\partial_{TT})\mathcal{E}_0 e_y \\ -\partial_{XZ}\mathcal{E}_0 e_x - \partial_{YZ}\mathcal{E}_0 e_y \end{pmatrix} e^{i(kz-\omega t)}$$

$$- \begin{pmatrix} \partial_X(\partial_X\mathcal{E}_0 e_x + \partial_Y\mathcal{E}_0 e_y) \\ \partial_Y(\partial_X\mathcal{E}_0 e_x + \partial_Y\mathcal{E}_0 e_y) \\ -2k'\partial_T(\partial_X\mathcal{E}_0 e_x + \partial_Y\mathcal{E}_0 e_y) \end{pmatrix} e^{i(kz-\omega t)}$$

$$- 2ik(\partial_Z + k'\partial_T)\mathcal{E}_1 \mathbf{e} e^{i(kz-\omega t)} - 3\frac{\omega^2}{c^2}\widehat{\chi_3}(\omega, -\omega, \omega)|\mathcal{E}_0|^2\mathcal{E}_0\mathbf{e} e^{i(kz-\omega t)}$$

$$+ \text{ third order harmonics} + \text{c.c.} \tag{1.2.37}$$

The solvability condition reads

$$- 2ik(\partial_Z + k'\partial_T)\mathcal{E}_1 - (\partial_{XX} + \partial_{YY} + \partial_{ZZ} - \alpha(\omega)\partial_{TT})\mathcal{E}_0$$

$$- 3\frac{\omega^2}{c^2}\widehat{\chi_3}(\omega, -\omega, \omega)|\mathcal{E}_0|^2\mathcal{E}_0 = 0. \tag{1.2.38}$$

Combining (1.2.35) and (1.2.38), defining $\mathcal{E} = \mathcal{E}_0 + \epsilon\mathcal{E}_1$, and using (1.2.35) to evaluate the $O(\epsilon)$ terms originating from (1.2.38), we get ($\Delta_\perp = \partial_{XX} + \partial_{YY}$)

$$2ik(\partial_Z + k'\partial_T)\mathcal{E}$$

$$- \varepsilon\left((\Delta_\perp - kk''\partial_{TT})\mathcal{E} - 3\frac{\omega^2}{c^2}\widehat{\chi_3}(\omega, -\omega, \omega)|\mathcal{E}|^2\mathcal{E}\right) = 0. \tag{1.2.39}$$

The small parameter ε can then be eliminated by introducing the delayed time $\tau = T - k'Z = \varepsilon(t - k'z)$ and rescaling the propagation coordinate $\zeta = \varepsilon Z = \varepsilon^2 z$. We get

$$i\partial_\zeta\mathcal{E} + \frac{1}{2k}\Delta_\perp\mathcal{E} - \frac{k''}{2}\partial_{\tau\tau}\mathcal{E} + \frac{3\omega^2}{2c^2}\widehat{\chi_3}(\omega, -\omega, \omega)|\mathcal{E}|^2\mathcal{E} = 0 \tag{1.2.40}$$

which in the absence of time dependence reproduces (1.2.15).

In the case of the so-called anomalous group velocity dispersion ($k'' < 0$), (1.2.40) reduces to the canonical elliptic NLS equation in three dimensions. Experimental requirements for self-focusing of an optical pulse in this context are described by Silberberg (1990). The case of normal dispersion ($k'' > 0$) deserves a special discussion, which is detailed in Section 9.2. In situations where it is weak, its effect can be viewed as a perturbation to the canonical two-dimensional NLS equation and is amenable to an asymptotic analysis.

1.3   Basic Dynamical Effects

1.3.1   Modulational Instability

The NLS equation

$$i\psi_t + D\psi + \gamma|\psi|^2\psi = 0, \tag{1.3.1}$$

where $D = \alpha_{ij}\partial_i\partial_j$ (with summation over the repeated indices), admits exact solutions in the form $\psi = \psi_0 e^{i\gamma|\psi_0|^2 t}$, corresponding in the original problem to a plane wave whose frequency is slightly shifted by the nonlinearity. We study the linear stability of this solution by writing $\psi = \psi_0(1 + \widetilde{\psi})e^{i(\gamma|\psi_0|^2 t + \widetilde{\phi})} \approx \psi_0(1 + \widetilde{\psi} + i\widetilde{\phi})e^{i\gamma|\psi_0|^2 t}$ with infinitesimal perturbations $\widetilde{\psi}$ and $\widetilde{\phi}$ of the amplitude and phase. We get

$$\widetilde{\psi}_t + D\widetilde{\phi} = 0, \tag{1.3.2}$$

$$\widetilde{\phi}_t - D\widetilde{\psi} - 2\gamma|\psi_0|^2\widetilde{\psi} = 0, \tag{1.3.3}$$

or

$$(\partial_{tt} + D^2 + 2\gamma|\psi_0|^2 D)\widetilde{\psi} = 0. \tag{1.3.4}$$

Looking for harmonic perturbations $\widetilde{\psi}$ and $\widetilde{\phi}$ proportional to $e^{i\boldsymbol{\kappa}\cdot\boldsymbol{x}}e^{\sigma t}$, this leads to the relation

$$\sigma^2 = 2\gamma|\psi_0|^2\alpha_{ij}\kappa_i\kappa_j - (\alpha_{ij}\kappa_i\kappa_j)^2. \tag{1.3.5}$$

If $\gamma\alpha_{ij}\kappa_i\kappa_j$ is negative, σ is imaginary and the wave amplitude remains bounded. When it is positive and $2|\psi_0|^2 > \alpha_{ij}\kappa_i\kappa_j/\gamma$, the perturbation amplitude is exponentially amplified. This instability of a wave train of wave number **k** relative to disturbances in the form of a large-scale long-time modulation, or equivalently in the form of two "side-band" modes $\mathbf{k} \pm \epsilon\boldsymbol{\kappa}$, is known as the modulational, or Benjamin–Feir instability after the discoverers of this effect in the water-wave context (Benjamin and Feir 1967).

Let us specify the stability condition to the case of the "elliptic" NLS equation where the operator D is of the form $D = \alpha\Delta$, a situation addressed in detail in this monograph. When the coefficients α and γ have the same signs (attracting nonlinearity), disturbances of wave number κ obeying $\kappa^2 < \frac{2\gamma}{\alpha}|\psi_0|^2$ are linearly unstable, while larger wave numbers are stabilized by the dispersion. When the two coefficients have opposite sign (repulsing nonlinearity), constant-amplitude solutions are modulationally stable.

Another important example concerns the two-dimensional "hyperbolic" NLS equation where $D = \partial_{xx} - \partial_{yy}$ and which arises for example in the description of surface gravity waves on deep water. In this case $\sigma^2 = (\kappa_1^2 - \kappa_2^2)[2\gamma|\psi_0|^2 - (\kappa_1^2 - \kappa_2^2)]$, and the Fourier modes located in the domain of the (κ_1, κ_2)-plane bounded by the hyperbola $\kappa_1^2 - \kappa_2^2 = 2\gamma|\psi_0|^2$ and

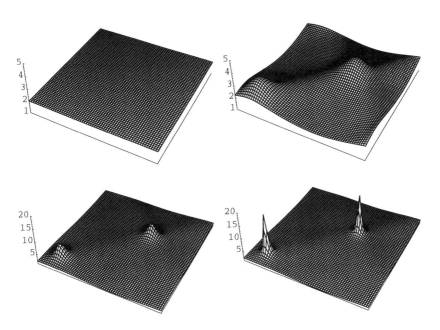

FIGURE 1.1. Modulational instability and wave collapse for the two-dimensional elliptic NLS equation (see text for details).

its asymptotes $\kappa_1 = \pm\kappa_2$ are unstable. Note that this domain includes arbitrarily large wave numbers.

Experimentally, the modulational instability has been observed in various physical contexts, including surface waves on deep water (Benjamin 1967a, Yuen and Lake 1975, 1980, Remoissenet 1996) and light waves in optical fibers (Tai, Hasegawa, and Tomita 1986).

The nonlinear development of this instability depends on the space dimension. Numerical simulations of the focusing one-dimensional cubic NLS equation in a periodic domain has revealed a Fermi–Pasta–Ulam (1955) recurrent behavior of the solution (Yuen and Ferguson 1978a, Hafizi 1981). Janssen (1981) constructed time periodic solutions by formal perturbative computations near threshold. Based on the argument that recurrence is related to the effective excitation of a finite number of modes on which the energy is perpetually redistributed (Thyragaraja 1979), models retaining a few Fourier modes were considered by Infeld (1981). A treatment based on the assumption that the linearly stable modes are phase locked (or slaved) to the unstable ones is given by Rowlands (1980). A theoretical interpretation of this recurrent behavior is given in Ablowitz and Clarkson (1991) and references therein, based on the existence of homoclinic orbits associated to the unstable eigendirections of the linear problem. Recurrence was also observed with the hyperbolic two-dimensional NLS equation (Yuen and Ferguson 1978b, Yuen & Lake 1980), a question discussed in Section 9.2.2. It is also reported in the context of one-dimensional Langmuir waves in the situation where the dynamics are roughly adiabatic (see Chapter 13), using finite-size particle codes (Tajima, Goldman, Leboeuf, and Dawson 1981). In contrast, for the periodic one-dimensional NLS with quintic nonlinearity or in the case of the two-dimensional elliptic NLS equation with cubic nonlinearities, numerical evidence of finite-time blowup was reported (see, e.g., Sulem, Sulem, and Frisch 1983, Sulem, Sulem, and Patera 1984). The nonlinear development of the modulational instability into wave collapse is illustrated in Fig.1.1, in the case of an initial wave amplitude $\psi_0(\mathbf{x}) = 2 + 0.01 \sin(x + \frac{\pi}{4}) \sin(y + \frac{\pi}{4})$, evolving according to the NLS equation $i\psi_t + \Delta\psi + |\psi|^2\psi = 0$. The wave amplitude is displayed within the periodic domain $[2\pi \times 2\pi]$ at times $t = 0$, $t = 1.4$, $t = 1.55$, and $t = 1.5813$. A standard pseudo-spectral method with $(256)^2$ collocation points is used. The phenomenon of wave collapse is actually delicate to analyze theoretically in a periodic geometry and was mostly considered in the context of localized solutions in the entire space. This phenomenon is extensively discussed in the forthcoming chapters.

1.3.2   Solitons in One Space Dimension

We consider in this section the one-dimensional cubic Schrödinger equation with an attracting nonlinearity, in the canonical form

$$i\psi_t + \psi_{xx} + |\psi|^2\psi = 0, \qquad (1.3.6)$$

in the entire space **R**. When the nonlinearity is repulsive, the nonlinearity enhances the linear dispersion, and nontrivial dynamics require that the solution be nonzero at infinity (see Kivshar and Luther-Davies 1998 for a recent review on "dark solitons").

   We look for a localized traveling wave solution of the form

$$\psi = e^{i(rx - st)}v(\xi) \qquad (1.3.7)$$

with $\xi = x - Ut$, r and s denoting real constants and the function v being assumed real and decaying at infinity. Substituting in the Schrödinger equation after choosing $r = U/2$ in order to eliminate the derivative term v', and defining $\alpha = r^2 - s$, we obtained the equation prescribing the shape of the profile

$$v'' - \alpha v + v^3 = 0. \qquad (1.3.8)$$

After multiplication by v, the left-hand side of the equation rewrites as a derivative. Assuming that v and its derivatives vanish at infinity, we have

$$v'^2 = \alpha v^2 - \frac{1}{2}v^4. \qquad (1.3.9)$$

For solutions vanishing at infinity, the quartic term is negligible at large distance and the existence of such solutions requires that α be taken positive. Dividing by v^4 and defining $w = 1/v$, we have

$$w'^2 = \alpha w^2 - \frac{1}{2}, \tag{1.3.10}$$

which implies [4] $w = \frac{1}{(2\alpha)^{1/2}} \cosh(\alpha^{1/2}\xi)$. It follows that

$$v = (2\alpha)^{1/2} \frac{1}{\cosh(\alpha^{1/2}\xi)}. \tag{1.3.11}$$

We thus obtain the existence of localized solutions in the form of a solitary waves

$$\psi(x,t) = (2\alpha)^{1/2} \frac{1}{\cosh(\alpha^{1/2}(x - Ut - x_0))} e^{i\left(\frac{U}{2}x + (\alpha - \frac{U^2}{4})t + \varphi_0\right)}, \tag{1.3.12}$$

where the constants x_0 and φ_0 reflect the invariance of the Schrödinger equation by space translation and by phase shift, respectively. The velocity U is associated to the invariance by Galilean transformation (see Section 2.3).

The above solutions, which are localized in space, preserve their form and propagate with a constant velocity. They are, in fact, (bright) solitons, which after they undergo collisions emerge with the same shape and velocity. The cubic Schrödinger equation in one space dimension is indeed integrable by the inverse scattering transform (IST). The description of this method, which reduces the resolution of the initial value problem to that of an inverse scattering problem for an associated linear eigenvalue equation, is beyond the scope of this monograph. The reader is referred to the original articles (Zakharov and Shabat 1972, 1974, Ablowitz, Kaup, Newell, and Segur 1974), to Chapter 6 of Newell and Moloney (1992) or for a general introduction to IST to the review of Scott, Chu, and McLaughlin (1973) and to the several books devoted this subject, among them Ablowitz and Segur (1981), Eckhaus and Van Harten (1981), Dodd, Eilbeck, Gibbon, and Morris (1982), Novikov, Manakov, Pitaevskii, and Zakharov (1984), Newell (1985), Drazin and Johnson (1989). The integrability by IST leads in particular to the conclusion that for a large class of initial conditions vanishing at infinity, the solution of the focusing cubic NLS equation in one space dimension evolves asymptotically in time to a finite set of solitons with some radiation that escapes to infinity.

The utilization of the dependence of the refractive index on the wave intensity to balance the broadening effect of dispersion and make possible the transmission of optical pulses without distortion in dielectric fiber

---

[4]Note that if the nonlinearity is taken to be repulsive, one has $w'^2 = \alpha w^2 + \frac{1}{2}$. A localized solution is then associated to $w = (2\alpha)^{-1/2} \sinh(\alpha^{1/2}\xi)$, leading to a singularity at $\xi = 0$ for the amplitude $v = 1/w$, which is not acceptable.

waveguides was pointed out by Hasegawa and Tappert (1973). The resulting "optical temporal solitons "[5] were observed by Mollenauer, Stollen & Gordon (1980) (see also the review of Segev & Stegeman 1998).

Shape-preserving localized solutions can also be constructed in higher dimensions (or for higher nonlinearities), and are usually referred to as standing or solitary waves. The conditions for their existence and stability relative to perturbations are discussed in Chapter 4. Even when stable, in contrast with the solitons, they do not however enjoy collision stability, specific of the cubic nonlinearity in one space dimension.

1.3.3 Soliton Instability for Transverse Perturbations

The effect of an infinitesimal long-wavelength transverse perturbation on a soliton solution of the one-dimensional Schrödinger equation, was addressed by Zakharov and Rubenchik (1974) and Yajima (1974), and reviewed in Ablowitz and Segur (1981, pp. 271–274). A general discussion of soliton stability is presented in Kuznetsov, Rubenchik, and Zakharov (1986).

We consider the equation

$$i\psi_t + \psi_{xx} + \lambda\psi_{yy} + |\psi|^2\psi = 0, \tag{1.3.13}$$

with $\lambda = \pm 1$. In the comoving frame, the soliton is given by

$$\psi_0(x,t) = \Psi(x)e^{i\alpha t}, \tag{1.3.14}$$

where $\alpha > 0$ and

$$\Psi_{xx} - \alpha\Psi + \Psi^3 = 0, \tag{1.3.15}$$

or

$$\Psi(x) = \frac{\sqrt{2\alpha}}{\cosh(\alpha^{1/2}x)}. \tag{1.3.16}$$

This solution is perturbed in the form

$$\psi = \Psi(1 + \widetilde{\chi})e^{i(\alpha t + \widetilde{\phi})} \tag{1.3.17}$$

where $\widetilde{\chi}$ and $\widetilde{\phi}$ are infinitesimal disturbances that slowly depend on the transverse variable y. In the linear approximation,

$$\psi = \psi_0 + (f + ig)e^{i\alpha t}, \tag{1.3.18}$$

where $f = \widetilde{\chi}\Psi$ and $g = \widetilde{\phi}\Psi$ are real quantities. After linearization about the soliton solution, the NLS equation is replaced by the system

$$L_R f = g_t - \lambda f_{yy}, \tag{1.3.19}$$

---

[5]As already seen, when eq. (1.3.6) is considered in optics, the variable t holds for the coordinate along the direction of propagation, while the second derivative ψ_{xx} holds for the time-dispersion.

$$L_I g = -f_t - \lambda g_{yy}, \tag{1.3.20}$$

where

$$L_R = \frac{d^2}{dx^2} - \alpha + 3\Psi^2, \tag{1.3.21}$$

and

$$L_I = \frac{d^2}{dx^2} - \alpha + \Psi^2. \tag{1.3.22}$$

Restricting to long-wavelength transverse perturbations, we introduce the slow variables $Y = \epsilon y$ and $T = \epsilon t$, and expand the perturbation in the form

$$f = f_0 + \epsilon f_1 + \epsilon^2 f_2 + \cdots, \tag{1.3.23}$$
$$g = g_0 + \epsilon g_1 + \epsilon^2 g_2 + \cdots. \tag{1.3.24}$$

At leading order we have

$$L_R f_0 = 0, \tag{1.3.25}$$
$$L_I g_0 = 0, \tag{1.3.26}$$

and thus

$$f_0 = a(Y,T)\Psi'(x) \tag{1.3.27}$$
$$g_0 = b(Y,T)\Psi(x). \tag{1.3.28}$$

The functions a and b are easily interpreted by noticing that to leading order, the perturbed solution reads $\psi \approx (\psi_0 + a\Psi' + b\Psi)e^{i\alpha t}$, which can be viewed as the expansion of a soliton $\Psi(x + a(X,T))e^{i(\alpha t + b(Y,T))}$, where both amplitude and phase are weakly modulated.

At order ϵ, we get

$$L_R f_1 = \partial_T g_0 = b_T \Psi(x), \tag{1.3.29}$$
$$L_I g_1 = -\partial_T f_0 = -a_T \Psi'(x). \tag{1.3.30}$$

This system is always solvable in the form

$$f_1 = b_T \frac{d\Psi}{d\alpha}, \tag{1.3.31}$$
$$g_1 = -\frac{a_T}{2} x \Psi(x). \tag{1.3.32}$$

At order ϵ^2, we have

$$L_R f_2 = \partial_T g_1 - \lambda \partial_{YY} f_0, \tag{1.3.33}$$
$$L_I g_2 = -\partial_T f_1 - \lambda \partial_{YY} g_0. \tag{1.3.34}$$

This system requires the solvability conditions

$$\int f_0 \, \partial_T g_1 \, dx - \lambda \int f_0 \, \partial_{YY} f_0 \, dx = 0, \tag{1.3.35}$$

$$-\int g_0 \, \partial_T f_1 \, dx - \lambda \int g_0 \, \partial_{YY} g_0 \, dx = 0, \tag{1.3.36}$$

where the integrals are taken from $-\infty$ to $+\infty$. We get

$$-\frac{1}{2}a_{TT}\int x\Psi\Psi'\,dx - \lambda a_{YY}\int \Psi'^2\,dx = 0, \qquad (1.3.37)$$

$$-b_{TT}\int \Psi\frac{d\Psi}{d\alpha}\,dx - \lambda b_{YY}\int \Psi^2\,dx = 0. \qquad (1.3.38)$$

We then compute the coefficients $\int \Psi^2 dx = 4\alpha^{1/2}$, $\int x\Psi\Psi' dx = -2\alpha^{1/2}$, $\int \Psi\frac{d\Psi}{d\alpha}dx = \alpha^{-1/2}$, $\int \Psi'^2 dx = \frac{4}{3}\alpha^{3/2}$, and finally get

$$a_{TT} - \frac{4\alpha\lambda}{3}a_{YY} = 0, \qquad (1.3.39)$$

$$b_{TT} + 4\alpha\lambda b_{YY} = 0. \qquad (1.3.40)$$

It follows that since α is positive, the soliton is always unstable relative to long-wavelength transverse perturbations, whatever the elliptic ($\lambda > 0$) or hyperbolic ($\lambda < 0$) character of the linear operator of the NLS equation. In the former case, the unstable mode is even in the x-variable, while in the latter, it is odd (see (1.3.27) and (1.3.28)). A similar analysis in the context of a nonlocal NLS equation governing the subsonic dynamics of Langmuir oscillations, (where the ion dynamics are adiabatic), is presented in Zakharov (1984). A numerical study in the general context is found in Denavit, Pereira, and Sudan (1994) and Degtyarev, Zakharov, and Rudakov (1975). A stability analysis of the NLS soliton for transverse perturbations not restricted to be at asymptotically large scale is presented in Janssen and Rasmussen (1983), who obtained that for $\lambda > 0$, there exists a critical value for the transverse wave number above which the soliton is stable relative to the corresponding perturbation. The case $\lambda < 0$ has been the object of discussions (Rubenchik and Zakharov 1986), but a similar conclusion holds (Rypdal and Rasmussen 1989 and references therein). A discussion is presented by Kuznetsov, Rubenchik, and Zakharov (1986).

In optical systems with negligible time dispersion (where t holds for the coordinate along the direction of propagation), the transverse instability can be arrested by spatially modifying the refractive index so that it provides waveguiding in one transverse direction, while self-trapping occurs in the perpendicular one. For this purpose, the scale of the waveguiding must be smaller than the width of the soliton (Barthelemy, Maneuf and Froehly 1985, Aitchison, Weiner, Silberberg, Oliver, Jackel, Leaird, Vogel, and Smith 1990, Maillote, Monneret, Barthelemy and Froehly 1994, see also Soljačić, Sears, and Segev 1998, Segev and Stegeman 1998). A comprehensive review of the transverse instability of solitary waves was recently presented by Kivshar and Pelinovsky (1999).

1.4   Fluid-Dynamical Form of the NLS Equation

We conclude this chapter with a representation of the NLS equation, formally close to the equations of hydrodynamics. For this purpose, starting with the elliptic NLS equation

$$i\psi_t + \Delta\psi + g|\psi|^2\psi = 0, \tag{1.4.1}$$

we write $\psi = \sqrt{\rho}e^{i\varphi}$ in terms of the phase and amplitude. This transformation is usually called Madelung's transformation and was originally introduced in the context of the linear Schrödinger equation for quantum mechanics (see, e.g., Spiegel 1980 for historical references, including attempts of classical interpretations of quantum mechanics). After substitution in the NLS equation and separation of the real and imaginary parts of the equation, one obtains (after rescaling time by a factor 2)

$$\rho_t + \nabla \cdot (\rho\nabla\varphi) = 0, \tag{1.4.2}$$

$$\varphi_t + \frac{1}{2}|\nabla\varphi|^2 - \frac{g}{2}\rho = \frac{1}{2\sqrt{\rho}}\Delta\sqrt{\rho}. \tag{1.4.3}$$

Looking at ρ as a density and at φ as an hydrodynamic potential, (1.4.2)–(1.4.3) identify for $g < 0$ with the equations for an irrotational barotropic gas, up to the additional term in the right-hand side of (1.4.3). Special interest was thus devoted to the above system in the context of the NLS equation with a repulsing nonlinearity and solutions of finite amplitude at infinity. This model is indeed relevant for a description of superfluid helium II at zero temperature, viewed as an imperfect Bose condensate with local interactions (Pitaevskii 1961, Gross 1963). In this context, where the NLS equation[6] is often named the Gross–Pitaevskii equation, the right-hand side of (1.4.3) corresponds to a "quantum-mechanical pressure." A recent discussion of the NLS equation for superfluid hydrodynamics is given by Coste (1998).

Madelung's transformation is, however, singular when ψ vanishes. Since this field is complex, these "topological defects" (defined as the common zeros of two real functions) are generically located on points in two space dimensions and on lines in three dimensions. In the context of superfluidity, they are known as "quantum vortices" (Ginzburg and Pitaevskii 1958, Donnelly 1991). The circulation of the velocity $\mathbf{v} = \nabla\varphi$ around each of them is $\pm 2\pi$. Quantification of vorticity in a superfluid was suggested in a footnote of a celebrated paper by Onsager (1949) on classical vortex flows and turbulence. Experimental observations of these vortices were reported early (see Yamchuk, Gordon, and Packard 1979 and references given in

---

[6]The NLS equation is then sometimes written with an additional linear term proportional to ψ, corresponding to the elimination of a linear phase in the time variable.

Coste 1998), and numerical simulations are presented in Frisch, Pomeau, and Rica (1992) and Nore, Brachet, and Fauve (1993).

The quantum-mechanical pressure induces a dispersive effect when (1.4.2)–(1.4.3) are linearized about the solution $\rho = 1$, $\nabla \phi = 0$. As a consequence, a long-wavelength reductive perturbation expansion leads to the Korteweg–de Vries equation in one space dimension and to a Kadomtsev–Petviashvili equation (KP) in higher dimensions (Zakharov and Kuznetsov 1986a, Kuznetsov and Turitsyn 1988). This approach was used to address the stability of one-dimensional dark solitons and of two-dimensional lumps (Kuznetsov and Rasmussen 1995). The nonlinear development of this instability is presented as a mechanism for the generation of vortices in a superfluid by Josserand and Pomeau (1995).

The quantum-mechanical pressure becomes negligible in the "semiclassical" or WKB limit, where ∇ and ∂_t scale like ϵ as $\epsilon \to 0$, and the initial condition has the form $\psi_0(\mathbf{x}, 0) = \Psi_0(\mathbf{x}) e^{i \frac{S_0(\mathbf{X})}{\epsilon}}$. In this limit, the Euler equations for an isentropic compressible flow are recovered. This was proved rigorously by Jin, Levermore, and McLaughlin (1994) in one dimension using the inverse scattering technique and by Grenier (1995) for higher dimensions in situations where no vortices are involved. In two dimensions, the result was recently extended by Lin and Xin (1997) to situations with a finite number of vortices of degree ± 1, with the condition that their sum is zero when the problem is considered in the entire space.

In the regime where ∂_t scales like ϵ^2, the Euler equations for an incompressible fluid are asymptotically recovered away from the vortices, while the vortex motion obeys the classical Kirchhoff law for fluid vortices (Neu 1990, Lund 1991, Creswick, and Morrison 1980, Ercolani and Montgomery 1993 and references therein, Ovchinnikov and Sigal 1998a, Colliander and Jerrard 1998). Application to the numerical simulation of classical fluids was proposed by Nore, Abid and Brachet (1994).

2
Structural Properties

This chapter, which assumes localized solutions vanishing at infinity, deals with the Lagrangian and Hamiltonian structures of the NLS equation and with the derivation of conservation laws, some of them being related through the Noether theorem to invariance properties of the action (see, e.g., Rasmussen and Rypdal 1986 and, for a comprehensive discussion, Ginibre 1998, where a systematic approach is presented). The so-called variance identity, or virial theorem which together with the conservation of the Hamiltonian plays a central role in proving finite–time singularities for the elliptic NLS equation, is also derived.

2.1 Variational Formulation

2.1.1 Lagrangian Structure

Let ψ denote a smooth solution of the NLS equation written in the canonical form

$$i\psi_t + \nabla \cdot \mathcal{D}\nabla\psi + f(|\psi|^2)\psi = 0. \tag{2.1.1}$$

The symmetric matrix $\mathcal{D}$ is defined as one half of the frequency Hessian matrix (see Section 1.1.1). When it is positive definite (elliptic NLS equation), it will be taken as the identity matrix (through a simple coordinate transformation), thus reducing the linear operator in the NLS equation to a Laplacian. We assume the problem defined in $\mathbf{R}^d$, with ψ and its derivatives vanishing at infinity. The nonlinearity f is a smooth function of $|\psi|^2$

and one defines

$$F(\lambda) = \int_0^\lambda f(\lambda)d\lambda. \tag{2.1.2}$$

A Lagrangian density $\mathcal{L}$ associated to (2.1.1) can be written in terms of the real and imaginary parts u and v of ψ, or equivalently in terms of ψ and ψ^* viewed as independent variables, in the form

$$\mathcal{L} = \frac{i}{2}(\psi^*\psi_t - \psi\psi_t^*) - \frac{1}{2}(\nabla\psi \cdot \mathcal{D}\nabla\psi^* + \nabla\psi^* \cdot \mathcal{D}\nabla\psi) + F(|\psi|^2). \tag{2.1.3}$$

One considers the action

$$S\{\psi, \psi^*\} = \int_{t_0}^{t_1} \int_{\mathbf{R}^d} \mathcal{L} \, d\mathbf{x} \, dt \tag{2.1.4}$$

as a functional on all admissible regular functions satisfying the prescribed conditions $\psi(\mathbf{x}, t_0) = \psi_0(\mathbf{x})$ and $\psi(\mathbf{x}, t_1) = \psi_1(\mathbf{x})$. Its variation

$$\delta S = S\{\psi + \delta\psi, \psi^* + \delta\psi^*\} - S\{\psi, \psi^*\} \tag{2.1.5}$$

for infinitesimal $\delta\psi$ and $\delta\psi^*$ reads

$$\delta S = \int_{t_0}^{t_1} \int_{\mathbf{R}^d} \left[\frac{\partial\mathcal{L}}{\partial\psi}\delta\psi + \frac{\partial\mathcal{L}}{\partial\nabla\psi} \cdot \nabla\delta\psi + \frac{\partial\mathcal{L}}{\partial\psi_t}\delta\psi_t\right] d\mathbf{x} \, dt + \text{c.c.}, \tag{2.1.6}$$

where $\frac{\partial\mathcal{L}}{\partial\nabla\psi}$ denotes the vector of components $\frac{\partial\mathcal{L}}{\partial(\partial_i\psi)}$ with $i = 1, \dots, d$. After integrations by parts, one gets

$$\delta S = \int_{t_0}^{t_1} \int_{R^d} \left[\frac{\partial\mathcal{L}}{\partial\psi} - \nabla \cdot \left(\frac{\partial\mathcal{L}}{\partial\nabla\psi}\right) - \partial_t\left(\frac{\partial\mathcal{L}}{\partial\psi_t}\right)\right] \delta\psi \, d\mathbf{x} \, dt$$

$$+ \left[\frac{\partial\mathcal{L}}{\partial\psi_t}\delta\psi\right]_{t_0}^{t_1} + \text{c.c.} \tag{2.1.7}$$

A necessary and sufficient condition for a function $\psi(\mathbf{x}, t)$ to lead to an extremum for the action S among the functions with prescribed values $\psi(., t_0)$ and $\psi(., t_1)$ at times t_0 and t_1, thus reduces to the Euler–Lagrange equations

$$\frac{\partial\mathcal{L}}{\partial\psi} = \nabla \cdot \left(\frac{\partial\mathcal{L}}{\partial\nabla\psi}\right) + \partial_t\left(\frac{\partial\mathcal{L}}{\partial\psi_t}\right), \tag{2.1.8}$$

which, when the Lagrangian (2.1.3) is used, reduces to the NLS equation. This system is easily rewritten in terms of the real fields $u = \frac{1}{2}(\psi + \psi^*)$ and $v = \frac{1}{2i}(\psi - \psi^*)$ as

$$\frac{\partial\mathcal{L}}{\partial u} = \nabla \cdot \left(\frac{\partial\mathcal{L}}{\partial\nabla u}\right) + \partial_t\left(\frac{\partial\mathcal{L}}{\partial u_t}\right), \tag{2.1.9}$$

$$\frac{\partial\mathcal{L}}{\partial v} = \nabla \cdot \left(\frac{\partial\mathcal{L}}{\partial\nabla v}\right) + \partial_t\left(\frac{\partial\mathcal{L}}{\partial v_t}\right). \tag{2.1.10}$$

The derivation of the NLS equation from a variational principle is the starting point of a heuristic apprach called the method of "collective co-ordinates," the "variational approach" or the "Rayleigh–Ritz optimization principle," where the solution is assumed to maintain a prescribed approximate profile (often Gaussian). Such methods strongly simplify the problem, reducing it to a system of ordinary differential equations for the evolution of a few characteristic parameters (Cooper, Lucheroni, and Shepard 1992 and Bergé 1998 for review). They must, however, be viewed as preliminary estimates unable, for example, to capture the delicate balance associated with critical collapse.

2.1.2 Hamiltonian Structure

As usual, a Hamiltonian structure is easily derived from the existence of a Lagrangian. Writing $\psi = u + iv$ in order to deal with real fields, the Hamiltonian density $\mathcal{H} = \frac{i}{2}(\psi^*\partial_t\psi - \psi\partial_t\psi^*) - \mathcal{L}$ becomes

$$\mathcal{H} = v\partial_t u - u\partial_t v - \mathcal{L}. \qquad (2.1.11)$$

Introducing the canonical variables

$$q_1 \equiv u \quad , \quad p_1 \equiv \frac{\partial \mathcal{L}}{\partial(\partial_t q_1)}, \qquad (2.1.12)$$

$$q_2 \equiv v \quad , \quad p_2 \equiv \frac{\partial \mathcal{L}}{\partial(\partial_t q_2)}, \qquad (2.1.13)$$

it takes the form

$$\mathcal{H} = \sum_i p_i \partial_t q_i - \mathcal{L}. \qquad (2.1.14)$$

One also defines

$$\boldsymbol{\rho}_i \equiv \frac{\partial \mathcal{L}}{\partial(\nabla q_i)}, \qquad (2.1.15)$$

and rewrites the Euler–Lagrange equations as

$$\frac{\partial \mathcal{L}}{\partial q_i} = \nabla \cdot \boldsymbol{\rho}_i + \partial_t p_i. \qquad (2.1.16)$$

Using that

$$\partial_t \mathcal{L} = \sum_i \frac{\partial \mathcal{L}}{\partial q_i}\partial_t q_i + \frac{\partial \mathcal{L}}{\partial(\nabla q_i)}\partial_t \nabla q_i + \frac{\partial \mathcal{L}}{\partial(\partial_t q_i)}\partial_{tt} q_i, \qquad (2.1.17)$$

and the Euler–Lagrange equations, one gets

$$\partial_t \mathcal{H} = -\nabla \cdot \sum_i \boldsymbol{\rho}_i \partial_t q_i, \qquad (2.1.18)$$

which ensures the conservation of the Hamiltonian $H = \int \mathcal{H}d\mathbf{x}$.

Similarly, from the variation of the Lagrangian density

$$\delta\mathcal{L} = \sum_i \frac{\delta\mathcal{L}}{\delta q_i}\delta q_i + \frac{\delta\mathcal{L}}{\delta(\nabla q_i)}\nabla\delta q_i + \frac{\delta\mathcal{L}}{\delta(\partial_t q_i)}\partial_t\delta q_i, \tag{2.1.19}$$

the Euler–Lagrange equations, and the definition of p_i one obtains the variation of the Hamiltonian H, in the form

$$\delta H = \sum_i \int (\partial_t q_i \delta p_i - \partial_t p_i \delta q_i)dx, \tag{2.1.20}$$

which leads to the Hamilton equations

$$\frac{\partial q_i}{\partial t} = \frac{\delta H}{\delta p_i}, \qquad \frac{\partial p_i}{\partial t} = -\frac{\delta H}{\delta q_i}, \tag{2.1.21}$$

or in complex form,

$$i\partial_t\psi = \frac{\delta H}{\delta\psi^*}. \tag{2.1.22}$$

2.2   The Noether Theorem

Conservation laws are of great importance for the analysis of the NLS equation. Some of them are related to the invariance of the action relative to several groups of transformations through the Noether theorem that we derive in this section, following the presentation given in Gelfand and Fomin (1963). We also refer to Bluman and Kumei (1989) for a detailed discussion.

Let us consider the action

$$S\{\psi\} = \int_{t_0}^{t_1}\int \mathcal{L}(\psi, \nabla\psi, \psi_t, \psi^*, \nabla\psi^*, \psi_t^*)d\mathbf{x}\, dt. \tag{2.2.1}$$

To simplify the notations, we introduce $\xi = (t, \mathbf{x}) = (\xi_0, \xi_1, \ldots, \xi_d)$, $\partial_0 = \partial_t$, $\partial = (\partial_0, \partial_1, \ldots, \partial_d)$, and $\Psi = (\Psi_1, \Psi_2) = (\psi, \psi^*)$. Furthermore, in this section, we denote by $\mathcal{D}$ the space–time integration domain.

We now consider a group of transformations T^ϵ depending on a parameter ϵ, such that under this transformation,

$$\xi \mapsto \widetilde{\xi}(\xi, \Psi, \epsilon), \qquad \Psi \mapsto \widetilde{\Psi}(\xi, \Psi, \epsilon), \tag{2.2.2}$$

where $\widetilde{\xi}$ and $\widetilde{\Psi}$ are assumed to be differentiable with respect to ϵ. For $\epsilon = 0$, the transformation reduces to the identity. Taking ϵ infinitesimal, we write

$$\widetilde{\xi} = \xi + \delta\xi, \qquad \widetilde{\Psi} = \Psi + \delta\Psi, \tag{2.2.3}$$

where $\delta\xi$ and $\delta\Psi$ are proportional to ϵ. By the transformation T^ϵ, $\Psi(\xi)$ is transformed into $\widetilde{\Psi}(\widetilde{\xi})$ and the domain $\mathcal{D}$ into the domain $\widetilde{\mathcal{D}}$. The action

$$S\{\psi\} = \int_{\mathcal{D}} \mathcal{L}(\Psi, \partial\Psi)d\xi, \tag{2.2.4}$$

where ∂ refers to differentiation with respect to ξ, becomes

$$\widetilde{S}\{\widetilde{\psi}\} = \int_{\widetilde{\mathcal{D}}} \mathcal{L}(\widetilde{\Psi}, \widetilde{\partial}\widetilde{\Psi})d\widetilde{\xi}, \tag{2.2.5}$$

where $\widetilde{\partial}$ denotes the differentiation with respect to $\widetilde{\xi}$. We are here interested in transformations leaving S invariant. Note that $\delta\xi$ and $\delta\psi$ depend on ξ and ψ, so that one can prescribe the bounds t_0 and t_1 of the integral defining the action to remain unchanged by the transformation. We define

$$\delta S = \widetilde{S}\{\widetilde{\psi}\} - S\{\psi\}, \tag{2.2.6}$$

and first notice that in the limit of infinitesimal ϵ, we can expand the Jacobian $\frac{\partial(\widetilde{\xi}_0,...,\widetilde{\xi}_d)}{\partial(\xi_0,...,\xi_d)} \approx 1 + \sum_{\mu=0}^{d} \frac{\partial\delta\xi_\mu}{\partial\xi_\mu}$. It follows that

$$\delta S = \int_{\mathcal{D}} \left[\mathcal{L}(\widetilde{\Psi}, \widetilde{\partial}\widetilde{\Psi}) - \mathcal{L}(\Psi, \partial\Psi) \right] d\xi + \int_{\mathcal{D}} \mathcal{L}(\Psi, \partial\Psi) \sum_{\mu=0}^{d} \frac{\partial\delta\xi_\mu}{\partial\xi_\mu} d\xi, \tag{2.2.7}$$

where in the last term of the right-hand side, $\mathcal{L}(\widetilde{\Psi}, \widetilde{\partial}\widetilde{\Psi})$ was replaced to leading order by $\mathcal{L}(\Psi, \partial\Psi)$. Again, to leading order, we have

$$\mathcal{L}(\widetilde{\Psi}, \widetilde{\partial}\widetilde{\Psi}) - \mathcal{L}(\Psi, \partial\Psi) = \frac{\partial\mathcal{L}}{\partial\Psi_i}(\widetilde{\Psi}_i(\widetilde{\xi}) - \Psi_i(\xi)) + \frac{\partial\mathcal{L}}{\partial(\partial_\mu\Psi_i)}(\widetilde{\partial}_\mu\widetilde{\Psi}_i(\widetilde{\xi}) - \partial_\mu\Psi_i(\xi)), \tag{2.2.8}$$

where summation is performed on the repeated indices. We also define

$$\delta\widehat{\Psi}_i \equiv \widetilde{\Psi}_i(\widetilde{\xi}) - \Psi_i(\xi) = \partial_\mu\Psi_i\,\delta\xi_\mu + \delta\Psi_i(\xi) \tag{2.2.9}$$

which includes the variations due to both the perturbation of the independent variable ξ and of the function Ψ. Similarly,

$$\widetilde{\partial}_\mu\widetilde{\Psi}_i(\widetilde{\xi}) - \partial_\mu\Psi_i(\xi) = (\widetilde{\partial}_\mu - \partial_\mu)\widetilde{\Psi}_i(\widetilde{\xi}) + \partial_\mu(\widetilde{\Psi}_i(\widetilde{\xi}) - \Psi(\xi)), \tag{2.2.10}$$

with

$$\partial_\mu = \frac{\partial\widetilde{\xi}_\nu}{\partial\xi_\mu}\widetilde{\partial}_\nu = (\delta_{\mu\nu} + \frac{\partial\delta\xi_\nu}{\partial\xi_\mu})\widetilde{\partial}_\nu = \widetilde{\partial}_\mu + \frac{\partial\delta\xi_\nu}{\partial\xi_\mu}\widetilde{\partial}_\nu. \tag{2.2.11}$$

We thus get

$$\mathcal{L}(\widetilde{\Psi}, \widetilde{\partial}\widetilde{\Psi}) - \mathcal{L}(\Psi, \partial\Psi) = \frac{\partial\mathcal{L}}{\partial\Psi_i}(\partial_\mu\Psi_i\delta\xi_\mu + \delta\Psi_i(\xi))$$
$$+ \frac{\partial\mathcal{L}}{\partial(\partial_\mu\Psi_i)}\left(-\frac{\partial\delta\xi_\nu}{\partial\xi_\mu}\widetilde{\partial}_\nu\widetilde{\Psi}_i + \partial_\mu(\partial_\nu\Psi_i\delta\xi_\nu + \delta\Psi_i) \right). \tag{2.2.12}$$

We finally use

$$\frac{\partial}{\partial\xi_\mu}(\mathcal{L}\,\delta\xi_\mu) = \mathcal{L}\frac{\partial\,\delta\xi_\mu}{\partial\xi_\mu} + \frac{\partial\mathcal{L}}{\partial\Psi_i}\partial_\mu\Psi_i\,\delta\xi_\mu + \frac{\partial^2\mathcal{L}}{\partial(\partial_\nu\Psi_i)}\partial^2_{\mu\nu}\Psi_i\,\delta\xi_\mu \tag{2.2.13}$$

and

$$\frac{\partial \mathcal{L}}{\partial(\partial_\mu \Psi_i)}\partial_\mu\, \delta\Psi_i = \frac{\partial}{\partial\xi_\mu}\left(\frac{\partial \mathcal{L}}{\partial(\partial_\mu \Psi_i)}\delta\Psi_i\right) - \frac{\partial}{\partial\xi_\mu}\left(\frac{\partial \mathcal{L}}{\partial(\partial_\mu \Psi_i)}\right)\delta\Psi_i. \quad (2.2.14)$$

This leads to

$$\mathcal{L}(\tilde{\Psi},\partial\tilde{\Psi}) - \mathcal{L}(\Psi,\partial\Psi) = \frac{\partial \mathcal{L}}{\partial \Psi_i}\delta\Psi_i + \partial_\mu\left(\mathcal{L}\,\delta\xi_\mu\right) - \mathcal{L}\frac{\partial\,\delta\xi_\mu}{\partial\xi_\mu}$$
$$+\partial_\mu(\frac{\partial \mathcal{L}}{\partial(\partial_\mu \Psi_i)}\delta\Psi_i) - \partial_\mu\left(\frac{\partial \mathcal{L}}{\partial(\partial_\mu \Psi_i)}\right)\delta\Psi_i.$$
$$(2.2.15)$$

Substituting in the variation δS of the action, we obtain

$$\delta S = \int_\mathcal{D}\left[\frac{\partial \mathcal{L}}{\partial \Psi_i} - \frac{\partial}{\partial\xi_\mu}\left(\frac{\partial \mathcal{L}}{\partial(\partial_\mu \Psi_i)}\right)\right]\delta\Psi_i d\xi$$
$$+ \int_\mathcal{D}\frac{\partial}{\partial\xi_\mu}\left(\mathcal{L}\delta\xi_\mu + \frac{\partial \mathcal{L}}{\partial(\partial_\mu \Psi_i)}\delta\Psi_i\right)d\xi, \quad (2.2.16)$$

where on the right-hand side the first integral vanishes because of the Euler–Lagrange equations. Since the domain $\mathcal{D}$ is arbitrary, this leads to the following result.

Theorem 2.1. *If the action (2.2.1) is invariant by the infinitesimal transformation of the dependent and independent variables $\psi \mapsto \psi + \delta\psi$, $\xi \mapsto \xi + \delta\xi$ where $\xi = (t, x_1, \ldots, x_d)$, the following conservation law holds:*

$$\frac{\partial}{\partial\xi_\mu}\left(\mathcal{L}\delta\xi_\mu + \frac{\partial \mathcal{L}}{\partial(\partial_\mu \Psi_i)}\delta\Psi_i\right) = 0, \quad (2.2.17)$$

or in terms of $\delta\widehat{\Psi}_i$ defined in (2.2.9),

$$\frac{\partial}{\partial\xi_\mu}\left[\mathcal{L}\delta\xi_\mu + \frac{\partial \mathcal{L}}{\partial(\partial_\mu \Psi_i)}\left(\delta\widehat{\Psi}_i - \frac{\partial\Psi_i}{\partial\xi_\nu}\delta\xi_\nu\right)\right] = 0. \quad (2.2.18)$$

By integration on the space variables, we get the following result.

Theorem 2.2. *If the action is invariant by the infinitesimal transformation*

$$t \mapsto \tilde{t} = t + \delta t(\mathbf{x}, t, \psi), \quad (2.2.19)$$
$$\mathbf{x} \mapsto \tilde{\mathbf{x}} = \mathbf{x} + \delta\mathbf{x}(\mathbf{x}, t, \psi), \quad (2.2.20)$$
$$\psi(\mathbf{x}, t) \mapsto \tilde{\psi}(\tilde{t}, \tilde{\mathbf{x}}) = \psi(t, \mathbf{x}) + \delta\psi(t, \mathbf{x}), \quad (2.2.21)$$

then

$$\int\left[\frac{\partial \mathcal{L}}{\partial\psi_t}\left(\partial_t\psi\,\delta t + \nabla\psi\cdot\delta\mathbf{x} - \delta\psi\right) + \frac{\partial \mathcal{L}}{\partial\psi_t^*}\left(\partial_t\psi^*\,\delta t + \nabla\psi^*\cdot\delta\mathbf{x} - \delta\psi^*\right) - \mathcal{L}\delta t\right]d\mathbf{x}$$
$$(2.2.22)$$

is a conserved quantity.

2.3 Invariances and Conservation Laws

In the context of the NLS equation , we have $\frac{\partial \mathcal{L}}{\partial \psi_t} = \frac{i}{2}\psi^*$ and $\frac{\partial \mathcal{L}}{\partial \psi_t^*} = -\frac{i}{2}\psi$, and several conservation laws result from the Noether theorem.

Invariance by Phase Shift or Gauge Invariance

One easily checks that the action associated to the NLS equation is invariant by the phase shift transformation $\widetilde{\psi} = e^{is}\psi$, which for infinitesimal s gives $\delta\psi = is\psi$ with $\delta t = \delta\mathbf{x} = 0$. Equation (2.2.18) becomes in this case

$$\partial_t |\psi|^2 + \nabla \cdot \{i(\psi\mathcal{D}\nabla\psi^* - \psi^*\mathcal{D}\nabla\psi)\} = 0. \tag{2.3.1}$$

This leads to the conservation of the *wave energy*, also called *mass, wave action, plasmon number*, or in optics *wave power*,

$$N = \int |\psi|^2 d\mathbf{x}. \tag{2.3.2}$$

In the following, we shall use any of these terms, according to the context.

Invariance by Time Translation

The action is invariant by the infinitesimal time translation $t \mapsto t + \delta t$ with $\delta\mathbf{x} = \delta\psi = \delta\psi^* = 0$. The matrix $\mathcal{D}$ being symmetric, (2.2.18) becomes

$$\partial_t(\nabla\psi \cdot \mathcal{D}\nabla\psi^* - F(|\psi|^2)) - \nabla \cdot (\psi_t \mathcal{D}\nabla\psi^* + \psi_t^* \mathcal{D}\nabla\psi) = 0. \tag{2.3.3}$$

This leads to the conservation of the *Hamiltonian* [1]

$$H = \int \left(\nabla\psi \cdot \mathcal{D}\nabla\psi^* - F(|\psi|^2) \right) d\mathbf{x}. \tag{2.3.4}$$

Invariance by Space Translation

The action is invariant by the infinitesimal space translation $\mathbf{x} \mapsto \mathbf{x} + \delta\mathbf{x}$ with $\delta t = \delta\psi = \delta\psi^* = 0$. In this case, eq. (2.2.18) becomes (I denotes the identity matrix)

$$\partial_t\{i(\psi\nabla\psi^* - \psi^*\nabla\psi)\}$$
$$+ \nabla \cdot \{2(\mathcal{D}\nabla\psi^* \otimes \nabla\psi + \mathcal{D}\nabla\psi \otimes \nabla\psi^* + \mathcal{L}I)\} = 0, \tag{2.3.5}$$

where, in expressing ψ_t and ψ_t^* by means of the NLS equation, the Lagrangian density $\mathcal{L}$ reads

$$\mathcal{L} = -\frac{1}{2}\nabla \cdot (\psi^*\mathcal{D}\nabla\psi + \psi\mathcal{D}\nabla\psi^*) - f(|\psi|^2)|\psi|^2 + F(|\psi|^2). \tag{2.3.6}$$

---

[1]We are using here a terminology adapted to waves. In a quantum mechanical context (where $\mathcal{D} = I$), this quantity corresponds to the energy, the Hamiltonian being defined as the operator $-\Delta - f(|\psi|^2)$.

This leads to the conservation of the *linear momentum*

$$\mathbf{P} = i \int (\psi \nabla \psi^* - \psi^* \nabla \psi) d\mathbf{x}. \tag{2.3.7}$$

Note that the *mass center*

$$\mathbf{X} = \frac{1}{N} \int \mathbf{x} |\psi|^2 d\mathbf{x} \tag{2.3.8}$$

obeys

$$N \frac{d\mathbf{X}}{dt} = \int \mathbf{x} \partial_t |\psi|^2 d\mathbf{x} = - \int \mathbf{x} \nabla \cdot \{ i(\psi \mathcal{D} \nabla \psi^* - \psi^* \mathcal{D} \nabla \psi) \} d\mathbf{x}$$

$$= \int i(\psi \mathcal{D} \nabla \psi^* - \psi^* \mathcal{D} \nabla \psi) d\mathbf{x} = \mathcal{D} \mathbf{P}, \tag{2.3.9}$$

and thus moves at a constant speed. For the elliptic NLS equation where $\mathcal{D} = I$, one recovers the usual definition of the momentum of a mechanical system.

We also remark that the energy (or mass) current density arising in (2.3.1)

$$\mathcal{J} = i(\psi \mathcal{D} \nabla \psi^* - \psi^* \mathcal{D} \nabla \psi) \tag{2.3.10}$$

is related to the linear momentum density

$$\mathcal{P} = i(\psi \nabla \psi^* - \psi^* \nabla \psi) \tag{2.3.11}$$

considered in (2.3.5) by

$$\mathcal{J} = \mathcal{D} \mathcal{P}. \tag{2.3.12}$$

This is an important property for the derivation of the variance identity (see Section 2.4).

Invariance by Space Rotation for Elliptic NLS Equation

For $\mathcal{D} = I$, the action is invariant by a rotation of angle $\delta\theta$ around an axis $\mathbf{n}$. In this case, $\delta t = \delta\psi = \delta\psi^* = 0$ and $\delta\mathbf{x} = \delta\theta\, \mathbf{n} \times \mathbf{x}$. This leads to the conservation of the *angular momentum*

$$\mathbf{M} = i \int \mathbf{x} \times (\psi^* \nabla \psi - \psi \nabla \psi^*) d\mathbf{x}. \tag{2.3.13}$$

Galilean Invariance

The NLS equation is invariant by the Galilean transformation

$$\mathbf{x} \mapsto \mathbf{x}' = \mathbf{x} - \mathcal{D}\mathbf{c}\, t, \tag{2.3.14}$$

$$t \mapsto t' = t, \tag{2.3.15}$$

$$\psi(\mathbf{x}, t) \mapsto \psi'(\mathbf{x}', t') = e^{-i[\frac{1}{2}\mathbf{c}\cdot\mathbf{x}' + \frac{1}{4}\mathbf{c}\cdot\mathcal{D}\mathbf{c}\, t']} \psi(\mathbf{x}' + \mathcal{D}\mathbf{c}\, t', t'), \tag{2.3.16}$$

which also keeps the action invariant. For an infinitesimal velocity $\mathbf{c}$, $\delta\mathbf{x} = -\mathcal{D}\mathbf{c}t$, $\delta t = 0$, and $\widehat{\delta\psi} = \psi'(\mathbf{x}', t') - \psi(\mathbf{x}, t) = -\frac{i}{2}\mathbf{c}\mathbf{x}\psi(\mathbf{x}, t)$. After integration on the space variables, (2.2.18) then leads to the conservation law (2.3.9), which expresses that the velocity of the wave-packet mass center is uniform.

Scale Invariance for Power Law Nonlinearities

In the case of a power law nonlinearity

$$f(|\psi|^2) = q|\psi|^{2\sigma}, \qquad (2.3.17)$$

the NLS equation is invariant by the scale transformation

$$\mathbf{x} \mapsto \mathbf{x}' = \lambda^{-1}\mathbf{x}, \qquad (2.3.18)$$

$$t \mapsto t' = \lambda^{-2}t, \qquad (2.3.19)$$

$$\psi(\mathbf{x}, t) \mapsto \psi'(\mathbf{x}', t') = \lambda^{1/\sigma}\psi(\lambda\mathbf{x}', \lambda^2 t'). \qquad (2.3.20)$$

Note that under this transformation, the action scales like $\lambda^{\frac{2}{\sigma} - \frac{1}{d}}$. So does the wave energy $\int |\psi|^2 d\mathbf{x}$. It follows that for $\sigma d = 2$ which corresponds to the criticality at the level of the L^2-norm (see Section 3.2.1), the action is invariant by the scale transformation keeping the equation invariant. It turns out that the condition $\sigma d = 2$ also corresponds to criticality for the possible existence of blowup solutions (see Sections 3.2.2 and 5.1) and is usually referred to as the *critical dimension*. In this special case, the invariance by scale transformation leads to an additional conservation law. For an infinitesimal transformation, $\lambda = 1 - \epsilon$ with $|\epsilon| \ll 1$, one has $\delta\mathbf{x} = \epsilon\mathbf{x}$, $\delta t = 2\epsilon t$ and $\widehat{\delta\psi} = -\frac{\epsilon}{\sigma}\psi(\mathbf{x}, t)$. Equation (2.2.18) then leads to

$$\int \{2\epsilon t\mathcal{L} - \frac{i}{2}\psi^*(\frac{\epsilon}{\sigma}\psi + \epsilon\mathbf{x}\cdot\nabla\psi + 2\epsilon t\psi_t) + \frac{i}{2}\psi(\frac{\epsilon}{\sigma}\psi^* + \epsilon\mathbf{x}\cdot\nabla\psi^* + 2\epsilon t\psi_t^*)\}d\mathbf{x} = \epsilon C_1,$$
$$(2.3.21)$$

or

$$i\int \mathbf{x}\cdot(\psi\nabla\psi^* - \psi^*\nabla\psi)d\mathbf{x} - 4Ht = 2C_1, \qquad (2.3.22)$$

where C_1 is a constant.

Pseudo-Conformal Invariance at Critical Dimension

As noted by Talanov (1970) in the case of the elliptic NLS equation, there is an additional symmetry when the nonlinearity is a power law $q|\psi|^{2\sigma}\psi$ at the critical dimension $\sigma d = 2$. Although this invariance is not restricted to the elliptic NLS equation, we first discuss it in this case for the sake of simplicity. Defining

$$\ell(t) = \frac{t_* - t}{t_0}, \quad t < t_*, \qquad (2.3.23)$$

where t_0 appears as an arbitrary time unit, the NLS equation at critical dimension is invariant by the "pseudo-conformal" transformation

$$\mathbf{x} \mapsto \mathbf{x}' = \frac{\mathbf{x}}{\ell(t)}, \tag{2.3.24}$$

$$t \mapsto t' = \int_0^t \frac{1}{\ell^2(s)} ds = \frac{t_0^2 t}{t_*(t_* - t)}, \tag{2.3.25}$$

$$\psi \mapsto \psi'(\mathbf{x}', t') = \ell^{d/2} \psi(\mathbf{x}, t) \exp\left(-i\frac{\ell_t}{\ell}\frac{|\mathbf{x}|^2}{4}\right),$$

$$= \ell^{d/2} \psi(\mathbf{x}, t) \exp\left(i\frac{a|\mathbf{x}|^2}{4\ell^2}\right), \tag{2.3.26}$$

with

$$a = -\ell\frac{d\ell}{dt} = -\frac{1}{\ell}\frac{d\ell}{dt'} = \frac{t_* - t}{t_0^2}. \tag{2.3.27}$$

A consequence of this symmetry is the existence for the *elliptic* NLS equation at critical dimension of an explicit solution singular at some finite time t_* (see Section 5.2.2).

When expressing ψ' in terms of ψ, the transformed Hamiltonian

$$H' = \int \left(|\nabla'\psi'|^2 - \frac{q}{\sigma+1}|\psi'|^{2\sigma+2}\right) d\mathbf{x}' \tag{2.3.28}$$

becomes

$$H' = \int \left(\left|\ell\nabla\psi + \frac{i}{2}\mathbf{x}\psi\right|^2 - \frac{\ell^2 q}{\sigma+1}|\psi|^{2\sigma+2}\right) d\mathbf{x} \tag{2.3.29}$$

and is conserved. Choosing $t_* = 0$, this yields the conservation of

$$C_2 = \int \left(|\mathbf{x}\psi + 2it\nabla\psi|^2 - \frac{4t^2 q}{\sigma+1}|\psi|^{2\sigma+2}\right) d\mathbf{x}. \tag{2.3.30}$$

As noticed by Kuznetsov and Turitsyn (1985), the above transformation preserves the action, and the invariant given by (2.3.30) is in fact a consequence of the Noether theorem. This is seen by considering a transformation close to identity by choosing $t_0 = t_*$ and by assuming that $\delta\lambda = 1/t_*$ is infinitely small. In this case, $\delta\mathbf{x} = \mathbf{x}t\delta\lambda$, $\delta t = t^2\delta\lambda$, and $\delta\widehat{\psi} = (-\frac{d}{2}t + \frac{i}{4}|\mathbf{x}|^2)\psi\delta\lambda$. Equation (2.2.18) then leads after space integration to

$$\int \left[t^2\left(-|\nabla\psi|^2 + \frac{q}{\sigma+1}|\psi|^{2\sigma+2}\right) + \frac{i}{2}t\mathbf{x}\cdot(\psi\nabla\psi^* - \psi^*\nabla\psi) - \frac{1}{4}|\mathbf{x}|^2|\psi|^2\right] d\mathbf{x} = C_2, \tag{2.3.31}$$

or equivalently to (2.3.30).

In optics, the pseudo-conformal transformation is known as the "lens transformation". In this context, the NLS equation is written in the form

$$i\psi_z + \Delta_\perp\psi + |\psi|^2\psi = 0 \tag{2.3.32}$$

with $\Delta_\perp = \partial_{xx} + \partial_{yy}$. It is also convenient to write

$$\ell = 1 + \frac{z}{f} \qquad (2.3.33)$$

where f is a constant. With these notations, the transformation then becomes

$$(x, y) \mapsto (\xi, \eta) = \left(\frac{x}{\ell(z)}, \frac{y}{\ell(z)} \right) \qquad (2.3.34)$$

$$z \mapsto \zeta = \int_0^z \frac{1}{\ell^2(s)} ds = \frac{z}{1 + z/f} \qquad (2.3.35)$$

$$\psi \mapsto \psi'(\xi, \eta, \zeta) = \ell \psi(x, y, z) \exp\left(-i \frac{\ell_z}{\ell} \frac{x^2 + y^2}{4} \right). \qquad (2.3.36)$$

The name "lens transformation" originates from the observation that

$$\frac{1}{\zeta} = \frac{1}{z} + \frac{1}{f}, \quad \sqrt{\xi^2 + \eta^2} = (\sqrt{x^2 + y^2})/\ell, \qquad (2.3.37)$$

and that a solution with initial condition $\psi_0(x, y)$ is transformed into a solution with initial condition $\psi_0(x, y) e^{i \frac{x^2 + y^2}{4f}}$. The transformation can be viewed as the effect of a thin lens with a focal distance f, located at $z = 0$. Note that the invariance by the lens transform holds in any dimension in the case of the linear Schrödinger equation, but only at critical dimension in the presence of a power-law nonlinearity.

As mentioned by Kuznetsov and Turitsyn (1985) the pseudo-conformal transformation also holds in the case of the hyperbolic NLS equation

$$i\psi_t + \frac{\partial^2 \psi}{\partial x^2} - \frac{\partial^2 \psi}{\partial y^2} + |\psi|^2 \psi = 0, \qquad (2.3.38)$$

This equations arises for example in the description of gravity wave at the surface of a liquid of infinite depth (see the Remark in Section 12.1.3 for the case of a liquid of finite depth). In this case one should replace y by iy in the definition of the transformation (2.3.26) and in the resulting conservation laws (2.3.30) (where $\frac{\partial}{\partial y}$ is also changed into $-i\frac{\partial}{\partial y}$).

2.4 Variance Identity

The square width of the wave packet is estimated by the quantity

$$v \equiv \frac{1}{N} \int (\mathbf{x} - \mathbf{X})^2 |\psi|^2 d\mathbf{x} = \frac{1}{N} \int \left(|\mathbf{x}|^2 - |\mathbf{X}|^2 \right) |\psi|^2 d\mathbf{x}, \qquad (2.4.1)$$

where $\mathbf{X}$ defined in (2.3.8) obeys (2.3.9). We rewrite

$$v = \frac{V}{N} - |\mathbf{X}|^2, \qquad (2.4.2)$$

where

$$V = \int |\mathbf{x}|^2 |\psi|^2 d\mathbf{x} \tag{2.4.3}$$

is called the "variance" (or the "momentum of inertia" in a context where N is referred to as the mass of the wave packet).

Due to the conservation of the wave energy N and of the linear momentum $\mathbf{P}$,

$$\frac{d^2 v}{dt^2} = \frac{1}{N} \frac{d^2 V}{dt^2} - 2 \frac{|\mathbf{P}|^2}{N^2}. \tag{2.4.4}$$

2.4.1   Elliptic NLS Equation

The key point to compute $\frac{d^2 V}{dt^2}$ is to notice that when $\mathcal{D} = I$, the energy (or mass) current density $\mathcal{J}$ and the linear momentum density $\mathcal{P}$ are equal. It follows that (Vlasov, Petrishchev, and Talanov 1971, Rypdal, Rasmussen, and Thomsen 1985)

$$\partial_{tt} |\psi|^2 = \nabla \cdot \nabla \cdot \mathcal{S} \tag{2.4.5}$$

with

$$\mathcal{S} = 2 \left(\nabla \psi^* \otimes \nabla \psi + \nabla \psi \otimes \nabla \psi^* + \mathcal{L} I \right). \tag{2.4.6}$$

We thus have

$$\frac{d^2 V}{dt^2} = \int |\mathbf{x}^2| \partial_i \partial_j \mathcal{S}_{ij} \, d\mathbf{x} \tag{2.4.7}$$

and after integration by parts

$$\frac{d^2 V}{dt^2} = 2 \int \sum_j \mathcal{S}_{jj} d\mathbf{x} = 4 \int \left(2|\nabla \psi|^2 + d\mathcal{L} \right) d\mathbf{x}$$

$$= 4 \int \left\{ 2|\nabla \psi|^2 + d\, F(|\psi|^2) - d\, |\psi|^2 f(|\psi|^2) | \right\} d\mathbf{x}$$

$$= 8H - 4 \int \left(d\, |\psi|^2 f(|\psi|^2) - (d+2)\, F(|\psi|^2) \right) d\mathbf{x}. \tag{2.4.8}$$

In the special case of a power nonlinearity $f(\lambda) = q\lambda^\sigma$, we get

$$\frac{d^2 V}{dt^2} = 8H - 4 \frac{d\sigma - 2}{\sigma + 1} q \int |\psi|^{2\sigma + 2} \, d\mathbf{x}. \tag{2.4.9}$$

This identity is central for proving finite-time blowup in dimension $d\sigma \geq 2$ for solutions with negative Hamiltonian since eq. (2.4.8) implies that the variance of smooth solutions (which is strictly positive) vanishes in a finite time (see Chapter 5). An interesting analogy with the collapse taking place in the classical n-body problem is discussed by Berkshire and Gibbon (1983) (see also Velo 1996).

Note that this "collapse" does not necessarily corresponds to a singularity for the primitive wave equation, but only to a breakdown of the modulational analysis leading to the (cubic when $\sigma = 1$) NLS equation. A simple example is provided by the Sine–Gordon equation (1.1.35) which remains globally well-posed. Regularization were proposed by retaining higher order saturating terms in the expansion, corresponding to higher order nonlinearity or to the breaking of the paraxial approximation (see Chapter 9 and also Xin (1998) for a detailed modeling of the saturation in the context of the Sine–Gordon equation).

At critical dimension, (2.4.9) reduces to

$$\frac{d^2 V}{dt^2} = 8H, \tag{2.4.10}$$

leading to

$$V(t) = 4Ht^2 + V'(0)t + V(0). \tag{2.4.11}$$

It is noticeable that the two integration constants $V(0)$ and $V'(0)$ arising in the variance identity (2.4.11) are directly related to the conserved quantities associated to the invariance of the action by scale transformation and pseudo-conformal transformation at critical dimension. Indeed,

$$V'(t) = 2i \int \mathbf{x} \cdot (\psi \nabla \psi^* - \psi^* \nabla \psi) \, d\mathbf{x}, \tag{2.4.12}$$

leading to $V'(0) = 4C_1$, where C_1 is the invariant defined in (2.3.22). On the other hand, (2.3.30) taken at $t = 0$ reduces to $C_2 = V(0)$.

2.4.2   Extension to the Non-Elliptic NLS Equation

By a simple change of coordinates, we consider the non-elliptic NLS equation of the form

$$i\psi_t + \sum_{i=1}^{d} \varepsilon_i \frac{\partial^2 \psi}{\partial x_i^2} + f(|\psi|^2)\psi = 0, \tag{2.4.13}$$

where $\varepsilon_i = \pm 1$ and $\varepsilon_j \neq \varepsilon_k$ for some (j, k). More precisely, we assume that in the $\mathcal{D}$ matrix, which is now diagonal, the first d_+ diagonal elements are $+1$ and the $d_- = d - d_+$ others are -1.

Since, as already mentioned, the mass current density $\mathcal{J} = i(\psi \mathcal{D} \nabla \psi^* - \psi^* \mathcal{D} \nabla \psi)$ is related to the linear momentum density $\mathcal{P} = i(\psi \nabla \psi^* - \psi^* \nabla \psi)$ by $\mathcal{J} = \mathcal{D}\mathcal{P}$, it follows that

$$\partial_{tt} |\psi|^2 + \nabla \cdot \nabla \cdot (-\mathcal{Q}\mathcal{D}) = 0, \tag{2.4.14}$$

with the momentum stress tensor

$$\mathcal{Q} = 2(\mathcal{D}\nabla\psi^* \otimes \nabla\psi + \mathcal{D}\nabla\psi \otimes \nabla\psi^* + \mathcal{L}I). \tag{2.4.15}$$

One then defines the partial variances

$$V_{\pm} = \int |\mathbf{x}_{\pm}|^2 |\psi|^2 d\mathbf{x}, \tag{2.4.16}$$

where $|\mathbf{x}_+|^2 = \sum_1^{d_+} |x_i|^2$ and $|\mathbf{x}_-|^2 = \sum_{d_++1}^{d} |x_i|^2$. One has

$$\frac{d^2}{dt^2} V_+ = \int |\mathbf{x}_+|^2 \partial_i \partial_j \mathcal{R}_{ij} = \sum_{i=1}^{d_+} \mathcal{R}_{ii}, \tag{2.4.17}$$

$$\frac{d^2}{dt^2} V_- = \int |\mathbf{x}_-|^2 \partial_i \partial_j \mathcal{R}_{ij} = \sum_{i=d_++1}^{d_++d_-} \mathcal{R}_{ii}, \tag{2.4.18}$$

where $\mathcal{R} = 2\mathcal{Q}\mathcal{D}$ or

$$\mathcal{R}_{ij} = 4(\mathcal{D}_{il}\mathcal{D}_{mj}\partial_l\psi^*\partial_m\psi + \text{c.c.} + \mathcal{L}\mathcal{D}_{ij}). \tag{2.4.19}$$

It follows that

$$\frac{d^2}{dt^2} V_{\pm} = 4 \int (2|\nabla_{\pm}\psi|^2 \pm d_{\pm}\mathcal{L})\, d\mathbf{x}, \tag{2.4.20}$$

where ∇_+ and ∇_- denotes the gradient operators with respect to the variables $(x_1, \cdots, x_{d_+})$ and $(x_{d_++1}, \cdots, x_d)$ respectively.

Finally, using the NLS equation to express ψ_t, one rewrites

$$\mathcal{L} = -\frac{1}{2}\nabla \cdot \{\psi^*\mathcal{D}\nabla\psi + \psi\mathcal{D}\nabla\psi^*\} - f(|\psi|^2)|\psi|^2 + F(|\psi|^2). \tag{2.4.21}$$

Consequently,

$$\frac{d^2}{dt^2} V_{\pm} = 4 \int \{2|\nabla_{\pm}\psi|^2 \pm d_{\pm}[F(|\psi|^2) - f(|\psi|^2)|\psi|^2]\}d\mathbf{x} \tag{2.4.22}$$

and for the total variance $V = \int |\mathbf{x}|^2 |\psi|^2 d\mathbf{x} = V_+ + V_-$,

$$\frac{d^2}{dt^2} V = 4 \int \{2|\nabla\psi|^2 + (d_+ - d_-)[F(|\psi|^2) - f(|\psi|^2)|\psi|^2]\}d\mathbf{x}. \tag{2.4.23}$$

In the special case $f(|\psi|^2) = |\psi|^{2\sigma}$,

$$F(|\psi|^2) - f(|\psi|^2)|\psi|^2 = -\frac{\sigma}{\sigma+1}|\psi|^{2(\sigma+1)}, \tag{2.4.24}$$

leading to

$$\frac{d^2 V_{\pm}}{dt^2} = 8|\nabla_{\pm}\psi|_{L^2}^2 \mp \frac{4d_{\pm}\sigma}{\sigma+1}|\psi|_{L^{2\sigma+2}}^{2\sigma+2}, \tag{2.4.25}$$

$$\frac{d^2 V}{dt^2} = 8|\nabla\psi|_{L^2}^2 - \frac{4(d_+ - d_-)\sigma}{\sigma+1}|\psi|_{L^{2\sigma+2}}^{2\sigma+2}. \tag{2.4.26}$$

In the case of the non-elliptic NLS equation, no definite conclusion on the possibility of wave collapse was reached from the above variance identities.

Part II

Rigorous Theory

3
Existence and Long-Time Behavior

This chapter together with the two following ones provides a survey of the extended mathematical literature dealing with the NLS equation. The present chapter concentrates on local and global existence of solutions as well as on the long-time behavior of global solutions. Many of these results involve delicate analysis, and we often review them without giving detailed proofs. A comprehensive review of the basic results is presented by Cazenave (1989), Strauss (1989), and Ginibre (1998), who also gives an introduction to various methods of functional analysis used in the mathematical theory of the NLS equation.

Before turning to a precise description of the results, we give a general overview of the method. Although the nonlinearity of the NLS equation obtained as an amplitude equation is generically cubic, it is useful in the following to extend the discussion of the elliptic NLS equation to the more general setting $(g = \pm 1)$

$$i\psi_t + \Delta\psi + g|\psi|^{2\sigma}\psi = 0, \tag{3.0.1}$$

$$\psi(\mathbf{x}, 0) = \varphi(\mathbf{x}). \tag{3.0.2}$$

When the initial value problem is considered in the entire space $\mathbf{R}^d$, the problem is conveniently rewritten in the integral form

$$\psi(t) = U(t)\varphi + ig \int_0^t U(t - t')|\psi(t')|^{2\sigma}\psi(t')\, dt', \tag{3.0.3}$$

where the "Schrödinger operator" $U(t) = e^{it\Delta}$ defines a one-parameter unitary group.

The existence of a solution for small enough t (local existence) is proved by a fixed point method for (3.0.3), using that as a result of the dispersion properties of the linear operator, this equation defines a contraction in a suitable Banach space of functions $\psi(\mathbf{x}, t)$ for small enough t. The result was first established for functions valued in $H^1(\mathbf{R}^d)$ and then extended to functions in $L^2(\mathbf{R}^d)$ by taking into account the local regularizing properties of the Schrödinger operator. Proving local existence is more delicate when the NLS equation is considered in a periodic domain where dispersive properties do not exist. We briefly discuss this case in Section 3.4.3.

Existence for all time (global existence) holds in the case where the local solutions can be continued in the large in time, by means of a priori estimates for the norms of the solution in the corresponding spaces, resulting from conservation laws and Sobolev-type inequalities. A natural question then concerns the asymptotic behavior of the solution as $t \to \infty$. Since the solutions of the linear problem converge locally to 0 as $t \to \infty$, one can ask under what conditions the dynamics of the nonlinear are mostly dispersive and the solutions of the NLS equation behave for long times like solutions of the linear problem.

When the problem is considered in the entire space $\mathbf{R}^d$, we denote by $L^p(\mathbf{R}^d)$ with $p \geq 1$ the space of functions $u(\mathbf{x})$ with $\mathbf{x} \in \mathbf{R}^d$, equipped with the norm $|u|_{L^p} = \left(\int_{\mathbf{R}^d} |u(\mathbf{x})|^p dx \right)^{1/p}$, and by $H^s(\mathbf{R}^d)$ the Sobolev space of the functions that together with their derivatives up to order s are square integrable. The associated norm is defined by $|u|_{H^s}^2 = \sum_{0 \leq |\alpha| \leq s} |D^\alpha u|_{L^2}^2$, where $D^\alpha = \partial_{x_1}^{\alpha_1} \cdots \partial_{x_d}^{\alpha_d}$ and $|\alpha| = \alpha_1 + \cdots + \alpha_d$. The Sobolev spaces H^s are also considered for fractional powers s. In the entire space, it is convenient to define Sobolev norms in terms of the Fourier transform $\hat{u}$ of the function u by $|u|_{H^s}^2 = \int (1 + |\mathbf{k}|^2)^s |\hat{u}(\mathbf{k})|^2 d\mathbf{k}$. In some instances, we shall write the variable as an index or its domain in parentheses to avoid confusion. For example, $L^p_{x_1} = L^p(\mathbf{R})$. Also, $L^q([0, T), L^r(\mathbf{R}^d))$ denotes the space of functions u in $L^q([0, T))$ in the t-variable with values in $L^r(\mathbf{R}^d)$.

3.1   The Linear Schrödinger Operator

This section briefly reviews a few estimates for the solutions of the free Schrödinger equation in the entire space,

$$i\psi_t + \Delta\psi = 0, \quad \mathbf{x} \in \mathbf{R}^d, \tag{3.1.1}$$

$$\psi(\mathbf{x}, 0) = \varphi(\mathbf{x}). \tag{3.1.2}$$

This equation is solved as $\psi(\mathbf{x}, t) = U(t)\varphi$, where the free Schrödinger operator $U(t) = e^{it\Delta}$ given by

$$U(t)\varphi(\mathbf{x}) = \left(\frac{1}{4\pi i t} \right)^{d/2} \int e^{i\frac{|\mathbf{X} - \mathbf{X}'|^2}{4t}} \varphi(\mathbf{x}') \, d\mathbf{x}' \tag{3.1.3}$$

defines a unitary transformation group in L^2.

The estimates discussed in this section play an important role in the proof of well-posedness of the NLS equation in $H^1(\mathbf{R}^d)$ or $L^2(\mathbf{R}^d)$. They are in fact very general and extend to cases where the operator $i\Delta$ is replaced by any skew-Hermitian operator for which the L^∞-norm of the kernel element behaves like $t^{-d/2}$.

3.1.1 Basic Estimates

The conservation of the L^2-norm $|\psi(t)|_{L^2} = |\varphi|_{L^2}$, together with the classical estimate $|\psi(\mathbf{x},t)| \leq (4\pi|t|)^{-d/2} |\varphi|_{L^1}$, leads, by the Riesz–Thorin interpolation theorem (Bergh and Löfström 1976, p. 2) to the following result.

Theorem 3.1. (Decay estimates) *For conjugate p and p' ($\frac{1}{p} + \frac{1}{p'} = 1$), with $2 \leq p \leq \infty$, and $t \neq 0$, the transformation $U(t)$ maps continuously $L^{p'}(\mathbf{R}^d)$ into $L^p(\mathbf{R}^d)$ and*

$$|U(t)\varphi|_{L^p} \leq (4\pi|t|)^{-d(\frac{1}{2} - \frac{1}{p})} |\varphi|_{L^{p'}}. \tag{3.1.4}$$

More refined space-time estimates known as the *Strichartz inequalities* show that in addition to the decay of the solution as $t \to \infty$, a small gain of spatial regularity occurs for $t > 0$. They were named after Strichartz (1977), who proved

$$|U(t)\varphi|_{L^{2+4/d}(\mathbf{R}\times\mathbf{R}^d)} \leq C|\varphi|_{L^2}. \tag{3.1.5}$$

This inequality was generalized by Ginibre and Velo (1985a) in the form of Theorem 3.3 below. Extension to the solutions of the inhomogeneous linear Schrödinger equation (Theorem 3.4) is due to Yajima (1987), and Cazenave and Weissler (1988).

Definition 3.2. The pair (q,r) of real numbers is said to be admissible if $\frac{2}{q} = \frac{d}{2} - \frac{d}{r}$ with $2 \leq r < \frac{2d}{d-2}$ when $d > 2$, or $2 \leq r \leq \infty$ when $d = 1$ or 2.

Theorem 3.3. *For every $\varphi \in L^2(\mathbf{R}^d)$ and every admissible pair (q,r), the function $t \mapsto U(t)\varphi$ belongs to $L^q(\mathbf{R}, L^r(\mathbf{R}^d)) \cap C(\mathbf{R}, L^2(\mathbf{R}^d))$, and there exists a constant C depending only on q such that*

$$|U(\cdot)\varphi|_{L^q(\mathbf{R},L^r(\mathbf{R}^d))} \leq C|\varphi|_{L^2}. \tag{3.1.6}$$

Proof. As shown in Cazenave (1989), in order to establish that $t \mapsto U(t)\varphi$ belongs to $L^q(\mathbf{R}, L^r(\mathbf{R}^d))$ and satisfies (3.1.6), it is sufficient by duality and density to prove that for a test function θ in $C^\infty(\mathbf{R}^{d+1})$ with compact support, one has

$$\left|(U(t)\varphi, \theta)_{L^2(\mathbf{R}^{d+1})}\right| \leq C|\varphi|_{L^2}|\theta|_{L^{q'}(\mathbf{R},L^{r'})}. \tag{3.1.7}$$

To establish the latter estimate, one writes

$$\left| \int_{-\infty}^{\infty} (U(s)\varphi, \theta)_{L^2(\mathbf{R}^d)} ds \right| = \left| \left(\varphi, \int_{-\infty}^{\infty} U(-s)\theta(s) ds \right)_{L^2(\mathbf{R}^d)} \right|$$

$$\leq |\varphi|_{L^2(\mathbf{R}^d)} \left| \int U(-s)\theta(s) ds \right|_{L^2(\mathbf{R}^d)} \qquad (3.1.8)$$

where

$$\left| \int_{-\infty}^{\infty} U(-s)\theta(s) ds \right|_{L^2(\mathbf{R}^d)}^2 = \left(\int_{-\infty}^{\infty} U(-s)\theta(s) ds, \int_{-\infty}^{\infty} U(-t)\theta(t) dt \right)_{L^2(\mathbf{R}^d)}$$

$$= \left(\int \theta(t) dt, \int U(t-s)\theta(s) ds \right)_{L^2(\mathbf{R}^d)}$$

$$\leq |\theta|_{L^{q'}(\mathbf{R},L^{r'})} \left| \int_{-\infty}^{\infty} U(\cdot - s)\theta(s) ds \right|_{L^q(\mathbf{R},L^r)}.$$

$$(3.1.9)$$

The last term of (3.1.9) is estimated by using Theorem 3.1 in the form

$$\left| \int_{-\infty}^{\infty} U(t-s)\theta(s) ds \right|_{L^r} \leq \int_{-\infty}^{\infty} |U(t-s)\theta(s)|_{L^r} ds$$

$$\leq \int_{-\infty}^{\infty} |t-s|^{-d(\frac{1}{2}-\frac{1}{r})} |\theta(s)|_{L^{r'}} ds$$

$$\leq \int_{-\infty}^{\infty} |t-s|^{-\frac{2}{q}} |\theta(s)|_{L^{r'}} ds. \qquad (3.1.10)$$

Finally, the Hardy–Littlewood–Sobolev inequality for integrals of the form $\int_{-\infty}^{\infty} |t-s|^{\alpha-1} g(s) ds$ with $0 < \alpha < 1$ (Riesz potentials in dimension one, see Stein 1970, p. 119) reads

$$\left| \int_{-\infty}^{\infty} |t-s|^{\alpha-1} g(s) ds \right|_{L^Q} \leq C|g|_{L^P} \qquad (3.1.11)$$

with $\frac{1}{Q} = \frac{1}{P} - \alpha$. When applied to the integral in the right-hand side of (3.1.10) with $\alpha - 1 = -2/q$ and $Q = q$, it leads to

$$\left| \int U(\cdot - s)\theta(s) ds \right|_{L^q(\mathbf{R};L^r)} \leq C|\theta|_{L^{q'}(\mathbf{R};L^{r'})}, \qquad (3.1.12)$$

which implies (3.1.7).

Theorem 3.4. *Let t_0 belong to the closure $\bar{I}$ of an interval $I \subset \mathbf{R}$. Let (γ, ρ) be an admissible pair and $f \in L^{\gamma'}(I, L^{\rho'})$, where γ' and ρ' denote the conjugates of γ and ρ, respectively. Then, for every admissible pair (q, r), the function*

$$t \in I \mapsto \Phi f(t) = \int_{t_0}^{t} U(t-s) f(s) ds \qquad (3.1.13)$$

belongs to $L^q(I, L^r(\mathbf{R}^d)) \cap C(\bar{I}, L^2(\mathbf{R}^d))$. Furthermore, there exists a constant C depending only on q and γ such that

$$|\Phi f|_{L^q(I,L^r)} \le C|f|_{L^{\gamma'}(I,L^{\rho'})}. \tag{3.1.14}$$

Proof. For simplicity, we assume $I = [0, T_0)$ for some T_0, and $t_0 = 0$. It is also convenient, for the rest of the estimates, to introduce the operators Ψ and Θ_t where $t \in [0, T_0)$:

$$\Psi f(s) = \int_s^{T_0} U(s-t)f(t)\,dt \quad, \quad \Theta_t f(s) = \int_0^t U(s-s')f(s')\,ds'. \tag{3.1.15}$$

Step 1. Following the lines of the proof of Theorem 3.3, one shows that Φ defined in (3.1.13) is continuous from $L^{q'}([0,T_0), L^{r'}(\mathbf{R}^d))$ to $L^q([0,T_0), L^r(\mathbf{R}^d))$ and satisfies

$$|\Phi f|_{L^q([0,T_0),L^r)} \le C|f|_{L^{q'}([0,T_0),L^{r'})}. \tag{3.1.16}$$

The same argument leads to the same result for the operators Ψ and Θ_t.

Step 2. The operators Φ, Ψ, and Θ_t are continuous from $L^{q'}([0,T_0), L^{r'}(\mathbf{R}^d))$ to $C([0,T_0), L^2(\mathbf{R}^d))$. Indeed,

$$\begin{aligned}
|\Phi f(t)|_{L^2}^2 &= \int_0^t \int_0^t (U(t-s)f(s), U(t-s')f(s'))_{L^2}\,ds'ds \\
&= \int_0^t \int_0^t (f(s), U(s-s')f(s'))_{L^2}\,ds'ds \\
&= \int_0^t (f(s), \Theta_t f(s))_{L^2}\,ds.
\end{aligned} \tag{3.1.17}$$

Using the Hölder inequality in space and in time, one gets

$$\begin{aligned}
|\Phi f(t)|_{L^2}^2 &\le |f|_{L^{q'}(0,T_0;L^{r'})}|\Theta_t f|_{L^q(0,T_0;L^r)} \\
&\le C|f|_{L^{q'}(0,T_0;L^{r'})}^2.
\end{aligned} \tag{3.1.18}$$

Step 3. Φ is continuous from $L^1(0,T_0; L^2(\mathbf{R}^d))$ to $L^q(0,T_0; L^r(\mathbf{R}^d))$. Indeed,

$$\begin{aligned}
\int_0^{T_0} (\Phi f(t), \theta(t))_{L^2}\,dt &= \int_0^{T_0} \int_0^t (U(t-s)f(s), \theta(t))_{L^2}\,ds\,dt \\
&= \int_0^{T_0} \int_s^{T_0} (f(s), U(s-t)\theta(t))_{L^2}\,ds\,dt \\
&= \int_0^{T_0} (f(s), \Psi\theta(s))_{L^2}\,ds.
\end{aligned} \tag{3.1.19}$$

By the Cauchy–Schwarz inequality,

$$\int_0^{T_0} (f(s), \Psi\theta(s))_{L^2}\,ds \le |f|_{L^1(0,T_0;L^2)}|\Psi\theta|_{L^\infty(0,T_0;L^2)}. \tag{3.1.20}$$

Using Step 2, it follows that

$$\left| \int_0^{T_0} (\Phi f(t), \theta(t))_{L^2} dt \right| \le C |f|_{L^1(0,T_0:L^2)} |\theta|_{L^{q'}(0,T_0:L^{r'})}. \qquad (3.1.21)$$

Step 4. The purpose of this paragraph is to extend the result of Step 1 and show the continuity of Φ from $L^{\gamma'}([0,T_0), L^{\rho'}(\mathbf{R}^d))$ to $L^q([0,T_0), L^r(\mathbf{R}^d))$ for any admissible pair (γ, ρ). Two cases are to be considered depending on $r \le \rho$ or $r > \rho$.

When $r \le \rho$, one defines $\mu \in [0,1]$ such that

$$\frac{1}{r} = \frac{\mu}{\rho} + \frac{1-\mu}{2}. \qquad (3.1.22)$$

Since (q,r) and (γ, ρ) are admissible pairs, one also has $\frac{1}{q} = \frac{\mu}{\gamma}$, and one can apply the Hölder inequality $|g|_{L^{p_1}} \le |g|_{L^{p_2}} |g|_{L^{p_3}}$ to Φf first in space (with $p_1 = r$, $p_2 = \rho$, and $p_3 = 2$), then in time (with $p_1 = q$, $p_2 = \gamma$, and $p_3 = \infty$) Using Steps 1 and 2, one gets

$$|\Phi f|_{L^q(0,T_0;L^r)} \le |\Phi f|_{L^\gamma(0,T_0;L^\rho)}^{1-\mu} |\Phi f|_{L^\infty(0,T_0;L^2)}^{\mu}$$

$$\le |f|_{L^{q'}(0,T_0;L^{r'})}. \qquad (3.1.23)$$

Therefore, Φ is continuous from $L^{\gamma'}(0,T_0; L^{\rho'}(\mathbf{R}^d))$ to $L^q(0,T_0; L^r(\mathbf{R}^d))$.

When $r > \rho$, one defines $\mu \in [0,1]$ by

$$\frac{1}{\rho'} = \frac{\mu}{2} + \frac{1-\mu}{r'} \quad \text{and} \quad \frac{1}{\gamma'} = \frac{\mu}{1} + \frac{1-\mu}{q'}. \qquad (3.1.24)$$

From Steps 1 and 3, Φ is a continuous function from $L^{q'}(0,T_0;L^{r'}(\mathbf{R}^d))$ to $L^q(0,T_0;L^r(\mathbf{R}^d))$ and from $L^1(0,T_0;L^2(\mathbf{R}^d))$ to $L^q(0,T_0;L^r(\mathbf{R}^d))$. By interpolation (see Bergh and Löfström 1976, p. 107), it is also continuous from $L^{\gamma'}(0,T_0;L^{\rho'}(\mathbf{R}^d))$ to $L^q(0,T_0;L^r(\mathbf{R}^d))$, where (γ, ρ) are defined in (3.1.24).

Remark. Improvements of the Strichartz inequalities are presented in a series of recent papers by Bourgain (1998a), Merle and Vega (1998), Moyua, Vargas, and Vega (1998). These results are used to prove the existence of solutions of the NLS equation in spaces of functions with lower regularity than those used in the more classical functional framework reviewed in Section 3.2.

3.1.2 Linear Schrödinger Equation with Potential

An important problem is to understand how the decay properties and the Strichartz inequalities extend to the linear Schrödinger equation

$$i\frac{\partial \psi}{\partial t} + H\psi = 0, \qquad (3.1.25)$$

$$\psi(\mathbf{x}, 0) = \varphi(\mathbf{x}), \qquad (3.1.26)$$

where $H = -\Delta + V(\mathbf{x})$ includes a real potential $V(\mathbf{x})$. Assuming that H has no bound states or zero resonance,[1] the first results are in dimension $d \geq 3$ in the form of time decay estimates for weighted L^2-norm when the potential V is bounded and decays fast enough at infinity (Rauch 1978, Jensen and Kato 1979, Jensen 1984). More recent results in the form of Theorems 3.5 and 3.6 below are due to Journé, Sogge, and Soffer (1991). They assume some regularity properties on V, namely that for any $f \in H^s$, $(1 + |\mathbf{x}|^2)^{\alpha/2} V(\mathbf{x}) f$ is also in H^s for some $\alpha > d + 4$ and $s > 0$, or in other words that the multiplication by $(1 + |\mathbf{x}|^2)^{\alpha/2} V(\mathbf{x})$ sends the Sobolev space H^s into itself. They also assume that the Fourier transform $\widehat{V}$ of V belongs to $L^1(\mathbf{R}^d)$. Denoting by P_c the projection on the continuous part of the spectrum of H, one has the following result.

Theorem 3.5. *Let $d \geq 3$ and the potential V satisfy the above assumption. If 0 is neither an eigenvalue nor a resonance of H, one has*

$$|e^{itH} P_c f|_{L^{p'}} \leq C t^{-\frac{1}{d}(\frac{1}{p} - \frac{1}{2})} |f|_{L^p} \qquad (3.1.27)$$

for p and p' conjugate. In particular, if φ is orthogonal to the bound states of H,

$$|\psi(t)|_{L^{p'}} \leq C t^{-\frac{1}{d}(\frac{1}{p} - \frac{1}{2})} |\varphi|_{L^p}. \qquad (3.1.28)$$

Theorem 3.6. *Suppose that there are no bound states or zero resonances. Then, for solutions of the inhomogeneous equation*

$$i\frac{\partial \psi}{\partial t} + H\psi = g, \quad \psi(\mathbf{x}, 0) = \varphi(\mathbf{x}), \qquad (3.1.29)$$

one has

$$|\psi|_{L^{q'}(\mathbf{R}^d \times \mathbf{R})} \leq C(|\varphi|_{L^2} + |g|_{L^q(\mathbf{R}^d \times \mathbf{R})}, \quad q = \frac{2(d+2)}{d+4}. \qquad (3.1.30)$$

Furthermore, if $g = 0$, estimate (3.1.30) also holds if H has bound states, provided that 0 is neither an eigenvalue nor a resonance, and that φ is orthogonal to the bound states.

3.1.3 Smoothing Properties

A simple smoothing effect of the Schrödinger operator is that for any $\varphi \in L^1(\mathbf{R}^d)$ with compact support, $U(t)\varphi$ is analytic in $\mathbf{R}^d \times (0, +\infty)$. Also, $U(t)$ maps functions that decay at infinity into smooth functions.

For a multi-index $\alpha = (\alpha_1, \ldots, \alpha_d)$ where the components α_k are nonnegative integers, and $\mathbf{x} = (x_1, \ldots, x_d) \in \mathbf{R}^d$, we denote by $D^\alpha = \frac{\partial^{|\alpha|}}{\partial x_1^{\alpha_1} \ldots \partial x_d^{\alpha_d}}$

---

[1]There exists a zero resonance if there is a distributional solution of $Hf = 0$ that belongs to the space $L^2((1 + |\mathbf{x}|^2)^{-\tau/2} d\mathbf{x})$ for some $\tau \neq 0$. No resonances exist for $d \geq 5$.

a multi-index derivative of order $|\alpha| = \alpha_1 + \alpha_2 + \cdots + \alpha_d$. We also define the monomial $\mathbf{x}^\alpha = \prod_1^d x_j^{\alpha_j}$ and the operator $P_\alpha = \prod_1^d P_j^{\alpha_j}$, where $P_j u = x_j u + 2it\partial_j u$.

Theorem 3.7. *Let $\varphi \in L^2$ with $\mathbf{x}^\alpha \varphi \in L^2$, and $\psi = U(t)\varphi$. Then,*

$$D^{|\alpha|}\big(e^{\frac{i\mathbf{X}^2}{4t}}\psi\big) \in C\big(\mathbf{R} - \{0\}; L^2\big), \tag{3.1.31}$$

and for $t \neq 0$,

$$(2|t|)^{|\alpha|}|D^\alpha\big(e^{\frac{i|\mathbf{X}|^2}{4t}}\psi(t)\big)|_{L^2} = |\mathbf{x}^\alpha \varphi|_{L^2}. \tag{3.1.32}$$

Proof. Define

$$P_\alpha u = (2it)^{|\alpha|}e^{i\frac{|\mathbf{X}|^2}{4t}}D^\alpha\Big(e^{-i\frac{|\mathbf{X}|^2}{4t}}u\Big). \tag{3.1.33}$$

If ψ is a solution of (3.1.1), $\psi_\alpha = P_\alpha\psi$ is also a solution of this equation. Thus, $\psi_\alpha(t) = U(t)\mathbf{x}^\alpha\varphi$, and (3.1.32) follows.

Theorem 3.8. *Let $\varphi \in L^2$ and $\mathbf{x}^\alpha\varphi \in L^2$ for any multi-index α. Then, $\psi = U(t)\varphi \in C^\infty(\mathbf{R} - \{0\} \times \mathbf{R}^d)$.*

The above theorems extend to initial conditions in the space of tempered distributions $S'(\mathbf{R}^d)$. For example, the Dirac distribution $\varphi(\mathbf{x}) = \delta(\mathbf{x})$ has all its moments finite, and one thus recovers that the kernel associated to the free Schrödinger equation is C^∞ for all $t \neq 0$.

The next result refers to local smoothing properties proved by Sjölin (1987), Vega (1987), and Constantin and Saut (1988) for a large class of dispersive equations. In the case of the linear Schrödinger equation, these results express that there is, locally in space, a gain of half a derivative:

Theorem 3.9. *Assume $\varphi \in L^2$. Then $U(t)\varphi \in H_{loc}^{1/2}(\mathbf{R}^d)$ for almost all t. Furthermore, if ψ obeys $i\psi_t + \Delta\psi = f$ with $f \in L^2(-T, T; L^2(\mathbf{R}^d))$, with initial condition $\psi(\mathbf{x}, 0) = \varphi(\mathbf{x})$, then $\psi \in L^2\big(-T, T; H^{1/2}(B)\big)$ for any open set $B \subset \mathbf{R}^d$.*

These results of local smoothing effects are important in proving local well-posedness of the NLS equation for a large class of nonlinearities including first-order derivatives (Kenig, Ponce, and Vega 1993, 1997). They also lead to analyticity properties in both space and time for smooth (but not necessarily analytic) initial data rapidly decaying at infinity (Hayashi and Saitoh 1990, Hayashi and Kato 1997). Global existence of small radially symmetric solutions were obtained in dimension $d \geq 3$ by Hayashi and Ozawa (1995).

Remarks.

(i) The decay estimates (Theorem 3.1) and the Strichartz inequalities (Theorem 3.2) as well as the smoothing properties (Theorem 3.6), hold for more general Schrödinger operators e^{itL} where $L = \sum a_{ij}\partial_{x_i}\partial_{x_j}$ with the coefficients a_{ij} real and the quadratic form (a_{ij}) nondegenerate (Ghidaglia and Saut 1990).

(ii) Regularity properties of the kernel for the linear Schrödinger equation

$$i\frac{\partial \psi}{\partial t} + \sum_{i,j=1}^{d} \frac{\partial}{\partial x_i}\left(a_{ij}(\mathbf{x})\frac{\partial \psi}{\partial x_j}\right) + V(\mathbf{x})\psi = 0, \qquad (3.1.34)$$

with variable dispersion coefficients and possibly including a potential V, were studied by Craig, Kappeler, and Strauss (1995), who gave a relation between the microlocal regularity of the kernel and the bicharacteristic flow associated to the principal symbol $a(\mathbf{x}, \boldsymbol{\xi}) = \Sigma_{i,j} a_{ij}(\mathbf{x})\xi_i\xi_j$.

3.2 Existence Properties

3.2.1 Local Existence

Although most of the theorems quoted below hold for more general nonlinearities, we restrict ourselves to the elliptic NLS equation (3.0.1)–(3.0.2) in $\mathbf{R}^d$, with an attracting ($g = +1$) or repulsive ($g = -1$) power-law nonlinearity. The equation is replaced by its integral form

$$\psi(\mathbf{x}, t) = U(t)\varphi - i\int_0^t U(t-s)|\psi|^{2\sigma}\psi(s)\,ds, \qquad (3.2.1)$$

where U is the linear Schrödinger operator, and existence theorems are proved in this formulation.

Theorem 3.10. (Solutions in H^1) *For $0 \leq \sigma < \frac{2}{d-2}$ (no condition on σ when $d = 1$ or 2) and an initial condition $\varphi \in H^1(\mathbf{R}^d)$, there exists, locally in time, a unique maximal solution ψ in $C((-T^*, T^*), H^1(\mathbf{R}^d))$, where maximal means that if $T^* < \infty$, then $|\psi|_{H^1} \to \infty$ as t approaches T^*. In addition, ψ satisfies the energy and Hamiltonian conservation laws*

$$N(\psi) \equiv \int |\psi|^2 d\mathbf{x} = N(\varphi), \qquad (3.2.2)$$

$$H(\psi) \equiv \int \left(|\nabla\psi|^2 - \frac{g}{\sigma+1}|\psi|^{2\sigma+2}\right)d\mathbf{x} = H(\varphi), \qquad (3.2.3)$$

and depends continuously on the initial condition φ in H^1.

If in addition, the initial condition φ belongs to the space $\Sigma = \{f, f \in H^1(\mathbf{R}^d), |\mathbf{x}f(\mathbf{x})| \in L^2(\mathbf{R}^d)\}$ of the functions in H^1 with finite variance, the above maximal solution belongs to $C((-T^*, T^*), \Sigma)$. The variance $V(t) = \int |\mathbf{x}|^2|\psi|^2 d\mathbf{x}$ belongs to $C^2(-T^*, T^*)$ and satisfies the identity

$$\frac{d^2V}{dt^2} = 8H - 4\frac{d\sigma - 2}{\sigma + 1}g\int |\psi|^{2\sigma+2}d\mathbf{x}. \qquad (3.2.4)$$

The existence theorem, given by Baillon, Cazenave, and Figueira (1977), Lin and Strauss (1978), Ginibre and Velo (1978, 1979a), Cazenave (1979), and in a more general context by Kato (1987, 1989), is based on a fixed point theorem applied to (3.2.1). It makes use of the dispersive properties of U and of the $L^p - L^{p'}$ estimate (3.1.4). The condition $\sigma < \frac{2}{d-2}$ ensures that in the Hamiltonian, the potential term $|\psi|_{L^{2\sigma+2}}$ is controlled by $|\psi|_{H^1}$. In this case, the nonlinear potential $|\psi|^{2\sigma}$ is said to be *subcritical* for the initial problem in H^1.

More generally, one defines *criticality at the level of H^k*, as follows (Ginibre, Tsutsumi, and Velo, 1997). The exponent σ is said to be *critical* for the initial value problem in H^k if both the equation and the highest derivative terms in the H^k-norm are invariant under the scaling transformation

$$\psi(\mathbf{x}, t) \rightarrow \psi_\lambda(\mathbf{x}, t) = \lambda^\theta \psi(\lambda \mathbf{x}, \lambda^2 t) \qquad (3.2.5)$$

for the same value of the parameter θ. The value of θ keeping the NLS equation invariant is $\theta = 1/\sigma$, and the value which ensures the invariance of the $|(-\Delta)^{\frac{k}{2}} \psi|_{L^2}$ is $\theta = d/2 - k$. Criticality at the level of H^k thus corresponds to $\sigma = \frac{2}{d-2k}$. The exponent σ is thus subcritical at the level of H^k (resp. supercritical) when $\sigma < \frac{2}{d-2k}$ (resp. $\sigma > \frac{2}{d-2k}$). This requires the condition $k < d/2$. In classical existence theory, the nonlinearity is considered as a perturbation of the linear problem, and contraction methods can be applied for initial conditions in H^k only when the exponent is subcritical or critical. A detailed discussion is given by Ginibre (1998). As seen in the theorems below, the notion of criticality thus prescribes the functional framework for local existence. Note that this concept of criticality refers here to the minimal smoothness required for local existence and is distinct from the notion of criticality for existence of blowup solutions discussed in the forthcoming chapters. Nevertheless, since at the critical dimension for blowup ($\sigma d = 2$), the L^2 norm of the solution is preserved by the scaling transformation, criticality for blowup identifies with criticality at the level of L^2.

Theorem 3.11. (Solutions in L^2) *For $0 \leq \sigma < \frac{2}{d}$ and an initial condition $\varphi \in L^2(\mathbf{R}^d)$, there exists a unique solution ψ in $C((-T^*, T^*), L^2(\mathbf{R}^d)) \cap L^q((-T^*, T^*), L^{2\sigma+2}(\mathbf{R}^d))$ with $q = \frac{4(\sigma+1)}{d\sigma}$, satisfying the L^2-norm conservation law (3.2.2).*

Proof. This theorem, due to Tsutsumi (1987), results from smoothing properties of the linear Schrödinger operator expressed by the Strichartz inequalities (3.1.6). Note that the condition $\sigma < 2/d$ corresponds to an exponent σ subcritical for the initial problem in L^2. We describe here its main steps and refer to the original paper for details.
Step 1. One regularizes the initial condition by defining a sequence $\varphi_{0j} = h_j \star \varphi$ by convolution of the initial condition φ with the mollifier $h_j(\mathbf{x}) = j^d h(j\mathbf{x})$, where h is a positive even C^∞-function with compact support such that $\int h(\mathbf{x})d\mathbf{x} = 1$. From the existence theorem in H^1, the solution $\psi_j(t)$ of the NLS equation with initial condition $\varphi_{0j} \in H^1(\mathbf{R}^d)$ exists in $C((-T_0, T_0), H^1(\mathbf{R}^d))$.

Step 2. The Strichartz inequality (3.1.6) with $r = 2\sigma + 2$ and $q = \frac{4(\sigma+1)}{d\sigma}$ reads

$$|U(\cdot)v|_{L^q(\mathbf{R}, L^r(\mathbf{R}^d))} \leq \delta(d, \sigma)|v|_{L^2}, \tag{3.2.6}$$

where the constant δ depends on d and σ. One then proves that if the time T_1 is small enough, then for all j, $\psi_j(t) \in M$, where

$$M = \{v \in L^\infty((-T_1, T_1), L^2) \cap L^q((-T, T), L^r),$$
$$|v|_{L^\infty((-T_1,T_1),L^2)} \leq |\varphi|_{L^2}, \ |v|_{L^q((-T_1,T_1),L^r)} \leq 2\delta|\varphi|_{L^2}\}. \tag{3.2.7}$$

The ingredients to prove this uniform estimate are the Strichartz inequality above, the $L^p - L^{p'}$ estimates, and Young-type inequalities. The smallness condition on T_1 is given by $T_1 \leq C(\delta|\varphi|_{L^2})^{-p}$ with $p = \frac{4\sigma}{2-\sigma d} > 0$.

Step 3. One proves that the sequence $\{\psi_j\}$ converges to a solution ψ in $L^q((-T_1, T_1), L^{2\sigma+2}) \cap C([-T_1, T_1], H^{-1})$ that satisfies $|\psi(t)|_{L^2} \leq |\varphi|_{L^2}$. Uniqueness also follows from Strichartz estimates and implies that $|\psi(t)|_{L^2} = |\varphi|_{L^2}$.

Cazenave and Weissler (1989) extended Theorems 3.10 and 3.11 by including the border values $\sigma = 2/(d-2)$ for the problem in $H^1(\mathbf{R}^d)$ and $\sigma = 2/d$ for the problem in $L^2(\mathbf{R}^d)$:

Theorem 3.12. *Let $d \geq 3$ and $\sigma = 2/(\sigma - 2)$, then, for all $\varphi \in H^1(\mathbf{R}^d)$, there exists a unique maximal solution $\psi \in C([0, T^*), H^1(\mathbf{R}^d))$. At critical dimension $\sigma = 2/d$, for all $\varphi \in H^1(\mathbf{R}^d)$, there exists a unique maximal solution $\psi \in C([0, T^*), L^2(\mathbf{R}^d)) \cap L^{2\sigma+2}_{\mathrm{loc}}([0, T^*), L^{2\sigma+2}(\mathbf{R}^d))$.*

Theorem 3.13 below given by Cazenave and Weissler (1990) interpolates between these two results using Sobolev spaces $H^s(\mathbf{R}^d)$ with fractional order $0 \leq s \leq 1$. The treatment of fractional derivatives requires the use of the Besov spaces B^s_{pq} that can be defined in terms of the moduli of continuity in L^p

$$\omega^1_p(f, h) = \sup_{|\eta|<h} |f(x + \eta) - f(x)|_{L^p}, \tag{3.2.8}$$

$$\omega^2_p(f, h) = \sup_{|\eta|<h} |f(x + \eta) - 2f(x) + f(x - \eta)|_{L^p}. \tag{3.2.9}$$

Two cases are to be distinguished according as s is or is not an integer. If s is not an integer, let $s = S + \sigma$, with S an integer and $0 < \sigma < 1$, and $\nu = 1$. If s is an integer, let $S = s - 1$, $\sigma = 1$ and $\nu = 2$. One defines

$$|f|_{B^s_{pq}} \equiv |f|_{L^p} + \sum_{|\alpha|=S} \left|\frac{1}{h^\sigma}\omega^\nu_p(D^\alpha f, h)\right|_{L^q(\mathbf{R}^+, \frac{dh}{h})} \tag{3.2.10}$$

where $\alpha = (\alpha_1, \ldots, \alpha_d)$ is a multi-index and $D^\alpha = \partial^{\alpha_1}_{x_1} \cdots \partial^{\alpha_d}_{x_d}$. Other equivalent definitions involving the Fourier transform are given in Brenner, Thomée, and Wahlbin (1975) and Bergh and Löfström (1976).

Theorem 3.13. *Let $0 < s < d/2$ and $0 < \sigma \leq 2/(d-2s)$. If 2σ is not an even integer, suppose also that the integer part $[s]$ is less than 2σ if s is not an integer, and $s - 1 < 2\sigma$ if s is an integer. For $\varphi \in H^s(\mathbf{R}^d)$, there exists a solution $\psi \in C([-T^*, T^*), H^s(\mathbf{R}^d))$ that satisfies the conservation of the L^2-norm and, if $s \geq 1$, of the Hamiltonian. This solution is unique in $L^\gamma(0, T; B^s_{\rho,2}(\mathbf{R}^d))$ for any $T < T^*$, where $\gamma = \frac{2(2\sigma+2)}{\sigma(d-2s)}$, $\rho = \frac{2\sigma+2}{1+2\sigma s/d}$, and $B^s_{\rho,2}$ denotes the associated Besov space.*

Pecher (1997) improved the lower bound condition on σ. He removed the condition $[s] < 2\sigma$ if $1 < s < 2$ and replaced it by $s - 2 < 2\sigma$ when $2 \leq s < 4$ and by $s - 3 < 2\sigma >$ when $s \geq 4$.

Local solutions in H^s with $s > d/2$ holds for any nonlinearity $|\psi|^{2\sigma}\psi$ if σ is an integer, and under the additional assumption $2\sigma + 1 > s$ if σ is not an integer to keep the differentiability of the nonlinearity near $\psi = 0$. The border line $s = d/2$ was recently studied by Nakamura & Ozawa (1998) for nonlinearities with exponential growth. Their analysis covers the case of power nonlinearities with $\sigma = 2$ in dimension 1, $\sigma = 1$ in dimension 2 and 3 and $\sigma = 1/2$ in dimension $d \geq 4$.

3.2.2 Global Existence for Large Initial Conditions

Theorem 3.14.   (Global existence in H^1) *Assume $0 \leq \sigma < 2/(d-2)$ if $g < 0$ (attracting nonlinearity), or $0 \leq \sigma < 2/d$ if $g > 0$ (repulsive nonlinearity). For any $\varphi \in H^1(\mathbf{R}^d)$, there exists a unique solution ψ in $C(\mathbf{R}, H^1(\mathbf{R}^d))$. It satisfies the conservation laws (3.2.2) and (3.2.3) and depends continuously on initial conditions in $H^1(\mathbf{R}^d)$.*

Proof. When $g = -1$, the conservation of the L^2-norm and of the Hamiltonian leads to a uniform upper bound for $|\psi|_{H^1}$. This estimate enables one to repeat the local existence theorem and to obtain a solution for all time. (Lin and Strauss 1978, Ginibre and Velo 1979a).

When $g = 1$, the conservation of the Hamiltonian leads to

$$|\nabla\psi(t)|^2_{L^2} \leq |H| + \frac{1}{\sigma+1}|\psi|^{2\sigma+2}_{L^{2\sigma+2}}. \tag{3.2.11}$$

The Gagliardo–Nirenberg inequality

$$|f|^{2\sigma+2}_{L^{2\sigma+2}} \leq C_{\sigma,d}|\nabla f|^{\sigma d}_{L^2}|f|^{2+\sigma(2-d)}_{L^2} \tag{3.2.12}$$

and the conservation of the L^2-norm then implies

$$|\nabla\psi(t)|^2_{L^2} \leq |H| + \frac{1}{\sigma+1}C_{\sigma,d}|\varphi|^{2+\sigma(2-d)}_{L^2}|\nabla\psi|^{\sigma d}_{L^2}, \tag{3.2.13}$$

which provides a uniform bound for $|\nabla\psi|_{L^2}$, when $\sigma d < 2$.

Remark. Recent results concern the growth in time of the successive Sobolev norms $|\psi(t)|_{H^s}$ of global solutions whose H^1-norm is uniformly bounded. Staffilani (1997) proved that in dimension 1 with polynomial non-linearity and in dimension 2 with cubic nonlinearity, $|\psi(t)|_{H^s} < |t|^{s-1}$. A stronger result holds in dimension 3 with cubic nonlinearity, where Bourgain (1997a) established that $\sup_t |\psi(t)|_{H^s_{\text{loc}}}$ is uniformly bounded. This improvement from the usual exponential growth estimates given by Gronwall's lemma is a consequence of the smoothing properties of the linear Schrödinger operator.

Theorem 3.15. (Global existence in L^2, Tsutsumi 1987) *For* $0 \le \sigma < 2/d$ *and* $\varphi \in L^2(\mathbf{R}^d)$, *there exists a unique solution* ψ *in* $C(\mathbf{R}, L^2(\mathbf{R}^d)) \cap L^q_{\text{loc}}(\mathbf{R}, L^{2\sigma+2}(\mathbf{R}^d))$ *with* $q = \frac{4(\sigma+1)}{d\sigma}$ *that satisfies the L^2-norm conservation (3.2.2) and depends continuously on the initial condition in L^2.*

3.2.3 Global Regularity for Small Initial Conditions

Theorem 3.16. *For a nonlinearity exponent* $\frac{2}{d} < \sigma < \frac{2}{d-2}$, *and an initial condition* φ *in* H^s *with* $s \ge d/2 - 1/\sigma \ge 0$, *and* $|(-\Delta)^{s/2}\varphi|_{L^2}$ *sufficiently small, the solution remains in H^s for all time.*

Global existence for a small condition in H^1 was proved by Strauss (1974, 1981) for $g = 1$ (repulsive nonlinearity). It is a consequence of (3.2.13). Indeed, defining $X(t) = |\nabla\psi(t)|^2_{L^2}$, and $\delta = \frac{1}{\sigma+1}C_{\sigma,d}|\varphi|^{2+\sigma(2-d)}_{L^2}$, (3.2.13) becomes

$$X(t) \le |H| + \delta X^{\sigma d/2}(t). \tag{3.2.14}$$

The function $X(t)$ will remain uniformly bounded if $X(0)$ and δ are sufficiently small.

One of the difficulties with fractional spaces is that the conservation laws do not provide a priori estimates in the space where local existence is proved. The theorem above was proven by Cazenave and Weissler (1990) in Sobolev spaces H^s of fractional order for $s = d/2 - 1/\sigma$ and recently extended by Pecher (1997) to $s > d/2 - 1/\sigma$.

At critical dimension $\sigma d = 2$ and for $g = 1$, Theorem 3.14 also holds when the initial condition has a sufficiently small L^2-norm, namely if $C^{2\sigma+2}_{\sigma,d}|\varphi|^{2+\sigma(2-d)}_{L^2} < 1$, where the optimal constant in the Gagliardo–Nirenberg inequality(3.2.12) can be calculated in terms of the L^2-norm of the ground state R, defined as the positive solution of

$$\Delta R - R + R^{2\sigma+1} = 0. \tag{3.2.15}$$

Theorem 3.17. (Weinstein 1983) *Assume* $g = 1$ *and* $\sigma d = 2$. *If* $\varphi \in H^1(\mathbf{R}^d)$ *with* $|\varphi|_{L^2} < |R|_{L^2}$, *there exists a unique global solution of (3.0.1) in $H^1(\mathbf{R}^d)$. If in addition, the initial condition φ belongs to the space Σ of functions in H^1 with finite variance, then for all* $2 \le p < \frac{2d}{d-2}$,

$$|\psi(t)|_{L^p} \le C|t|^{d/p-d/2}. \tag{3.2.16}$$

Proof. For initial condition $\varphi \in H^1(\mathbf{R}^d)$, there exists a unique solution in $H^1(\mathbf{R}^d)$ during a finite time. For this solution, the mass (or wave energy) (3.2.2) and the Hamiltonian (3.2.3) are conserved. From the Gagliardo–Nirenberg inequality (3.2.12), one has

$$|\nabla\psi(t)|^2_{L^2} \leq H + \frac{1}{\sigma + 1} C_{\sigma,d} |\psi|^{2\sigma}_{L^2} |\nabla\psi(t)|^2_{L^2}. \qquad (3.2.17)$$

The optimal constant $C_{\sigma,d}$ is obtained by minimizing the functional

$$J^{\sigma,d}(f) = \frac{|\nabla f|^{\sigma d}_{L^2} |f|^{2+\sigma(2-d)}_{L^2}}{|f|^{2\sigma+2}_{L^{2\sigma+2}}} \qquad (3.2.18)$$

among all functions $f \in H^1(\mathbf{R}^d)$. It is found to be $C_{\sigma,d} = (\sigma + 1)/|R|^{2\sigma}_{L^2}$. (See Section 4.2.2.) Thus, for initial condition φ such that $|\varphi|_{L^2} < |R|_{L^2}$, we have a uniform bound for $|\nabla\psi(t)|_{L^2}$ for all time.

Theorem 3.17 is a sharp result. Indeed, there exist explicit "minimal blowup solutions" with an L^2-norm equal to $|R|_{L^2}$ that blow up in a finite time. They have the form

$$\psi(\mathbf{x}, t) = \frac{1}{(t^* - t)^{d/2}} R\left(\frac{\mathbf{x}}{t^* - t}\right) e^{\frac{i|\mathbf{X}|^2}{4(t^* - t)}} e^{\frac{i}{t^* - t}}. \qquad (3.2.19)$$

The second part of the theorem states that solutions in Σ with an L^2-norm smaller than $|R|_{L^2}$ (the critical mass for existence) disperse to zero at the rate of the free Schrödinger equation.

Theorem 3.18. *For σ and d such that $\frac{1}{\sigma}(1 + \frac{1}{2\sigma}) < d$, and an initial condition $\varphi \in H^s(\mathbf{R}^d) \cap W^{s,p}(\mathbf{R}^d)$ where $s > 2 + d/2$ and $p = \frac{4\sigma+2}{4\sigma+1}$ such that $|\varphi|_{H^s} + |\varphi|_{W^{s,p}}$ is small enough, there exists a unique solution in $C(\mathbf{R}, H^s(\mathbf{R}^d) \cap W^{s,p}(\mathbf{R}^d))$ that as $t \to \infty$ satisfies*

$$|\psi(t)|_{L^{2\sigma+2}} = O(t^{-\frac{d\sigma}{2\sigma+1}}). \qquad (3.2.20)$$

This result is due to Strauss (1981). It was also proved by Shatah (1982) and Klainerman and Ponce (1983) for more general nonlinearities, possibly including derivatives. The analysis actually applies to a large class of nonlinear dispersive equations. The proof combines a local existence theorem and $L^p - L^{p'}$ estimate for the solution of the linear equation (see Theorem 3.1). The result originates from the observation that if the initial conditions are small enough, the dispersive effects of the linear term will take place before the nonlinearity becomes important. To implement this idea, one needs a "high" nonlinearity and a "high" dimension, which explains the condition on σ and d in the theorem. We illustrate this result by giving below a simplified proof in the case $d = 3$, $\sigma = 1$ (see also Bardos 1980 and Strauss 1989 for other cases).

The solution of the NLS equation being written in the integral form (3.2.1), we define

$$M(t) = \sup_{0 \leq s \leq t} \left(|\psi(s)|_{H^2} + (1+s)^{3/2} |\psi(s)|_{L^\infty} \right), \qquad (3.2.21)$$

where the second term is motivated by the long-time behavior of the linear solution. Standard estimates imply that

$$|\psi(t)|_{H^2} \leq C + \int_0^t \frac{M^3(s)}{(1+s)^3} ds \leq C + CM^3(t). \qquad (3.2.22)$$

The L^∞-estimate is obtained using the basic $L^1 - L^\infty$ inequality for the operator U. To avoid difficulties for small t, we write

$$|\psi(t)|_{L^\infty} \leq (1+t)^{-3/2} (|\varphi|_{L^1} + |\varphi|_{H^2})$$
$$+ \int_0^t (1+t-s)^{-3/2} \left(\left| |\psi|^2 \psi(s) \right|_{L^1} + \left| |\psi|^2 \psi(s) \right|_{H^2} \right) ds$$
$$\leq \delta (1+t)^{-3/2} + M^3(t) \int_0^t (1+t-s)^{-3/2} (1+s)^{-3/2} ds,$$

$$(3.2.23)$$

where $\delta = |\varphi|_{L^1} + |\varphi|_{H^2}$. The integral in the last inequality is bounded from above by $C(1+t)^{-3/2}$. It results that

$$M(t) \leq \delta + C\, M^3(t). \qquad (3.2.24)$$

The assumption that δ is small enough ensures that $M(t)$ remains bounded independently of t. Global existence follows.

Other methods such as the methods of normal forms (Shatah 1985a), of invariant norms (Klainerman 1986), and of conformal mappings (Christodoulou 1986) were developed in the same context for various dispersive equations (see Strauss 1989 for a review).

3.2.4 Self-Similar Solutions

In spite of its importance in the theory of the NLS equation, the H^1 space is not adapted to the study of self-similar solutions of the form $\psi(\mathbf{x}, t) = \frac{1}{t^{p/2}} f(\frac{\mathbf{x}}{\sqrt{t}})$ with $p \in \mathbf{C}$, $\Re p = 1/\sigma$. Indeed, $\psi|_{L^2}^2 = t^{d/2 - 1/\sigma} |f|_{L^2}^2$, and except in the case $\sigma d = 2$, a self-similar solution cannot be a classical H^1-solution, since such a solution should satisfy the L^2-norm conservation. One is thus led to look for self-similar solutions for which $f \notin L^2$ but $|\nabla f|_{L^2}$ and $|f|_{L^{2\sigma+2}}$ are finite. Since $t^\beta |\psi|_{L^{2\sigma+2}} = |f|_{L^{2\sigma+2}}$ with $\beta = \frac{1}{2\sigma} - \frac{d}{4(\sigma+1)}$, Cazenave and Weissler (1998a) introduced the space

$$X_\sigma = \{ u \in L^\infty_{\text{loc}}((0, \infty), L^{2\sigma+2}) : \sup_t t^\beta |u(t)|_{L^{2\sigma+2}} < \infty \}. \qquad (3.2.25)$$

As a consequence of Theorem 3.1, for an initial condition taken in the space of tempered distributions, the solution of the linear Schrödinger equation belongs to X_α when $2d\sigma^2 + (d-2)\sigma - 2 > 0$. In the nonlinear theory, one also needs $\sigma < \frac{2}{d-2}$. This leads one to restrict the nonlinearity to the range $\sigma_0 < \sigma < \frac{2}{d-2}$, where σ_0 is the positive zero of the above quadratic polynomial. In this context, assuming that σ and β satisfy the conditions above and an initial condition φ taken as a tempered distribution such that $\sup_{t>0} |t|^\beta |U(t)\varphi|_{L^{2\sigma+2}}$ is small enough (a precise condition is given in the original paper), Cazenave and Weissler (1998a) prove the existence for all $t \geq 0$ of a unique solution of the NLS equation in X_α. Furthermore, if Φ is a finite linear combination of functions of the form $P_k(\mathbf{x})|\mathbf{x}|^{-p-k}$ where P_k is a homogeneous polynomial of degree k, then $\sup_{t>0} |t|^\beta |U(t)\Phi|_{L^{2\sigma+2}} < \infty$, so that for c small enough, the solution of the NLS equation with $\varphi = c\Phi$ is global and self-similar.

The existence of a class of self-similar solutions with higher regularity is established in Cazenave and Weissler (1998b). The results are valid for a range of values of σ that is different from that previously considered in spite of some overlap. More specifically, they assume $d \geq 3$ and $\sigma_1 < \sigma < \frac{2}{d-4}$, where σ_1 is the positive root of $2(d-2)\sigma^2 + (d-4)\sigma - 2 = 0$. In the case $d = 3$ or 4, $\sigma_1 < \sigma < \infty$. They also define $\mu = \frac{2-(d-4)\sigma}{4\sigma(\sigma+1)}$ and $\theta = \frac{2d(\sigma+1)}{d+2\sigma}$. For $1 \leq r < d$, they introduce the Banach space

$$\widetilde{W}^{1,r}(\mathbf{R}^d) = \{u \in L^{\frac{dr}{d-r}}(\mathbf{R}^d), \nabla u \in L^r(\mathbf{R}^d)\}. \tag{3.2.26}$$

In this case, assuming the above conditions and taking the initial condition φ as a tempered distribution such that $U(t)\varphi \in \widetilde{W}^{1,\theta}(\mathbf{R}^d)$ for almost all $t > 0$ with $\sup_{t>0} |t|^\mu |\nabla U(t)\varphi|_{L^\theta}$ sufficiently small, there exists a unique solution $\psi(t) \in \widetilde{W}^{1,\theta}(\mathbf{R}^d)$ for almost all $t > 0$ with $\sup_{t>0} |t|^\mu |\nabla \psi(t)|_{L^\theta}$ finite. As in the previous functional framework, this result allows one to construct global self-similar solutions.

A result concerning convergence of an H^1-solution to a self-similar form (asymptotic self-similarity) is also given in Cazenave and Weissler (1998a). Suppose $\sigma_0 < \sigma < \frac{2}{d-2}$. Let φ be a p-homogeneous function (i.e., a function such that $\varphi(\mathbf{x}) = \lambda^p \varphi(\lambda \mathbf{x})$) with $\Re p = 1/\sigma$, such that $\sup_{t>0} t^\beta |U(t)\varphi|_{L^{2\sigma+2}} < \infty$ and $\varphi = \varphi_1 + \varphi_2$, where $\varphi_1 \in L^{\frac{2\sigma+2}{2\sigma+1}}(\mathbf{R}^d)$ and $\varphi_2 \in H^1(\mathbf{R}^d)$. Multiply φ by a sufficiently small constant to ensure the global existence of the self-similar solution $u(\mathbf{x}, t) = t^{-p/2} f\left(\frac{\mathbf{x}}{\sqrt{t}}\right)$ with initial condition φ and of the H^1-solution $v(t)$ with initial condition φ_2 and set $w(\mathbf{x}, t) = v(\mathbf{x}, t) - t^{-p/2} f\left(\frac{\mathbf{x}}{\sqrt{t}}\right)$. It follows that for all $\epsilon > 0$,

$$|w(t)|_{L^{2\sigma+2}} = O(t^{-\frac{d\sigma}{2\sigma+2}+\epsilon}) \tag{3.2.27}$$

and

$$|f - t^{p/2} v(\mathbf{x}\sqrt{t}, t)|_{L^{2\sigma+2}} \leq C_\epsilon t^{-\frac{d(2\sigma+1)}{4(\sigma+1)}+\frac{1}{\alpha}+\epsilon} \tag{3.2.28}$$

as $t \to \infty$. Both converge to 0 as $t \to \infty$ if ϵ is sufficiently small.

Finally, the existence of the above self-similar solutions as $t \to \infty$ can be used to construct blowup solutions that are asymptotically self-similar at the singularity time taken at $t = 0$. This is done by using the pseudo-conformal transformation ($\mathbf{y} \in \mathbf{R}^d$, $s > 0$)

$$v(\mathbf{y}, s) = s^{-d/2} e^{i \frac{|\mathbf{y}|^2}{4s}} \psi \left(\frac{\mathbf{y}}{s}, -\frac{1}{s} \right),$$ (3.2.29)

which interchanges the behavior at $t = 0$ and $t = \infty$. If ψ is a solution of the NLS equation, then $\mathbf{v}$ obeys a nonautonomous NLS equation where the nonlinear term includes a $s^{d\sigma - 2}$ factor, for which existence of solutions can be established for all $s > 0$ in the same $L^{2\sigma + 2}$ functional framework as above. This approach could provide a rigorous setting to the numerically observed blowup of H^1 solutions of the supercritical NLS equation that near the singularity display a self-similar behavior limited to a range of finite extension and continued at large distance by a more rapidly decaying profile (see Section 7.1).

3.3 Scattering Properties

This section deals with the long-time behavior and the scattering of globally smooth solutions of the elliptic NLS equation (3.0.1)-(3.0.2) in $\mathbf{R}^d$, $d \geq 2$, with $g = \pm 1$ and $0 \leq \sigma < 2/(d-2)$ if $d > 2$, and no condition on σ otherwise. A detailed review of these results is found in the monographs of Strauss (1989, Chapter 6), Cazenave (1989, Chapter 7), and Ginibre (1998, Chapters 10–12). Extensions to a class of NLS equations including a linear potential (see Section 3.1.2) are given by Journé, Sogger, and Sogge (1991).

We are here interested in the so-called "dispersive solutions" that for large time behave like solutions of the linear Schrödinger equation. The "asymptotic states" ψ_+ and ψ_- are defined as

$$\psi^{\pm} = \lim_{t \to \pm \infty} U(-t)\psi(t).$$ (3.3.1)

The maps

$$\Omega^{\pm} : \psi^{\pm} \to \psi(0) = \varphi$$ (3.3.2)

are called "wave operators." The functions $\psi^{\pm}$ can be seen as initial conditions. We are looking for a solution ψ of the NLS equation that behave asymptotically like $U(t)\psi^{\pm}$ as $t \to \pm \infty$.

The property that all solutions (with arbitrary large initial conditions in a given space X_0) are dispersive is referred to as "asymptotic completeness." The wave operators are then one-to-one on X_0 and one can define the scattering operator $S = (\Omega^+)^{-1} \circ \Omega^-$ on the space X_0. The results cover a large class of nonlinearities, including nonlocal ones. Here we restrict the discussion to power laws.

There has been an extensive literature on the subject, and two important issues emerge from these studies. The first one concerns the space in which the wave operators are defined. A natural space is the Sobolev space H^1. The space Σ of the functions in H^1 with finite variance was also considered in order to take advantage of the pseudo-conformal law (Proposition 3.19 below), which is identical to the variance identity at critical dimension and was independently derived by Ginibre and Velo (1979b). The second issue is whether the scattering properties require a smallness condition on the solutions.

Existence of wave operators $\Omega^\pm$ and asymptotic completeness in Σ were first proved by Lin and Strauss (1978) for $\frac{5}{6} < \sigma < 2$ in dimension 3 following methods introduced by Morawetz and Strauss (1972) for the nonlinear Klein–Gordon equation. They were extended by Ginibre and Velo (1979b) to any dimension $d \geq 2$ obeying $\frac{2}{d} < \sigma < \frac{2}{d-2}$. Strauss (1981) proved asymptotic completeness for solutions in H^1 (with no assumption of finite variance), provided that the solutions are "small", and he extended the range of possible power nonlinearities by replacing the lower bound $\sigma > 2/d$ by $\sigma > \sigma_0(d)$ with $\sigma_0(d) = \frac{2-d+\sqrt{d^2+12d+4}}{4d}$ (see also Cazenave and Weissler 1998 for an interpretation of this lower bound in terms of dilatation invariance properties).

The smallness assumption for solutions in H^1 was removed by Ginibre and Velo (1985b) under the condition $2/d < \sigma < 2/(d-2)$ in dimension $d \geq 3$. For solutions in Σ, Tsutsumi (1985) extended the validity range of σ to $\sigma > \sigma_0(d)$, and Cazenave and Weissler (1992) included the lower limit $\sigma = \sigma_0(d)$. Note that $1/d < \sigma_0(d) < 2/d$. They reduced the lower limit to $2/(d+2)$ for small solutions in Σ.

Proposition 3.19. (Pseudo-conformal law) *If the initial condition φ belongs to the space Σ of functions in H^1 with finite variance, then*

$$\int |(\mathbf{x} + 2it\nabla)\psi|^2 d\mathbf{x} - \frac{4t^2 g}{\sigma + 1} \int |\psi|^{2\sigma+2} d\mathbf{x} = \int |\mathbf{x}|^2 |\varphi|^2 d\mathbf{x} - \int_0^t s\theta(s) ds,$$

$$(3.3.3)$$

where

$$\theta(t) = 4\frac{d\sigma - 2}{\sigma + 1} \int |\psi|^{2\sigma+2} d\mathbf{x}. \qquad (3.3.4)$$

Proof. Denoting by $h(t)$ the left-hand side of (3.3.3), we have

$$h(t) = \int |\mathbf{x}\psi|^2 d\mathbf{x} - 4t \, \Im\left(\int \psi^* \mathbf{x} \cdot \nabla\psi d\mathbf{x} + 4t^2 H\right). \qquad (3.3.5)$$

Taking into account that

$$\frac{d}{dt} \int |\mathbf{x}\psi|^2 d\mathbf{x} = 4 \, \Im\left(\int \psi^* \mathbf{x} \cdot \nabla\psi d\mathbf{x}\right), \qquad (3.3.6)$$

and using the variance identity (3.2.4), we have

$$h'(t) = -t\left(\frac{d^2}{dt^2}\int |\mathbf{x}\psi|^2 d\mathbf{x} - 8H\right)$$

$$= 4t\frac{d\sigma - 2}{\sigma + 1}\int |\psi|^{2\sigma+2} d\mathbf{x} = t\theta(t). \tag{3.3.7}$$

Equality (3.3.3) follows by time integration. Notice that at critical dimension $d\sigma = 2$, it reproduces the conservation law (2.3.30).

3.3.1 The Case of Repulsive Nonlinearity

The pseudo-conformal law (3.3.3) is used to characterize the long-time behavior of the solutions (Ginibre and Velo 1979b, Tsutsumi 1985).

Theorem 3.20. (Rate of decay). *For an initial condition* $\varphi \in \Sigma$, *let* ψ *be the maximal solution of (3.0.1) with* $g < 0$. *Let* r *satisfy* $2 \le r \le \frac{2d}{d-2}$ *if* $d > 2$ *or* $2 \le r \le \infty$ *if* $d = 1$ *or* 2.

(i) *If* $\frac{2}{d} \le \sigma < \frac{2}{d-2}$, *there exists a constant* C *such that for all* $t \in \mathbf{R}$,

$$|\psi(t)|_{L^r} \le C|t|^{-(d/2-d/r)}. \tag{3.3.8}$$

(ii) *If* $\sigma < \frac{2}{d}$, *there exists a constant* C *such that for all* $t \in \mathbf{R}$,

$$|\psi(t)|_{L^r} \le C|t|^{(-d/2-d/r)(1-\eta(r))}, \tag{3.3.9}$$

where $\eta(r) = 0$ *if* $2 \le r \le 2\sigma + 2$ *and* $\eta = \frac{(r-2\sigma-2)(2-d\sigma)}{(r-2)(2\sigma+2-d\sigma)}$ *if* $r > 2\sigma + 2$.

Remark. For $\sigma \ge 2/d$ the solution decays in all spaces L^r like the solution of the linear Schrödinger equation, while for $\sigma < 2/d$ this is only the case for $r < 2\sigma + 2$.

Theorem 3.21. (Scattering properties). *Let* $\sigma_0(d) = \frac{2-d+\sqrt{d^2+12d+4}}{4d} \le \sigma < \frac{2}{d-2}$, $\varphi \in \Sigma$, *and let* ψ *be the maximal solution of (3.0.1). There exists* $\psi^\pm \in \Sigma$ *such that*

$$|U(-t)\psi - \psi^\pm|_\Sigma \to 0 \quad as \quad t \to \pm\infty. \tag{3.3.10}$$

In addition, $|\psi^+|_{L^2} = |\psi^-|_{L^2} = |\varphi|_{L^2}$ *and* $\int |\nabla\psi^+|^2 d\mathbf{x} = \int |\nabla\psi^-|^2 d\mathbf{x} = H(\varphi)$.

This result shows that as $t \to \pm\infty$, the solution is asymptotic to a solution of the linear Schrödinger equation. It also defines

$$U^\pm : \varphi \mapsto \psi^\pm \tag{3.3.11}$$

as continuous mappings from Σ to Σ. One has the formulas

$$\psi^\pm = \varphi - i\int_0^{\pm\infty} U(-\tau)f(\psi(\tau))\,d\tau \tag{3.3.12}$$

and

$$\psi(t) = U(t)\psi^{\pm} + i \int_{\mp\infty}^{t} U(t - \tau) f(\psi(\tau)) \, d\tau. \qquad (3.3.13)$$

Theorem 3.22. (Existence of wave operators). *Let $\sigma_0(d) \le \sigma < \frac{2}{d-2}$. For every $\psi^+ \in \Sigma$ there exists a unique $\varphi \in \Sigma$ such that the maximal solution $\psi \in C(\mathbf{R}, H^1)$ with initial condition φ satisfies*

$$|U(-t)\psi(t) - \psi^+|_{\Sigma} \to 0 \quad as \quad t \to +\infty. \qquad (3.3.14)$$

Similarly, for every $\psi^- \in \Sigma$ there exists a unique $\varphi \in \Sigma$ such that the maximal solution $\psi \in C(\mathbf{R}, H^1)$ with initial condition φ satisfies

$$|U(-t)\psi(t) - \psi^-|_{\Sigma} \to 0 \quad as \quad t \to -\infty. \qquad (3.3.15)$$

The wave operators $\Omega^+ : \psi^+ \to \varphi$ and $\Omega^- : \psi^- \to \varphi$ are continuous one-to-one operators from Σ into Σ, and $(\Omega^{\pm})^{-1} = U^{\pm}$, which ensures asymptotic completeness in Σ.

Theorem 3.23. *For every $\psi^- \in \Sigma$ there exists a unique $\psi^+ \in \Sigma$ satisfying the following properties: There exists a unique $\varphi \in \Sigma$ such that the maximal solution $\psi \in C(\mathbf{R}, H^1)$ with initial condition φ satisfies*

$$U(-t)\psi(t) \to \psi^{\pm} \quad in \ \Sigma \quad as \quad t \to \pm\infty. \qquad (3.3.16)$$

The scattering operator

$$S : \psi^- \in \Sigma \to \psi^+ \in \Sigma \qquad (3.3.17)$$

is continuous and one-to-one, and its inverse is also continuous. In addition, $|\psi^+|_{L^2} = |\psi^-|_{L^2} = |\varphi|_{L^2}$ and $\int |\nabla\psi^+|^2 dx = \int |\nabla\psi^-|^2 dx = H(\varphi)$.

Theorem 3.24. *For $1/d < \sigma < 2/(d-2)$, and $\varphi \in \Sigma$, there exist unique asymptotic states $\psi^{\pm} \in L^2(\mathbf{R}^d)$ such that the solution of eq. (3.0.1) with initial condition φ satisfies*

$$|\psi(t) - U(-t)\psi^{\pm}|_{L^2} \to 0 \quad as \quad t \to \pm\infty. \qquad (3.3.18)$$

This result, due to Tsutsumi and Yajima (1984) defines the mappings $U^{\pm}$ from Σ into L^2. The proof is based on the observation that if $\psi(\mathbf{x}, t)$ is a solution of (3.0.1), then u defined by

$$\psi(\mathbf{x}, t) = (2it)^{-d/2} e^{i|\mathbf{x}|^2/4t} u^* \left(\frac{\mathbf{x}}{2t}, \frac{1}{2t} \right)$$

satisfies, in the new variables,

$$iu_t + \Delta u + g|t|^{\sigma d - 2} |u|^{2\sigma} u = 0. \qquad (3.3.19)$$

The theorem is thus equivalent to the existence of $u(0)$ such that

$$|u(t) - u(0)|_{L^2} \to 0 \text{ as } t \to 0^{\pm}.$$

Theorem 3.25. (Cazenave and Weissler 1992) *For $\frac{2}{d+2} < \sigma < \frac{2}{d-2}$ (or $\sigma > 1$ if $d = 1$) the wave operators $\Omega^{\pm}$ are defined on all Σ. In addition, the scattering operator S is defined from a neighborhood of 0 in Σ onto a neighborhood of 0.*

Remark. The lower bound $1/d$ in Theorem 3.24 is in a sense optimal. Indeed, when $0 < \sigma \leq 1/d$, scattering operators cannot be defined: as shown by Strauss (1974), for any solution $\psi(t) \in \Sigma$, the quantity $U(-t)\psi(t)$ does not have a strong limit as $t \to \infty$, even in L^2.

Note the gap in the lower bounds for σ given in Theorems 3.24 and 3.25. Ginibre, Ozawa, and Velo (1994) recently bridged the gap in dimension $d = 3$ and reduced it in dimension $d \geq 4$ by proving the following theorem.

Theorem 3.26. *For $d = 3$, and $\frac{1}{3} < \sigma < \frac{2}{3}$, the results of Theorem 3.25 hold when Σ is replaced by $\Sigma_\rho = \{u, u \in H^\rho, \mathbf{x}^\rho u \in L^2\}$ with $\frac{3}{2} \leq \rho < 2$. In dimension $d \geq 4$, the lower bound on σ reads $\sigma(d + 4\sigma + 2) > 2$.*

The above results concern solutions with a finite variance. A natural question is whether one can define the wave operators and prove asymptotic completeness in a larger space where global existence holds, that is, in H^1 instead of Σ. A partial answer is given by the following result.

Theorem 3.27. (Ginibre and Velo 1985b) *If $d \geq 3$ and $2/d < \sigma < 2/(d-2)$, asymptotic completeness holds for arbitrary initial conditions in $H^1(\mathbf{R}^d)$.*

3.3.2 The case of Attracting Nonlinearity

Theorem 3.28. (Cazenave and Weissler 1992) *For $\frac{2}{d+2} < \sigma < \frac{2}{d-2}$ (or $\sigma > 1$ if $d = 1$), the scattering operator S is defined from a neighborhood of 0 in Σ onto a neighborhood of 0. Furthermore, for $\frac{2}{d+2} < \sigma < \frac{2}{d}$ (or $\sigma > 1$ if $d = 1$), the wave operators $\Omega^{\pm}$ can be defined on the whole space Σ.*

When dealing with solutions whose variance is infinite, we have the following theorem.

Theorem 3.29. (Strauss 1981) *If $d \geq 3$ and $2/d < \sigma < 2/(d-2)$, asymptotic completeness holds for small enough initial conditions in $H^1(\mathbf{R}^d)$.*

Remark. The lower bounds $\sigma > 2/(d+2)$ for the theory in Σ (Theorem 3.28) and $\sigma > 2/d$ for the theory in $H^1(\mathbf{R}^d)$ (Theorem 3.29) as well as the upper bound $\sigma < 2/d$ in the second part of Theorem 3.28 are optimal (see Remarks 4.4 and 4.7 of Cazenave and Weissler 1992).

3.4   Further Results

3.4.1   Generalized NLS Equations

Consider the generalized Schrödinger equation

$$i\psi_t + D\psi + P(\psi, \psi^*, \nabla_{\mathbf{x}}\psi, \nabla_{\mathbf{x}}\psi^*) = 0, \qquad (3.4.1)$$

$$\psi(\mathbf{x}, 0) = \varphi(x), \qquad (3.4.2)$$

where D is a nondegenerate second-order operator with constant coefficients

$$D = \sum_{j \leq k} \partial_{x_j}^2 - \sum_{j > k} \partial_{x_j}^2, \text{ for } 1 \leq k \leq d, \qquad (3.4.3)$$

and P a polynomial with no constant or linear terms :

$$P(z) = P(z_1, z_2, \ldots, z_{2d+2}) = \sum_{\ell_0 \leq |\alpha| \leq \ell_1} a_\alpha z^\alpha, \quad \ell_0 \geq 2. \qquad (3.4.4)$$

Theorem 3.30. (Kenig, Ponce, and Vega 1997) *For $\ell_0 \geq 3$, there exists $s = s(d, P) > 0$ depending on the dimension d and the nonlinear terms P such that for all initial conditions $\varphi \in H^s(\mathbf{R}^d)$, the initial value problem has a unique solution defined in a finite interval $[0, T]$ with $T = T(|\varphi|_{H^s})$, satisfying $\psi \in C([0, T], H^s(\mathbf{R}^d))$, with the "local smoothing property"*

$$\sup_{\mu \in \mathbf{Z}^d} \int_0^T \int_{Q_\mu} |(1 - \Delta)^{s/2+1/4}\psi|^2 d\mathbf{x} \, dt < \infty, \qquad (3.4.5)$$

where $\{Q_\mu\}_{\mu \in \mathbf{Z}^d}$ is a family of unit cubes of side one with disjoint interiors covering $\mathbf{R}^d$.

For $\ell_0 = 2$, there exist $s = s(d, P) > 0$ and $m = m(d, P) > 0$ such that the same result holds with the space $H^s(\mathbf{R}^d))$ replaced by the weighted space $H^s(\mathbf{R}^d)) \cap L^2(\mathbf{R}^d, |\mathbf{x}|^{2m} dx)$.

Kenig, Ponce, and Vega (1993) first established this result in the case of small initial conditions. The restriction was removed in one space dimension by Hayashi and Ozawa (1994a) and in any dimension for the operator $D = \Delta$ by Chihara (1996) and Hayashi and Kaikina (1998). The proof is based on a detailed analysis of the smoothing properties of linear Schrödinger operators with nonconstant coefficients.

When $D = \Delta$ and the nonlinear term P is quadratic, with the condition

$$|\partial_\psi P| + |\partial_{\psi^*} P| < C|\nabla\psi|, \qquad (3.4.6)$$

small analytic solutions exist globally in time (Hayashi and Kato 1997). A result of global existence of radially symmetric solutions with small initial conditions is given by Hayashi and Ozawa (1995).

Note that in one space dimension, the so-called derivative nonlinear Schrödinger equation

$$\frac{\partial b}{\partial t} \pm \partial_x(|b|^2 b) + i\partial_{xx}b = 0, \tag{3.4.7}$$

which is not an amplitude equation but rather describes the long-wavelength dynamics of dispersive Alfvén waves propagating along an ambient magnetic field (Mjølhus 1976, Mio, Ogino, Minami and Takeda 1976, see also Mjølhus and Hada 1997 for a recent review), is integrable by the inverse scattering technique (Kaup and Newell 1978, Lee 1989). The Cauchy problem and the analyticity properties of global solutions are considered by Hayashi and Ozawa (1992). The effect of perturbations of the form $\lambda|\psi|^p\psi$ are discussed by Hayashi and Ozawa (1994b) and Ozawa (1996).

A physical example of an NLS equation with a nonpolynomial nonlinearity including derivatives arises in the context of the continuous limit of a cubic lattice of classical spins processing in a magnetic field created by their closest neighbors. Their dynamics are governed by

$$\frac{\partial \mathbf{S}}{\partial t} = \mathbf{S} \times \Delta\mathbf{S}, \qquad \mathbf{x} = (x_1, x_2, x_3) \in \mathbf{R}^3, \tag{3.4.8}$$

$$\mathbf{S}(\mathbf{x}, 0) = \mathbf{S}_0(\mathbf{x}), \tag{3.4.9}$$

where $\mathbf{S}_0(\mathbf{x})$ is a field of vectors in $\mathbf{R}^3$ of unit length. This equation is rewritten using the stereographic projection of the unit sphere $\mathcal{S}$ on the plane $x_3 = 1$. Each point $\mathbf{S}$ of the unit sphere, except the south pole $\mathbf{P} = (0, 0, -1)$, has an image $Q = (\alpha, \beta, 1)$ obtained as the intersection of the straight line $\mathbf{SP}$ with the plane $x_3 = 1$. Defining $z = \alpha + i\beta$, one gets

$$i\frac{\partial z}{\partial t} + \Delta z = 2\frac{z^*(\nabla z)^2}{4 + |z|^2}. \tag{3.4.10}$$

Existence and long-time asymptotic freedom for initial conditions corresponding to small spin deviations are established in Sulem, Sulem, and Bardos (1986). An extension of this model including a coupling to a mean field was name after Ishimori (1984) and considered by various authors (see among recent publications Hayashi and Naumkin 1998a, Kenig, Ponce, and Vega 1998 and references therein). Other extensions were recently reviewed in Myrzakulov, Vijayalakshmi, Syzdykova, and Lakshmanan (1998).

A special example of the NLS equation including first and second derivatives in the nonlinear term and arising in the context of the propagation of an ultrashort high-intensity laser pulse in a plasma is considered by de Bouard, Hayashi, and Saut, (1997) who established local existence and uniqueness of solutions with small initial conditions. For this purpose, they rewrite the equation as a system for the solution and its complex conjugate where second derivatives do not appear in the nonlinearities. The dispersion becomes nonlinear, or equivalently, a nonlinear factor appears in front of the time derivative.

We finally mention the NLS equation with a Hartree potential

$$i\psi_t + \Delta\psi = (V \star |\psi|^2)\psi = 0, \quad \mathbf{x} \in \mathbf{R}^d. \tag{3.4.11}$$

Here $V(\mathbf{x}) = \lambda|\mathbf{x}|^{-1}$ is the Coulomb potential. The Cauchy problem and the long-time behavior were first addressed by Chadam and Glassey (1975) and Glassey (1977a). Extensions to more general potentials can be found in Ginibre and Velo (1980), Hayashi and Tsutsumi (1987), Hayashi and Ozawa (1987), and Nawa and Ozawa (1992). Sharp estimates in L^∞ and existence of modified scattering states are derived for small initial conditions in weighted Sobolev spaces (Hayashi and Naumkin 1998b, Hayashi, Naumkin, and Ozawa 1998).

3.4.2 Defocusing NLS with Nonzero Condition at Infinity

In the case of a repulsive nonlinearity, the nonlinearity amplifies the broadening effects of dispersion and diffraction. Nonlinear dynamics are thus not possible with a localized wave packet, and nonzero boundary conditions corresponding to an unperturbed monochromatic plane wave are required at infinity. As mentioned in Section 1.4, it turns out that such an equation also describes the dynamics of an imperfect Bose condensate and was the object of several studies in this field. Here we only briefly summarize a few recent results.

Up to a phase translation, the equation is written

$$i\psi_t + \Delta\psi + (1 - |\psi|^2)\psi = 0. \tag{3.4.12}$$

When prescribing the general boundary condition

$$|\psi(\mathbf{x}, t)| \to 1 \quad \text{as} \quad |\mathbf{x}| \to \infty, \tag{3.4.13}$$

which corresponds to a phase whose value at infinity depends on the direction, the initial value problem has to be addressed in the framework of local L^r_{loc} or H^s_{loc} spaces. This is a delicate situation for which no rigorous result are to our knowledge presently available. Results were obtained by Ginibre and Velo (1996, 1997a, 1997b) in the context of the complex & Ginzburg–Landau equation, but their extension to the nondissipative regime associated to the nonlinear Schrödinger equation remains open.

The situation where the boundary condition reduces to $\psi \to 1$ as $|\mathbf{x}| \to \infty$, independently of the direction, corresponds in terms of the hydrodynamic interpretation (Section 1.4) to a fluid at rest with an unperturbed density at infinity. This case can be addressed in the usual Sobolev framework. The (conserved) Hamiltonian

$$H = \int \left(|\nabla\psi|^2 + \frac{1}{2}(1 - |\psi|^2)^2 \right) d\mathbf{x} \tag{3.4.14}$$

is finite and the problem globally well-posed in two space dimensions (Bethuel and Saut 1998).

The zeros of the complex function ψ are called vortices or defects. Their dynamics were first studied by Neu (1990), who formally showed that the vortex centers satisfy a Hamiltonian equation at leading order in r^{-1}, where r measures the minimum distance between the centers (Kirchhoff law). A rigorous approach is given by Lin and Xin (1997) and Colliander and Jerrard (1998). Radiation effects that occur at higher orders are retained by Ovchinnikov and Sigal (1998a), who considered long-time behavior of solutions with initial conditions corresponding to two vortices of the same charge ±1 or opposite charges +1 and −1. They proved that in the former case, the vortices radiate while rotating around each other and asymptotically move apart at a rate proportional to $t^{1/6}$ asymptotically. In the latter case, the radiation is absent for large vortex separation, while for small or moderate separations, the two vortices produce a shock wave and eventually collapse, annihilating each other.

Special interest was devoted to traveling waves $\psi(\mathbf{x}, t) = v(\xi, y)$ with $\xi = x - ct$. The equation for v becomes

$$ic\partial_\xi v = (\partial_{\xi\xi} + \partial_{yy})v + (1 - |v|^2)v \tag{3.4.15}$$

with $v \to 1$ at infinity. Formal constructions of solutions and analysis of their stability were presented both in two dimensions (a model for thin helium films) and in threedimensional axisymmetric geometries (Jones and Roberts 1982, Jones, Putterman, and Roberts 1986, Kuznetsov, and Rasmussen 1995, and references therein). Rigorous results were recently given by Bethuel and Saut (1998). Existence of solutions was proved for small enough velocity $c > 0$ and is not expected to hold when c exceeds the velocity of sound. Near this threshold, the amplitudes of the travelling waves are small, and a standard reductive perturbative expansion reduces the NLS equation to a Kadomtsev–Petviashvili (KPI) equation that admits "lump solitons." The density well at the center of the soliton decreases with the velocity. Below the velocity for which the density vanishes, the solution bifurcates and splits into two separated zeros in the direction transverse to the direction of propagation.

For $c = 0$, there are no nontrivial solutions with finite energy. Time-independent solutions have thus a phase at infinity that depends on the direction. They satisfy

$$\Delta u + (1 - |u|^2)u = 0, \quad \text{with} \quad |u| \to 1 \quad \text{as} \quad |\mathbf{x}| \to \infty, \tag{3.4.16}$$

an equation extensively considered in two space dimensions (Brezis, Merle, and Rivière 1994, Bethuel, Brezis, and Hélein 1994, Ovchinnikov and Sigal 1997). It admits "vortex solutions" in the form

$$u^{(n)}(r, \theta) = f^{(n)}(r)e^{in\theta}, \tag{3.4.17}$$

where $f^{(n)}$ is a unique function monotonically increasing from 0 to 1 as r goes from 0 to ∞, and n is the winding number. The stability of these solutions was addressed by Weinstein and Xin (1996) and Gustafson (1997a), who proved that in dimension 2, the vortices associated to $n = \pm 1$ are marginally stable in the spectral sense. Extension of vortices to higher dimensions (called monopoles in three dimensions) has been considered by Gustafson (1997b). He proved the existence and uniqueness of a spherically symmetric solution with nonzero topological degree. It is a degree-one solution of the form $\psi = f(|\mathbf{x}|)\mathbf{x}$ with the function f being real-valued, and it is stable in the spectral sense.

Stability of magnetic vortices (which are solutions of (3.4.16) coupled with a magnetic field) were recently studied by Gustafson and Sigal (1998). A recent survey including an extended bibliography is given by Ovchinnikov and Sigal (1998b).

3.4.3 Periodic Boundary Conditions and Invariant Measures

As already discussed, the local existence theory for solutions of the NLS equations in $\mathbf{R}^d$ uses the dispersive effect of the free Schrödinger operator, in the form of the Strichartz inequalities. When the domain is periodic, such inequalities do not hold. Using refined properties of trigonometric series, Bourgain (1993, 1994a) developed an analogue to these estimates, by defining the temporal norm on a finite time interval and dealing with the projection of the linear Schrödinger equation on spaces spanned by a finite number of Fourier modes. This leads to estimates similar to classical Strichartz inequalities, except that they involve constants that increase with the number of retained Fourier modes, at a rate that grows with the space dimension and the order of the norm. A detailed review is presented by Ginibre (1996). The main steps are as follows. For a periodic function $u \in L^2(\mathbf{T}^d)$, one defines

$$f_N(\mathbf{x},t) = \sum_{\mathbf{m}\in\mathbf{Z}^d, |m_i|\leq N} \widehat{u}(m)e^{i(\mathbf{m}\cdot\mathbf{x}-|\mathbf{m}|^2 t)}, \tag{3.4.18}$$

which is identical to the projection of $U(t)u$ on the subspace spanned by N Fourier modes. The question is then to find the best constant $K_{q,d}(N)$ such that

$$|f_N|_{L^q(\mathbf{T}^{d+1})} \leq K_{q,d}(N)|u|_{L^2(\mathbf{T}^d)}. \tag{3.4.19}$$

By comparison with the problem in $\mathbf{R}^d$ and using some counterexamples, Bourgain conjectured that

$$K_{q,d}(N) \leq \begin{cases} C_q N^{\beta(q)} & \text{for } q > r_0, \\ CN^\epsilon \;\; \forall \epsilon, & \text{for } q = r_0, \\ C_q & \text{for } q < r_0, \end{cases} \tag{3.4.20}$$

where $\beta(q) = \frac{d}{2} - \frac{d+2}{q}$, while $r_0 = 2 + \frac{4}{d}$ is the exponent of the Strichartz inequality in the entire space. He proved the conjecture when

$$
\begin{aligned}
d &= 1, \quad 2 \le q \le 4, \quad q \ge 6 = r_0 \\
d &= 2, \quad q \ge 4 = r_0, \\
d &= 3, \quad q > 4 > r_0.
\end{aligned}
\tag{3.4.21}
$$

Furthermore, since this approach involves Fourier transforms in both space and time, with t belonging to a finite interval $[-T, T]$ with small $T < 1$, the integral form of the NLS equation is rewritten as

$$
\psi(t) = \chi_{T_0} U(t)\varphi + \chi_T(t) \int_0^t U(t' - t)|\psi(t')|^{2\sigma}\psi(t')dt'
\tag{3.4.22}
$$

with $T_0 \le 1$ and $\chi_T = \chi_1(\frac{t}{T})$, where χ_1 is in $C^\infty(\mathbf{R})$ and satisfies $\chi_1(t) = 1$ if $|t| \le 1$ and $\chi_1(t) = 0$ if $|t| \ge 2$. The free evolution contribution to the integral equation is then eliminated by defining the spaces X such that $|u|_X = |U(-t)u|_H$ where H holds for a classical functional space, like the Sobolev space $H^{s,b}$ in the $\mathbf{x}$ and t variables, defined as $H^{s,b} = \{u/(1 + |k|^2)^{\frac{s}{2}}(1 + |\tau|^2)^{\frac{b}{2}}|u|_{L^2} < \infty\}$. We quote below some of the results.

For $d = 1$, the NLS equation on the one-dimensional torus $\mathbf{T}$ is locally well-posed in $H^s(\mathbf{T})$, provided that $\sigma < 2/(1 - 2s)$. If $\sigma \le 1$, it is globally well-posed in the space $L^4(\mathbf{T} \times \mathbf{R}_{\mathrm{loc}})$ for initial data $\varphi \in L^2(\mathbf{T})$. If $\sigma = 1$, the solution is in $C(\mathbf{R}, H^s(\mathbf{T}))$ for all $\varphi \in H^s(\mathbf{T})$, $s \ge 0$. This corresponds to the case where the NLS equation is integrable. Its spectral properties in periodic geometry are addressed for both attracting and repulsive nonlinearities by Ma and Ablowitz (1981).

For $d = 2$, the NLS equation is locally well-posed in $H^s(\mathbf{T})$, provided that $1 \le \sigma < 1/(1 - s)$, $0 < s \le 1$. It is globally well-posed in $H^1(\mathbf{T}^2)$ for $\sigma = 1$ and initial condition $\varphi \in H^1$ with sufficiently small L^2-norm. The same result holds for all $\sigma \ge 1$ if $|\varphi|_{H^1}$ is small enough.

For $d = 3$, the NLS equation is globally well-posed in $H^1(\mathbf{T}^d)$ with $1 \le \sigma < 2$ and sufficiently small initial data in H^1.

For $d \ge 4$, the NLS equation is locally well-posed in $H^s(\mathbf{T}^d)$ when $1 \le \sigma < 2/(d - 2s)$ and $3d/(d + 4) < s \le d/2$. In dimension $d = 4$, and for $\sigma = \frac{1}{2}$, the NLS equation is globally well-posed in $H^2(\mathbf{T}^4)$ for an initial condition $\varphi \in H^2$ with sufficiently small L^2-norm.

An important issue addressed in the context of periodic boundary conditions concerns the "statistical mechanics" of the NLS equation. Since this topic is outside the scope of the present monograph which is mostly concerned with classical solutions, we give only a brief discussion and refer to the original papers and to Bourgain (1996a) for a review. The method consists in replacing the study of individual solutions by that of appropriate probability measures on the phase space of the system. It can be viewed as an extension to a continuous system of the statistical-mechanics approach of an ensemble of particles at thermal equilibrium.

For a system of n particles confined in a compact spatial region Ω, the Gibbs probability distribution for finding a system of n particles confined in a compact spatial region Ω in a neighborhood dX_n of the microscopic state X_n is given by

$$\mu(dX_n) = \mathcal{Z}_n^{-1} \exp[-\beta H(X_n)] dX_n, \qquad (3.4.23)$$

where H is the Hamiltonian of the system, β the inverse temperature, and $\mathcal{Z}_n$ a normalization constant that to be finite requires the Hamiltonian to be bounded from below. The extension to a continuous field is challenging not only because of the infinite number of degrees of freedom already present in a finite volume, but also because as is the case for the NLS equation with attracting nonlinearity, the Hamiltonian may be unbounded from below. In one space dimension, the problem was first addressed by Lebowitz, Rose, and Speer (1988, 1989) (see also Bourgain 1997b for a brief description of the results), who show that a normalized measure can be defined for any subcritical nonlinearity by restricting the Gibbs measure to any ball in the L^2-space. For the critical nonlinearity, it can also be defined, provided that this ball is small enough. No normalized measure is obtained in supercritical dimension. In other words, the conditions for the existence of a normalized measure are essentially the same as those for the existence of a global solution for the NLS equation. This normalized measure was shown to be invariant under the dynamics (Bourgain 1994b, McKean 1995, Bidégaray 1995, Zhidkov 1995, Bourgain 1996b, Brydges and Slade 1996), which allows one to apply the Poincaré recurrence theorem. The existence of invariant measures in the case of an interval with Dirichlet boundary conditions was shown by Zhidkov (1991). This involves the proof of existence of a well-defined flow associated to initial data with low regularity in the support of the Gibbs measure. Extension to nonlocal potential of Hartree type is obtained in Bourgain (1997b).

In relation to the existence of invariant measures, there arises the problem of the persistence of time periodic or quasi-periodic solutions of the linear Schrödinger equation with a smooth real potential under Hamiltonian perturbations. The subject is related to the KAM theory for invariant tori of finite dimension in an infinite-dimensional phase space. In the case of a finite interval with Dirichlet boundary conditions, the problem was solved by Kuksin (1993). The analytical difficulties are due to the presence of small divisors similar to the case of pertubation theory for near Hamiltonian integrable systems. An additional difficulty arises with periodic boundary conditions because of the presence of resonances as well. The problem was addressed by Craig and Wayne (1994) using a Lyapunov–Schmidt decomposition, in the case of time-periodic solutions in one space dimension. The result was extended to quasi-periodic solutions in one and two space dimensions by Bourgain (1994b, 1998b). A recent review of these problems is given by Craig (1998).

4

Standing Wave Solutions

The elliptic nonlinear Schrödinger equation

$$i\psi_t + \Delta\psi + |\psi|^{2\sigma}\psi = 0, \quad \mathbf{x} \in \mathbf{R}^d, \qquad (4.0.1)$$

with an attracting nonlinearity and the condition $\psi \to 0$ as $|\mathbf{x}| \to \infty$, admits special solutions called "standing waves", "bound states", "solitary waves", or "wave guides", of the form $\psi(\mathbf{x}, t) = e^{i\lambda^2 t}\Phi(\mathbf{x})$, where the profile Φ is time-independent. Because of the resulting time-invariance of the wave intensity (which corresponds in quantum mechanics to the probability of finding a particle at a given location), these solutions are also referred to as stationary. The function Φ satisfies

$$\Delta\Phi - \lambda^2\Phi + |\Phi|^{2\sigma}\Phi = 0, \quad \mathbf{x} \in \mathbf{R}^d, \qquad (4.0.2)$$

where the coefficient λ^2 is to be positive in order to ensure that Φ vanishes at infinity. More general solutions are constructed using invariance properties of the NLS equation, among them the Galilean invariance.

It is easily seen that Φ is a solution of the variational problem

$$\delta\{H + \lambda^2 N\} = 0, \qquad (4.0.3)$$

where $N = \int |\Phi|^2 d\mathbf{x}$ and $H = \int (|\nabla\Phi|^2 - \frac{1}{\sigma+1}|\Phi|^{2\sigma+2})d\mathbf{x}$.

Defining $\phi(\mathbf{x}) = \lambda^{-1/\sigma}\Phi(\lambda^{-1}\mathbf{x})$, we have $N = \lambda^{2/\sigma-d}N_0$ with $N_0 = \int |\phi|^2 d\mathbf{x}$, and

$$\Delta\phi - \phi + |\phi|^{2\sigma}\phi = 0. \qquad (4.0.4)$$

In one dimension, there exists a unique solution of (4.0.4) of the form

$$\phi(x) = \frac{(\sigma+1)^{1/2\sigma}}{\cosh^{1/\sigma}(\sigma x)} \qquad (4.0.5)$$

that satisfies the zero boundary conditions at infinity. This is not the case in higher dimensions where there exists a denumerable set of solutions that can be analyzed in details when isotropy is assumed (Anderson and Derrick 1970).

As precisely stated in Section 4.2, there exists for fixed λ^2 a unique solution g of (4.0.2) that is positive and radially symmetric. It is an extremum of H at fixed N and also minimizes the action $\frac{1}{2}(H + \lambda^2 N)$ among all nontrivial solutions of (4.0.2). By analogy with the problem of a quantum particle in a potential (see e.g. Lieb and Loss 1997, Chapter 11), this solution is referred to as the "ground state", while the other solutions are called "excited states" and are characterized in the isotropic case by their number of nodes. The ground state associated to $\lambda^2 = 1$ will be denoted by R.

An important question concerns the conditions of stability or instability of the standing waves and in particular, of that associated to the ground state we refer to as ground-state standing wave. In the case $d = 2$, $\sigma = 1$ (especially relevant in optics), the latter is usually called "Townes soliton."

4.1   A Heuristic Approach

Before reviewing rigorous results concerning the existence and stability of the standing-wave solutions, we present in this section a linear analysis initiated by Zakharov (1968a) and Vakhitov and Kolokolov (1973), and also a heuristic argument based on an analogy with Hamiltonian systems with a finite number of degrees of freedom, as reviewed by Kuznetsov, Rubenchik, and Zakharov (1986) and Kuznetsov (1996) (see also Laedke, Spatschek, and Stenflo 1983, Laedke, and Spatschek 1985). A discussion of the various types of stability is given in Pelinovsky and Grimshaw (1997).

4.1.1   Linear Stability Analysis

For fixed λ^2, let g denote the ground state of (4.0.2) and $g(\mathbf{x})e^{i\lambda^2 t}$ the corresponding standing-wave solution of the NLS equation. When subject to infinitesimal perturbations of the amplitude and phase, this solution becomes

$$\psi(\mathbf{x}, t) = g(\mathbf{x})(1 + r(\mathbf{x}, t))e^{i(\lambda^2 t + s(\mathbf{x}, t))}, \qquad (4.1.1)$$

or after expansion,

$$\psi(\mathbf{x}, t) = (g(\mathbf{x}) + u(\mathbf{x}, t) + iv(\mathbf{x}, t))e^{i\lambda^2 t}, \qquad (4.1.2)$$

where we have introduced the real functions $u = gr$ and $v = gs$. When linearizing the nonlinear Schrödinger equation about the ground-state standing wave, we get

$$\partial_t \begin{pmatrix} u \\ v \end{pmatrix} = \mathbf{N} \begin{pmatrix} u \\ v \end{pmatrix} \qquad (4.1.3)$$

with

$$\mathbf{N} = \begin{pmatrix} 0 & L_0 \\ -L_1 & 0 \end{pmatrix} \qquad (4.1.4)$$

and

$$L_0 = -\Delta + \lambda^2 - g^{2\sigma}, \qquad (4.1.5)$$
$$L_1 = -\Delta + \lambda^2 - (2\sigma + 1)g^{2\sigma}. \qquad (4.1.6)$$

For perturbations $u, v \propto e^{i\Omega t}$, we have

$$\Omega^2 u = L_0 L_1 u. \qquad (4.1.7)$$

The operators L_0 and L_1 are self-adjoint. Using the fact that g satisfies $\Delta g - \lambda^2 g + g^{2\sigma+1} = 0$, one easily checks that the operator L_0 can be rewritten as

$$L_0 = -\frac{1}{g}\text{div}\left(g^2\text{grad}\left(\frac{1}{g}\cdot\right)\right). \qquad (4.1.8)$$

As a consequence, $\int u L_0 u d\mathbf{x} = \int |\nabla(\frac{u}{g})|^2 g^2 d\mathbf{x} \geq 0$, and the operator L_0 is nonnegative. Furthermore, g belongs to the kernel of L_0 and ∇g to the kernel of L_1. The ground state g, which is positive, is also spherically symmetric about some point O and decreases with respect to the radial coordinate about this point. As a consequence, ∇g has a single node at the point O. So, as proved by Weinstein (1985), zero is the second eigenvalue of L_1, which has exactly one negative eigenvalue. For solutions u of (4.1.7) orthogonal to g, the minimum value of Ω^2 is given by

$$\Omega_m^2 = \min \frac{\langle u|L_1|u\rangle}{\langle u|L_0^{-1}|u\rangle}, \qquad (4.1.9)$$

where we have used the notation $\langle a|L|b\rangle = \int aLb\, d\mathbf{x}$. For this class of functions, L_0 is positive definite, so the sign of Ω_m^2 is prescribed by that of $\langle u|L_1|u\rangle$. The minimum value of $\langle u|L_1|u\rangle$ is attained for a function u (orthogonal to g), which satisfies the spectral problem

$$L_1 u = \mu u + \alpha g, \qquad (4.1.10)$$

where α is an undetermined Lagrange multiplier and $\mu = \langle u|L_1|u\rangle$. Let $\mu_0, \mu_1, \ldots$ be the eigenvalues of L_1. Since L_1 has only one negative eigenvalue μ_0 and since the eigenvector ∇g associated to the second eigenvalue

$\mu_1 = 0$ is orthogonal to g, (4.1.10) can be written as

$$u = \alpha(L_1 - \mu)^{-1}g \qquad (4.1.11)$$

if $\mu_0 < \mu < \mu_2$, where μ_2 is the first positive eigenvalue. Taking the scalar product with g and using the orthogonality condition $\langle g|u \rangle = 0$, the above equation is replaced by

$$f(\mu) \equiv \langle g|(L_1 - \mu)^{-1}|g \rangle = 0. \qquad (4.1.12)$$

When μ increases from μ_0 to μ_2, $f(\mu)$ varies monotonically from $-\infty$ to $+\infty$, and consequently passes through zero for $\mu = \mu_{\min} \in (\mu_0, \mu_2)$. In order to determine the sign of $\mu_{\min}$, it is enough to consider the sign of $f(0) = \langle g|L_1^{-1}|g \rangle$. If it is positive, $\mu_{\min}$ is negative, while if it is negative, $\mu_{\min}$ is positive. To calculate this quantity, one differentiates the equation obeyed by g with respect to λ^2, and gets

$$L_1 \frac{\partial g}{\partial \lambda^2} + g = 0. \qquad (4.1.13)$$

Hence,

$$f(0) = -\left\langle g \left| \frac{\partial g}{\partial \lambda^2} \right. \right\rangle = -\frac{1}{2} \frac{\partial N}{\partial \lambda^2}, \qquad (4.1.14)$$

where

$$N = \int g^2 d\mathbf{x}. \qquad (4.1.15)$$

It follows that $\partial N/\partial \lambda^2$ prescribes the sign of $\mu_{\min}$. When $\partial N/\partial \lambda^2 < 0$, a condition that corresponds to $d\sigma > 2$, then $\mu_{\min}$ and thus Ω_m^2 are negative. In this case, harmonic pertubations increase exponentially and the ground-state standing wave is unstable. In the other case, for $\partial N/\partial \lambda^2 > 0$ or equivalently $d\sigma < 2$, the ground-state standing wave is linearly stable. At the critical dimension $d\sigma = 2$, the above criterion cannot be used.

Another way to derive the criterion for instability (Laedke, Spatschek, and Stenflo 1983) is to introduce the function

$$\xi = \langle \xi_-|g \rangle L_1^{-1}g - \langle g|L_1^{-1}|g \rangle \xi_-, \qquad (4.1.16)$$

where ξ_- is such that $\langle \xi_-|L_1|\xi_- \rangle < 0$. Such a function ξ_- exists, since L_1 has a negative eigenvalue. Note that $L_1^{-1}g = -\frac{\partial g}{\partial \lambda^2}$ and that $\langle \xi, g \rangle = 0$. Calculating $\langle \xi|L_1|\xi \rangle$, we get

$$\langle \xi|L_1|\xi \rangle = -\langle g|L_1^{-1}|g \rangle \langle \xi_-|g \rangle^2 + \langle g|L_1^{-1}|g \rangle^2 \langle \xi_-|L_1|\xi_- \rangle. \qquad (4.1.17)$$

If $\langle g|L_1^{-1}|g \rangle > 0$ or equivalently $\partial N/\partial \lambda^2 < 0$, then $\langle \xi|L_1|\xi \rangle < 0$ and thus the minimal value Ω_m^2 defined in (4.1.9) is strictly negative. The ground-state standing wave is then unstable.

4.1.2   The Case of Finite Amplitude Perturbations

Consider the transformation

$$\Psi'(r) = \frac{1}{a^{d/2}} \Psi\left(\frac{r}{a}\right), \tag{4.1.18}$$

which preserves the mass N and transforms the Hamiltonian into

$$H(a) = \frac{1}{a^2} \int |\nabla\Psi|^2 dx - \frac{1}{a^{\sigma d}} \frac{1}{(\sigma+1)} \int |\Psi|^{2\sigma+2} dx. \tag{4.1.19}$$

Since N is preserved with varying a, (4.0.3) implies that for a solution Φ of (4.0.2), $\partial H/\partial a|_{a=1} = 0$ or equivalently, $X = \frac{\sigma d}{2(\sigma+1)}Y$, where $X = \int |\nabla\Phi|^2 dx$ and $Y = \int |\Phi|^{2\sigma+2} dx$. Since on the other hand, $X = -\lambda^2 N + Y$, one has

$$X = \frac{\sigma d \lambda^{2+2/\sigma-d} N_0}{2\sigma+2-\sigma d}, \quad Y = \frac{2(\sigma+1)\lambda^{2+2/\sigma-d} N_0}{2\sigma+2-\sigma d}, \tag{4.1.20}$$

$$H = \frac{(\sigma d - 2)\lambda^{2+2/\sigma-d} N_0}{2\sigma+2-\sigma d}, \tag{4.1.21}$$

where $N_0 = \int |\phi|^2 dx$. It follows that the Hamiltonian of the ground state vanishes at the critical dimension $\sigma d = 2$ and is positive for $\sigma d < 2$. It is negative in dimension $d = 1$ when $\sigma > 2$ or in dimension $d > 2$ when $2/d < \sigma < 2/(d-2)$. Higher nonlinearities are not permitted, since they do not preserve the positivity of X and Y (to be compared to the condition for the existence of a solution of the initial value problem given in Chapter 3).

Furthermore, $\partial^2 H/\partial a^2|_{a=1} = 2(2 - \sigma d)X$, indicating that the ground state realizes a minimum of $H(a)$ when $\sigma d < 2$ and a maximum when $\sigma d > 2$ (with the same condition on σ as above when $d > 2$).

Another transformation preserving N is provided by the gauge transformation

$$\Psi'(r) = \Psi(r)e^{i\chi(r)}, \tag{4.1.22}$$

which changes the Hamiltonian H into

$$H' = H + \int (\nabla\chi)^2 \Psi^2 dx > H. \tag{4.1.23}$$

For $d\sigma > 2$, the ground state is thus a saddle point of the Hamiltonian functional. One can thus conjecture by analogy with Hamiltonian systems with a finite number of degrees of freedom that the ground-state standing wave will be stable for $\sigma d < 2$ and unstable for $\sigma d > 2$. The case $d\sigma = 2$ is critical.

4.2   Existence and Variational Approach

The existence of solutions to equations of the form $-\Delta u = f(u)$ has been extensively studied (see, e.g. Evans 1998, Chapter 8). The first results concern the case $f(u) = -mu + |u|^p u$ $(m > 0)$, which is associated to standing-wave solutions of the NLS and Klein–Gordon equations (Pohozaev 1971 in a bounded domain of $\mathbf{R}^d$, Coffman 1972 in $\mathbf{R}^3$). These results were later generalized to larger classes of functions f (Strauss 1977, Coleman, Glazer, and Martin 1978, Berestycki and Lions 1983 in dimensions $d \geq 3$, Esteban 1980 and Berestycki, Gallouët, and Kavian 1983 in dimension 2). Here, we restrict the discussion to solutions $g \in H^1(\mathbf{R}^d)$ of

$$\Delta g - \lambda^2 g + g^{2\sigma+1} = 0. \tag{4.2.1}$$

This problem is amenable to variational formulations. To obtain a solution of (4.2.1), one looks for the critical points on $H^1(\mathbf{R}^d)$ of the Lyapunov functional, also called action,

$$S(u) = \frac{1}{2}(H(u) + \lambda^2 N(u)). \tag{4.2.2}$$

Other approaches include an ODE method that consists in looking directly for radially symmetric solutions (Berestycki, Lions, and Peletier 1981), and also a "local" method that solves (4.2.1) in a ball of finite radius with Dirichlet boundary conditions and then passes to the limit (see Berestycki and Lions 1980 for details).

4.2.1   Necessary Conditions for Existence in H^1

Lemma 4.1. (Pohozaev identities) *Any solution of (4.2.1) in $H^1(\mathbf{R}^d) \cap L^{2\sigma+2}(\mathbf{R}^d)$ satisfies the Pohozaev identities*

$$\int |\nabla g|^2 d\mathbf{x} = \frac{\sigma d}{2(\sigma+1)} \int |g|^{2\sigma+2} d\mathbf{x}, \tag{4.2.3}$$

$$\frac{\lambda^2}{d} \int |g|^2 d\mathbf{x} = \left(\frac{1}{d} - \frac{\sigma}{2(\sigma+1)}\right) \int |g|^{2\sigma+2} d\mathbf{x}. \tag{4.2.4}$$

Consequently, there are no solutions in $H^1(\mathbf{R}^d)$ of (4.2.1) when $\sigma > \frac{2}{d-2}$.

Proof. These identities are direct consequences of (4.1.20)–(4.1.21). They can also be directly established as follows. Multiplying (4.2.1) by g and integrating over the whole space, one gets

$$\int \left(-|\nabla g|^2 - \lambda^2|g|^2 + |g|^{2\sigma+2}\right) d\mathbf{x} = 0. \tag{4.2.5}$$

Similarly, multiplying (4.2.1) by $x_i \partial_i g$ and integrating over $\mathbf{R}^d$, one obtains

$$\int \left((\frac{d}{2} - 1)|\nabla g|^2 + \frac{\lambda^2 d}{2}|g|^2 - \frac{d}{2(\sigma+1)}|g|^{2\sigma+2}\right) d\mathbf{x} = 0. \tag{4.2.6}$$

Identities (4.2.3)–(4.2.4) follow by combining (4.2.5)–(4.2.6).

4.2.2 Existence Results

Theorem 4.2. (Existence of a positive solution or ground state) *Suppose $d \geq 2$ and $\sigma < 2/(d-2)$ (no condition on σ if $d = 2$). Equation (4.2.1) has a positive, spherically symmetric solution $g \in C^2(\mathbf{R}^d)$. In addition, g and its derivatives up to order 2 have an exponential decay at infinity. This solution minimizes the action, among all $H^1(\mathbf{R}^d)$-solutions of (4.2.1).*

Theorem 4.3. (Uniqueness of positive solutions) *For $0 < \sigma < \frac{2}{d-2}$, the positive solution obtained in Theorem 4.2 is unique.*

Theorem 4.3, due to Kwong (1989), ensures uniqueness of the ground state. Since the solution is radial, the equation reduces to an ordinary differential equation. Kwong's result improves several previous contributions. Coffman (1972) studied the special case $d = 3$, $\sigma = 1$, and proved the existence and uniqueness of a positive radially symmetric solution. McLeod and Serrin (1987) extended Coffman's result to any dimension and power nonlinearity satisfying the conditions $\sigma < \infty$ if $1 \leq d \leq 2$, $\sigma \leq \frac{1}{d-2}$, if $2 < d \leq 4$, and $\sigma < \frac{8-d}{2d}$ if $4 < d < 8$.

Theorem 4.4. *Let $u(\mathbf{x})$ be a C^2 positive solution vanishing at infinity of the equation $\Delta u + f(u) = 0$ in $\mathbf{R}^d$, $d \geq 2$, with $f \in C^{1+\mu}(\mathbf{R}^d)$, $\mu > 0$, $f(0) = 0$, and $f'(0) < 0$. Then $u(\mathbf{x})$ is spherically symmetric about some point in $\mathbf{R}^d$ and $\frac{\partial u}{\partial r} < 0$ for $r > 0$, where r is the radial coordinate about that point. Furthermore,*

$$\lim_{r \to \infty} r^{(d-1)/2} e^r u(r) = \text{const} > 0. \tag{4.2.7}$$

This result, due to Gidas, Ni, and Nirenberg (1981), ensures that any smooth positive solution of (4.2.1) is radially symmetric. We quoted it here in the form of the remark following Theorem 2 in their paper (p. 371).

Theorem 4.5. (Existence of infinitely many solutions or bound states) *Under the hypotheses of Theorem 4.2, there exists an infinite number of radially symmetric solutions g_k of class C^2. Each g_k has exactly k nodes as a function of $|\mathbf{x}| = r$ and decays exponentially at infinity. In addition, $S(g_k)$ and $|\nabla g_k|_{L^2} \to \infty$ as $k \to \infty$.*

Existence of bound states was established by Strauss (1977) and Berestycki and Lions (1983) for a large class of nonlinearities using critical point theory. This result was also obtained by means of other approaches consisting in proving existence of radial solutions with a prescribed number of nodes. They were developed by Jones, Küpper, and Plakties (1988) and Grillakis (1990) using methods of degree theory.

4.2.3 Variational Approaches

In this section we sketch the proof for Theorem 4.2 for existence of the ground state given in Berestycki and Lions (1983) and then briefly review other minimization procedures developed in this context.

Define the functionals

$$T(u) = \int |\nabla u|^2 d\mathbf{x}, \tag{4.2.8}$$

$$V(u) = \int \left(\frac{1}{2\sigma + 2}|u|^{2\sigma+2} - \frac{\lambda^2}{2}|u|^2 \right) d\mathbf{x}, \tag{4.2.9}$$

$$S(u) = \frac{1}{2}T(u) - V(u), \tag{4.2.10}$$

$$K(u) = \int \left(\frac{d-1}{2d}|\nabla u|^2 + \frac{\lambda^2}{2}|u|^2 - \frac{1}{2(\sigma+1)}|u|^{2\sigma+2} \right) d\mathbf{x}. \tag{4.2.11}$$

A consequence of the Pohozaev identities is that if g is a solution of (4.2.1), then

$$S(g) = \frac{1}{d} T(g) > 0. \tag{4.2.12}$$

To obtain a solution of (4.2.1), one looks for the critical points of the action S on $H^1(\mathbf{R}^d)$ which can be proved to be a C^1-functional on $H^1(\mathbf{R}^d)$. They can equivalently be viewed as the critical points of the Hamiltonian $H = \int \left(|\nabla u|^2 - \frac{1}{\sigma+1}|u|^{2\sigma+2} \right) d\mathbf{x}$ under the constraint $N(u) = \int |u|^2 d\mathbf{x} =$ constant.

One can also look for a solution of the constrained minimization problem

$$\min\{T(w), w \in H^1(\mathbf{R}^d), V(w) = 1\} \tag{4.2.13}$$

(Coleman, Glazer, and Martin 1978, Berestycki and Lions 1983). We first note that the solution obtained by this procedure has the property of minimizing the action among all solutions of (4.2.1).

Let us prove that

$$S(g) \le S(v) \tag{4.2.14}$$

for all $v \in H^1(\mathbf{R}^d)$ satisfying (4.2.1). Indeed, if $\bar{g}$ solves (4.2.13), then there exits a Lagrange multiplier θ such that $T'(\bar{g}) = \theta V'(\bar{g})$, that is to say,

$$-\Delta \bar{g} = \theta(\bar{g}^{2\sigma+1} - \lambda^2 \bar{g}). \tag{4.2.15}$$

Thus the function $g(\mathbf{x}) = \bar{g}(\mathbf{x}/\sqrt{\theta})$ satisfies (4.2.1), and from the Pohozaev equalities (4.2.3) and (4.2.4), $T(g) = \frac{2d}{d-2}V(g)$. We have

$$T(g) = \theta^{\frac{d-2}{2}}T(\bar{g}) \quad \text{and} \quad V(g) = \theta^{d/2}V(\bar{g}) = \theta^{d/2}. \tag{4.2.16}$$

It follows that $\theta = \frac{d-2}{2d}T(\bar{g})$ and

$$S(g) = \frac{1}{d}\left(\frac{d-2}{2}\right)^{\frac{d-2}{2d}} T(\bar{g})^{d/2}. \tag{4.2.17}$$

Now let v be another solution of (4.2.1). Again, $T(v) = \frac{2d}{d-2}V(v)$. The rescaled function $v_\tau(\mathbf{x}) = v(\mathbf{x}/\sqrt{\tau})$ satisfies $V(v_\tau) = 1$. This implies that

$$\tau = V(v)^{-1/d} = \left(\frac{d-2}{2d}\right)^{-1/d} T(v)^{-1/d}. \tag{4.2.18}$$

Furthermore, $T(v_\tau) = \tau^{2-d}T(v)$. Thus,

$$T(v) = \left(\frac{d-2}{2d}\right)^{(d-2)/2} T(v_\tau)^{d/2}. \tag{4.2.19}$$

Using (4.2.12),

$$S(v) = \frac{1}{d}\left(\frac{d-2}{2d}\right)^{(d-2)/2} T(v_\tau)^{d/2}. \tag{4.2.20}$$

From the minimization problem, we know that

$$T(v_\tau) \geq T(\bar{g}). \tag{4.2.21}$$

Comparison of (4.2.17) and (4.2.20) then leads to (4.2.14).

The above approach has to be modified in dimension $d = 2$, since in this case, a solution of (4.2.1) satisfies $V(g) = 0$. As shown by Berestycki, Gallouët, and Kavian (1983), the minimization problem (4.2.13) is then replaced by

$$\min\{T(w), w \in H^1(\mathbf{R}^d), V(w) = 0\}. \tag{4.2.22}$$

To solve the minimization problem (4.2.13) or (4.2.22), one constructs an appropriate minimizing sequence which, by Schwarz symmetrization (Lieb 1977, Lieb and Loss 1997 Chapter 3) can be chosen nonnegative, spherically symmetric and nonincreasing. Key points for the passage to the limit are the uniform decay at infinity of radial functions in $H^1(\mathbf{R}^d)$ (Lemma 5.6) and a compactness lemma, both due to Strauss (1977).

Other minimization procedures can be used to construct the ground state. Keller (1983) and Shatah (1985b) consider

$$\inf\left\{\frac{1}{d}T(u),\ K(u) = 0\right\}, \tag{4.2.23}$$

which is the same as

$$\inf\left\{\frac{1}{d}T(u),\ K(u) \leq 0\right\}, \tag{4.2.24}$$

where the functional K is defined in (4.2.11). Note that from (4.2.6), any solution of (4.2.1) satisfies $K(u) = 0$. The minimum value is also equal to

$$\inf\{S(u),\ T(u) = T(g)\} \tag{4.2.25}$$

because if u satisfies $\int |\nabla u|^2 d\mathbf{x} = \int |\nabla g|^2 d\mathbf{x}$, then by (4.2.24), $K(u) \geq 0$. Thus $S(u) = \frac{1}{d}T(u) + K(u) \geq S(g)$.

Using a different approach, Weinstein (1983) minimizes

$$J^{\sigma,d}(u) = \frac{|\nabla u|_{L^2}^{\sigma d}|u|_{L^2}^{2+\sigma(2-d)}}{|u|_{L^{2\sigma+2}}^{2\sigma+2}} \tag{4.2.26}$$

among all $u \in H^1(\mathbf{R}^d)$, $0 < \sigma < \frac{2}{d-2}$. It turns out that $J^{\sigma,d}(u)$ reaches its minimum α at a function $R \in H^1(\mathbf{R}^d) \cap C^\infty(\mathbf{R}^d)$ that is positive, radially symmetric, and is a solution of

$$\frac{\sigma d}{2}\Delta R - \left(1 + \frac{\sigma}{2}(2-d)\right)R + R^{2\sigma+1} = 0 \tag{4.2.27}$$

with the minimal L^2-norm (ground state).

An important consequence of Weinstein's approach is the following corollary, which is used for the derivation of sharp conditions for existence of solutions for the nonlinear Schrödinger equation with critical nonlinearity (see Theorem 3.17).

Corollary 4.6. *The best constant $C_{\sigma,d}$ for the Gagliardo–Nirenberg inequality*

$$|f|_{L^{2\sigma+2}}^{2\sigma+2} \le C_{\sigma,d}|\nabla f|_{L^2}^{\sigma d}|f|_{L^2}^{2+\sigma(2-d)} \tag{4.2.28}$$

with $f \in H^1(\mathbf{R}^d)$, $0 < \sigma < \frac{2}{d-2}$ and $d \ge 2$ is given by

$$C_{\sigma,d} = (\sigma+1)\frac{2(2+2\sigma-\sigma d)^{-1+\sigma d/2}}{(\sigma d)^{\sigma d/2}}\frac{1}{|R|_{L^2}^{2\sigma}}, \tag{4.2.29}$$

where R is the positive solution of $\Delta R - R + R^{2\sigma+1} = 0$.

4.3   Stability/Instability Conditions

We investigate the stability of the standing waves $\psi(\mathbf{x},t) = e^{i\lambda^2 t}g(\mathbf{x})$ constructed above. The first theorem of this section concerns the *orbital stability* in the subcritical case $\sigma < 2/d$. Orbital stability refers to stability up to the transformations keeping the equation invariant (translation and phase shift). It was first proved by Cazenave and Lions (1982) using the concentration–compactness methods of Lions (1984). Another proof based on Lyapunov functionals was given by Weinstein (1986a).

Theorems 4.8 and 4.9, due to Weinstein (1983) and Berestycki and Cazenave (1981), concern the critical and supercritical cases, respectively. All the standing waves in the former case and the ground-state standing waves in the latter one are shown to be unstable due to the existence of solutions with arbitrarily close initial conditions that blow up in a finite time.

In Section 4.4 we discuss an alternative approach initiated by Shatah and Strauss (1985), where the stability/instability properties of the ground-state standing waves are characterized by the convexity/concavity of the action restricted to the family of ground states, viewed as a function of the Lagrange multiplier. This general approach extends to solitary waves of general Hamiltonian systems that are invariant under a group of transformations (Grillakis, Shatah, and Strauss 1987, 1990, Grillakis 1988). A dynamical system approach to the study of linear instability of standing waves is presented in Jones (1988).

4.3.1 Subcritical Dimension: Orbital Stability

Theorem 4.7. *Assume $\sigma < 2/d$ and let g be the ground state of (4.2.1). Then, for any $\epsilon > 0$, there exists $\delta > 0$ such that if the initial condition φ obeys*

$$\inf_{\theta \in \mathbf{R}, y \in \mathbf{R}^d} |\varphi(\cdot) - e^{i\theta} g(\cdot + y)|_{H^1} < \delta, \tag{4.3.1}$$

then the solution $\psi(x, t)$ of (3.0.1)–(3.0.2) satisfies

$$\inf_{\theta \in \mathbf{R}, y \in \mathbf{R}^d} |\psi(x, \cdot) - e^{i\theta} g(\cdot + y)|_{H^1} < \epsilon. \tag{4.3.2}$$

Proof. We only sketch the main steps of the proof given by Weinstein (1986a) and refer to this reference for details. The proof follows the main lines of the linear stability analysis, except that it deals with finite perturbations and estimates the remainders.

The orbit of a function u is defined as

$$\mathcal{G}_u = \{u(\cdot + \mathbf{x_0})e^{i\gamma}, \forall \mathbf{x_0} \in \mathbf{R}^d, \gamma \in [0, 2\pi)\}. \tag{4.3.3}$$

The deviation of the solution ψ of the NLS equation from the orbit of the ground state g is measured by the norm

$$\rho(\psi(t), \mathcal{G}_g) = \inf_{\mathbf{x_0}, \gamma} \left\{ \left| \nabla \psi(\cdot + \mathbf{x_0}, t)e^{i\gamma} - \nabla g \right|_{L^2} + \lambda^2 \left| \psi(\cdot + \mathbf{x_0}, t)e^{i\gamma} - g \right|_{L^2} \right\}. \tag{4.3.4}$$

Step 1 : Lyapunov functional. We consider the Lyapunov functional

$$\mathcal{S}(\psi) = H(\psi) + \lambda^2 N(\psi) \tag{4.3.5}$$

(here without the factor $1/2$), and write the solution $\psi(\mathbf{x} + \mathbf{x_0}, t)e^{i\gamma} = g + w$, with $w = u + iv$. The conservation of the L^2-norm and of the Hamiltonian together with the scale invariance yield

$$\delta \mathcal{S} \equiv \mathcal{S}(\varphi) - \mathcal{S}(g) = \mathcal{S}(g + w) - \mathcal{S}(g). \tag{4.3.6}$$

Expanding the functional near g, the first variation vanishes (because g satisfies (4.2.1)), while the second variation gives

$$\delta \mathcal{S} = (L_0 v, v) + (L_1 u, u) + W, \tag{4.3.7}$$

where the operators L_0 and L_1 are given by (4.1.5) and (4.1.6), and the remainder W can be bounded from below by

$$W \geq -C_1 |w|_{H^1}^{2+\theta} - C_2 |w|_{H^1}^6, \tag{4.3.8}$$

with $\theta > 0$.

Step 2 : Spectral properties of L_0 and L_1. If $\mathbf{x}_0 = \mathbf{x}_0(t)$ and $\gamma = \gamma(t)$ are chosen to minimize $\rho(\psi(t), \mathcal{G}_g)$, then

$$\int g^{2\sigma} \frac{\partial g}{\partial x_j} u(\mathbf{x}, t) d\mathbf{x} = 0, \qquad (4.3.9)$$

$$\int g^{2\sigma+1} v(\mathbf{x}, t) d\mathbf{x} = 0. \qquad (4.3.10)$$

To estimate the first two terms of (4.3.7), one has to consider the spectral properties of the linear operators L_0 and L_1. As seen in Section 4.1.1, L_0 is a nonnegative operator, since $L_0 g = 0$ and $g > 0$. Also, the infimum of $(Lv, v)/(v, v)$ under the constraint (4.3.10) can be zero only if $v = g$, but this contradicts (4.3.10). Thus

$$(L_0 v, v) > c|v|_{L^2}^2, \qquad (4.3.11)$$

and also

$$(L_0 v, v) \geq c_1 |v|_{H^1}^2. \qquad (4.3.12)$$

The analysis of the second term of (4.3.7) is slightly more delicate. One proves that

$$(L_1 u, u) \geq C_1 |u|_{H^1}^2 - C_2 |\nabla w|_{L^2} |w|_{L^2}^2 - C_3 |w|_{L^2}^4. \qquad (4.3.13)$$

It is done in two steps. One first gets the bound under the restriction that the perturbed solution has the same L^2 norm as the ground state. Then, for general perturbations, given the initial condition φ, one rescales the ground state g in the form $\widetilde{g}(\mathbf{x}) = \lambda^{1/\sigma} g(\lambda \mathbf{x})$ such that $\widetilde{g}$ has the same L^2-norm as φ, and $|\widetilde{g} - g|_{H^1} < \epsilon/2$. Therefore,

$$|\psi(\cdot + \mathbf{x}_0, t) e^{i\gamma} - g|_{H^1} \leq |\psi(\cdot + \mathbf{x}_0, t) e^{i\gamma} - \widetilde{g}|_{H^1} + |\widetilde{g} - g|_{H^1}. \qquad (4.3.14)$$

The operator L_1 has exactly one negative eigenvalue, and its null-space is

$$Ker(L_1) = \text{span} \left\{ \frac{\partial g}{\partial x_j}, j = 1, \ldots, d \right\} \qquad (4.3.15)$$

(Weinstein 1985). Let us first suppose that the initial condition φ has the same L^2 norm as the ground state. To obtain the lower bound (4.3.13), one decomposes

$$u = u_{\parallel} + u_{\perp} \qquad (4.3.16)$$

with $u_{\parallel} = (u, g)g$ and $u_{\perp} = u - u_{\parallel}$. The hypothesis $|\psi|_{L^2} = |g|_{L^2}$ implies that

$$(u, g) = -\frac{1}{2}((u, u) + (v, v)). \qquad (4.3.17)$$

One establishes lower bounds for $(L_1 u_{\perp}, u_{\perp})$, $(L_1 u_{\parallel}, u_{\parallel})$, and $(L_1 u_{\perp}, u_{\parallel})$. It is in the estimate for $(L_1 u_{\perp}, u_{\perp})$ that the hypothesis of subcriticality is used. Indeed, the second variation of the functional $J^{\sigma, d}$ (defined in (4.2.26)) about its minimum g is nonnegative. On functions satisfying $(f, g) = 0$, one has

$$(L_1 f, f) + k^2(\sigma d - 2)(f, \Delta g)^2 \geq 0. \tag{4.3.18}$$

For $\sigma d \leq 2$, $(L_1 f, f) \geq 0$ for all f orthogonal to g. Since $L_1 \nabla g = 0$ and $(\nabla g, g) = 0$,

$$\inf_{(f, g) = 0} (L_1 f, f) = 0, \qquad \text{if} \quad \sigma d \leq 2. \tag{4.3.19}$$

If $\sigma d < 2$, the infimum is attained at $f = c \cdot \nabla g$, for $c \in \mathbf{R}^d$. But this violates the condition (4.3.9). Thus

$$(L_1 u_\perp, u_\perp) \geq d(u_\perp, u_\perp). \tag{4.3.20}$$

Step 3 : A lower bound for δS. Since

$$\sqrt{\min(1, \lambda^2)} |w|_{H^1} \leq \rho(\psi(t), \mathcal{G}_g) \leq \sqrt{\max(1, \lambda^2)} |w|_{H^1}, \tag{4.3.21}$$

it follows that (4.3.7), (4.3.8), (4.3.12), and (4.3.13) imply

$$\delta S \geq h(\rho(\psi(t), \mathcal{G}_g)), \tag{4.3.22}$$

where $h(s) = cs^2(1 - as^\theta - bs^6)$, with $a, b, c, \theta > 0$.
Step 4 : Stability result. Let $\epsilon > 0$. By continuity of S in H^1, there exists $\eta(\epsilon)$ such that if $\rho(\psi_0, \mathcal{G}_g) < \eta$, then

$$\delta S(0) < h(\epsilon). \tag{4.3.23}$$

Since S is constant in time, it follows from (4.3.22) that

$$h(\rho(\psi(t), \mathcal{G}_g)) < h(\epsilon) \tag{4.3.24}$$

for all $t > 0$ and, by continuity of $\rho(\psi(t), \mathcal{G}_g)$ as a function of t, that

$$\rho(\psi(t), \mathcal{G}_g) < \epsilon, \tag{4.3.25}$$

which expresses the orbital stability of the ground-state standing wave.
Remark. Note that except for the one-dimensional case where the solution is unique, it is still unknown whether the bound states associated to the excited states, are stable.

4.3.2 Critical Case: Instability by Blowup

Theorem 4.8. *At critical dimension $\sigma = 2/d$, all the H^1 solutions of (4.2.1) are unstable for the NLS equation in the following sense: Let $g \in H^1(\mathbf{R}^d)$ be a solution of (4.2.1). For any $\epsilon > 0$, there exists a function $\varphi \in H^1$, with $|\varphi - g|_{L^2} < \epsilon$, such that the solution ψ of (3.0.1) with initial condition φ, satisfies $\lim_{t \to T} |\nabla \psi|_{L^2} = \infty$, for some $0 < T < \infty$.*

Proof. The proof uses that standing-wave solutions have a zero Hamiltonian. One constructs an initial condition $\varphi_\epsilon = (1 + \epsilon)g(\mathbf{x})$ that is "close" to g

in H^1-norm but has a Hamiltonian $H(\varphi_\epsilon) = -2\epsilon|g|^2_{L^2} + O(\epsilon^2)$, which is negative for small enough ϵ. Theorem 5.1 then ensures that the corresponding solution will blow up in a finite time.

4.3.3 Supercritical Case: Instability by Blowup

Theorem 4.9. *Let $\sigma > 2/d$ and let g be the ground state of (4.2.1). For any $\epsilon > 0$ there exists a function $\varphi \in H^1$, with $|\varphi - g|_{H^1} < \epsilon$, such that the solution ψ of (3.0.1) with initial condition φ satisfies*

$$\lim_{t \to T} |\nabla\psi|_{L^2} = \infty, \qquad (4.3.26)$$

for some $0 < T < \infty$.

Proof. The proof uses the variance identity (3.2.4), which for any solution ψ of the NLS equation is written in the form

$$\frac{1}{8}\frac{d^2}{dt^2}V(t) = \int \left(|\nabla\psi|^2 - \frac{d\sigma}{2\sigma+2}|\psi|^{2\sigma+2}\right)d\mathbf{x}. \qquad (4.3.27)$$

Define

$$Q(u) = \int \left(|\nabla u|^2 - \frac{d\sigma}{2\sigma+2}|u|^{2\sigma+2}\right)d\mathbf{x} \qquad (4.3.28)$$

and the set $M = \{u \in H^1(\mathbf{R}^d),\ Q(u) = 0\}$.

The purpose is to find a sequence of initial conditions φ_α such that (i) $Q(\varphi_\alpha) < 0$, (ii) φ_α converges to the ground state g in H^1 as $\alpha \searrow 1$, (iii) during the time they exist, the solutions ψ_α of the NLS equation with initial conditions φ_α also satisfy $Q(\psi_\alpha) < 0$ (note that Q is not a conserved quantity). The proof is in several steps (see Cazenave 1989, Chapter 8, for details).

Step 1: If u solves (4.2.1), it satisfies $Q(u) = 0$. Furthermore, u is a ground state of (4.2.1) if and only if it minimizes the action among all the functions in M.

Step 2: Define $\varphi_\alpha(\mathbf{x}) = \alpha^{d/2}g(\alpha\mathbf{x})$ and let ψ_α be the solutions of the NLS equation associated with these initial conditions. From the conservation of the L^2-norm and of the Hamiltonian, the action satisfies

$$\mathcal{S}(\psi_\alpha) = \mathcal{S}(\varphi_\alpha). \qquad (4.3.29)$$

In addition,

$$Q(\varphi_\alpha) < 0 \qquad (4.3.30)$$

for $\alpha > 1$. To prove (4.3.30), one computes the derivative of $\mathcal{S}(\varphi_\alpha)$ with respect to α and obtains

$$\frac{d\mathcal{S}(\varphi_\alpha)}{d\alpha} = \alpha\int|\nabla g|^2 d\mathbf{x} - \frac{d\sigma}{2\sigma+2}\alpha^{\sigma d-1}\int|g|^{2\sigma+2}d\mathbf{x} = \frac{1}{\alpha}Q(\varphi_\alpha). \qquad (4.3.31)$$

Using that $\mathcal{Q}(g) = 0$, one gets

$$\frac{d\mathcal{S}(\varphi_\alpha)}{d\alpha} = \alpha(1 - \alpha^{d\sigma - 2}) \int |\nabla g|^2 dx. \tag{4.3.32}$$

For $\alpha > 1$, $\frac{d\mathcal{S}(\varphi_\alpha)}{d\alpha} < 0$. Inequality (4.3.30) is then satisfied, and $\mathcal{S}(\varphi_\alpha) < \mathcal{S}(g)$.

Step 3: By continuity, $\mathcal{Q}(\psi_\alpha) < 0$ for small $|t|$. In addition, one proves that if $m = \inf\{\mathcal{S}(u), u \in M\} = \mathcal{S}(g)$, then for all $u \in H^1$ such that $\mathcal{Q}(u) < 0$, one has $\mathcal{Q}(u) < \mathcal{S}(u) - m$. It follows that for small t,

$$\mathcal{Q}(\psi_\alpha) < \mathcal{S}(\psi_\alpha) - m = \mathcal{S}(\varphi_\alpha) - m = -\delta < 0. \tag{4.3.33}$$

By continuity, (4.3.33) holds during the time $(-T^*(\varphi_\alpha), T^*(\varphi_\alpha))$ of existence of ψ_α.

Step 4: The variance identity for ψ_α gives

$$\frac{d^2}{dt^2} V(t) = 8\, \mathcal{Q}(\psi_\alpha) < -8\delta \quad \text{for} \quad t \in (-T^*, T^*). \tag{4.3.34}$$

Since $\varphi_\alpha \to \varphi$ in $H^1(\mathbf{R}^d)$ as $\alpha \searrow 1$, Theorem 5.1 (presented in the next chapter) for the existence of a finite-time blowup, applies. Theorem 4.9 follows, since $\varphi_\alpha \to \varphi$ in $H^1(\mathbf{R}^d)$ as $\alpha \searrow 1$.

4.4   A General Approach for Hamiltonian Systems

4.4.1   Setting of the Problem and Main Results

The results discussed in the previous section can, in fact, be viewed as special cases of stability properties of solitary waves for general Hamiltonian systems of the form

$$\frac{du}{dt} = JE'(u(t)), \tag{4.4.35}$$

where following the notations of the original papers, E' denotes the functional derivative of the Hamiltonian E and J a skew-symmetric linear operator that are invariant under a representation $T(.)$ of a group G. We also assume the existence of an additional conserved quantity $Q(u)$.

The present approach was first used in the context of the Klein–Gordon equations (Shatah 1983, 1985b) and then extended by Shatah and Strauss (1985) and Grillakis, Shatah, and Strauss (1987) to a general context, including traveling waves of nonlinear wave equations, standing waves in the presence of a potential or in in an optical wave-guide, and solitary waves of generalized KdV equations. In particular, this analysis holds in situations where solitary wave instability does not imply blowup of the solution of the initial value problem.

Following the notation of Grillakis, Shatah, and Strauss (1987), the ground state equation is written in the form

$$\Delta g_\omega - \omega g_\omega + g_\omega^{2\sigma+1} = 0 \qquad (4.4.36)$$

with $\omega > 0$. The orbit of g_ω is the set $\{T(\lambda)g_\omega, \lambda \in G\}$. In the context of the NLS equation, $G = \mathbf{R}$, and the transformation $T(\cdot)$ is multiplication by $e^{i\lambda}$. In terms of the previously used notation, the Hamiltonian is $E(u) = \frac{1}{2}H(u)$, and the additional conserved quantity is $Q(u) = -\frac{1}{2}N(u)$.

A quantity that plays an important role in the study of the stability of the solution $u(t) = T(\omega t)g_\omega$ of (4.4.35) is the scalar function

$$\delta(\omega) = E(g_\omega) - \omega Q(g_\omega), \qquad (4.4.37)$$

which is the action

$$\mathcal{S}(u) = E(u) - \omega Q(u), \qquad (4.4.38)$$

restricted to the family of ground states $\{g_\omega\}$. In the following, we will say that $\delta(\omega_0)$ is convex in ω_0, if $\delta(\omega) - \delta(\omega_0) \geq (\omega - \omega_0)\delta'(\omega_0)$ or equivalently, if its second derivative at this point $\delta''(\omega_0)$ is strictly positive.

The functional framework is provided by the space $X = H_r^1$ of radially symmetric complex functions in H^1. We define the identification map $I : X \to X^*$ by

$$\langle I(u), v \rangle = \Re \int uv^* dx, \qquad (4.4.39)$$

where $\langle\, ,\, \rangle$ refers to the duality between X and X^*, and the skew-symmetric operator J entering (4.4.35) is given by $J = -iI^{-1}$.

Theorem 4.10.   *The two statements, $\delta(\omega_0)$ is convex at ω_0, and the Hamiltonian $E(u) = \frac{1}{2}\int \left(|\nabla u|^2 - \frac{1}{\sigma+1}|u|^{2\sigma+2}\right) dx$ restricted to the manifold $M_0 = \{u \in H_r^1, Q(u) = Q(g_{\omega_0})\}$ has a local minimum at $u = g_{\omega_0}$, are equivalent.*

On the other hand, when $\delta(\omega)$ is strictly concave at ω_0, g_{ω_0} is a saddle point for E, as expressed in the following proposition.

Proposition 4.11.   *Fix ω_0 and consider the curve*

$$\omega \mapsto q(\omega, \mathbf{x}) = g_\omega\left(\frac{\mathbf{x}}{\lambda(\omega)}\right), \qquad (4.4.40)$$

where $\lambda(\omega)^d = \frac{Q(g_{\omega_0})}{Q(g_\omega)}$. Then $Q(q(\omega)) = Q(g_{\omega_0})$. One has:

(i) $\partial_\omega^2 E(q(\omega_0)) \leq \delta''(\omega_0)$.
(ii) *If $\delta(\omega)$ is strictly concave at ω_0, then $E(q(\omega)) < E(g_{\omega_0})$ for $\omega \neq \omega_0$ near ω_0.*

Theorem 4.12.   *If the action $\delta(\omega)$ is convex at ω_0, the associated ground-state standing wave $g_{\omega_0}e^{i\omega_0 t}$ is orbitally stable. Conversely, if $\delta(\omega)$ is concave at ω_0, the orbit of g_{ω_0} is unstable.*

In the case of the NLS equation, the first and second derivatives of δ read $\delta'(\omega) = \frac{1}{2}\omega^b \int |R|^2 dx$ and $\delta''(\omega) = \frac{b}{2}\omega^{b-1} \int |R|^2 dx$ with $b = 1/\sigma - d/2$, and the function R being the unique positive solution of $\Delta R - R + R^{2\sigma+1} = 0$. Thus, the ground-state standing wave $g_\omega e^{i\omega t}$ is orbitally stable if $\sigma < 2/d$ (subcritical case) and unstable if $\sigma > 2/d$ (supercritical case).

When $\delta(\omega) \neq 0$, Stubbe (1989) proved the equivalence between the linear stability (that is, the stability for the equation linearized near the ground-state standing wave) and the orbital stability.

4.4.2   Main Steps of the Proofs

Following Shatah and Strauss (1985), we restrict the discussion to the problem in dimension $d \geq 2$. A more general approach is found in Grillakis, Shatah, and Strauss (1987).

Proof of Theorem 4.10.

First, note that

$$\delta(\omega) = S(g_\omega) = \frac{1}{d} \int |\nabla g_\omega|^2 dx. \tag{4.4.41}$$

Also, g_ω is a critical point of the action and thus satisfies

$$E'(g_\omega) - \omega Q'(g_\omega) = 0. \tag{4.4.42}$$

Consequently,

$$\delta'(\omega) = \left\langle E'(g_\omega) - \omega Q'(g_\omega), \frac{dg_\omega}{d\omega} \right\rangle - Q(g_\omega) = -Q(g_\omega). \tag{4.4.43}$$

Using the notation $g_0 \equiv g_{\omega_0}$, one defines

$$M_0 = \{u \in X, Q(u) = Q(g_0)\}. \tag{4.4.44}$$

Step 1: Assume $\delta(\omega)$ convex at ω_0. For any $u \in M_0$ in a small neighborhood of g_0, there exists ω such that $\int |\nabla u|^2 dx = \int |\nabla g_\omega|^2 dx$. This follows from the continuity of δ near g_0 and from the fact that $\delta(\omega_0) \neq 0$. Now,

$$E(u) = S(u) + \omega Q(g_0) = S(u) - \omega \delta'(\omega_0). \tag{4.4.45}$$

From (4.2.25), g_ω minimizes S among all functions u such that $\int |\nabla u|^2 dx = \int |\nabla g_\omega|^2 dx$. Thus,

$$E(u) \geq \delta(\omega) - \omega \delta'(\omega_0). \tag{4.4.46}$$

The convexity of δ at ω_0 implies $\delta(\omega) - \omega \delta'(\omega_0) \geq \delta(\omega_0) - \omega_0 \delta'(\omega_0)$. Thus,

$$E(u) \geq \delta(\omega_0) - \omega_0 \delta'(\omega_0) = E(g_0) \tag{4.4.47}$$

for all $u \in M_0$ close to g_0.

Step 2: Assume conversely that $E(g_0)$ is a local minimum. Consider the curve $q(\omega, \mathbf{x})$ defined in Proposition 4.11 and compute

$$E(q(\omega)) = \mathcal{S}(q(\omega)) + \omega Q(g_0)$$

$$= \omega Q(g_0) + \left(\frac{\lambda^{d-2}}{2} + \left(\frac{1}{d} - \frac{1}{2} \right) \lambda^d \right) \int |\nabla g_\omega|^2 dx. \quad (4.4.48)$$

A simple analysis of the function $f(\lambda) = \frac{1}{2}\lambda^{d-2} + (\frac{1}{d} - \frac{1}{2})\lambda^d$ shows that if $d \geq 2$, then $f(\lambda) \leq f(1) = 1/d$. Thus,

$$E(q(\omega)) \leq -\omega \delta'(\omega_0) + \delta(\omega). \quad (4.4.49)$$

By assumption, $E(q(\omega)) \geq E(g_0) = \delta(\omega_0) - \omega_0 \delta'(\omega_0)$. Thus,

$$\delta(\omega_0) - \omega_0 \delta'(\omega_0) \leq -\omega \delta'(\omega_0) + \delta(\omega), \quad (4.4.50)$$

and δ is convex at ω_0.

Proof of Proposition 4.11.

Let $\alpha(\omega) = E(q(\omega)) - \delta(\omega) + \omega \delta'(\omega_0)$. Clearly, $\alpha(\omega_0) = 0$, and from (4.4.49), $\alpha(\omega) \leq 0$. Therefore, $\alpha(\omega)$ has a maximum at ω_0, and $\alpha''(\omega_0) \leq 0$. This proves (i). Again from (4.4.49) we get that if δ is strictly concave at ω_0, then $E(q(\omega)) < E(g_0)$ for ω near ω_0 with $\omega \neq \omega_0$, which proves (ii).

Proof of Theorem 4.12.

Let U_ϵ be a neighborhood of the orbit $e^{i\gamma} g_0$. For $u \in U_\epsilon$, one defines $\gamma(u)$ as the value of γ for which $|e^{-i\gamma} u - g_0|_{H^1}$ is minimal. Under the assumption that δ is convex at ω_0, one proves a result stronger than (4.4.47), namely (Grillakis, Shatah, and Strauss 1987)

$$E(u) \geq E(g_0) + c|e^{-i\gamma(u)} u - g_0|_{H^1} \quad (4.4.51)$$

for all $u \in U_\epsilon$ and $Q(u) = Q(g_0)$. The proof of this inequality uses in an essential way that the linearized operator $L_1 = -\Delta + \omega - (2\sigma + 1)g_\omega^{2\sigma}$ has exactly one negative eigenvalue.

The proof of the stability result in Theorem 4.12 proceeds by contradiction. Let η be an arbitrary positive number and suppose that there exists a sequence of initial conditions $\psi_n(0)$ such that

$$\inf_s |\psi_n(0) - e^{is} g_0|_{H^1} \to 0, \quad \text{and} \quad \sup_t \inf_s |\psi_n(t) - e^{is} g_0|_{H^1} \geq \eta, \quad (4.4.52)$$

where $\psi_n(t)$ is the solution of NLS equation with initial condition $\psi_n(0)$. One defines t_n as the first time for which

$$\inf_s |\psi_n(t_n) - e^{is} g_0|_{H^1} = \eta, \quad (4.4.53)$$

the solution existing in the interval $[0, t_n]$. As $n \to \infty$, $Q(\psi_n(t_n)) \to Q(g_0)$ and $E(\psi_n(t_n)) \to E(g_0)$. Now choose a sequence v_n such that $Q(v_n) = Q(g_0)$ and $|v_n - \psi_n(t_n)|_{H^1} \to 0$. Because of the continuity of E, $E(v_n) \to E(g_0)$. Choosing η sufficiently small, one uses (4.4.51) and gets

$$c|v_n - e^{i\gamma(v_n)} g_0|_{H^1} \leq E(v_n) - E(g_0) \to 0. \quad (4.4.54)$$

Thus, $|\psi_n(t_n) - e^{i\gamma(v_n)}g_0|_{H^1} \to 0$, which contradicts (4.4.53). This concludes the proof of the stability condition given in Theorem 4.12.

The proof of the instability result in Theorem 4.12 involves several steps we only sketch.

Lemma 4.13. *Define y_0 as $\partial q / \partial \omega$ evaluated at $\omega = \omega_0$. One has:*

(i) $\langle (E''(g_0) - \omega_0 Q''(g_0))y_0, y_0 \rangle \le \delta''(\omega_0)$,
(ii) $\langle Q'(g_0), y_0 \rangle = 0$,
(iii) $\int \nabla g_0 \cdot \nabla y_0 dx < 0$ if $\delta''(\omega_0) < 0$.

Proof.   (i) Differentiating E along the curve $q(\omega)$, one has

$$\frac{d}{d\omega} E(q(\omega)) = \langle E'(q(\omega)), q'(\omega) \rangle \qquad (4.4.55)$$

$$\frac{d^2}{d\omega^2} E(q(\omega)) = \langle E''(q(\omega))q', q' \rangle + \langle E'(q(\omega)), q'' \rangle. \qquad (4.4.56)$$

Since Q is constant along the curve $q(\omega)$,

$$Q''(q(\omega)) = \langle Q''(q(\omega))q', q' \rangle + \langle Q'(q(\omega)), q'' \rangle = 0. \qquad (4.4.57)$$

Combining (4.4.56) and (4.4.57) at $\omega = \omega_0$, one gets

$$\frac{d^2}{d\omega^2} E(q(\omega))|_{\omega=\omega_0} = \langle (E''(g_0) - \omega_0 Q''(g_0))y_0, y_0 \rangle, \qquad (4.4.58)$$

since $E'(g_0) - \omega_0 Q'(g_0) = 0$. Using Corollary 4.11, one gets part (i) of the lemma.
(ii) Since $Q(q(\omega)) = Q(g_0)$, one has $\langle Q'(g_0), y_0 \rangle = 0$.
(iii) Using (4.4.40), one has

$$\int |\nabla q(\omega)|^2 dx = \lambda(\omega)^{d-2} |\nabla g_\omega)|^2 dx. \qquad (4.4.59)$$

Differentiation with respect to ω gives at ω_0

$$2 \int \nabla g_0 \cdot \nabla y_0 dx = (d-2)\lambda'(\omega_0)|\nabla g_\omega)|^2 dx + d\delta'(\omega_0). \qquad (4.4.60)$$

But $\delta'(\omega) < 0$ and $\lambda^d(\omega) = Q(g_0)/Q(g_\omega) = \delta'(\omega_0)/\delta'(\omega)$. Thus $\lambda'(\omega_0)$ and $\delta''(\omega_0)$ have the same sign, which is negative. This concludes the proof of part (iii) of the lemma.

Proposition 4.14.   (Shatah and Strauss 1985) *Define*

$$A(u) = -\langle Jy_0, e^{i\gamma(u)}u \rangle, \qquad (4.4.61)$$

where $\gamma(u)$ is the value of γ for which $|e^{-i\gamma}u - g_0|_{H^1}$ is minimal. Let $\delta''(\omega_0) < 0$. There exists $\epsilon > 0$ such that for all u in a neighborhood of the orbit of g_0, with $u \ne g_0 e^{i\theta}$ and $Q(u) = Q(g_0)$, there is a $\lambda \in (1 - \epsilon, 1 + \epsilon)$ such that

$$E(g_0) < E(u) + (\lambda - 1)P(u), \qquad (4.4.62)$$

where

$$P(u) = \langle E'(u), -JA'(u) \rangle. \tag{4.4.63}$$

Furthermore, when restricted to the curve $q(\omega)$, the functional P changes sign as ω passes ω_0

Consider now the evolution equation (4.4.35) with an initial condition $u(0) = u_0$. Let K denote the orbit of g_0 and let $N = \mathcal{O} - K$ be a small tubular neighborhood around the orbit. Define $E_0 = E(g_0)$.

Lemma 4.15. *The sets*

$$S_1 = \{u \in N, E(u) < E_0, Q(u) = Q(g_0), P(u) > 0\}, \tag{4.4.64}$$
$$S_2 = \{u \in N, E(u) < E_0, Q(u) = Q(g_0), P(u) < 0\} \tag{4.4.65}$$

are invariant under the flow (4.4.35).

Lemma 4.16. *Let $u_0 \in S_1$ (resp. S_2) and let $T_0 = \sup\{t, \ u(s) \in N, 0 \le s < t\}$ be the exit time. There is $\epsilon_0 > 0$ such that $P(u(t)) > \epsilon_0$ (resp. $P(u(t)) < -\epsilon_0$) for $t < T_0$.*

The proof of Theorem 4.12 is completed by showing that if $u_0 \in S_1 \cup S_2$, the solution of the evolution problem with initial condition u_0 exits the set N in a finite time $(T_0 < \infty)$. One has

$$\frac{dA}{dt} = \left\langle \frac{du}{dt}, A'(u(t)) \right\rangle. \tag{4.4.66}$$

From (4.4.35),

$$\left\langle \frac{du}{dt}, A'(u(t)) \right\rangle = \langle JE'(u(t)), A'(u(t)) \rangle. \tag{4.4.67}$$

By (4.4.63),

$$\frac{d}{dt} A(u(t)) = P(u(t)). \tag{4.4.68}$$

By the previous lemma, $|P(u(t))| > \epsilon_0$ as long as $u \in N$. Thus,

$$|A(u(t)) - A(u_0)| > \epsilon_0 t. \tag{4.4.69}$$

Since N is a bounded set and A is bounded on N, the solution must exit the neighborhood N of the ground-state orbit in a finite time.

4.4.3 Extension to Abstract Hamiltonian Systems

More generally, consider the abstract Hamiltonian system (4.4.35) that is invariant under the action T of a Lie group G. A solitary or standing wave is a solution of the form $T(e^{t\omega})g_\omega$, where ω now belongs to the Lie algebra $\mathcal{G}$ of G. We also need the functionals Q_ω for $\omega \in \mathcal{G}$ defined by $Q'_\omega = J^{-1}T_\omega$ where T_ω denotes the differential of T. We define the linear operator $L_\omega = E''(g_\omega) - Q''_\omega(g_\omega)$ and the function $\delta(\omega) = E(g_\omega) - Q_\omega(g_\omega)$. Furthermore, δ'' denotes the Hessian of δ. Grillakis, Shatah, and Strauss (1990) proved the following result.

Theorem 4.17. *Suppose that δ'' is nondegenerate. The ground-state standing wave $T(e^{t\omega})g_\omega$ is orbitally stable if the number $n(L_\omega)$ of negative eigenvalues of L_ω is equal to the number $p(\delta'')$ of positive eigenvalues of δ''. It is unstable if the difference between $n(L_\omega)$ and $p(\delta'')$ is equal to an odd positive integer.*

In the framework of the NLS equation $n(L_\omega) = 1$, the nondegeneracy reduces to $\delta''(\omega) \neq 0$, and the number of positive eigenvalues of δ'' is 1 if $\delta''(\omega) > 0$ and 0 if $\delta''(\omega) < 0$. Note also that the instability result allows $n(L_\omega) - p(\delta'')$ to be an odd integer other than one.

4.5 Further Stability Results

4.5.1 Asymptotic Stability Results

Stronger results referred to as "asymptotic stability" of solitary waves of the one-dimensional NLS equation $i\psi_t + \psi_{xx} + f(|\psi|^2)\psi = 0$ have been proved by Buslaev and Perel'man (1993, 1995). Assuming a nonlinearity $f(s) \sim s^q$ for s large, with $q < 2$ and $f(s) \sim -s^p$, $p \geq 4$, for small s, together with some conditions on the spectrum of the linearized operator near the solitary waves, the solution, initially close to a solitary wave, can be decomposed as t tends to ∞ into a sum of a solitary wave, a dispersive part, and a remainder that tends to 0.

In higher dimensions, asymptotic stability results were proved by Soffer and Weinstein (1990, 1992) for the NLS equation

$$i\psi_t = -\Delta\psi + V(\mathbf{x})\psi + \lambda|\psi|^{2\sigma}\psi, \qquad (4.5.1)$$

where the potential V decreases fast enough at infinity and the operator $-\Delta + V(\mathbf{x})$ has exactly one bound state in L^2 with a strictly negative eigenvalue. Furthermore, the nonlinearity and/or the initial conditions are such that solutions exist in the large. Existence of solitary waves for the above equation was studied by Strauss (1977), Berestycki and Lions (1983), and Rose and Weinstein (1988). Soffer and Weinstein (1990, 1992) proved that for initial conditions in a neighborhood of a particular solitary wave $g_{E_0} e^{i\gamma_0}$ of energy E_0 and phase γ_0, the asymptotic behavior of the solution as $t \to \pm\infty$ is given by a solitary wave of nearby energy and phase plus a remainder that disperses to 0. This result was recovered by Pillet and Wayne (1997) using a center manifold approach. It shows in particular that the decomposition into a localized part (solitary waves) and a dispersive contribution is not restricted to equations that are completely integrable by the inverse scattering method (such as the Korteweg–de Vries or the one-dimensional cubic NLS equation).

4.5.2   Ground-State Orbital Stability and Global Existence

When looking at the condition of global existence for the initial value problem associated to the elliptic NLS equation with an attracting nonlinearity (Section 3.2.2) and the orbital stability of the ground-state standing waves (Section 4.3.1), one notices a striking connection between them. Indeed, when $\sigma < 2/d$, global existence for the NLS equation holds for arbitrary initial conditions in H^1, and on the other hand, ground-state standing waves are orbitally stable. When $\sigma \geq 2/d$, solutions to the NLS equation with an attracting nonlinearity may blow up in finite time, and ground-state standing waves are unstable. Stubbe (1991) proved that this relationship holds for a large class of NLS equations. Let us write the elliptic NLS equation with a general nonlinearity in the form

$$i\psi_t + \Delta\psi + f(|\psi|^2)\psi = 0. \tag{4.5.2}$$

The ground states g_ω are the positive solutions of

$$\Delta g_\omega - \omega g_\omega + f(g_\omega^2)g_\omega = 0. \tag{4.5.3}$$

Define also $N(g) = \int |g|^2 d\mathbf{x}$.

Theorem 4.18. *Let (ω_-, ω_+) be the maximal interval for which ground states exist in H^1. Define $N^* = \liminf_{\omega \nearrow \omega_+} N(g_\omega) \in [0, \infty]$. If $N^* > 0$, all solutions to the NLS equation (4.5.2) with initial conditions $\varphi \in H^1$ satisfying $N(\varphi) < N^*$ are global in time. In particular, if $N^* = \infty$, all solutions with initial conditions on H^1 are global in time.*

Let us apply this theorem to the NLS equation with power nonlinearity $f(s) = s^\sigma$. The ground states g_ω exist for all $\omega > 0$ and $N(g_\omega) = \omega^{1/\sigma - d/2} N(g_1)$. Therefore, $N^* = \infty$ if $\sigma < 2/d$, and $N^* = N(g_1)$ if $\sigma = 2/d$. Notice that we recover the sharp existence condition (Theorem 3.17) for the critical NLS equation. The proof uses the variational characterization of the ground states and the existence of invariant regions under the NLS flow (see Lemma 4.16).

5
Blowup Solutions

This chapter is devoted to rigorous results concerning solutions of the elliptic NLS equation with an attracting nonlinearity

$$i\partial_t \psi + \Delta \psi + |\psi|^{2\sigma}\psi = 0 \qquad (5.0.1)$$

$$\psi(\mathbf{x}, 0) = \varphi(\mathbf{x}), \qquad (5.0.2)$$

at critical ($\sigma d = 2$) and supercritical ($\sigma d > 2$) dimensions that become infinite in a finite time (see Cazenave 1994 for a review). This is the phenomenon of "blowup" or "wave collapse", which has important physical consequences in that it corresponds to a violent energy transfer from large to small scales where dissipative processes can act efficiently. In the case of rapidly decaying initial solutions, the blow up results are direct consequences of the variance identity discussed in Section 2.4.1. In some instances like that of isotropic solutions, the proof was extended to solutions with infinite variance. Other estimates concern the rate of blowup and, at critical dimension, the phenomenon of L^2-norm (or mass) concentration near the singularity.

5.1 Finite-Time Blowup

5.1.1 Case of Finite Variance

Classical blowup results are based on the "variance identity", also known as the "virial theorem" (see Section 2.4.1). Defining the variance $V(t) = \int |\mathbf{x}|^2 |\psi|^2 d\mathbf{x}$, we have the identity (Vlasov, Petrishchev, and Talanov 1971,

Zakharov 1972)

$$\frac{1}{8}\frac{d^2}{dt^2}V(t) = H - \frac{d\sigma - 2}{2\sigma + 2}\int |\psi|^{2\sigma+2}d\mathbf{x}. \tag{5.1.1}$$

Theorem 5.1. *Suppose that $d\sigma \geq 2$. Consider an initial condition $\varphi \in H^1$ with $V(0)$ finite that satisfies one of the conditions below:*

(i) $H(\varphi) < 0,$
(ii) $H(\varphi) = 0,$ *and* $\Im \int \mathbf{x} \cdot \varphi^* \nabla \varphi d\mathbf{x} < 0,$
(iii) $H(\varphi) > 0$ *and* $\Im \int \mathbf{x} \cdot \varphi^* \nabla \varphi d\mathbf{x} \leq -4\sqrt{H(\varphi)}|\mathbf{x}\varphi|_{L^2}.$

Then, there exists a time $t_* < \infty$ such that

$$\lim_{t \to t_*} |\nabla \psi|_{L^2} = \infty \quad \text{and} \quad \lim_{t \to t_*} |\psi|_{L^\infty} = \infty. \tag{5.1.2}$$

Proof. If $d\sigma \geq 2,$

$$\frac{d^2}{dt^2}V(t) \leq 8H, \tag{5.1.3}$$

and by time integration,

$$V(t) \leq 4Ht^2 + V'(0)t + V(0), \tag{5.1.4}$$

where $V'(0) = 4\,\Im \int \varphi^* \mathbf{x} \cdot \nabla \varphi d\mathbf{x}$ and $V(0) = \int |\mathbf{x}|^2 |\varphi|^2 d\mathbf{x}$. Under any of the hypotheses (i)–(iii) of the above theorem, there exists a time t_0 such that the right–hand side of (5.1.1) vanishes, and thus also $t_1 \leq t_0$ such that

$$\lim_{t \to t_1} V(t) = 0. \tag{5.1.5}$$

Furthermore, from the equality,

$$\int |f|^2 d\mathbf{x} = \frac{1}{d}\int (\nabla \cdot \mathbf{x})|f|^2 d\mathbf{x}$$
$$= -\frac{1}{d}\int \mathbf{x} \cdot \nabla(|f|^2)d\mathbf{x}, \tag{5.1.6}$$

one gets the "uncertainty principle"

$$|f|_{L^2}^2 \leq \frac{2}{d}|\nabla f|_{L^2}|\mathbf{x}f|_{L^2}. \tag{5.1.7}$$

When this inequality is applied to a solution ψ, one gets from (5.1.5) and from the conservation of $|\psi|_{L^2}^2$, that there exists a time $t_* \leq t_1$ such that $\lim_{t \to t_*} |\nabla \psi|_{L^2} = \infty$ (Tsutsumi 1978). The conservation of H then ensures that $\lim_{t \to t_*} |\psi|_{L^{2\sigma+2}}^{2\sigma+2} = \infty$, and since $|\psi|_{L^2}^2$ is conserved, this implies that $\lim_{t \to t_*} |\psi|_{L^\infty} = \infty$.

Remark. The vanishing of the variance that corresponds to a global collapse of the wave packet at the focus provides only an upper bound for the time of blowup that could in fact occur before.

Sharper Results in Supercritical Dimensions

Equation (5.1.4) is optimal at critical dimension, where it becomes an equality. Noticing that in this case the Hamiltonian $H(R)$ of the ground state R is zero, the above theorem then ensures that a sufficient condition for blowup at critical dimension is $H(\varphi) < H(R)$. An analogous condition, involving, nevertheless, a lower bound on $|\nabla\varphi|_{L^2}$ was derived in supercritical dimension $\sigma d > 2$, by estimating the last term in (5.1.1). It involves the (positive) Hamiltonian of the ground state with the same L^2-norm as that of the initial condition of the initial-value problem (Turitsyn 1993, Kuznetsov, Rasmussen, Rypdal, and Turitsyn 1995, Lushnikov 1995).

Theorem 5.2. *Assume $\sigma d > 2$ and an initial condition $\varphi \in H^1(\mathbf{R})$. Let R_N be the ground state having the same mass $N = |\varphi|^2_{L^2}$ as that of the initial condition. Define $X_N = |\nabla R_N|^2_{L^2}$ and let H_N be the associated Hamiltonian. For an initial condition φ that satisfies $X_0 \equiv \int |\nabla\varphi|^2 dx > X_N$ with a Hamiltonian H such that $H < H_N$, there exists a time $t_* < \infty$ such that*

$$\lim_{t \to t_*} |\nabla\psi|_{L^2} = \infty \quad \text{and} \quad \lim_{t \to t_*} |\psi|_{L^\infty} = \infty. \tag{5.1.8}$$

Proof. By the Gagliardo–Nirenberg inequality (4.2.28), we have

$$H \geq |\nabla\psi(t)|^2_{L^2} - \frac{C_{\sigma,d}}{\sigma+1}|\varphi|_{L^2}^{2+2\sigma-\sigma d}|\nabla\psi(t)|^{\sigma d}_{L^2}, \tag{5.1.9}$$

where the optimal constant $C_{\sigma,d}$ is given by (4.2.29). We rewrite (5.1.9) in the form

$$H \geq F(X) \equiv X - C_{\sigma,d}N^{1+\sigma-\sigma d/2}X^{\sigma d/2}, \tag{5.1.10}$$

where $X = X(t) = |\nabla\psi(t)|^2_{L^2}$. As a function of X, F reaches its maximum exactly when X takes the value

$$X_N = \left(\frac{2(\sigma+1)}{\sigma d C_{\sigma,d}}\frac{1}{N^{1+\sigma-\sigma d/2}}\right)^{2/(\sigma d-2)}. \tag{5.1.11}$$

The maximum $F(X_N)$ is equal to the Hamiltonian

$$H_N = \frac{\sigma d - 2}{\sigma d}X_N, \tag{5.1.12}$$

associated to the ground state R_N, solution of

$$\Delta R_N - \lambda^2(N)R_N + R_N^{2\sigma+1} = 0, \tag{5.1.13}$$

with the coefficient $\lambda(N)$ prescribed by the condition $\lambda(N)^{\frac{2}{\sigma}-d} = \frac{N}{|R|^2_{L^2}}$ (see Section 4.1.2). In this formula, R denotes, as usual, the ground state associated to $\lambda = 1$. Furthermore, $F(X)$ vanishes for $X = \left(\frac{\sigma d}{2}\right)^{2/(\sigma d-2)}X_N$.

During the time where it exists, the solution satisfies the identity (5.1.1), which can be rewritten as

$$\frac{1}{4}\frac{d^2}{dt^2}V(t) = \sigma dH - (d\sigma - 2)\int |\nabla\psi|^2 dx$$

$$= \sigma d(H - H_N) + \sigma dH_N - (d\sigma - 2)X(t). \qquad (5.1.14)$$

If initially $X(0) > X_N$ and $0 < H < H_N$, then $X(t) > X_N$ as long as the solution exists, since the opposite would contradict the fact that $H > F(X)$. Thus, (5.1.14) becomes

$$\frac{1}{4}\frac{d^2}{dt^2}V(t) \leq \sigma d(H - H_N) + \sigma dH_N - (d\sigma - 2)X_N$$

$$= \sigma d(H - H_N). \qquad (5.1.15)$$

It follows that

$$V(t) \leq 2\sigma d(H - H_N)t^2 + V'(0)t + V(0). \qquad (5.1.16)$$

The condition $H < H_N$ ensures that V will vanish in a finite time, and the proof proceeds as in the previous theorem.

Alternative Proof of Blowup in Supercritical Dimension

When the Hamiltonian is negative and the dimension supercritical, an alternative proof of finite time singularity was developed by Glassey (1977b), who concentrates on the blowup of the variance time-derivative rather than on the vanishing of the variance and established the following result.

Theorem 5.3. *Assume $\sigma d > 2$ and an initial condition $\varphi \in H^1$ with a finite variance and a negative Hamiltonian. There exists a time $t_* < \infty$ such that*

$$y(t) \equiv -\frac{1}{4}\frac{dV}{dt} = -\Im\int \psi^* \mathbf{x} \cdot \nabla\psi \, d\mathbf{x} \to \infty. \qquad (5.1.17)$$

As a consequence, $|\nabla\psi(t)|_{L^2}$ also blows up in a finite time.

Proof. Equation (5.1.1) can be rewritten as

$$\frac{dy}{dt} = (d\sigma - 2)\int |\nabla\psi|^2 d\mathbf{x} - d\sigma H. \qquad (5.1.18)$$

Assuming a negative Hamiltonian, in supercritical dimension $d\sigma > 2$ one writes

$$\frac{dy}{dt} \geq (d\sigma - 2)\int |\nabla\psi|^2 d\mathbf{x}. \qquad (5.1.19)$$

On the other hand, from the definition of y, one has

$$y(t) \leq V(t)^{1/2}|\nabla\psi|_{L^2}. \qquad (5.1.20)$$

Suppose that the solution exists for all time. Then $y(t)$ is an increasing function that, if not initially positive, will become so after a time t_0. After this time, $V(t)$ is a decreasing function and thus has an upper bound d_0^2. We write

$$y(t) \leq d_0|\nabla\psi|_{L^2} \qquad (5.1.21)$$

and thus

$$\frac{dy}{dt} \geq Ay^2 \qquad (5.1.22)$$

with $A = \frac{d\sigma-2}{d_0^2}$ and $y(t_0) > 0$. It follows that on the time interval $t_0 \leq t < t_0 + \frac{1}{Ay(t_0)}$,

$$y(t) \geq \frac{y(t_0)}{1 - Ay(t_0)(t - t_0)}, \qquad (5.1.23)$$

which leads to the estimate

$$|\nabla\psi|_{L^2} \geq \frac{1}{d_0}\frac{y(t_0)}{1 - Ay(t_0)(t - t_0)}. \qquad (5.1.24)$$

Hence, for some time t_* such that $t_0 \leq t_* < t_0 + \frac{1}{Ay(t_0)}$,

$$\lim_{t \to t_*} |\nabla\psi|_{L^2} = \infty, \qquad (5.1.25)$$

and because of the conservation of the Hamiltonian,

$$\lim_{t \to t_*} |\psi|_{L^\infty} = \infty. \qquad (5.1.26)$$

Note that the variance can remain strictly positive in the above interval.

This approach based on the blowup of the variance derivative does not extend to the critical dimension. For $d\sigma = 2$, one has

$$\frac{dy}{dt} = 2|H|, \qquad (5.1.27)$$

indicating a *linear* growth of $y(t)$ and thus a quadratic variation of the variance, as described at the beginning of this chapter.

Further Results

(i) Kavian (1987) gave sufficient conditions for blowup in the case of the NLS equation in a smooth domain $\Omega \in \mathbf{R}^d$ with Dirichlet boundary conditions. When Ω is star-shaped, $H < 0$ and $\sigma d \geq 2$, solutions become singular in a finite time. Sufficient conditions for blowup are more complicated when Ω is a general smooth domain of $\mathbf{R}^d$. Examples of blowup solutions with Neumann or periodic boundary conditions are also presented.

(ii) In a periodic domain, a negative Hamiltonian cannot be a sufficient condition for blowup as exemplified by the particular solution $\psi = ae^{ia^2t}$. In one dimension and quintic nonlinearity ($\sigma = 2$), sufficient conditions for blowup were given by Ogawa and Tsutsumi (1990). In addition to a negative Hamiltonian, these conditions essentially express that the initial field φ is sufficiently peaked to make $\frac{1}{3}|\varphi|_{L^6}^6$ much larger than $|\varphi|_{L^2}^6$ (see the original paper for details).

(iii) In the case of the NLS equation with a linear dissipation

$$i(\psi_t + \gamma\psi) + \Delta\psi + |\psi|^{2\sigma}\psi = 0, \tag{5.1.28}$$

then for $d > 2/\sigma$ and negative initial Hamiltonian, blowup may occur for small values of γ (Tsutsumi 1984), while large enough values of γ ensure that solutions exist for all time in both critical and supercritical dimensions.

5.1.2   Extensions to Solutions with Infinite Variance

Progress has been made these last few years concerning the existence of blowup solutions in $H^1(\mathbf{R}^d)$ without the restriction of finite variance.

Theorem 5.4.   (Ogawa and Tsutsumi 1991a) *In the case of dimension $d = 1$ and critical exponent $\sigma = 2$, for an initial condition $\varphi \in H^1(\mathbf{R})$ such that the Hamiltonian $H(\varphi) < 0$, the solution blows up in a finite time in H^1.*

Theorem 5.5.   (Ogawa and Tsutsumi 1991b) *Assume $d \geq 2$ and $\frac{2}{d} \leq \sigma \leq \min(\frac{2}{d-2}, 2)$. If the initial condition $\varphi \in H^1(\mathbf{R}^d)$ is radially symmetric with a Hamiltonian $H(\varphi) < 0$, the solution blows up in a finite time in H^1.*

Remark. In the supercritical case, Martel (1997) extends the existence of a finite-time blowup to a class of initial conditions that in some directions have their moments $\int x_i^2|\psi|^2 d\mathbf{x}$ bounded and in the other directions are radially symmetric (see the original paper for the precise conditions).

The idea of the proofs is to modify the definition of the function $y(t)$ (or of the variance $V(t)$) by replacing the $|\mathbf{x}|^2$ factor by a suitable weight function $p(\mathbf{x})$ with the same behavior at small $|\mathbf{x}|$ but whose large-distance behavior eliminates the divergence. This, in turn, introduces additional terms in the estimate of $\frac{dy}{dt}$ that are to be controlled. The restriction to radially symmetric solutions in dimension $d > 1$ originates from the use of an upper bound for the sup-norm of a radial function away from the origin in terms of its H^1-norm given by the following lemma.

Lemma 5.6.   (Strauss 1977) *Let v be a radially symmetric function in $H^1(\mathbf{R}^d)$ with $d \geq 2$. Then, for any $A > 0$,*

$$|v|_{L^\infty(|\mathbf{x}|>A)}^2 \leq CA^{-d+1}|\nabla v|_{L^2(|\mathbf{x}|>A)}|v|_{L^2(|\mathbf{x}|>A)}, \tag{5.1.29}$$

where C is a constant independent of A and v.

In order to monitor the magnitude of the weight function mentioned above and of its derivatives, it is convenient to define

$$p(r) = \int_0^r h(s)ds \tag{5.1.30}$$

where $h(r) \in C^3$ denotes a radially symmetric function to be specified later, and to construct the scaled functions

$$h_m = mh\left(\frac{r}{m}\right), \quad p_m = m^2 p\left(\frac{r}{m}\right). \qquad (5.1.31)$$

We then define

$$y_m(t) = -\Im \int (\nabla p_m \cdot \nabla \psi) \psi^* d\mathbf{x}. \qquad (5.1.32)$$

The function h is specified differently in extending Glassey's proof in supercritical dimension and in dealing with a modified variance identity at critical dimension. In the former case only y_m should be kept finite, while in the latter approach one also uses the local variance

$$V_m(t) = \int p_m |\psi|^2 d\mathbf{x}, \qquad (5.1.33)$$

related to $y_m(t)$ by

$$y_m(t) = -\frac{1}{2} \frac{dV_m}{dt}(t). \qquad (5.1.34)$$

In both cases, one uses the following lemma.

Lemma 5.7. *Suppose that the function p belongs to C^4. The solution of the NLS equation satisfies*

$$\frac{d}{dt}\Im \int (\nabla p \cdot \nabla \psi) \psi^* d\mathbf{x} = 2 \, \Re \int \partial_i \partial_j p \partial_i \psi^* \partial_j \psi dx - \frac{\sigma}{\sigma+1} \int \Delta p |\psi|^{2\sigma+2} dx$$
$$-\frac{1}{2} \int \Delta^2 p |\psi|^2 d\mathbf{x}, \qquad (5.1.35)$$

or when all the functions are radially symmetric,

$$\frac{d}{dt}\Im \int \partial_r p \partial_r \psi \psi^* d\mathbf{x}$$
$$= d\sigma H + (2 - d\sigma) \int |\partial_r \psi|^2 dx - 2 \int (1 - \partial_{rr} p) |\partial_r \psi|^2 dx$$
$$+ \frac{\sigma}{\sigma+1} \int (d - \Delta p) |\psi|^{2\sigma+2} dx - \frac{1}{2} \int \Delta^2 p |\psi|^2 dx. \qquad (5.1.36)$$

Proof. We have

$$\partial_t (p(\mathbf{x}) |\psi|^2) = -2p(\mathbf{x}) \Im \partial_j (\partial_j \psi \ \psi^*) \qquad (5.1.37)$$

and

$$\partial_t \Im(\partial_j \psi \ \psi^*) = \partial_i \left(2\Re(\partial_i \psi^* \partial_j \psi) - \partial_{ij} |\psi|^2 \right) + \frac{\sigma}{\sigma+1} \partial_j |\psi|^{2\sigma+2}. \qquad (5.1.38)$$

Thus, multiplying the last equation by $p(\mathbf{x})\partial_j$ and integrating over $\mathbf{R}^d$ gives (5.1.35), which reduces to (5.1.36) when all the functions are radially symmetric.

Proof of Theorem 5.5. It follows from Lemma 5.7 that $y_m(t)$ obeys

$$y'_m(t) = -d\sigma H - (2 - d\sigma) \int |\partial_r \psi|^2 dx + 2 \int (1 - \partial_{rr} p_m) |\partial_r \psi|^2 dx$$

$$- \frac{\sigma}{\sigma + 1} \int (d - \Delta p_m) |\psi|^{2\sigma + 2} dx + \frac{1}{2} \int \Delta^2 p_m |\psi|^2 dx, \quad (5.1.39)$$

where the weight functions are to be adequately chosen in order to control the resulting additional contributions. Near the origin, $p(r)$ should behave like r^2 and thus $h(r)$ like r. At large distance, $h(r)$ is required to be zero only at critical dimension, where one has to deal with the local variance itself. Following Merle (1995), we use in supercritical dimensions

$$h(r) = r \text{ for } 0 \le r \le 1,$$
$$h(r) = r - (r-1)^4 \text{ for } r \ge 1 \text{ near } 1,$$
$$h(r) = \frac{3}{2} \text{ for } r \ge 3,$$
$$h'(r) < 1 \text{ for } r > 1, \quad (5.1.40)$$

while when in critical dimension, we choose

$$h(r) = r \text{ for } 0 \le r \le 1,$$
$$h(r) = r - (r-1)^3 \text{ for } 1 \le r \le 1 + 1/\sqrt{3},$$
$$h(r) \text{ smooth and } h'(r) < 0 \text{ for } 1 + 1/\sqrt{3} \le r \le 2,$$
$$h(r) = 0 \text{ for } r \ge 2. \quad (5.1.41)$$

Both definitions ensure that $1 - \partial_{rr} p > 0$ and $-\Delta p + d \ge 0$.
(i) *Supercritical dimensions $d\sigma > 2$*
    With the definition (5.1.40) for h, adapted to the supercritical dimensions, one has $(1 - \partial_{rr} p_m) \ge 0$, and

$$\int \Delta^2 p_m |\psi|^2 \le \frac{C}{m^2} |\varphi|_{L^2}^2. \quad (5.1.42)$$

Also, $d - \Delta p_m = 0$ for $r \le m$, and $0 \le d - \Delta p_m < c$ otherwise. Thus, one has the following upper bound for y'_m:

$$y'_m(t) \ge -d\sigma H + (d\sigma - 2) \int |\partial_r \psi|^2 - c|\varphi|_{L^2}^2 |\psi|_{L^\infty(r \ge m)}^{2\sigma} - \frac{C}{m^2} |\varphi|_{L^2}^2. \quad (5.1.43)$$

From Lemma 5.6, we have

$$|\psi|_{L^\infty(r \ge m)}^{2\sigma} \le m^{-\sigma(d-1)} |\psi|_{L^2}^\sigma |\nabla \psi|_{L^2}^\sigma \quad (5.1.44)$$

and by Young's inequality, since $\sigma < 2$,

$$|\psi|_{L^\infty(r \ge m)}^{2\sigma} \le \epsilon |\nabla \psi|_{L^2}^2 + C(\epsilon) m^{-\frac{2\sigma(d-1)}{2-\sigma}} |\psi|_{L^2}^{\frac{4+2\sigma}{2-\sigma}}. \quad (5.1.45)$$

It follows that

$$y'_m(t) \geq -d\sigma H + (d\sigma - 2 - \epsilon) \int |\partial_r \psi|^2 - C(\epsilon) m^{-\frac{2\sigma(d-1)}{2-\sigma}} |\varphi|_{L^2}^{\frac{4+2\sigma}{2-\sigma}} - \frac{C}{m^2} |\varphi|_{L^2}^2.$$
$$(5.1.46)$$

In supercritical dimension $d\sigma > 2$, we choose ϵ such that $\alpha \equiv d\sigma - 2 - \epsilon > 0$. Since $H < 0$, it follows that for m large enough,

$$y'_m(t) \geq \frac{d\sigma}{2} |H| + \alpha \int |\partial_r \psi|^2 d\mathbf{x}. \qquad (5.1.47)$$

Suppose that the solution exists in H^1 for all time. Then,

$$y_m(t) \geq y_m(0) + \frac{d\sigma}{2} |H| t + \alpha \int_0^t |\partial_r \psi|_{L^2}^2 ds. \qquad (5.1.48)$$

From the last inequality we see that there exists a time t_1 such that for $t \geq t_1$, $y_m(t) > 0$. We also have, by the Cauchy–Schwarz inequality, that

$$y_m(t) \leq |\partial_r p_m|_{L^\infty} |\varphi|_{L^2} |\partial_r \psi|_{L^2}. \qquad (5.1.49)$$

Equation (5.1.48) now becomes, for $t > t_1$,

$$y_m(t) \geq y_m(0) + \frac{d\sigma}{2} |H| t + A \int_0^t y_m^2(s) ds, \qquad (5.1.50)$$

with $A = C|\varphi|_{L^2}$, or

$$y_m(t) \geq A \int_{t_1}^t y_m^2(s) ds, \qquad (5.1.51)$$

and by integration over $[t_1, t]$,

$$y_m(t) \geq \frac{y_m(t_1)}{1 - A y_m(t_1)(t - t_1)}. \qquad (5.1.52)$$

It follows that $y_m(t)$ blows up in a finite time. From (5.1.49), so does $|\partial_r \psi(t)|_{L^2}$, which leads to a contradiction.
(ii) *Critical dimension $d\sigma = 2$*
   In this case,

$$y'_m(t) = -2H + 2 \int (1 - \partial_{rr} p_m) |\partial_r \psi|^2$$
$$- \frac{\sigma}{\sigma + 1} \int (d - \Delta p_m) |\psi|^{2\sigma + 2} + \frac{1}{2} \int \Delta^2 p_m |\psi|^2. \quad (5.1.53)$$

Here again $|\Delta^2 p|_{L^\infty} \leq C/m^2$, and (5.1.53) becomes

$$y'_m(t) \geq -2H + 2 \int_{r>m} a(r) |\partial_r \psi|^2 - \frac{2}{d+2} \int_{r>m} b(r) |\psi|^{4/d+2} - \frac{C}{m^2} |\varphi|_{L^2}^2,$$
$$(5.1.54)$$

with $a(r) = 1 - \partial_{rr} p_m \geq 0$ and $b(r) = d - \Delta p_m \geq 0$. In (5.1.54), the third term of the right-hand side can be bounded as follows

$$
\frac{2}{d+2} \int_{r>m} b(r)|\psi|^{4/d+2} \leq \frac{2}{d+2} \, |\varphi|_{L^2}^2 \, |b^{\frac{d}{4}}\psi|_{L^\infty(r>m)}^{\frac{4}{d}}
$$

$$
\leq \frac{2C}{d+2}|\varphi|_{L^2}^2 m^{-\frac{2(d-1)}{d}} \, |b^{\frac{d}{4}}\psi|_{L^2}^{\frac{2}{d}} \, |\nabla(b^{\frac{d}{4}}\psi)|_{L^2(r>m)}^{\frac{2}{d}}
$$

$$
\leq \epsilon \, |\nabla(b^{\frac{d}{4}}\psi)|_{L^2(r>m)}^2 + C(\epsilon)m^{-2} \, |\varphi|_{L^2}^{\frac{2d}{d-1}} \, |b^{\frac{d}{4}}\psi|_{L^2}^{\frac{2}{d-1}}
$$

$$
\leq 2\epsilon \, |\nabla(b^{\frac{d}{4}})\psi|_{L^2(r>m)}^2 + 2\epsilon \, |b^{\frac{d}{4}}\nabla\psi|_{L^2(r>m)}^2
$$

$$
+ C(\epsilon)m^{-2} \, |b|_{L^\infty}^{\frac{2d}{d-1}} \, |\varphi|_{L^2}^{\frac{2(d+1)}{d-1}}, \tag{5.1.55}
$$

since $d \geq 2$. We then note that

$$
|\nabla(b^{\frac{d}{4}})| \leq \frac{C}{m} \quad \text{for} \quad r > m. \tag{5.1.56}
$$

Therefore, from eqs. (5.1.54)–(5.1.56), we have

$$
y_m'(t) \leq 2|H| - Cm^{-2}\,|\varphi|_{L^2}^2 - C(\epsilon)\,m^{-2}\,|b|_{L^\infty}^{\frac{2d}{d-1}}\,|\varphi|_{L^2}^{\frac{2(d+1)}{d-1}}
$$

$$
+ 2 \int_{r>m} \left(a - \epsilon b^{\frac{d}{2}}\right)|\nabla\psi|^2. \tag{5.1.57}
$$

We can choose ϵ small enough so that for $r > m$,

$$
a - \epsilon b^{\frac{d}{2}} \geq 0. \tag{5.1.58}
$$

Indeed, for $r > 1 + 1/\sqrt{3}$,

$$
a(r) > 1 \quad \text{and} \quad b(r) < C, \tag{5.1.59}
$$

where C is a constant independent of m. For $1 < r < 1 + 1/\sqrt{3}$,

$$
a(r) - \epsilon b(r)^{\frac{d}{2}} = 3\left(\frac{r}{m} - 1\right)^2 - \epsilon 3\left(\frac{r}{m} - 1\right)^d\left(1 + \frac{d-1}{3}(1 - \frac{r}{m})\right)^{d/2}
$$

$$
\geq 3\left(\frac{r}{m} - 1\right)^2\left(1 - \epsilon(1 - \frac{r}{m})^{d-2}(1 + \frac{d-1}{3\sqrt{3}})\right). \tag{5.1.60}
$$

Thus, $a(r) - \epsilon b(r)^{\frac{d}{2}} > 0$ if $\epsilon\left(1 + \frac{d-1}{3\sqrt{3}}\right)^{d/2} < 1$.

For ϵ fixed, we now choose m large enough so that

$$
y_m'(t) \geq \eta|H| \tag{5.1.61}
$$

for some $\eta > 0$. By integration, we get

$$
y_m(t) \geq y_m(0) + \eta|H|\,t. \tag{5.1.62}
$$

Now, the truncated variance $V_m(t) = \int p_m|\psi|^2 dx$ is well-defined. It is positive, and satisfies

$$
\frac{dV_m}{dt} = -2y_m(t). \tag{5.1.63}
$$

It follows that

$$V_m(t) \leq -\frac{\eta}{2}|H|t^2 - y_m(0)t + V_m(0). \tag{5.1.64}$$

Assuming that the solution exists for all time, the last equation shows that V_m has to vanish in a finite time, leading to a contradiction.

Both in critical and supercritical dimensions, it is still unknown whether a negative Hamiltonian provides a sufficient condition to ensure finite-time blowup of non-radially symmetric solutions with infinite variances. Nevertheless, one has the following result.

Theorem 5.8. (Nawa 1993, 1998a) *At critical dimension, if the H^1 norm of the solution is uniformly bounded in time on $[0,\infty)$, the Hamiltonian is nonnegative. Equivalently, if the Hamiltonian is strictly negative, there exists a time T_m finite or infinite such that the corresponding solution satisfies*

$$\lim_{t \to T_m} |\nabla\psi|_{L^2} = \lim_{t \to T_m} |\psi|_{L^\sigma} = \infty. \tag{5.1.65}$$

5.2 Analysis of the Blowup

5.2.1 Rate of Blowup

This section provides lower bounds for the rate of blowup of Sobolev norms both at critical and supercritical dimensions.

Theorem 5.9. *At critical dimension $d = 2/\sigma$, assume that the solution of the NLS equation blows up in H^1 at a finite time t_*. Then there exists a positive constant M depending on the initial conditions, such that*

$$|\nabla\psi(t)|_{L^2} \geq M(t_* - t)^{-1/2}. \tag{5.2.1}$$

Proof. The above result is given in Cazenave and Weissler (1990) in a more general context. We present here a simpler proof, following Merle (1996a).

Define for fixed $t < t_*$,

$$\psi_1(\mathbf{x}, s) = \frac{1}{\lambda^{d/2}(t)}\psi\left(\frac{\mathbf{x}}{\lambda(t)}, t + \frac{s}{\lambda^2(t)}\right) \tag{5.2.2}$$

with $\lambda(t) = |\nabla\psi(t)|_{L^2}$. We have, for all $s \in [0, \lambda(t)^2(t_* - t))$,

$$i\partial_s\psi_1 + \Delta\psi_1 + |\psi_1|^{2\sigma}\psi_1 = 0 \tag{5.2.3}$$

and $|\nabla\psi_1(s)|_{L^2} \to \infty$ as $s \to \lambda^2(t)(t_* - t)$. Moreover, $|\psi_1(0)|_{L^2} = |\varphi|_{L^2}$, and the definition of $\lambda(t)$ implies that $|\nabla\psi_1(0)|_{L^2} = 1$. Thus, $|\psi_1(0)|_{H^1}^2 = |\varphi|_{L^2}^2 + 1$.

On the other hand, by the classical existence theory of the NLS equation in H^1 (see, for example, Kato 1987), for all $c_1 > 0$, there exists $t_1(c_1) > 0$

such that if $|\psi_1(0)|_{H^1} \leq c_1$, then there exists c_2 such that $|\psi_1(t)|_{H^1} < c_2$ in the interval $t \in [0, t_1]$. Applying this statement with $c_1 = |\varphi|_{L^2}^2 + 1$, we get that

$$\forall 0 < t < t_*, \quad \lambda^2(t)(t_* - t) \geq t_1, \tag{5.2.4}$$

from which the theorem follows.

In supercritical dimensions, one has the following result.

Theorem 5.10. *(Cazenave and Weissler 1990) Let $0 < s < d/2$ and $\frac{2}{d} < \sigma < \frac{2}{d-2s}$. If the solution obtained in Theorem 3.13 exists only for a finite time t_*, then $\lim_{t \to t_*} |(-\Delta)^{s/2}\psi(t)|_{L^2} = \infty$, and there is an estimate for the rate of blowup in the form*

$$|(-\Delta)^{s/2}\psi(t)|_{L^2} \geq \frac{C}{(t_* - t)^{1/(2\sigma)-(d-2s)/4}}. \tag{5.2.5}$$

A related result for $s = 1$ is given by Merle (1989) in the form of the following proposition.

Proposition 5.11. *Defining $J(\Theta) = \int_0^{t_*} |\nabla\psi(t)|_{L^2}^\Theta dt$, one has that $J(\Theta)$ is finite for $\Theta < 1$ and infinite for $\Theta > \Theta_2 = 4\sigma/(2\sigma + 2 - d\sigma)$,*

5.2.2   A Self-Similar Solution at Critical Dimension

From the standing-wave solution $\psi(\mathbf{x}, t) = R(|\mathbf{x}|)e^{it}$, a self-similar blowup solution is easily constructed in critical dimension, using the pseudo-conformal transformation (see Section 2.3). By means of this transformation, (choosing $t_0 = 1$ in (2.3.23)), the above solution is transformed into another solution, which up to a phase shift reads

$$\psi(\mathbf{x}, t) = \frac{1}{(t_* - t)^{d/2}} e^{\frac{i}{t_* - t}} R\left(\frac{|\mathbf{x}|}{t_* - t}\right) e^{-\frac{i|\mathbf{X}|^2}{4(t_* - t)}} \tag{5.2.6}$$

$$= \frac{1}{(t_* - t)^{1/\sigma}} e^{\frac{i}{t_* - t}} R\left(\frac{|\mathbf{x}|}{t_* - t}\right) e^{-\frac{i|\mathbf{X}|^2}{4(t_* - t)}}. \tag{5.2.7}$$

In this context, special attention has been paid to initial conditions with critical mass $|\varphi|_{L^2} = |R|_{L^2}$, where R is the positive solution of $\Delta R - R + R^{2\sigma+1} = 0$. Weinstein (1986b) proved the following result.

Theorem 5.12. *Let φ in $H^1(\mathbf{R}^d)$ such that $|\varphi|_{L^2} = |R|_{L^2}$, and suppose that the solution $\psi(t)$ blows up in a finite time t_*. Define the operator S_μ*

$$S_\mu f(\mathbf{x}, t) = \mu^{1/\sigma} f(\mu\mathbf{x}, t), \tag{5.2.8}$$

and let $\mu(t) = \frac{|\nabla R|_{L^2}}{|\nabla\psi|_{L^2}}$. Then, there exist two functions $y(t) \in \mathbf{R}^d$ and $\gamma(t) \in [0, 2\pi)$ such that

$$S_{\mu(t)}\psi(\cdot + y(t), t)e^{i\gamma(t)} \to R(\cdot) \quad \text{strongly in} \quad H^1 \tag{5.2.9}$$

as $t \to t_*$.

This result predicts that in the case of a special family of initial conditions whose mass is critical, up to a space translation and a phase shift, the solution near the singularity behaves like

$$\psi \sim \frac{1}{\mu(t)^{\frac{1}{\sigma}}} R\left(\frac{\mathbf{x}}{\mu(t)}\right). \tag{5.2.10}$$

A uniform characterization of the solution during the interval of existence of the solution is also given by Weinstein (1987, Theorem 4) in the case where the initial conditions not only have a critical mass, but are also "close" to the ground state. This requires the extension of the concept of orbit of the ground state, by taking dilations (i.e. rescalings of the spatial variable and of the solution amplitude, which at critical dimension keep the L^2-norm constant) as well as translations and phase shifts into account. Defining a convenient measure to estimate the distance to the (extended) ground-state orbit, the result states that for any $\epsilon > 0$, there is $\delta(\epsilon)$ such that if the distance of the initial conditions to the ground state is smaller than δ, then for all $t \in [0, T_*)$, the distance of the solution to the extended orbit remains smaller than ϵ.

A stability result up to space translations, phase shifts, and dilations is given at critical dimension by Laedke, Blaha, Spatschek, and Kuznetsov (1992) for initial conditions with a mass that exceeds the critical one. Assuming a negative Hamiltonian and defining the invariant set

$$S = \{\tilde{g}, \tilde{g} = g(\mathbf{x} - \mathbf{a}_1)e^{i(\lambda^2 t + a_0)}\} \tag{5.2.11}$$

with $\mathbf{a}_1 \in \mathbf{R}^d$ and $a_0 \in [0, 2\pi]$, where g is the ground state solution of $\Delta g - \lambda g + g^{2\sigma+1} = 0$, they state that in the interval $0 \le t < t_*$, for every ϵ, there exists δ_ϵ such that

$$|\varphi(\mathbf{x}) - \tilde{g}(\mathbf{x})|_{H^1} < \delta_\epsilon \Rightarrow |\psi(\mathbf{x}, t) - S_\mu \tilde{g}(\mathbf{x}, t)|_L^2 < \epsilon, \tag{5.2.12}$$

with $\mu(t) = |\nabla \psi|_{L^2}/|\nabla g|_{L^2}$ where λ is chosen by the condition $\mu(0) = 1$.

The rate of blowup for solutions with critical mass was first determined in the case of initial conditions that are radially symmetric with finite variance (Merle 1992a) and then extended to initial conditions lacking this symmetry (Merle 1993) as stated in the following result.

Theorem 5.13. *Let* $\varphi \in H^1(\mathbf{R}^d)$ *and* $|\mathbf{x}\varphi| \in L^2(\mathbf{R}^d)$ *(finite variance). Suppose in addition that* $|\varphi|_{L^2} = |R|_{L^2}$, *and that the solution* $\psi(t)$ *blows up in a finite time* t_*. *Then there exist constants* θ, ω, *and* $\boldsymbol{\xi}_0$ *and* $\mathbf{x}_1 \in \mathbf{R}^d$ *such that the solution is expressed in the form*

$$\psi(\mathbf{x}, t) = \frac{\omega^{d/2}}{(t_* - t)^{d/2}} R\left(\omega\left(\frac{\mathbf{x} - \mathbf{x}_1}{t_* - t} - \boldsymbol{\xi}_0\right)\right) e^{i\left(\theta + \frac{\omega^2}{t_* - t} - \frac{|\mathbf{X} - \mathbf{X}_1|^2}{4(t - t_*)}\right)}. \tag{5.2.13}$$

5.2.3   Solutions with Exactly k Blowup Points

At critical dimension, explicitly known solutions that blowup in a finite time are those given by

$$\psi(\mathbf{x}, t) = \frac{1}{(t_* - t)^{d/2}} R\left(\frac{\mathbf{x}}{t_* - t}\right) e^{-\frac{i|\mathbf{X}|^2}{4(t_* - t)}} e^{\frac{i}{t_* - t}}, \qquad (5.2.14)$$

where R is the unique positive solution of

$$\Delta R - R + R^{1 + 4/d} = 0. \qquad (5.2.15)$$

For these solutions, the only blowup point is taken at the origin. A question that can be asked is whether there exist blowup solutions that have more than one blowup point and, if so, what is their behavior. Given k arbitrary points $\mathbf{x}_1, \mathbf{x}_2, \ldots, \mathbf{x}_k$ in $\mathbf{R}^d$, Merle (1990) constructed a solution of the critical NLS equation that blows up in a finite time t_* at exactly these locations and displays the L^2-concentration property at the points $\mathbf{x}_i$.

Denoting by $R_0, R_1, \ldots$ the infinite sequence of radial solutions to the equation

$$\Delta R - R + R^{\frac{4}{d} + 1} = 0 \qquad (5.2.16)$$

such that R_k has exactly k nodes as a function of r and decreases exponentially at infinity, one has the following result.

Theorem 5.14. *Let the dimension be critical ($\sigma d = 2$) and $\mathbf{x}_1, \ldots, \mathbf{x}_k$ be given in $\mathbf{R}^d$. There is a constant ω_0 such that for any constants $\omega_1, \ldots, \omega_k$ all strictly larger than ω_0, there exists a solution ψ of the NLS equation that blows up in a finite time t_* such that:*

(i)   *The set of blowup points in $L^{2+4/d}$ and H^1 is $\{\mathbf{x}_1, \ldots, \mathbf{x}_k\}$,*
(ii) *For $i = 1, \ldots, k$ and all A such that the balls $B_i = B(\mathbf{x}_i, A)$ are disjoint, $\lim_{t \to t_*} |\psi(t)|_{L^2(B_i)} = |R_i|_{L^2}$,*
(iii) *$\lim_{t \to t_*} |\psi(t)|_{L^2(\bar{B})} = 0$, where $\bar{B} = \mathbf{R}^d - \cup_{i=1,\cdots,k} B_i$.*

In addition, there is a constant $\gamma > 0$ such that on $[0, t_*)$,

$$\left| \psi(t) - \sum_{i=1}^{k} \frac{1}{|(t_* - t)\omega_i|^{d/2}} e^{\frac{-i}{(t_* - t)\omega_i^2} + \frac{i|\mathbf{X}|^2}{4(t_* - t)}} R_i\left(\frac{\mathbf{x} - \mathbf{x}_i}{\omega_i(t_* - t)}\right) \right|_{L^{2+4/d}} \le e^{-\frac{\gamma}{t_* - t}}. $$
$$(5.2.17)$$

Finally, we mention a recent result by Nawa (1998b), who exhibited for the critical one-dimensional NLS equation a class of initial conditions such that the solutions blow up at two symmetric points and concentrate at least the critical mass at each of these points.

5.2.4 L^2-Norm Concentration at Critical Dimension

We consider blowing-up solutions in the critical case $\sigma d = 2$ and study their L^2-properties. The forthcoming theorems give a rigorous basis to the concept of "strong collapse" (Zakharov and Kuznetsov 1986b) for which as the singularity is approached, a finite amount of the L^2-norm (mass or energy) of the solution remains trapped near the blowup point.

Suppose again that the initial condition φ belongs to $H^1(\mathbf{R}^d)$.[1] We know that there exists a time t_* such that for all $t \in [0, t_*)$, there is a unique solution $\psi(t)$ in $H^1(\mathbf{R}^d)$ with either $t_* = +\infty$ or $\lim_{t \to t_*} |\psi(t)|_{H^1} = \infty$. In the case where t_* is finite, that is to say when the solution $\psi(t)$ blows up in a finite time in $H^1(\mathbf{R}^d)$, Merle and Tsutsumi (1990) proved the two following theorems:

Theorem 5.15. *Suppose that $\psi(t)$ becomes infinite in $H^1(\mathbf{R}^d)$ in a finite time t_*. Then $\psi(t)$ does not have a strong limit in L^2 as $t \to t_*$. In addition, there is no sequence t_n converging to t_* such that $\psi(t_n)$ converges in L^2 as $t_n \to t_*$.*

Proof. This result is a consequence of the conservation of the Hamiltonian and of the Gagliardo–Nirenberg inequality (3.2.12). It is proved by contradiction.

Suppose that $\psi(t)$ is a solution of (3.0.1) in $C([0, t_*), H^1)$, and that $\{t_n\}$ is a sequence such that $t_n \to t_*$ and $\psi(t_n)$ has a strong limit in L^2 as $n \to \infty$. It follows that $|\nabla \psi(t)|_{L^2}$ is bounded for all $t \in (0, t_*)$. Indeed,

$$|\nabla \psi(t_n)|_{L^2}^2 \leq H + C \int |\psi(t_n) - \psi(t_m)|^{2\sigma+2} d\mathbf{x} + C \int |\psi(t_m)|^{2\sigma+2} d\mathbf{x}$$

$$\leq H + |\psi(t_n) - \psi(t_m)|_{L^2}^2 (|\nabla \psi(t_n)|_{L^2}^2 + |\nabla \psi(t_m)|_{L^2}^2)$$

$$+ C \int |\psi(t_m)|^{2\sigma+2} d\mathbf{x}.$$

Since $\psi(t_n)$ converges strongly in L^2, there exists a positive integer m such that for $n > m$,

$$|\psi(t_n) - \psi(t_m)|_{L^2}^2 \leq \frac{1}{2}. \tag{5.2.18}$$

Thus, for all $n > m$,

$$|\nabla \psi(t_n)|_{L^2}^2 \leq \frac{1}{2} |\nabla \psi(t_n)|_{L^2}^2 + C_m. \tag{5.2.19}$$

This implies that the sequence $\psi(t_n)$ is uniformly bounded in H^1.

Remark. It is known that for initial condition in L^2, there exists a local solution in L^2 (Tsutsumi 1987). The previous result implies that the blowup solution cannot be extended beyond the singularity time in the strong topology of L^2. Equivalently, the singularity time is the same in H^1 and in L^2.

---

[1]For initial condition in L^2 (not in H^1), concentration properties of the L^2-norm were recently established by Bourgain (1998a) for solutions of the cubic Schrödinger equation in two space dimensions.

Assume that the initial condition φ is radially symmetric. Then, the solution $\psi(t)$ is also radially symmetric, and the blowup occurs at the origin, in the following sense:

$$\forall A > 0, \quad |\nabla\psi|_{L^2(|\mathbf{x}|<A)} \to \infty \quad \text{as} \quad t \to t_*, \tag{5.2.20}$$

and

$$\forall A > 0, \quad |\psi|_{L^\infty(|\mathbf{x}|<A)} \to \infty \quad \text{as} \quad t \to t_*. \tag{5.2.21}$$

Theorem 5.16. *If the solution is radially symmetric and blows up in H^1, then the origin is a blowup point, and for all $A > 0$,*

$$\liminf_{t \to t_*} |\psi(t)|_{L^2(B(0;A))} \geq |R|_{L^2}, \tag{5.2.22}$$

where $B(0; A)$ is the ball of radius A centered at the origin.

This is an optimal result in the sense that there exist explicit examples of blowing-up solutions with $\liminf_{t \to t_*} |\psi(t)|_{L^2(B(0;A))} = |R|_{L^2}$.

For solutions not necessarily radially symmetric but blowing up at one point, Weinstein (1989) proved the following result:

Theorem 5.17. *Let $\varphi \in H^1$ and suppose that the solution blows up in H^1 at a finite time t_*. For any sequence $t_k \to t_*$, there is a subsequence t_{k_j} such that for $2 < p < \frac{2d}{d-2}$*

$$S_{\mu(t)}\psi(\cdot + y(t_{k_j}), t_{k_j})e^{i\gamma(t_{k_j})} \to \Psi \quad \text{strongly in} \quad L^p, \tag{5.2.23}$$

where

$$S_\mu f(\mathbf{x}, t) = \mu^{1/\sigma} f(\mu\mathbf{x}, t), \tag{5.2.24}$$

$$\mu(t) = \frac{|\nabla R|_{L^2}}{|\nabla\psi|_{L^2}}, \tag{5.2.25}$$

and

$$|\Psi|_{L^2} \geq |R|_{L^2}. \tag{5.2.26}$$

Note that $|R|_{L^2}$ is a threshold value for the L^2 norm of the initial conditions. Indeed,

(i) When $|\varphi|_{L^2} < |R|_{L^2}$, solutions exist globally in time and disperse to 0 at the free Schrödinger rate, in the sense that $\sup_t |t|^{d/2 - d/p}|\psi(t)|_{L^p} < \infty$.

(ii) When $|\varphi|_{L^2} = |R|_{L^2}$, blowing-up solutions can be obtained explicitly with a rate of blowup $(t_* - t)^{-d/2}$ and a spatial structure given by the ground state R.

(iii) When $|\varphi|_{L^2} > |R|_{L^2}$, the L^2-norm of blowing-up solutions concentrates near the singularity by an amount which is bounded from below by $|R|_{L^2}$.

A detailed discussion of solutions blowing up at several points simultaneously is presented by Nawa (1990, 1994).

The rate of the L^2-norm concentration was obtained by Tsutsumi (1990) for radially symmetric solutions:

Proposition 5.18. *Define* $\lambda(t) = |\nabla \psi(t)|_{L^2}$. *If* $a(t)$ *is a decreasing function of* t *such that* $a(t) \to 0$ *as* $t \to t_*$, *and* $\lambda(t)a(t) \to \infty$, *then*

$$\liminf_{t \to t_*} |\psi(t)|_{L^2(|\mathbf{x}|<a(t))} \geq |R|_{L^2}, \tag{5.2.27}$$

and for all $\epsilon > 0$ *there exists* $K > 0$ *such that*

$$\liminf_{t \to t_*} |\psi(t)|_{L^2(|\mathbf{x}|<K/\lambda(t))} \geq (1 - \epsilon)|R|_{L^2}. \tag{5.2.28}$$

Proof.   Let ρ be a radially symmetric function in C_0^1 such that

$$\rho(\mathbf{x}) = \begin{cases} 1 & \text{if } |\mathbf{x}| < \dfrac{1}{2} \\ 0 & \text{if } |\mathbf{x}| > 1, \end{cases}$$

and $\rho_a(\mathbf{x}) = \rho(\mathbf{x}/a)$. Define also

$$\lambda_a(t) = |\nabla(\rho_a \psi(t))|_{L^2}, \tag{5.2.29}$$

and $q = 2 + 4/d$. From Lemma 5.6, we have

$$|\psi|_{L^q(|\mathbf{x}|>a/2)}^q \leq |\psi|_{L^\infty(|\mathbf{x}|>a/2)}^{4/d} N \leq \frac{C}{a^{2(d-1)/d}} \lambda(t)^{2/d} N^{1+1/d}. \tag{5.2.30}$$

The conservation of the Hamiltonian ensures that

$$\lambda(t)^2 \leq H + \frac{2}{q}|\psi|_{L^q(|\mathbf{x}|<a/2)}^q + \frac{C}{a^{2(d-1)/d}} \lambda(t)^{2/d} N^{1+1/d}. \tag{5.2.31}$$

Straightforward calculations then give

$$|\psi|_{L^q(|\mathbf{x}|<a/2)} \leq |\rho_a \psi|_{L^q}, \tag{5.2.32}$$

and

$$\begin{aligned} \lambda_a^2(t) &\leq (\lambda(t) + \frac{C}{a}N^{1/2})^2 \\ &\leq \lambda(t)^2 + \frac{C}{a}N^{1/2}\lambda(t) + \frac{C}{a^2}N. \end{aligned} \tag{5.2.33}$$

Finally, the variational characterization of the ground state R leads to

$$\frac{2}{q}|\rho_a \psi|_{L^q}^q \leq \left(\frac{|\rho_a \psi|_{L^2}}{|R|_{L^2}}\right)^{4/d} \lambda_a(t)^2. \tag{5.2.34}$$

Combining (5.2.31)–(5.2.34) gives

$$\begin{aligned} 1 - \frac{|\rho_a \psi|_{L^2}^{4/d}}{|R|_{L^2}^{4/d}} &\leq \frac{H}{\lambda_a^2} + \frac{C}{(a\lambda_a)^{2(d-1)/d}}\left(\frac{\lambda}{\lambda_a}\right)^{2/d} N^{1+1/d} \\ &\quad + \frac{C\lambda}{a\lambda_a^2}N^{1/2} + \frac{C}{(a\lambda_a)^2}N. \end{aligned} \tag{5.2.35}$$

From the assumption on the function $a(t)$ and the fact (proved below) that $\frac{\lambda(t)}{\lambda_a(t)}$ remains bounded as $t \to t_*$, it follows that in the limit $t \to t_*$,

$$\limsup_{t \to t_*} \left(1 - \frac{|\rho_a \psi|_{L^2}^{4/d}}{|R|_{L^2}^{4/d}} \right) \le 0, \tag{5.2.36}$$

or equivalently, (5.2.27) holds. Rewriting (5.2.35) with $a(t) = K/\lambda(t)$ gives (5.2.28) in the limit $t \to t_*$, if K is taken sufficiently large.

To complete the proof of the proposition, $\frac{\lambda(t)}{\lambda_a(t)}$ should be shown to remain bounded as $t \to t_*$. Equations (5.2.31) and (5.2.34) imply

$$\left(1 - \frac{CN^{1+1/d}}{(a\lambda)^{2-2/d}} \right) \lambda^2(t) \le H + C\lambda_a^2(t)N^{2/d}. \tag{5.2.37}$$

As t approaches t_*, $1/a\lambda \to 0$ and the factor in parentheses on the left-hand side of (5.2.37) is smaller than $\frac{1}{2}$. Thus, $\lambda_a(t) \to \infty$, and λ/λ_a remains bounded.

Theorem 5.19. *If $a(t)$ is a decreasing function of t such that $a(t) \to 0$ as $t \to t_*$ and $(t_* - t)^{1/2}/a(t) \to 0$, then*

$$\liminf_{t \to t_*} |\psi(t)|_{L^2(|\mathbf{x}|<a(t))} \ge |R|_{L^2}, \tag{5.2.38}$$

and for all $\epsilon > 0$, there exists $K > 0$ such that

$$\liminf_{t \to t_*} |\psi(t)|_{L^2(|\mathbf{x}|<K(t_*-t)^{1/2})} \ge (1 - \epsilon)|R|_{L^2}. \tag{5.2.39}$$

Corollary 5.20. *For any $p > 2$,*

$$|\psi(t)|_{L^p} \ge \frac{M}{(t_* - t)^\alpha} \quad \text{with} \quad \alpha = \left(\frac{1}{2} - \frac{1}{p} \right) \frac{d}{2}. \tag{5.2.40}$$

In particular, for $p = 2\sigma + 2$,

$$|\psi(t)|_{L^{2\sigma+2}}^{2\sigma+2} \ge \frac{M}{t_* - t}. \tag{5.2.41}$$

Proof. Using the Hölder inequality with $\frac{1}{p} + \frac{p-2}{2p} = \frac{1}{2}$, estimate (5.2.39) implies

$$C_1 \le |\psi(t)|_{L^2(|\mathbf{x}|<K(t_*-t)^{1/2})} \le C(t_* - t)^{\frac{d(p-2)}{4p}} |\psi(t)|_{L^p}, \tag{5.2.42}$$

and (5.2.40) follows.

Remark. Extension of the analysis to the critical NLS equation with a nonhomogeneous nonlinearity $k(\mathbf{x})|\psi|^{4/d}$ is considered by Merle (1996b). In particular, the existence of a minimal blowup solution depends on the form of the function $k(\mathbf{x})$ near the point where it reaches its maximum.

Finally, we mention a recent result by Nawa and Tsutsumi (1998):

Theorem 5.21. *If the solution is radially symmetric and blows up in H^1 in a finite time t_*, then the following properties are equivalent:*

(i) $\lim_{t \to t_*} |\psi(\mathbf{x}, t)|^2 dx = |\varphi|^2_{L^2} \delta_0(dx)$ *in the sense of measures where δ_0 is the Dirac measure at $0 \in \mathbf{R}^d$,*

(ii) $|\mathbf{x}|\varphi \in L^2(\mathbf{R}^d)$ *and* $\lim_{t \to t_*} \||\mathbf{x}|\psi(\mathbf{x}, t)|_{L^2} = 0$.

Such a situation where all the mass of the solution concentrates at the origin and the variance vanishes at the singularity time is realized by the particular blowup solutions (5.2.13). For more general collapsing solutions, numerical simulations indicate that the variance remains strictly positive at the time of blowup (see Chapter 6). In this case, only a fraction of the total L^2-norm (asymptotically equal to $|R|_{L^2}$) concentrates near the origin, while the rest forms wings of moderate amplitude. Rigorous results associated to this picture were recently given by Nawa (1998a). They express that at critical dimension, a blowup solution behaves like a finite superposition of dilated zero-Hamiltonian, zero-momentum, H^1-bounded solution accompanied by a dilated wave of the free Schrödinger equation. At the singularity time, an amount of the mass at least equal to that of the ground state R concentrates at each of the foci, and the solution cannot have a strong limit in L^2. More precisely, one has the following result.

Theorem 5.22. *Let $\psi(t)$ be a singular solution of the NLS equation at critical dimension such that $\lim\sup_{t \to t_*} |\nabla\psi|_{L^2} = \infty$ for some time $t_* \in (0, \infty]$. Let $\{t_n\}$ be any sequence such that as $n \to \infty$, $t_n \uparrow t_*$ and $\sup_{t \in [0, t_n]} |\nabla\psi(t)|_{L^2} = |\nabla\psi(t_n)|_{L^2}$. Define*

$$\lambda_n = |\nabla\psi(t_n)|_{L^2}^{-1} \qquad (5.2.43)$$

and, for $t \in [-\frac{t_* - t_n}{\lambda_n^2}, \frac{t_n}{\lambda_n^2})$ the scaled functions,

$$u_n(\mathbf{x}, t) = \lambda_n^{d/2}\psi(\lambda_n\mathbf{x}, t_n - \lambda_n^2 t). \qquad (5.2.44)$$

There exists a finite number of smooth nontrivial solutions $u^1, u^2, \ldots, u^L$ of the NLS equation at critical dimension with zero Hamiltonian and zero momentum, and sequences $\{\gamma_n^1\}, \ldots, \{\gamma_n^L\}$ in $\mathbf{R}^d$ with $\lim_{n \to \infty} |\gamma_n^j - \gamma_n^k| = \infty$ for $j \neq k$ (in the case of a radially symmetric initial condition, $L = 1$, and $\gamma_n^1 = 0$), such that for any $T > 0$,

$$\lim_{n \to \infty} \sup_{t \in [0,T]} |\nabla u_n(\mathbf{x}, t) - \sum_{j=1}^{L} \nabla u^j(\mathbf{x} - \gamma_n^j, t)|_{L^2(\mathbf{R}^d)} = 0 \quad (5.2.45)$$

$$\lim_{n \to \infty} \sup_{t \in [0,T]} |u_n(\mathbf{x}, t) - \sum_{j=1}^{L} u^j(\mathbf{x} - \gamma_n^j, t) - \phi_n(\mathbf{x}, t)|_{L^2(\mathbf{R}^d)} = 0, \quad (5.2.46)$$

where the function ϕ_n, which can be viewed as corresponding to a shoulder, solves the free Schrödinger equation

$$i\frac{\partial \phi_n}{\partial t} + \Delta \phi_n = 0, \tag{5.2.47}$$

$$\phi_n(\mathbf{x}, 0) = u_n(\mathbf{x}, 0) - \sum_{j=1}^{L} w^j(\mathbf{x} - \gamma_n^j, 0). \tag{5.2.48}$$

Furthermore, the initial condition satisfies

$$|\varphi|_{L^2}^2 \geq \sum_{j=1}^{L} |w^j(t)|_{L^2}^2 \geq L|R|_{L^2}^2, \tag{5.2.49}$$

where R denotes the ground state solution. In addition, for any $T > 0$ and any $f \in B = C(\mathbf{R}^d) \cap L^\infty(\mathbf{R}^d)$,

$$\lim_{n\to\infty} \sup_{t\in[t_n - \lambda_n^2 T, t_n]} \left| \int_{\mathbf{R}^d} \left(|\psi(\mathbf{x}, t)|^2 - \sum_{j=1}^{L} \left| \frac{1}{\lambda_n^{d/2}} w^j \left(\frac{\mathbf{x}}{\lambda_n} - \gamma_n^j, \frac{t_n - t}{\lambda_n^2} \right) \right|^2 \right. \right.$$
$$\left. \left. - \left| \frac{1}{\lambda_n^{d/2}} \phi_n \left(\frac{\mathbf{x}}{\lambda_n}, \frac{t_n - t}{\lambda_n^2} \right) \right|^2 \right) f(\mathbf{x}) d\mathbf{x} \right| = 0. \tag{5.2.50}$$

When the conditions for blowup are satisfied, then for any $T > 0$, one defines $s_n = t_n - \lambda_n^2 T$ (where $s_n \to \infty$ as $n \to \infty$). Then, there exists a subsequence of $\{s_n\}$ (still denoted by $\{s_n\}$) that satisfies the following properties: There is a finite number $L \in \mathbf{N}$, a family of points $\{a^1, \ldots, a^L\}$ in $\mathbf{R}^d$ such that $a^j = \lim_{n\to\infty} \lambda_n \gamma_n^j$, and a positive measure $\mu \in B'$ (the dual of B) such that

$$|\psi(\mathbf{x}, s_n)|^2 d\mathbf{x} \to \sum_{j=1}^{L} |w^j(0)|_{L^2}^2 \delta_{a^j}(d\mathbf{x}) + \mu(d\mathbf{x}), \tag{5.2.51}$$

as $n \to \infty$ in the weak topology of measures. In this formula, μ is the limit of

$$\frac{1}{\lambda_n^{d/2}} \phi_n \left(\frac{\mathbf{x}}{\lambda_n}, \frac{t_n - t}{\lambda_n^2} \right).$$

Part III

Asymptotic Analysis
near Collapse

6
Numerical Observations

Before discussing in Chapters 7 and 8 the asymptotic theory of the collapse for the nonlinear Schrödinger equation

$$i\partial_t \psi + \Delta\psi + |\psi|^{2\sigma}\psi = 0 \qquad (6.0.1)$$

in the supercritical and critical cases, respectively, we describe in this chapter the accurate numerical methods that play an important role in differentiating among various possible singular behaviors and suggesting the adequate asymptotic ansatz. As observed with all the simulations with generic initial conditions satisfying a sufficient condition for collapse (usually a negative Hamiltonian), the singularity takes place much before the variance vanishes, even at critical dimension.

Simulations of the cubic nonlinear Schrödinger equation ($\sigma = 1$) performed in dimension $d = 3$ for radially symmetric solutions with a unique maximum located at the origin early led to the conclusion that the collapse is in this case self-similar with a rate of blowup scaling like $(t_* - t)^{-1/2}$ (Budneva, Zakharov, and Synakh 1975, Goldman, Rypdal, and Hafizi 1980). For the Schrödinger equation at critical dimension $d = 2$, the asymptotic regime is established only extremely close to the singularity. It is thus much more delicate to simulate it numerically and to discriminate between the conflicting predictions proposed for critical collapse. A $(t_* - t)^{-2/3}$ scaling for a peak amplitude superimposed on a slowly varying shelf was proposed by Zakharov and Synakh (1976), a picture supported by simulations of the early phase of the collapse presented by Sulem, Sulem, and Patera (1984). Similar observations were reported for the quintic NLS equation at the critical dimension $d = 1$ (Sulem, Sulem, and Frisch 1983),

and a formal construction of these solutions was presented by Papanicolaou, McLaughlin, and Weinstein (1982). In a different direction, Vlasov,
Piskunova, and Talanov (1978), Wood (1984), and Rypdal and Rasmussen
(1986) suggested a $(|\ln(t_* - t)|/(t_* - t))^{1/2}$ law, while Landman, Papan-
icolaou, Sulem, and Sulem (1988) and LeMesurier, Papanicolaou, Sulem,
and Sulem (1988b) (see also Landman, LeMesurier, Papanicolaou, Sulem,
and Sulem 1989, Dyachenko, Newell, Pushkarev, and Zakharov 1992 and
Sulem and Sulem 1997) conclude that the rate of blowup is of the form
$(\ln|\ln(t_* - t)|/(t_* - t))^{1/2}$. This rate was previously predicted by Fraiman
(1985) (see also Smirnov and Fraiman 1991) on the basis of delicate physical
arguments. An unfortunate misprint in both the original version and the
English translation however, transformed the double log correction into
a simple log, and this result was not immediately recognized (Zakharov
and Kuznetsov 1986b). Laws of the form $|\ln(t_* - t)|^\gamma/(t_* - t)^{1/2}$, with
$0.35 \leq \gamma \leq 0.65$, depending on the initial conditions, were also considered
as consistent with the simulations (Zakharov and Shvets 1988, Shvets and
Zakharov 1990, Kosmatov, Shvets and Zakharov 1991).

Especially at critical dimension, accurate numerical simulations of the
blowup require a special adaptive mesh refinement leading to a strong
enhancement of the resolution near the singularity. A simple method suggested by the theoretical understanding that the solution is asymptotically
self-similar or quasi-self-similar near collapse is provided by the dynamic
rescaling method first described in the simplified case of radially symmetric
solutions in McLaughlin, Papanicolaou, Sulem, and Sulem (1986) and then
extended to more general initial conditions localized about a unique peak
in Landman, Papanicolaou, Sulem, Sulem, and Wang (1991). It consists
in rescaling the dependent and independent variables by factors prescribed
self-consistently in terms of a norm of the solution that blows up at the
singularity. The corresponding norm of the rescaled solution then remains
constant in time. As a consequence, the rescaled problem is no longer singular, and its numerical integration can be pushed very close to the blowup. A
main numerical observation concerns the stability of the isotropic collapse
towards which anisotropic solutions are seen to converge. Furthermore, the
collapse appears to be self-similar in dimension 3, while at the critical dimension $d = 2$, the self-similarity is weakly broken, with a rate of blowup
of the form $F(t_* - t)/(t_* - t)^{1/2}$, where the function $F(t_* - t)$ blows up
more slowly than any power of $\ln \frac{1}{t_* - t}$ (LeMesurier, Papanicolaou, Sulem,
and Sulem 1987). Very accurate computations recently performed using
an adaptive Galerkin finite element method directly on the NLS equation (Akrivis, Dougalis, Karakashian, and McKinney 1993, 1997) support
the log-log law for the far asymptotic regime. Descriptions of intermediate
regimes were proposed by Malkin (1993), Fibich, Malkin, and Papanicolaou
(1995), Fibich (1996a), Fibich and Papanicolaou (1997a).

6.1 Capturing the Blowup Structure

6.1.1 A Scale Transformation for Isotropic Solutions

The invariance of the nonlinear Schrödinger equation with a power law nonlinearity by the scale transformation suggests in the case of radially symmetric solutions the change of variables $(r = |\mathbf{x}|)$

$$\psi(r,t) = \frac{1}{L(t)^{1/\sigma}} u(\xi,\tau) , \quad \xi = \frac{r}{L(t)} , \quad \tau = \int_0^t \frac{ds}{L^2(s)} . \tag{6.1.1}$$

Substituting in (6.0.1), we get

$$iu_\tau + ia(\tau)\left(\xi u_\xi + \frac{u}{\sigma}\right) + \Delta u + |u|^{2\sigma} u = 0, \tag{6.1.2}$$

where

$$a = -L\frac{dL}{dt} = -\frac{d\ln L}{d\tau}. \tag{6.1.3}$$

The initial condition $\psi(r,0) = \varphi(r)$ becomes $u_0(\xi) = L(0)^{1/\sigma}\varphi(L(0)\xi)$.
  Under this scaling, the wave energy and the Hamiltonian take the form

$$N = L(\tau)^{d-2/\sigma} \int |u|^2 \xi^{d-1} d\xi, \tag{6.1.4}$$

$$H = L(\tau)^{-2-\frac{2}{\sigma}+d} \int \left(|\nabla u|^2 - \frac{1}{\sigma+1}|u|^{2\sigma+2}\right) \xi^{d-1} d\xi. \tag{6.1.5}$$

At this step, it is natural to look for radially symmetric singular solutions of self-similar type near collapse. This corresponds to setting $a(\tau) = a$ to be a constant and imposing that either $u(\xi,\tau)$ be independent of τ (Goldman, Rypdal, and Hafizi 1980) or, more generally, that $u(\xi,\tau) = e^{i\tau}Q(\xi)$, where the profile Q satisfies (Zakharov 1984)

$$Q_{\xi\xi} + \frac{d-1}{\xi}Q_\xi - Q + ia\left(\xi Q_\xi + \frac{Q}{\sigma}\right) + |Q|^{2\sigma}Q = 0, \quad \xi > 0,$$
$$Q_\xi(0) = 0, \ Q(\infty) = 0, \tag{6.1.6}$$

with the normalization $Q(0)$ real. This leads to a class of singular solutions of the form

$$\psi(\mathbf{x},t) = \frac{1}{(2a(t_* - t))^{1/2\sigma}} Q\left(\frac{|\mathbf{x} - \mathbf{x}_*|}{(2a(t_* - t))^{1/2}}\right) \exp\left(i\left(\theta + \frac{1}{2a}\ln\frac{t_*}{t_* - t}\right)\right), \tag{6.1.7}$$

where $\mathbf{x}_*$ and t_* are the location and the time of the blowup.
  Such solutions with $a \neq 0$ are possible in supercritical dimension (see Chapter 7). In contrast, at critical dimension, (6.1.6) has no admissible solutions for $a \neq 0$, and a more refined analysis is needed (Chapter 8).

6.1.2 Dynamic Rescaling

Analysis of the blowup of radially symmetric solutions is conveniently done using the so-called dynamic rescaling method, where the rescaled equations resulting from the scale transformation are complemented by the determination of the scaling factor $L(t)$. The latter should be chosen in such a way that it tends to 0 fast enough so that the rescaled time τ goes to infinity as the singularity time t_* is approached. This is done by imposing that the rescaled solution $u(\xi, \tau)$ remains smooth, uniformly in τ (McLaughlin, Papanicolaou, Sulem, and Sulem 1986, LeMesurier, Papanicolaou, Sulem, and Sulem 1987). Various constraints have been used with comparable efficiency. They consist in requiring that a spatial norm remains constant in time. The simplest choice is $L(t)^{1/\sigma} = |u_0(0)| |\psi(t)|_{L^\infty}^{-1}$ which ensures that $|u(\tau)|_{L^\infty}$ remains constant. It leads to

$$a(\tau) = -\frac{\sigma}{|u_0(0)|^2} \Im(u^* \Delta u)(0, \tau). \tag{6.1.8}$$

A related method involving the rescaling of the independent variables but not of the amplitude of the solution was used by Zakharov and Shvets (1988), Kosmatov, Petrov, Shvets, and Zakharov (1988) and Kosmatov, Shvets, and Zakharov (1991).

It is also possible to construct $L(t)$ using a spatial integral norm of ψ that goes to infinity at blowup, a choice that appears to be preferable for numerical accuracy. Choosing, for example, a scaling such that $|\nabla u(\tau)|_{L^2}$ remains constant leads to

$$L(t) = \left(\frac{|\nabla u_0|_{L^2}}{|\nabla \psi(t)|_{L^2}} \right)^{2/p}, \qquad p = 2 + \frac{2}{\sigma} - d. \tag{6.1.9}$$

Substitution into (6.1.3) and use of (6.1.1) then give

$$a(\tau) = -\frac{2}{p |\nabla u_0|_{L^2}^2} \int_0^\infty |u|^{2\sigma} \Im(u^* \Delta u) \xi^{d-1} d\xi. \tag{6.1.10}$$

Note that numerical errors do not accumulate in the computation of $|\nabla u(\tau)|_{L^2}$. Indeed, differentiating $|\nabla u(\tau)|_{L^2}$ with respect to τ and using (6.1.2) and (6.1.10), one gets

$$\frac{d}{d\tau} |\nabla u(\tau)|_{L^2}^2 = -pa \left(|\nabla u(\tau)|_{L^2}^2 - |\nabla u_0|_{L^2}^2 \right). \tag{6.1.11}$$

Since for singular solutions $L(t)$ is decreasing, it follows that a is positive, and consequently, $|\nabla u(\tau)|_{L^2}^2 = |\nabla u_0|_{L^2}^2$ is a stable solution.

Instead of fixing the L^2-norm of the gradient, higher order norms like $|\nabla u(t)|_{L^r}$ ($r > 2$) or $|\Delta u(t)|_{L^2}$ can also be used. Using a dual procedure, one can also define $L(t)$ by prescribing that a typical scale defined as the ratio

$$\frac{\int \xi^2 |u(\xi)|^{2p} \xi^{d-1} d\xi}{\int |u(\xi)|^{2p} \xi^{d-1} d\xi} \quad \text{with} \quad p \geq 2,$$

remain constant. This prescription appears to be well adapted to extension of the method to non-radially symmetric solutions.

Equation (6.1.2) together with the expression of $a(\tau)$ viewed as a functional of u (such as (6.1.10)) constitutes a closed system that is nonsingular and thus amenable to standard resolution algorithms. To reconstruct the original solution ψ, one computes the scaling factor $L(t)$ by solving (6.1.3) with $a(\tau)$ given in terms of u.

Note that (6.1.2) must be solved in the entire space $\xi > 0$. At large enough distance however, the solution being small, the nonlinear term is negligible in (6.1.2) and so are the second derivatives. The equation then reduces to the hyperbolic equation

$$u_\tau + a(\tau)\left(\xi u_\xi + \frac{u}{\sigma}\right) = 0, \qquad (6.1.12)$$

which can be solved exactly in the form

$$u(\xi,\tau) = u_0\left(\xi\frac{L(\tau)}{L(0)}\right)\frac{L^{1/\sigma}(\tau)}{L^{1/\sigma}(0)}. \qquad (6.1.13)$$

Kosmatov, Petrov, Shvets, and Zakharov, (1988) and Kosmatov, Shvets, and Zakharov (1991) suggested that this property be used in solving numerically the rescaled NLS equation. They solved (6.1.2) in the finite domain $[0, A]$ with the condition that at the boundary $\xi = A$, the solution takes the value given by (6.1.13) at each time step. This approximation allows a substantial reduction of the size of the spatial integration domain.

Within the finite region of integration $\xi < A$, the wave energy and the Hamiltonian

$$N_A = L(\tau)^{d-2/\sigma} \int_0^A |u(\tau)|^2 \xi^{d-1} d\xi,$$

$$H_A = L(\tau)^{-2-\frac{2}{\sigma}+d} \int_0^A \left(|\nabla u|^2 - \frac{1}{\sigma+1}|u|^{2\sigma+2}\right)\xi^{d-1}d\xi$$

are no longer constants of motion. However, the quantities

$$N_A + A^d \int_0^\tau L(\tau')^{d-2/\sigma} a(\tau')|u(A,\tau')|^2 d\tau' \qquad (6.1.14)$$

and

$$H_A + A^d \int_0^\tau L(\tau')^{-2-\frac{2}{\sigma}+d} a(\tau')\left(|\nabla u(A,\tau')|^2 - \frac{1}{\sigma+1}|u(A,\tau')|^{2\sigma+2}\right) d\tau'$$
$$\qquad (6.1.15)$$

are invariant.

An alternative method consists in mapping the spatial domain $[0, \infty)$ onto the interval $[-1, +1]$ by a transformation of the type $z = \frac{\ell-\xi}{\ell+\xi}$, where ℓ is an adjustable parameter. When a regular mesh is used for the z variable, half of the grid points are located between 0 and ℓ.

The dynamic rescaling method presented above enables numerical integrations up to times very close to the singularity with amplification factors of the maximum larger than 10^8, using moderate resolutions.

An alternative method recently used by Budd, Chen, and Russell (1999) involves discretization on a moving mesh in the interval $[0, L]$ (where L is taken large enough). The N grid points $R_i(t)$ are chosen to equidistribute a monitor function M of the solution $\psi(r, t)$, by imposing, in the simplest setting, the condition $\int_0^{R_i} M dr = \frac{i}{N} \int_0^L M dr$ for $i = 1, \ldots, n$. In more efficient schemes, the time evolution of the grid points is prescribed by an integro-differential equation of the form $\dot{R}_i = - \left(\int_{R_0}^{R_i} M dr - \frac{i}{N} \int_0^L M dr \right)$, $i = 1, \ldots, N$ (see Huang and Russell (1996) and Budd, Huang, and Russell (1996).

6.2   Simulation of Isotropic Collapse

6.2.1   Stability of Supercritical Self-Similar Solutions

The cubic NLS equation in three dimensions was integrated in McLaughlin, Papanicolaou, Sulem, and Sulem (1986) using the rescaled form (6.1.2) with the function $a(\tau)$ given by (6.1.10). This corresponds to a scaling factor $L(t)$ chosen such that the L^2-norm of the gradient of the rescaled solution $\int |\nabla u(\xi, \tau)|^2 \xi^2 d\xi$ remains constant (see (6.1.9)). Radially symmetric initial conditions associated to a negative Hamiltonian and to a finite or infinite variance were considered.

As τ tends to ∞ (or $t \to t_*$), $a(\tau)$ was observed to tend to a positive constant A, while $u(\xi, \tau)$ evolved up to a linear phase in τ to a fixed profile, as predicted by Zakharov (1984),

$$u(\xi, \tau) \to e^{iC\tau} S(\xi). \tag{6.2.1}$$

Consequently,

$$L(t) = A^{1/2}(t_* - t)^{1/2}, \tag{6.2.2}$$

and S satisfies

$$S_{\xi\xi} + \frac{2}{\xi} S_\xi - CS + iA(\xi S)_\xi + |S|^2 S = 0, \qquad \xi > 0. \tag{6.2.3}$$

Note that the asymptotic rate of blowup is rapidly reached and was observed for moderate values of the amplitude in direct numerical integration of the NLS equation (Budneva, Zakharov, and Synakh 1975, Goldman, Rypdal, and Hafizi 1980). Figure 6.1 illustrates for the initial condition $\psi_0 = 6 \exp(-|\mathbf{x}|^2)$, the evolution of $|u(\xi, \tau)|$ and its fast convergence to a limit profile decreasing monotonically with ξ.

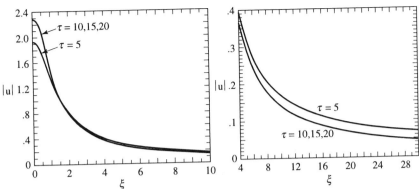

FIGURE 6.1. Convergence of the rescaled solution to a limit profile in three dimensions for $\sigma = 1$, at short and large distances (LeMesurier et al. 1987).

By rescaling $\tilde{\xi} = \sqrt{C}\xi$ and $\tilde{Q}(\tilde{\xi}) = \frac{1}{\sqrt{C}}S(\xi)$, this equation becomes, after dropping the tildes and defining $k = A/C$,

$$Q_{\xi\xi} + \frac{2}{\xi}Q_\xi - Q + ik(\xi Q)_\xi + |Q|^2 Q = 0, \qquad \xi > 0. \qquad (6.2.4)$$

This suggests that in the primitive variables, the solution near the singularity takes the asymptotic form

$$\psi(\mathbf{x}, t) = \frac{(2k)^{-\frac{1}{2}}}{(t_* - t)^{\frac{1}{2}}} Q\left(\frac{(2k)^{-\frac{1}{2}}|\mathbf{x}|}{(t_* - t)^{\frac{1}{2}}}\right) \exp\left(i\theta + i\frac{1}{2k}\ln\frac{t_*}{t_* - t}\right). \qquad (6.2.5)$$

where Q is a solution of (6.2.4) with a profile $|Q|$ decreasing monotonically. A remarkable fact is that for the different initial conditions considered in the numerical simulations, the ratio k always has the same value, $k \approx 0.917\ldots$. Other computations performed by LeMesurier, Sulem, Sulem, and Papanicolaou 1987, Kosmatov, Petrov, Shvets, and Zakharov (1988) and Kosmatov, Shvets, and Zakharov (1991) confirm the local form of the solution near the singularity and the value of the constant k. Properties of solutions of equation (6.1.6) for the self-similar profile are discussed in Chapter 7.

6.2.2 The NLS Equation at Critical Dimension

Numerical simulations were performed using the dynamic rescaling presented in Section 6.1.2 and various choices for the scaling factor $L(t)$. A simple condition is to prescribe that $|\nabla u(\tau)|_{L^q}$ with $q > d$, stays constant. It follows that the rescaled solution $|u(\tau)|_{L^\infty}$ remains bounded as a consequence of the L^2-norm conservation and the Gagliardo–Nirenberg inequality

$$|u|_{L^\infty} \leq C|\nabla u|_{L^q}^\alpha |u|_{L^2}^{1-\alpha}, \qquad \frac{1}{2} = \alpha\left(\frac{1}{2} - \frac{1}{q} + \frac{1}{d}\right), \qquad 0 < \alpha < 1. \qquad (6.2.6)$$

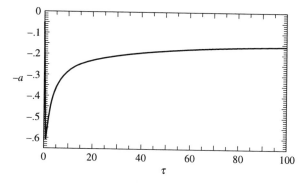

FIGURE 6.2. Typical evolution of $-a(\tau)$ at critical dimension $d = 2$, $\sigma = 1$ (LeMesurier *et al.* 1987).

In this case,

$$L(t) = \left(\frac{|\nabla u_0|_{L^q}}{|\nabla u(\tau)|_{L^q}}\right)^{q/p}, \qquad p = q + \frac{q}{\sigma} - d, \qquad (6.2.7)$$

and

$$a(\tau) = \frac{q}{p|\nabla u_0|_{L^q}^q} \int \{\nabla|\nabla u|^{q-2}\Im(\nabla u^*\Delta u) - |\nabla u|^{q-2}\Im(\nabla u^*u)\nabla(|u|^{2\sigma})\}\xi^{d-1}d\xi.$$

$$(6.2.8)$$

Numerical computations for the cubic NLS equation in two space dimensions with initial conditions corresponding to a negative Hamiltonian and a finite or infinite variance show that the quantity $a(\tau)$ remains positive and thus that $L(t)$ is decreasing. After an initial transient, $a(\tau)$ decays to zero, at a very slow rate, as illustrated in Fig. 6.2 in the case of the initial condition $\psi_0 = 4\exp(-|\mathbf{x}|^2)$. Figure 6.3 shows the corresponding evolution of $|u(\xi, \tau)|$, whose convergence to a fixed profile appears to be limited to

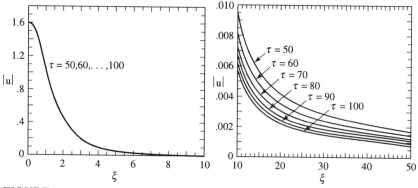

FIGURE 6.3. Evolution of the rescaled profile at critical dimension $d = 2$, $\sigma = 1$ showing a persistent slow time-dependency at large distance (LeMesurier *et al.* 1987).

relatively short distances. At larger distances, the profile evolves adiabatically while $a(\tau)$ continue to decrease. Furthermore, the phase of u at the
origin tends to become linear in τ.

The continuous and slow decay of $a(\tau)$ with no indication of convergence
to a nonzero value as the singularity is approached suggests that the rate
of blowup $L(t)$ displays a weak correction to the $(t_* - t)^{1/2}$ law and that
strict self-similarity is broken. An asymptotic analysis of this behavior is
presented in Chapter 8. It predicts a log-log corrective factor to the above
scaling together with the profile of the collapsing pulse. It is noticeable
that even at critical dimension, the singularity time obtained numerically
with generic initial conditions is significantly shorter than the upper bound
given by the variance identity. The reason is that not all the wave energy
gets localized near the focus when the singularity occurs.

6.2.3 An Adaptive Galerkin Finite Element Method

More recently, an alternative numerical method was introduced by Akrivis,
Dougalis, Karakashian, and McKinney (1993, 1997). The radially symmetric NLS equation is solved in the primitive variables, using an adaptive
scheme where both the mesh size and the time step are refined as the
singularity is approached. The solution is considered on a finite radial domain $0 \leq r \leq A$, beyond which the amplitude of the solution is negligible.
The spatial discretization is done with a Galerkin finite element method
that uses continuous piecewise polynomial functions in the radial variable.
The temporal discretization is made using an implicit Runge–Kutta scheme
of high order. A rigorous analysis of the stability and convergence of this
method in Cartesian coordinates in the absence of mesh refinement is given
in Akrivis, Dougalis, and Karakashian (1993) and Karakashian, Akrivis,
and Dougalis (1993). Anticipating that the solution blows up at the origin
as t approaches t_*, the above algorithm is in fact implemented in a code
where the spatial resolution near $r = 0$ and the temporal resolution are
drastically refined as the singularity is approached. When the criterion for
time step reduction is met, the time step is halved, and the spatial mesh
also refined by increasing the number of points in the vicinity of the origin as follows: Suppose that the spatial grid has already been refined N_s
times, then the interval $[0, A]$ is partitioned in $N_s + 2$ adjacent successive
intervals $I_0, I_1, \ldots, I_{N_s+1}$ such that the left-hand boundary of I_0 is 0, the
right-hand boundary of I_{N_s-1} is A, and on each I_j the mesh is uniform.
Specifically, if N and $M < 2N$ are given integers and $h = 1/N$ is the
initial mesh length, then I_0 consists of M subintervals of constant mesh
length $h/2^{N_s+1}$, while $I_1, I_2, \ldots, I_s$ consist of $M/2$ subintervals of constant
mesh length $h/2^{N_s}, h/2^{N_s+1}, \ldots, h/2$, respectively and I_{N_s+1} consists of
$N - M/2$ subintervals of mesh size h. Thus each time the grid is refined,
I_0 is cut in half into two new intervals I_0 and I_1, and all the other regions
are redefined so that I_1 becomes I_2, I_2 becomes I_3, etc. The approximated

solution is embedded in the new mesh by linear interpolation. For this purpose, the interval $[0, A]$ is partitioned into successive adjacent intervals. Suppose that it has already been split N_s times into $I_0, I_1, \ldots, I_{N_s}$ with $M, M/2, \ldots, M/2^{N_s}$ points in each of them, respectively.

Spatial grid refinement is implemented (i.e., N_s is replaced by N_s+1) if on the interval I_0, one has $|\psi|_{L^\infty(I_0)} h_{\min}^{1/2}/|\psi|_{L^2(I_0)} > c_h$ where $h_{\min} = h/2^{N_s+1}$. The spatial tolerance factor c_h was taken equal to 0.14 in dimension $d = 3$ and 0.12 in dimension $d = 2$. The time step is halved if between two successive time steps, the variation of the Hamiltonian normalized by the square L^2-norm of the radial gradient exceeds a value of order 10^{-8}.

When this method is implemented in two or three dimensions with $N_s = 35$ mesh refinements, the singularity is approached so closely that the amplitude at the origin has been amplified by a factor close to 10^{11}. In this case, the far asymptotic regime is reached, even at critical dimension. The authors then fitted the amplitude at the origin and several other norms of the computed solution or of its spatial gradient against theoretical predictions. For this purpose, they estimate the singularity time t_* by the largest value of t that a particular run can reach beyond which the adaptive mechanism is no longer able to refine the spatial mesh. In three space dimensions a blowup factor $L^{-1} \propto (t_* - t)^{-1/2}$ is easily confirmed (see also Tourigny and Sanz-Serna 1992).

At critical dimension $d = 2$, the computed solution was tested against several theoretical predictions. In particular, its L^∞-norm at the times t_i of the successive spatial refinements was fitted with scaling laws of the form

$$\frac{1}{L} = \left(\frac{F(t_* - t)}{t_* - t}\right)^\rho \tag{6.2.9}$$

with $F(s) = 1, |\ln s|, |\ln s|^{0.6}, |\ln s|^{0.5}, |\ln s|^{0.4}$, and $\ln|\ln s|$ in order to select the functional form for which the exponent ρ has minimal fluctuations. We reproduce below part of a table taken from their paper and showing this comparison. Denoting by A_i the value of $|\psi(0, t)|$ at the refinement times t_i, the exponent ρ is approximated by

$$\rho_i = \ln\left(\frac{A_i}{A_{i+1}}\right) \bigg/ \ln\left(\frac{F_i/(t_* - t_i)}{F_{i+1}/(t_* - t_{i+1})}\right), \tag{6.2.10}$$

where $F_i = F(t_* - t_i)$. Columns (a), (b),...,(f) are the values of ρ_i calculated respectively with the different functional form of F listed above at time t_i. The last two rows of each column show the mean value and the standard deviation. The rate of column (f), which corresponds to the double log, clearly stabilizes very close to $\frac{1}{2}$ better than any other law.

i	(a)	(b)	(c)
19	0.50553	0.48947	0.49577
20	0.50514	0.48976	0.49580
21	0.50479	0.49004	0.49583
22	0.50448	0.49030	0.49587
23	0.50422	0.49057	0.49594
24	0.50396	0.49081	0.49599
25	0.50372	0.49103	0.49603
26	0.50351	0.49126	0.49609
27	0.50323	0.49138	0.49605
28	0.50292	0.49145	0.49598
mean	0.50415	0.49061	0.49593
standard deviation	0.844×10^{-3}	0.698×10^{-3}	0.110×10^{-3}

i	(d)	(e)	(f)
19	0.49737	0.49898	0.50072
20	0.49733	0.49888	0.50060
21	0.49730	0.49878	0.50049
22	0.49729	0.49871	0.50039
23	0.49730	0.49867	0.50033
24	0.49730	0.49862	0.50026
25	0.49729	0.49856	0.50018
26	0.49731	0.49854	0.50013
27	0.49723	0.49842	0.49999
28	0.49712	0.49827	0.49982
mean	0.49729	0.49864	0.50029
standard deviation	0.672×10^{-3}	0.211×10^{-3}	0.276×10^{-3}

The authors conclude that the exponent $\rho = 1/2$ and a double logarithm
for the corrective factor F ensure a fit with high accuracy. These simulations thus provide a strong support to the asymptotic analysis presented
in Chapter 8, which predicts an asymptotic behavior given, up to finite
rescaling factors, by

$$\psi(r,t) \approx \frac{1}{L(t)} R\left(\frac{r}{L(t)}\right) \exp\left[i\tau(t) + \frac{iL_t(t)r^2}{4L(t)}\right], \qquad (6.2.11)$$

where R refers to the ground state and

$$L(t) \approx \left(\frac{t_* - t}{\ln|\ln(t_* - t)|}\right)^{1/2}, \quad \tau(t) \approx \frac{1}{2\lambda}|\ln(t_* - t)|\ln|\ln(t_* - t)|, \quad (6.2.12)$$

with $\lambda = \pi$. The numerics reported in Akrivis, Dougalis, Karakashian, and
McKinney (1997) reproduces this prediction, including an accurate value
of the constant λ when the next term is included in the expansion of $\tau(t)$
in the form

$$\tau(t) \approx \frac{1}{2\pi}|\ln(t_* - t)|(\ln|\ln(t_* - t)| + 4\ln\ln|\ln(t_* - t)|. \qquad (6.2.13)$$

6.3 Simulation of Non-Radially Symmetric Solutions

6.3.1 Anisotropic Dynamic Rescaling

In order to follow the formation of singularities for initial data that are not radially symmetric, the dynamic rescaling method was extended to anisotropic solutions. It remains, however, restricted to problems where the solution blows up at only one point. The method allows scaling, rotation, and translation of the coordinate axes, in such a way that in the transformed variables, the solution remains regular and only moderately anisotropic. The behavior of the scaling factors in various directions determine the associated rate of collapse and the nature of the singularity in the primitive variables.

Consider the NLS equation in dimension d. Let $D(t)$ be a $d \times d$ matrix, $\mathbf{x}_0(t)$ a vector function and $L(t)$ a nonnegative scalar function. We introduce the change of variables

$$\boldsymbol{\xi} = D^{-1}(t)(\mathbf{x} - \mathbf{x}_0), \quad \tau = \int_0^t \frac{1}{L^2(s)} ds, \quad u(\boldsymbol{\xi}, \tau) = L(t)^{1/\sigma} \psi(\mathbf{x}, t),$$

$$(6.3.1)$$

where the matrix $D(t)$ is written the form

$$D(t) = O^T \Lambda(t), \tag{6.3.2}$$

with $O(t)$ an orthogonal matrix and $\Lambda(t)$ a diagonal matrix whose diagonal elements are denoted by λ_i ($i = 1, \ldots, d$). Noticing that $D^T D = \Lambda^2$, the above change of variables leads to the rescaled NLS equation

$$i(u_\tau - \frac{1}{\sigma} L^{-1} L_\tau u + \mathbf{f}.\nabla u) + L^2 \sum_{i=1}^d \frac{1}{\lambda_i(t)^2} u_{\xi_i \xi_i} + |u|^{2\sigma} u = 0, \tag{6.3.3}$$

where

$$\mathbf{f} = -D^{-1} \frac{dD}{d\tau} \boldsymbol{\xi} - D^{-1} \frac{d\mathbf{x}_0}{d\tau}.$$

Note that D, L, and $\mathbf{x}_0$ may be viewed as functions of t or τ. They are chosen so that the rescaled solution u is nonsingular. This can be done in several ways. One possibility is the following. Let p a positive integer. We define $\mathbf{x}_0(t)$ as the centroid

$$x_{0i}(t) = \frac{\int x_i |\psi|^{2p} dx}{\int |\psi|^{2p} dx}, \tag{6.3.4}$$

where p is chosen to ensure accuracy in the numerical computation of the integrals (typically $p \geq 3$). This point approaches the blowup point (assumed to be unique) near the singularity time. We choose $D(t)$ such that the second moment of $|u|^{2p}$ is the identity matrix, i.e.,

$$\frac{\int \xi_i \xi_j |u|^{2p} d\xi}{\int |u|^{2p} d\xi} = \delta_{ij}, \tag{6.3.5}$$

or, equivalently, using (6.3.1),

$$D^{-1}S(D^{-1})^T = I, \tag{6.3.6}$$

where $S = (s_{ij})$ is defined by

$$s_{ij} = \frac{\int (x_i - x_{0i})(x_j - x_{0j})|\psi|^{2p} dx}{\int |\psi|^{2p} dx}. \tag{6.3.7}$$

We also prescribe

$$\sum \frac{1}{\lambda_i^2} = \frac{d}{L(t)^2}, \tag{6.3.8}$$

which makes the coefficients of the second-order derivative terms in (6.3.3) bounded. From (6.3.2) and (6.3.6) we have the decomposition $S = O^T \Lambda^2 O$. If S has distinct eigenvalues (in particular if ψ is not radially symmetric), O is unique. To see how $\mathbf{x}_0$, D, and L vary with time, we differentiate (6.3.3)–(6.3.6) with respect to the rescaled time τ. From (6.3.3), using (6.3.1), we get

$$\frac{d\mathbf{x}_0}{d\tau} = 2D\beta, \tag{6.3.9}$$

where $\beta = (\beta_j)(j = 1, \ldots, d)$ and

$$\beta_j = \frac{p \int \xi_j |u|^{2(p-1)} \Im(L^2 \sum_i \lambda_i(t)^{-2} u_{\xi_i \xi_i} u^*) d\xi}{\int |u|^{2p} d\xi}. \tag{6.3.10}$$

Differentiating (6.3.7) and using (6.3.3) and (6.3.1) give

$$\frac{dS}{d\tau} = -2DAD^T, \tag{6.3.11}$$

where $A = (a_{ij})(i, j = 1, \ldots, d)$ with

$$a_{ij} = \frac{p \int (\delta_{ij} - \xi_i \xi_j)|u|^{2(p-1)} \Im(L^2 \sum_i \lambda_i(t)^{-2} u_{\xi_i \xi_i} u^*) d\xi}{\int |u|^{2p} d\xi}. \tag{6.3.12}$$

But from (6.3.6), $S = DD^T = O^T \Lambda^2 O$, so that we have

$$\frac{dS}{d\tau} = \frac{d}{d\tau}(DD^T) = \dot{O}^T \Lambda^2 O + 2O^T \Lambda \dot{\Lambda} O + O^T \Lambda^2 \dot{O} \tag{6.3.13}$$

and thus

$$O\dot{S}O^T = O\dot{O}^T \Lambda^2 + 2\Lambda \dot{\Lambda} + \Lambda^2 \dot{O}O^T, \tag{6.3.14}$$

where the dot denotes the derivative with respect to τ. Then, using (6.3.11), we have

$$-2\Lambda A \Lambda = O\dot{O}^T \Lambda^2 + 2\Lambda \dot{\Lambda} + \Lambda^2 \dot{O}O^T. \tag{6.3.15}$$

From this equation we can derive a system of decoupled evolution equations for Λ and O. Setting

$$O\dot{O}^T = G = (g_{ij}), \tag{6.3.16}$$

which is skew-symmetric, we have $g_{ii} = 0$. Equating the diagonal elements of (6.3.15), we obtain

$$\dot{\lambda}_i = -\lambda_i a_{ii}. \tag{6.3.17}$$

Equating the off-diagonal elements, we get

$$-2\lambda_i\lambda_j a_{ij} = \lambda_j^2 g_{ij} - \lambda_i^2 g_{ij}, \tag{6.3.18}$$

so that

$$g_{ij} = \frac{2\lambda_i\lambda_j}{\lambda_i^2 - \lambda_j^2}a_{ij} \qquad \text{for} \quad \lambda_i \neq \lambda_j. \tag{6.3.19}$$

From (6.3.18), we have that if $\lambda_i = \lambda_j$ for $i \neq j$, then $a_{ij} = 0$ and

$$\lim_{\lambda_i \to \lambda_j} g_{ij} = \lim_{\lambda_i \to \lambda_j} \frac{2\lambda_i\lambda_j}{\lambda_i^2 - \lambda_j^2}a_{ij}. \tag{6.3.20}$$

It then follows from (6.3.16) that

$$\frac{dO}{d\tau} = -GO, \tag{6.3.21}$$

where G is given by (6.3.19).

Finally, we derive an evolution equation for L. We differentiate (6.3.8) with respect to τ and use (6.3.17), which gives

$$\frac{1}{L}\frac{dL}{d\tau} = -\frac{\sum a_{ii}/\lambda_i^2}{\sum 1/\lambda_i^2}. \tag{6.3.22}$$

Using (6.3.9), (6.3.17), (6.3.19) and (6.3.22), the vector $\mathbf{f}$ in (6.3.3) can be rewritten as

$$\mathbf{f} = B\boldsymbol{\xi} - 2\boldsymbol{\beta}, \tag{6.3.23}$$

where $B = (b_{ij}) = \frac{dD}{d\tau}D^{-1} = -\dot{\Lambda}\Lambda^{-1} + \Lambda^{-1}\dot{O}O^T\Lambda$, i.e.,

$$\begin{cases} b_{ii} = -\dot{\lambda}_i\lambda_i^{-1} = a_{ii}, \\ b_{ij} = -\lambda_i^{-1}g_{ji}\lambda_j = \dfrac{2\lambda_j^2}{\lambda_j^2 - \lambda_i^2}a_{ij} \ (i \neq j), \end{cases} \tag{6.3.24}$$

and $\boldsymbol{\beta}$ is given by (6.3.10).

In summary, we get a coupled system of evolution equations for the transformed solution u and the scaling factors λ_i in the form

$$iu_\tau = i(\frac{1}{\sigma}L^{-1}L_\tau u - \mathbf{f}\cdot\nabla u) - L^2\sum_{i=1}^d\frac{1}{\lambda_i(t)^2}u_{\xi_i\xi_i} - |u|^{2\sigma}u, \tag{6.3.25}$$

$$\frac{d\lambda_i}{d\tau} = -a_{ii}\lambda_i \ (i = 1,\ldots,d). \tag{6.3.26}$$

Here $1/L^2(t) = \frac{1}{d}\sum 1/\lambda_i^2$, or equivalently,

$$L^{-1}L_\tau = -\frac{\sum a_{ii}/\lambda_i^2}{\sum 1/\lambda_i^2}, \qquad (6.3.27)$$

and $\mathbf{f} = B\boldsymbol{\xi} - 2\boldsymbol{\beta}$, where B and $\boldsymbol{\beta}$ are given by (6.3.24) and (6.3.10), respectively, while a_{ij} is given by (6.3.12). We also have the system of evolution equations for the centroid $\mathbf{x}_0$ and the rotation matrix O,

$$\frac{d\mathbf{x}_0}{d\tau} = 2O^T\Lambda\boldsymbol{\beta} \qquad (6.3.28)$$

$$\frac{dO}{d\tau} = -GO, \qquad (6.3.29)$$

where G is given by (6.3.19).

We note that (6.3.25) and (6.3.26) form a closed system, so that u and λ_i are determined by these equations alone without requiring computations of the rotation O and of the centroid $\mathbf{x}_0$. This shows that the translation and rotation are not fundamental in the singularity formation. It is the local scaling that determines the collapse of the solution ψ. The quantities $\mathbf{x}_0$ and O can be computed from (6.3.28) and (6.3.29) once u and λ_i have been obtained.

In the radially symmetric case, we have $\mathbf{x}_0=0$ and $a_{ij} = 0$ for $i \neq j$, while the a_{ii} are all equal. We denote them by $a = a_{ii}$. Therefore $\lambda_i = \lambda$, $L = \sqrt{d/(\sum 1/\lambda_i^2)} = \lambda$, $L^{-1}L_\tau = -a$, and $\mathbf{f} = a\boldsymbol{\xi}$. Equation (6.3.25) becomes

$$iu_\tau + u_{\xi\xi} + \frac{d-1}{\xi}u_\xi + ia(\xi u_\xi + \frac{u}{\sigma}) + |u|^{2\sigma}u = 0, \qquad (6.3.30)$$

where ξ is the radial coordinate and a is given by

$$a = \frac{p\int(1-\xi^2)|u|^{2(p-1)}\Im(L^2\Lambda^{-2} : u\boldsymbol{\nabla}\boldsymbol{\nabla})u^*d\boldsymbol{\xi}}{\int|u|^{2p}d\boldsymbol{\xi}}. \qquad (6.3.31)$$

To solve (6.3.29) numerically, we express O in terms of the Euler angles ϕ, θ, ψ in the form

$$O = \begin{pmatrix} -\sin\phi\sin\theta + \cos\psi\cos\phi\cos\theta & \cos\phi\sin\theta + \cos\psi\sin\phi\cos\theta & -\sin\psi\cos\theta \\ -\sin\phi\cos\theta - \cos\psi\cos\phi\sin\theta & \cos\phi\cos\theta - \cos\psi\sin\phi\sin\theta & \sin\psi\cos\theta \\ \sin\psi\cos\phi & \sin\psi\sin\phi & \cos\psi \end{pmatrix}.$$

From this expression and (6.3.19), we solve for $\dot\phi, \dot\theta, \dot\psi$, and obtain

$$\dot\phi = \frac{2}{\sin\psi}\left(\frac{\lambda_1\lambda_3}{\lambda_1^2 - \lambda_3^2}a_{13}\sin\theta + \frac{\lambda_2\lambda_3}{\lambda_2^2 - \lambda_3^2}a_{23}\cos\theta\right),$$

$$\dot\theta = -\frac{2\lambda_1\lambda_2}{\lambda_1^2 - \lambda_2^2}a_{12} - \frac{2\cos\psi}{\sin\psi}\left(\frac{\lambda_1\lambda_3}{\lambda_1^2 - \lambda_3^2}a_{13}\sin\theta + \frac{\lambda_2\lambda_3}{\lambda_2^2 - \lambda_3^2}a_{23}\cos\theta\right),$$

$$\dot\psi = 2\left(\frac{\lambda_1\lambda_3}{\lambda_1^2 - \lambda_3^2}a_{13}\cos\theta - \frac{\lambda_2\lambda_3}{\lambda_2^2 - \lambda_3^2}a_{23}\sin\theta\right). \qquad (6.3.32)$$

In two dimensions, the rotation is defined by a single angle θ,

$$O = \begin{pmatrix} \cos\theta & \sin\theta \\ -\sin\theta & \cos\theta \end{pmatrix}, \tag{6.3.33}$$

and making use of (6.3.19), we get

$$\dot\theta = -\frac{2\lambda_1\lambda_2}{\lambda_1^2 - \lambda_2^2}a_{12}. \tag{6.3.34}$$

6.3.2   Stability of Isotropic Collapse

A main result of the numerical simulations of focusing cubic NLS equation in $\mathbf{R}^2$ and $\mathbf{R}^3$ with anisotropic initial conditions, using the dynamic rescaling method, is that initially anisotropic one-peak solutions rapidly become isotropic near the singularity both in two and three dimensions (Landman, Papanicolaou, Sulem, Sulem, and Wang 1991). In the latter case, this result confirms the numerical linear stability analysis of isotropic collapse with respect to anisotropic disturbances performed by Vlasov, Piskunova, and Talanov (1989). Anisotropic collapsing solutions were formally constructed (Pelletier 1987) but not observed in the simulations.

When considering a radially symmetric initial condition with a ring-type shape, one again observes that the collapse takes place at the origin. In this case, if the maximum of the initial conditions is sufficiently far from the origin, the curvature term in the Laplacian is initially negligible. For short times, the solution evolves "one-dimensionally," shedding wave packets. It is only when a sufficient amount of energy has moved to the neighborhood of the origin that the curvature term in the Laplacian becomes relevant and the singularity develops (Budneva, Zakharov and Synakh 1975, Sulem, Sulem, and Patera (1983), Akrivis, Dougalis, Karakashian, and McKinney 1997).

Initial conditions corresponding to distinct peaks of different powers were considered by Bergé, Schmidt, Rasmussen, Christiansen, and Rasmussen (1997) in a two-dimensional periodic domain, using spectral methods. It turns out that the condition for the structures to amalgamate (i.e., fuse and self-focus into a central lobe) depends not only on their mass and their separation but also on their relative phases.

Direct extension of the dynamic rescaling method to multipeak solutions seems delicate. The multidimensional adaptive codes of Karakashian and Plexousakis for non-radially-symmetric solutions of the NLS equation announced in Akrivis, Dougalis, Karakashian, and McKinney (1997) seem more promising. Preliminary results for initial conditions leading to the formation of several foci confirm the local isotropy of each structure (Dougalis, private communication).

6.3.3   An Iterative Grid Redistribution Method

Very recently, an algorithm for multi-dimensional singular problems in an arbitrary geometry was developed by Ren and Wang (1999) and implemented to the simulation of the NLS equation in two dimensions in a bounded domain. We review here this algorithm which combines a dynamic rescaling of the solution amplitude ψ and of the time variable t, in the form

$$u(\xi, \eta, \tau) = \lambda(t)\psi(x, y, t) \tag{6.3.35}$$

$$\frac{d\tau}{dt} = \frac{1}{\lambda^2(t)} \tag{6.3.36}$$

where the scaling factor $\lambda(t)$ is chosen as $\lambda(t) = 1/|\psi|_{L^\infty(|\Omega_p)}$, with an adaptive mesh redistribution method based on the transformation of the space coordinates

$$(x, y) \in \Omega_p \mapsto (\xi(x, y), \eta(x, y)) \in \Omega_c, \tag{6.3.37}$$

from the physical domain Ω_p to the computational domain Ω_c, (both chosen to be $[-1, +1] \times [-1, +1]$). This transformation is chosen to insure that the maximum amplitude of the gradient of the solution does not exceed a prescribed value M.

In the new variables, the NLS equation becomes

$$u_\tau - \frac{\lambda_\tau}{\lambda}u - i(\lambda^2\Delta_B u + |u|^2 u) = 0 \tag{6.3.38}$$

with

$$\lambda_\tau = \lambda^3 \Im(u^* \Delta_B u)\Big|_{(\xi_0, \eta_0)} \tag{6.3.39}$$

where (ξ_0, η_0) is the maximum point of $|u(\xi, \eta, \tau)|$. The operator Δ_B is defined by

$$\Delta_B u = \frac{1}{J}\left[\frac{\partial}{\partial\xi}\left(\frac{b_{22}u_\xi - b_{12}u_\eta}{J}\right) + \frac{\partial}{\partial\eta}\left(\frac{b_{11}u_\eta - b_{12}u_\xi}{J}\right)\right] \tag{6.3.40}$$

where

$$b_{11} = x_\xi^2 + y_\xi^2, \quad b_{12} = x_\xi x_\eta + y_\xi y_\eta, \quad b_{22} = x_\eta^2 + y_\eta^2, \tag{6.3.41}$$

and $J = x_\xi y_\eta - x_\eta y_\xi$ is the Jacobian of the coordinate transformation.

Choosing a monitor function of the form

$$w(x, y) = (1 + 2|\psi(x, y)|^2 + |\nabla\psi(x, y)|^2), \tag{6.3.42}$$

the transformation (6.3.37) is determined by solving

$$\nabla \cdot \left(\frac{1}{w}\nabla\xi\right) = 0 \tag{6.3.43}$$

$$\nabla \cdot \left(\frac{1}{w}\nabla\eta\right) = 0, \tag{6.3.44}$$

These equations are viewed as Euler-Lagrange equations whose solutions minimize the functional

$$E(\xi, \eta) = \int_{\Omega_c} \frac{1}{w} \left(|\nabla \xi|^2 + |\nabla \eta|^2 \right) dx dy, \qquad (6.3.45)$$

which insures an approximate "equidistribution" of the grid points in the rescaled variables. The solution of the above elliptic system is obtained as the steady state of the heat equation

$$\frac{\partial \xi}{\partial t} - \nabla \cdot \left(\frac{1}{w} \nabla \xi \right) = 0 \qquad (6.3.46)$$

$$\frac{\partial \eta}{\partial t} - \nabla \cdot \left(\frac{1}{w} \nabla \eta \right) = 0, \qquad (6.3.47)$$

Since the resulting mesh should concentrate more points in the regions of large variations, the transformed solution should also be better behave than the original function. In fact, this improvement may in some instances be very limited. The authors suggest to improve the result by defining an "iterative remeshing procedure" where the same procedure is repeated on the transformed solution, until a satisfactory form is obtained. The computations then proceeds as follows.

(i) Given an initial condition $\psi(x, y, 0)$, the initial grid transform is determined by the iterative remeshing, and the initial condition $u(\xi, \eta, 0)$ in the computational domain constructed.

(ii) The (rescaled) NLS equation is solved in the computational variables, with the grid transformation $x(\xi, \eta), y(\xi, \eta)$ being fixed, until some time τ_0 where the solution $u(\xi, \eta, \tau)$ cannot meet the criterion that its gradient remains smaller than M (usually taken equal to 5 or 7).

(iii) At this moment, a new mesh is generated by the iterative meshing, starting with $u(\xi, \eta, \tau_0)$. The interpolation is then used to move the solution on the new grids.

(iv) The integration is continued as described in (ii).

As noted by the authors, in most of the cases, only one remeshing iteration is needed when starting the iteration form $u(\xi, \eta, \tau)$ in step (iii).

This method which appears to be efficient and relatively easy to implement, was used to confirm results obtained by the usual dynamic rescaling such as stability of the isotropic collapse. It was also applied to simulate solutions which blow up at two different points, because of a symmetry which is accurately preserved by the simulation. It turns out that the structure of each singularity is the same as that of the solution with a single blowup point.

7
Supercritical Collapse

7.1 Self-Similar Blowup Solutions

As discussed in Section 6.2.1, the numerical simulations of the cubic NLS equation in three space dimensions clearly suggest that in supercritical dimension, blow-up solutions approach asymptotically self-similar solutions with radial symmetry. Up to a simple rescaling of the dependent and independent variables, one asymptotically has

$$\psi(\mathbf{x}, t) \approx \frac{1}{(2a(t_* - t))^{1/2\sigma}} Q\left(\frac{|\mathbf{x}|}{(2a(t_* - t))^{1/2}}\right) \exp\left(i\left(\theta + \frac{1}{2a}\ln\frac{t_*}{t_* - t}\right)\right),$$
$$(7.1.1)$$

where a is a strictly positive constant whose value depends only on d and σ. Furthermore, the (complex) function Q satisfies the ordinary differential equation

$$Q_{\xi\xi} + \frac{d-1}{\xi}Q_\xi - Q + ia\left(\frac{Q}{\sigma} + \xi Q_\xi\right) + |Q|^{2\sigma}Q = 0, \qquad \xi \in \mathbf{R}^+, \quad (7.1.2)$$

with the boundary conditions

$$Q_\xi(0) = 0 \qquad Q(\xi) \to 0 \quad \text{as} \quad \xi \to \infty. \qquad (7.1.3)$$

and the constraint that the profile $|Q(\xi)|$ decreases monotonically with ξ.

Here, we use the notation a for the constant denoted by k in the previous chapter. Noticing that (7.1.2) is phase invariant, we can fix the phase by imposing $Q(0)$ real.

7.1.1   Properties of the Profile

There is no complete mathematical theory of (7.1.2)–(7.1.3). In dimension $2 < d < 4$ with cubic nonlinearity, for any value of the parameter a and any prescribed value of $Q(0)$, (7.1.2)–(7.1.3) was shown to admit a unique non-trivial solution (Wang 1990, Budd, Chen, and Russell 1999). Furthermore, one has the following properties (LeMesurier, Papanicolaou, Sulem, and Sulem 1988a).

Proposition 7.1.   *The large-distance behavior* $(\xi \to \infty)$ *of the solutions* Q *of (7.1.2) is given by* $Q = \alpha Q_1 + \beta Q_2$, *where* $Q_1 \approx |\xi|^{-\frac{i}{a}-\frac{1}{\sigma}}$, $Q_2 \approx e^{-\frac{ia\xi^2}{2}} |\xi|^{\frac{i}{a}-d+\frac{1}{\sigma}}$ *and* α *and* β *denote complex numbers.*

Proof.   We write $Q(\xi) = X(\xi)Z(\xi)$ and choose the function X such that after substitution in (7.1.2), the resulting equation for Z does not contain first-order derivatives. It gives $X(\xi) = e^{-ia\xi^2/4}\xi^{(1-d)/2}$ and

$$- Z'' + \left(-\frac{a^2}{4}\xi^2 + 1 - ia\frac{d\sigma-2}{2\sigma} - |\xi|^{(1-d)\sigma}|Z|^{2\sigma} + \frac{(d-1)(d-3)}{4\xi^2}\right)Z = 0.$$
$$(7.1.4)$$

We then write $Z = e^w$ and get

$$w'' + w'^2 + \frac{a^2}{4}\xi^2 - 1 + ia\frac{d\sigma-2}{2\sigma} - \frac{(d-1)(d-3)}{4\xi^2} = O(e^{2\sigma w}\xi^{(1-d)\sigma}). \quad (7.1.5)$$

This equation has two solutions, which in the limit $\xi \to \infty$ are given by

$$w_1 \approx ia\frac{\xi^2}{4} - \frac{i}{a}\ln|\xi| - \left(\frac{1-d}{2} + \frac{1}{\sigma}\right)\ln|\xi|, \quad (7.1.6)$$

$$w_2 \approx -ia\frac{\xi^2}{4} + \frac{i}{a}\ln|\xi| + \left(\frac{1-d}{2} + \frac{1}{\sigma}\right)\ln|\xi|. \quad (7.1.7)$$

The proposition is obtained by substituting these asymptotic behaviors in $Q_i(\xi) = X(\xi)e^{w_i}$. Note that $Q_{1\xi}$ and Q_2 belong to $L^2(\mathbf{R}^d)$ but that Q_1 and $Q_{2\xi}$ do not.

   Another derivation of the behavior of solutions of (7.1.2) can be obtained by introducing $v = \xi^{(d-2)/2}Q$. The resulting equation for v has a regular singular point at infinity, and it is possible to write its solutions explicitly in terms of contour integrals (Lochak 1990). Johnson and Pan (1993) also derived the large-distance behavior of the solutions with a precise estimate of the corrective terms.

Proposition 7.2.   *If* Q *is a solution of (7.1.2) with* $Q_\xi \in L^2(\mathbf{R}^d)$ *and* $Q \in L^{2\sigma+2}(\mathbf{R}^d)$, *and* $d \neq 2 + 2/\sigma$, *its Hamiltonian vanishes:*

$$\int \left(|Q_\xi|^2 - \frac{1}{\sigma+1}|Q|^{2\sigma+2}\right)\xi^{d-1}d\xi = 0. \quad (7.1.8)$$

Equivalently, $P = e^{ia\xi^2/4}Q$ obeys

$$\int \left(|P_\xi|^2 - \frac{1}{\sigma+1}|P|^{2\sigma+2} + a\Im(\xi PP_\xi^*) + \frac{a^2\xi^2}{4}|P|^2 \right) \xi^{d-1}d\xi = 0. \quad (7.1.9)$$

Proof. Multiply (7.1.2) by ΔQ^*, take imaginary parts, and integrate. This gives

$$a\Re \int \left(\frac{Q}{\sigma} + \xi Q_\xi \right) \Delta Q^* \xi^{d-1}d\xi + \Im \int |Q|^{2\sigma}Q\Delta Q^*\xi^{d-1}d\xi = 0. \quad (7.1.10)$$

The first integral in (7.1.10) can be transformed into

$$a\left(\frac{d}{2}-1-\frac{1}{\sigma}\right) \int |Q_\xi|^2\xi^{d-1}d\xi. \quad (7.1.11)$$

After using (7.1.2) to express ΔQ^*, the second integral in (7.1.10) is rewritten in the form

$$-\frac{a}{\sigma+1}(\frac{d}{2}-1-\frac{1}{\sigma}) \int |Q|^{2\sigma+2}\xi^{d-1}d\xi. \quad (7.1.12)$$

Combining these two transformations leads to (7.1.8), provided that $d \neq 2 + 2/\sigma$.

Proposition 7.3. *If Q is a solution of (7.1.2) and $Q \in H^1(\mathbf{R}^d) \cap L^{2\sigma+2}(\mathbf{R}^d)$, then $Q \equiv 0$.*

Proof: Multiplying (7.1.2) by Q^*, taking the imaginary part, and integrating over $\mathbf{R}^d$, we get

$$a\left(\frac{1}{\sigma}-\frac{d}{2}\right) \int |Q|^2\xi^{d-1}d\xi = 0. \quad (7.1.13)$$

Since $a > 0$ and $d\sigma > 2$, Q has to be identically 0. In other words, solutions with finite Hamiltonian have an infinite L^2-norm.

Remark. We are interested in complex solutions Q of (7.1.2) with a monotonically decreasing amplitude $|Q|$ and zero Hamiltonian, which provide the limiting profiles of singular solutions of the NLS equation. We call such solutions "admissible solutions." They are of the form αQ_1 because Q_2 does not have a finite Hamiltonian. They depend on three real constants, $\Re\alpha$, $\Im\alpha$, and a, that are determined by three conditions at $\xi = 0$, namely $Q_\xi(0) = 0$ and $Q(0)$ real. This suggests that admissible solutions exist only for a discrete set of values of the parameter a. Equations (7.1.2)–(7.1.3) can be indeed viewed as a nonlinear eigenvalue problem that has finite-Hamiltonian solutions only for special values of a, given the dimension $d > 2/\sigma$. Numerically, this value appears to be unique. The problem is delicate especially because of the infinite domain. A discussion of the structure of the spectrum for a model equation of the form $-u'' + if(x,|u|^2) = \lambda u$, in the finite interval $0 < x < 1$, with homogeneous Dirichlet boundary conditions is given in Lochak (1990).

It was recently shown numerically by Budd, Chen, and Russell (1999) that (7.1.2)–(7.1.3) with cubic nonlinearity admit an infinite number of "multi-bump" solutions which decrease at infinity and lie on solution branches parametrized by the dimension d, such that as $d \to 2$, $a \to 0$ on each branch and $Q(0)$ tends either to zero or to the value $R(0) \approx 2.2062\ldots$ of the ground state at the origin. On each branch, the solutions are characterized by the number of oscillations of the profile $|Q(\xi)|$. From such functions Q, self-similar solutions with a non-monotone profile can be constructed for the NLS equation in supercritical dimension. Such multi-bump blowup solutions all appear to be unstable.

7.1.2 Spatial Extension of the Self-Similar Profile

In supercritical dimension, the self-similar solution whose L^2-norm is infinite cannot be extended to infinity. The boundedness of the L^2-norm of the NLS equation solution suggests that it is continued at large distance by a rapidly decaying function. Indeed, in this far asymptotic range, the NLS equation reduces to the linear equation

$$i\partial_t \psi + \Delta\psi = 0. \tag{7.1.14}$$

Looking for solutions of the form $\psi = e^{i\lambda^2 t} f(r)$, we obtain that f is an eigenfunction of the Laplace equation in spherical geometry. As $r \to \infty$, we have (Shvets, Kosmatov, and LeMesurier 1993)

$$f(r) \approx r^{(1-d)/2} e^{-\lambda r}, \tag{7.1.15}$$

which ensures the finite character of the wave energy.

A detailed analysis of the matching between the self-similar and the large-distance regimes is presented in Bergé and Pesme (1992, 1993a). The extension of the self-similar region is recovered from a simple argument. Noticing that the contribution to the mass N originating from the large-distance region is convergent since the solution belongs to L^2, we have

$$N = \int_{r<r_0} |\psi|^2 d\mathbf{x} + \int_{r>r_0} |\psi|^2 d\mathbf{x}, \tag{7.1.16}$$

where r_0 estimates the extension of the self-similar region which can a priori depend on time. We thus have

$$N \approx L(t)^{d-2/\sigma} \int_{\xi<r_0/L(t)} |Q(\xi)|^2 \xi^{d-1} d\xi + \int_{r>r_0} |\psi|^2 d\mathbf{x}, \tag{7.1.17}$$

where the first term of the right-hand side is written as $N_1 + N_2$ with

$$N_1 \approx L(t)^{d-2/\sigma} \int_{\xi<\xi_1} |Q(\xi)|^2 \xi^{d-1} d\xi, \qquad \xi_1 \text{ finite}, \tag{7.1.18}$$

and

$$N_2 \approx L(t)^{d-2/\sigma} \int_{\xi_1 < \xi < r_0/L(t)} |Q(\xi)|^2 \xi^{d-1} d\xi. \qquad (7.1.19)$$

In the range $\xi_1 < \xi < r_0/L(t)$, Q is estimated by its asymptotic behavior $|Q| \approx |\xi|^{-1/\sigma}$. It follows that

$$N_2 \approx \frac{r_0^{d-2/\sigma}}{d - 2/\sigma}. \qquad (7.1.20)$$

Asymptotically close to the singularity,

$$N \approx \frac{r_0^{d-2/\sigma}}{d - 2/\sigma} + \int_{r > r_0} |\psi|^2 d\mathbf{x}, \qquad (7.1.21)$$

which implies that in supercritical dimension, the self-similar region has a finite extension (depending on the initial condition) in the physical variables. In the rescaled variable ξ, the extension thus scales like $\xi(\tau) = 1/L(\tau)$, which grows exponentially in τ, a point confirmed by accurate numerical simulations (Shvets, Kosmatov, and LeMesurier 1993). The fast decay of the solution at the boundary of the self-similar zone observed in these simulations justifies a posteriori the estimate of N on the basis of the contribution of the sole self-similar region. It is important to notice that the mass of the collapsing core tends to zero, a property often referred to as "weak collapse", and that the mass of the wave packet is transferred to the tail, where the profile displays an algebraic decay. This situation contrasts with the critical case, where it was proved that a finite amount of mass concentrates in a neighborhood of the focus with a width of order $(t_* - t)^{1/2}$ ("strong collapse," see Section 5.2.4). Note also that solutions associated to strong collapse were constructed in supercritical dimensions by Zakharov and Kuznetsov (1986b) (see also Bergé 1998), but these solutions appear to be unstable in three dimensions (Zakharov, Litvak, Rakova, Sergeev, and Shvets 1988, Bergé 1994a).

7.1.3 Rate of Convergence to Self-Similar Solutions

Let us come back to the rescaled equation derived from the NLS equation by the scale transformation,

$$i u_\tau + i a(\tau) \left(\xi u_\xi + \frac{u}{\sigma} \right) + \Delta u + |u|^{2\sigma} u = 0, \qquad (7.1.22)$$

where $a(\tau)$ is prescribed as a suitable functional of u ensuring that this solution remains smooth. Up to simple rescalings depending on the initial conditions, we write the solution near the singularity in the form

$$u(\xi, \tau) \approx e^{i\varphi(\tau)} \varphi_\tau^{1/2\sigma} Q(\varphi_\tau^{1/2}\xi, k) \chi \left(\frac{\xi}{L(\tau)} \right), \qquad (7.1.23)$$

where φ_τ approaches a constant C, and k denotes the limit of the ratio $a(\tau)/\varphi_\tau$, which depends only on d and σ. Furthermore, χ is a smooth cut-off function such that $\chi(\xi) = 1$ if $0 \le \xi \le 1$ and $\chi(\xi) = 0$ if $\xi \ge 2$. Let us assume for simplicity that the scaling factor $L(t)$ is prescribed by imposing that the norm $|u|_{L^{2\sigma+2}}$ remains equal to its initial value I_0. Substituting the asymptotic behavior into the conserved norm, we get

$$
I_0 \approx \varphi_\tau^{1/2} \int_0^{\varphi_\tau^{1/2}/L(\tau)} |Q(\xi)|^{2\sigma+2} \xi^{d-1} d\xi
$$

$$
\approx \varphi_\tau^{1/2} \left(\int_0^\infty |Q|^{2\sigma+2} \xi^{d-1} d\xi - \int_{\frac{\varphi_\tau^{1/2}}{L(\tau)}}^\infty |Q|^{2\sigma+2} \xi^{d-1} d\xi \right). \quad (7.1.24)
$$

In the limit of large τ, the last integral is approximated by $\frac{L(\tau)}{\varphi_\tau^{1/2}}$. Expressing that asymptotically, $\varphi_\tau(\tau) = C(1 + f(\tau))$ with $f \ll 1$, we get

$$
C^{\frac{1}{2}}(1 + \frac{f}{2}) \int |Q|^{2\sigma+1} \xi^{d-1} d\xi = I_0 + \frac{L}{2}, \quad (7.1.25)
$$

which fixes the value of the constant C and prescribes that asymptotically f varies like L and thus decays exponentially in τ. It follows that φ_τ and thus $a(\tau)$ converge exponentially to their limit values.

7.2 Dissipation and Postcollapse Dynamics

In plasmas, wave collapse is an important mechanism of small-scale formation, permitting dissipative processes like Landau damping or ion cyclotron resonance to act and heat the medium. In optics, the dissipation associated to multiphoton absorption is usually taken to be nonlinear (Vlasov, Piskunova, and Talanov 1989). In this case, the NLS equation becomes

$$
i(\psi_t + \epsilon|\psi|^p\psi) + \Delta\psi + |\psi|^{2\sigma}\psi = 0 \quad (7.2.1)
$$

with $p = 2\sigma(1 + s)$, $s > 0$. For any $\epsilon > 0$, the solution exists for all times, and the L^2-norm of the solution that corresponds to the energy of the wave is a decreasing function

$$
\frac{d}{dt} \int |\psi|^2 d\mathbf{x} = -2\epsilon \int |\psi|^{p+2} d\mathbf{x}. \quad (7.2.2)
$$

Denoting by T a characteristic lifetime of the collapse, the purpose is to evaluate the energy dissipation

$$
\Delta N = -\int_0^T \frac{d}{dt}|\psi|_{L^2}^2 dt = 2\epsilon \int_0^T \left(\int |\psi|^{p+2} d\mathbf{x} \right) dt \quad (7.2.3)
$$

in the limit $\epsilon \to 0$.

The amplitude Ψ_M and the size l_d of the collapson at the moment where the dissipation enters into play is easily estimated by a balance argument, made precise by the self-similar character of the solution assumed to be preserved. We get $\Psi_M \propto \epsilon^{-1/2\sigma s}$ and $l_d \propto \epsilon^{1/2s}$. It follows that the energy N_{l_d} contained in the collapson and thus available for dissipation scales like

$$\Delta N \propto l_d^{\frac{\sigma d - 2}{\sigma}} \propto \epsilon^{\frac{\sigma d - 2}{2\sigma s}} . \tag{7.2.4}$$

In supercritical dimension ($d\sigma > 2$) and under the condition $p > 2\sigma$, ΔN tends to zero with ϵ, indicating that the collapse is *weak* (Zakharov and Kuznetsov 1986b, Kosmatov, Shvets and Zakharov 1991), a situation that contrasts with the *strong collapse* arising at critical dimension, where a finite amount of the energy remains trapped near the blowup point. This property associated to the L^2-norm concentration is discussed in Section 5.2.4. The distinction between strong and weak collapses is physically important because it determines the efficiency of the collapse as a nonlinear mechanism for energy dissipation.

In supercritical dimensions, numerical simulations show however that according to the values of d and s, the efficiency of the dissipation can be significantly enhanced by the existence of an energy flux from the wings of the Q profile, where most of the energy is contained, towards the dissipating region, making the dynamics quasi-stationary and leading to a so-called *long-lived* or *super-strong* collapse. This suggests to look for solutions that, in the outer region (outside the collapson) solve the steady nondissipative NLS equation and become singular near the origin (Zakharov, Kosmatov, and Shvets 1989, Malkin and Shapiro 1990).

In terms of phase and amplitude ($\psi = A e^{i\phi}$), the (nondissipative) NLS equation becomes

$$\partial_t A^2 + \nabla \cdot (A^2 \nabla \phi) = 0, \tag{7.2.5}$$

$$-A\partial_t \phi + \Delta A - |\nabla \phi|^2 A + A^{2\sigma+1} = 0. \tag{7.2.6}$$

Radial stationary solutions obey

$$\Delta A - \frac{P^2}{A^4 r^{2(d-1)}} A + A^{2\sigma+1} = 0, \tag{7.2.7}$$

where

$$P = \lim_{r \to 0} r^{d-1} A^2 \nabla \phi. \tag{7.2.8}$$

The phase ϕ is solved in terms of the amplitude as

$$\phi_r = \frac{P}{r^{d-1} A^2}. \tag{7.2.9}$$

Solutions that are singular near the origin and behave like a power law

$$A \sim \frac{C}{r^\alpha} \quad \text{with} \quad C > 0 \quad \text{and} \quad \alpha > 0 \tag{7.2.10}$$

satisfy

$$\frac{\alpha(\alpha+1)}{r^2} - \frac{\alpha(d-1)}{r^2} - \frac{P^2}{C^4 r^{2(d-1)-4\alpha}} + \frac{C^{2\sigma}}{r^{2\sigma\alpha}} = 0, \qquad (7.2.11)$$

where various dominant balances can be considered.

(i) When the nonlinear term balances the Laplacian and the flux term is negligible, $\alpha = 1/\sigma$ and $d < 2/\sigma + 2$. The constant C is then given by $C^{2\sigma} = \frac{1}{\sigma}(d-2-\frac{1}{\sigma})$, provided that $d > 2 + 1/\sigma$. Thus, if $2 + 1/\sigma < d < 2 + 2/\sigma$, then $A \sim C r^{-1/\sigma}$ and $C = \left(\frac{1}{\sigma}(d-2-\frac{1}{\sigma})\right)^{1/(2\sigma)}$.

(ii) When the flux term balances the nonlinear term and the Laplacian is negligible, one gets for $d > 2 + 2/\sigma$ that $A \sim \frac{1}{\sigma+2} r^{\frac{d-1}{\sigma+2}}$.

(iii) If $2 < d < 2 + 1/\sigma$, then $A \sim C r^{2-d}$.

Two limiting cases are also to be discussed. When $d = 2 + 2/\sigma$, then $A \sim \frac{C}{r^{1/\sigma}}$ with the constant C solving $C^4(C^{2\sigma} - \frac{1}{\sigma^2}) = P^2$. When $d = 2 + 1/\sigma$, one looks for a solution of the form $A \sim \frac{C}{r^{1/\sigma}}|\ln r|^\beta$. After substitution, one gets $\beta = -1/(2\sigma)$ and $C = \left(\frac{1}{2\sigma^2}\right)^{1/(2\sigma)}$. When $d = 2$, $A \sim C|\ln r|$. Finally, when $1 \le d < 2$, A is constant but $\nabla\phi$ is singular at the origin and behaves like $P^2 C^{-2} r^{1-d}$.

The question then arises of the stability of these solutions associated to the existence of a finite energy flux to the origin that balances the dissipation taking place in the collapson, in the form

$$\int_{r=r_0} A^2 \nabla\phi \cdot d\mathbf{S} + \epsilon \int_{r<r_0} A^{p+2} d\mathbf{x} = 0. \qquad (7.2.12)$$

Numerical simulations show that this regime is stable only when $d \ge 2+2/s$. In this case the amplitude at the origin plotted versus time displays a long plateau. For $d < 2+2/s$, these solutions are not observed and the collapse is weak, as numerically observed with the one-dimensional Schrödinger equation with a high nonlinearity $\sigma = 3$ (Zakharov, Litvak, Rakova, Sergeev, and Shvets 1988). Note that in dimension $d = 2$, a super-strong collapse cannot become established whatever the power of the nonlinearity (Shvets 1990). It follows in particular that, as stressed by Vlasov, Piskunov, and Talanov (1989), the postcollapse dynamics are not prescribed by the sole condition of criticality or supercriticality, but also depend on the space dimension, making one-dimensional models of supercritical collapse questionable.

It was also argued that for wave-packets with an anisotropic setting like lower hybrid or Langmuir waves in the presence of a weak magnetic field (see Section 13.1.3), the existence of an "effective dimension" that can be greater than 4 (with $\sigma = 1$) could make the regime of super-strong collapse physically possible (Kuznetsov and Zakharov 1998).

8
Critical Collapse

As discussed in Chapter 6, numerical simulations of the cubic NLS equation in two dimensions show that at critical dimension, the function $a(t)$ defined in (6.1.3) does not converge to a finite limit as the singularity is approached, but rather continues to decrease slowly. This reflects the presence of a correction to the self-similar rate of blowup. Consequently, the rescaled solution u of (6.1.2) does not tend uniformly to the self-similar profile given up to a simple rescaling, by $u(\xi, \tau) = e^{i\tau} Q(\xi, a)$ with a nonzero constant value of a.

The aim of this chapter is to construct asymptotic solutions near collapse at critical dimension. The construction involves several steps. Due to the slow variation of a, the idea is to keep also a slow variation in the profile of the solution at leading order through the presence of the parameter $a = a(\tau) > 0$. The construction of a profile associated to the instantaneous value of this parameter should take into account the relation between a and the space dimension or the power of the nonlinearity required for existence of a self-similar profile. The rate of blowup is then determined as a solvability condition in an expansion near this quasi self-similar blowup. This approach was developed in LeMesurier, Papanicolaou, Sulem, and Sulem (1988b) and Landman, Papanicolaou, Sulem, and Sulem (1988). A related approach avoiding the explicit variation of the space dimension or the degree of nonlinearity is given in Dyachenko, Newell, Pushkarev, and Zakharov (1992). A more heuristic approach was used by Fraiman (1985). A misprint, however, affects the ultimate expression of the scaling factor. A reformulation of this analysis is presented by Malkin (1993).

8.1 Self-Similar Profile near Critical Dimension

8.1.1 Constraints on the Self-Similar Profile

It is easily shown that the equation obeyed by the (radially symmetric) complex self-similar profile

$$Q_{\xi\xi} + \frac{d-1}{\xi} Q_\xi - Q + ia(\xi Q_\xi + \frac{Q}{\sigma}) + |Q|^{2\sigma}Q = 0, \quad \xi > 0, \qquad (8.1.1)$$

$$Q_\xi(0) = 0, \quad Q(\infty) = 0 \qquad (8.1.2)$$

has no admissible solution at critical dimension when a has a finite value. To wit, we rewrite $Q = We^{i\theta}$ in terms of (real) amplitude and phase given by

$$W_{\xi\xi} + \frac{d-1}{\xi} W_\xi - W + W^{2\sigma+1} - \theta_\xi(a\xi + \theta_\xi)W = 0, \qquad (8.1.3)$$

$$\frac{\partial}{\partial\xi}\left(\xi^{2/\sigma-1}W^2(\theta_\xi + \frac{a}{2}\xi)\right) + \frac{\sigma d - 2}{\sigma}\xi^{2/\sigma-2}\theta_\xi W^2 = 0. \qquad (8.1.4)$$

In the critical case $\sigma d = 2$, (8.1.4) can be solved explicitly, and this implies that regular even solutions have a quadratic phase

$$\theta(\xi) = -a\xi^2/4 . \qquad (8.1.5)$$

The derivation of the large ξ behavior of Q given in Section 7.1.1 is still valid when $\sigma d = 2$, and this leads to

$$Q \approx \frac{e^{-ia\xi^2/4}}{\xi^{1/\sigma}}\left(\lambda_+ e^{i(\frac{a\xi^2}{4} - \frac{1}{a}\ln\xi)} + \lambda_- e^{-i(\frac{a\xi^2}{4} - \frac{1}{a}\ln\xi)}\right), \qquad (8.1.6)$$

where λ_+ and λ_- are complex constants. Since smooth even solutions have a quadratic phase, they correspond to $\lambda_+ = \lambda_-^*$. Their amplitude thus displays an oscillatory decay at infinity. On the other hand, solutions corresponding to $\lambda_- = 0$ have a monotonic decay at infinity but do not have a quadratic phase, and thus cannot be smooth and even. There is consequently no admissible solutions to (8.1.1)–(8.1.2) at critical dimension $d = 2/\sigma$ if a is a nonzero constant.

8.1.2 A Nonlinear Eigenvalue Problem

Existence of solutions of (8.1.1)–(8.1.2) with monotonically decreasing profile $|Q|$ and zero Hamiltonian in dimension $d > 2/\sigma$ requires the adjustment of the value of a to the distance to criticality. In this section, we fix the nonlinearity and analyze how a and d should be related in the limit $a \to 0^+$, $d \downarrow 2/\sigma$. Varying continuously the dimension is possible, since in a radially symmetric geometry, the dimension appears only as a coefficient in the curvature term of the Laplacian. It is useful to write $Q = e^{-ia\xi^2/4}P$ with

P satisfying

$$\frac{d^2P}{d\xi^2} + \frac{d-1}{\xi}\frac{dP}{d\xi} - P + \frac{a^2\xi^2}{4}P - ia\frac{d\sigma-2}{2\sigma}P + |P|^{2\sigma}P = 0 \qquad (8.1.7)$$

and the boundary conditions

$$P_\xi(0) = 0, \quad P(\xi) \to 0 \quad \text{as} \quad \xi \to \infty, \quad P(0) \quad \text{real.} \qquad (8.1.8)$$

For $a = 0$, the equations for Q and P with the constraint of monotonic profiles, reduce to the equation for the ground state R:

$$\frac{d^2R}{d\xi^2} + \frac{d-1}{\xi}\frac{dR}{d\xi} - R + R^{2\sigma+1} = 0. \qquad (8.1.9)$$

Proposition 8.1.   *Assuming the existence of a solution Q of (8.1.1)–(8.1.2) with zero Hamiltonian, the function $d(a)$ is differentiable to all orders at $a = 0$ and*

$$\frac{d^p}{da^p}\Big(d(a) - \frac{2}{\sigma}\Big)\Big|_{a=0} = 0 \text{ for all } p = 0, 1, 2, \ldots \qquad (8.1.10)$$

This proposition states that as $a \to 0$, $d(a) \to 2/\sigma$ faster than any power law. It follows that a global analysis is needed to find the precise behavior of $d(a)$ as a approaches 0. In order to prove this proposition, we start with a few identities.

Lemma 8.2.

$$\int_0^\infty \Big(|R_\xi|^2 - \frac{1}{\sigma+1}|R|^{2\sigma+2}\Big)\xi^{\frac{2}{\sigma}-1}d\xi = 0, \quad (8.1.11)$$

$$\int_0^\infty \Big(R\rho - \frac{1}{8}\xi^2 R^2\Big)\xi^{\frac{2}{\sigma}-1}d\xi = 0, \quad (8.1.12)$$

$$\int_0^\infty \Big(-2\nu R + \frac{\xi^2}{2}\rho R + \rho_\xi^2 - (2\sigma+1)\rho^2 R^{2\sigma}\Big)\xi^{\frac{2}{\sigma}-1}d\xi = 0, \quad (8.1.13)$$

where R denotes the positive solution (ground state) of

$$\Delta R - R + R^{2\sigma+1} = 0 \qquad (8.1.14)$$

in dimension $d = 2/\sigma$ and ρ and ν satisfy respectively

$$\rho_{\xi\xi} + \frac{2/\sigma-1}{\xi}\rho_\xi - \rho + (2\sigma+1)R^{2\sigma}\rho = -\frac{1}{4}\xi^2 R \qquad (8.1.15)$$

and

$$\nu_{\xi\xi} + \frac{2/\sigma-1}{\xi}\nu_\xi - \nu + (2\sigma+1)R^{2\sigma}\nu = -\frac{1}{4}\xi^2\rho - \sigma(2\sigma+1)\rho^2 R^{2\sigma-1}. \quad (8.1.16)$$

Proof of Lemma 8.2 We consider the equation

$$P_{\xi\xi} + \frac{2/\sigma-1}{\xi} - P + \frac{a^2\xi^2}{4}P + |P|^{2\sigma}P = 0, \quad \xi > 0, \qquad (8.1.17)$$

at critical dimension $d = 2/\sigma$. Let

$$P^{(n)} = R + \frac{a^2}{2}P_2 + \cdots + \frac{a^{2n}}{(2n)!}P_{2n} \qquad (8.1.18)$$

be a sequence of approximate solutions of (8.1.17) with monotonic profiles that obey

$$P^{(n)}_{\xi\xi} + \frac{2/\sigma - 1}{\xi}P^{(n)} - P^{(n)} + \frac{a^2\xi^2}{4}P^{(n)} + |P^{(n)}|^{2\sigma}P^{(n)} = O(a^{2n+2}). \quad (8.1.19)$$

We have chosen $P(0; a)$ real, so $P^{(n)}(0; a)$ is real, and consequently $P^{(n)}(\xi; a)$ is also real. We have an estimate for the Hamiltonian of $P^{(n)}$ in the form

$$H(P^{(n)}) = \int_0^\infty \left(|P^{(n)}_\xi|^2 - \frac{1}{\sigma + 1}|P^{(n)}|^{2\sigma+2} + \frac{a^2\xi^2}{4}|P^{(n)}|^2\right)\xi^{2/\sigma-1}d\xi$$
$$= O(a^{2n+2}). \qquad (8.1.20)$$

It follows that when $H(P^{(n)})$ is expanded as a power series in a in the form $\Sigma_{j=0}^\infty m_{2j}a^{2j}$, the coefficients m_j, $j = 0, 2, 4, \ldots, 2n$, vanish. The explicit form of the first three coefficients m_1, m_2, m_3 are given by the left-hand side of (8.1.11)–(8.1.13) which are thus zero.

Proof of Proposition 8.1. The proof is done by induction. We recall that the function P satisfies (8.1.7), where the dimension d is viewed as a function of a with $d(0) = 2/\sigma$.

Step 1. We prove that $d'(0) = 0$. Differentiating the identity (7.1.9) where $d = d(a)$ with respect to a, and evaluating it at $a = 0$, we obtain

$$\int_0^\infty 2(-\Delta R - R^{2\sigma+1})\Re p_1\xi^{2/\sigma-1}d\xi$$

$$+ d'(0)\int_0^\infty (R_\xi^2 - \frac{1}{\sigma+1}R^{2\sigma+2})\xi^{2/\sigma-1}\ln\xi d\xi = 0, \,(8.1.21)$$

where $p_1 = \frac{dP}{da}|_{a=0}$ satisfies

$$\Delta p_1 - p_1 + (2\sigma + 1)R^{2\sigma}p_1 = -d'(0)\frac{R_\xi}{\xi} \qquad (8.1.22)$$

and $\Delta = \partial_{\xi\xi} + \frac{2/\sigma-1}{\xi}\partial_\xi$. Thus

$$\Re p_1 = d'(0)g, \qquad (8.1.23)$$

where g satisfies

$$\Delta g - g + (2\sigma + 1)R^{2\sigma}g = -\frac{R_\xi}{\xi}. \qquad (8.1.24)$$

We now rewrite (8.1.21) in the form

$$d'(0)\left(-2\int_0^\infty Rg\xi^{2/\sigma-1}d\xi + \int_0^\infty (R_\xi^2 - \frac{1}{\sigma+1}R^{2\sigma+2})\xi^{2/\sigma-1}\ln\xi d\xi\right) = 0.$$
(8.1.25)

To prove that $d'(0) = 0$, we have to show that the quantity

$$\gamma = -2\int_0^\infty Rg\xi^{2/\sigma-1}d\xi + \int_0^\infty (R_\xi^2 - \frac{1}{\sigma+1}R^{2\sigma+2})\xi^{2/\sigma-1}\ln\xi d\xi \quad (8.1.26)$$

in (8.1.25) does not vanish. This results from the identity

$$\gamma = \frac{1}{2}\int R^2 \xi^{2/\sigma-1}d\xi \qquad (8.1.27)$$

proved in Lemma 8.3 below.

From (8.1.23), we see that $\Re p_1 = 0$. In addition, $\Im p_1 = 0$ because $P(0;a)$ is chosen to be real.

Step 2. We proceed with the induction by assuming that

$$d^{(j)}(0) \equiv \left(\frac{d}{da}\right)^j d(a)\Big|_{a=0} = 0 \quad \text{for} \quad j = 1, 2, \ldots, n-1, \qquad (8.1.28)$$

$$\left(\frac{d}{da}\right)^j P(\xi;a)\Big|_{a=0} = 0 \quad \text{for} \quad j \text{ odd}, \quad j \leq n-1, \quad (8.1.29)$$

$$\Im\left(\frac{d}{da}\right)^j P(\xi;a)\Big|_{a=0} = 0 \quad \text{for} \quad j = 1, 2, \ldots, n-1, \qquad (8.1.30)$$

where $\frac{d}{da}$ denotes the total derivative with respect to a, taking into account the variation of the dimension $d(a)$. We will prove that (8.1.28)–(8.1.30) hold for $j = n$.

From (8.1.7) and (8.1.28) we have

$$\left(\frac{d}{da}\right)^j P(\xi;a)\Big|_{a=0} = P_j \text{ for } j \text{ even }, j < n. \qquad (8.1.31)$$

We then differentiate (8.1.7) n times with respect to a and set a to 0. Using the induction hypotheses (8.1.29) and (8.1.30), we obtain

$$\left(\frac{d}{da}\right)^n P(\xi;a)\Big|_{a=0} = 0, \quad n \text{ odd}, \qquad (8.1.32)$$

$$\Im\left(\frac{d}{da}\right)^n P(\xi;a)\Big|_{a=0} = 0, \quad n \text{ even}, \qquad (8.1.33)$$

and

$$\Re\left(\frac{d}{da}\right)^n P(\xi;a)\Big|_{a=0} = P_n + d^{(n)}(0)g, \quad n \text{ even}, \qquad (8.1.34)$$

where g is defined by (8.1.24). From the induction hypothesis (8.1.28), we have

$$\left(\frac{d}{da}\right)^{j}(\xi^{d(a)-1})\Big|_{a=0} = 0, \qquad j = 1, 2, \ldots, n-1, \qquad (8.1.35)$$

$$\left(\frac{d}{da}\right)^{n}(\xi^{d(a)-1})\Big|_{a=0} = d^{(n)}(0)\,\xi^{2/\sigma-1}\ln\xi. \qquad (8.1.36)$$

We now differentiate identity (7.1.9) n times with respect to a and set a to 0. We obtain

$$\int \left(\frac{d}{da}\right)^{n}\left(|P_{\xi}|^{2} - \frac{1}{\sigma+1}|P|^{2\sigma+2} + a\Im(\xi P\bar{P}_{\xi}) + \frac{a^{2}}{4}\xi^{2}|P|^{2}\right)\Big|_{a=0}\xi^{2/\sigma-1}d\xi$$

$$+ d^{(n)}(0)\int\left(R_{\xi}^{2} - \frac{1}{\sigma+1}|R|^{2\sigma+2}\right)\xi^{2/\sigma-1}\ln\xi = 0 \,.(8.1.37)$$

From the identities $m_{j} = 0$, for $j = 0, 2, 4, \ldots$, established at the end of the proof of Lemma 8.3 it follows that (8.1.37) reduces to $\gamma d^{(n)}(0) = 0$. Since $\gamma > 0$, we have $d^{(n)}(0) = 0$, and the induction is complete.

Lemma 8.3.   *The quantity γ defined in (8.1.26) by*

$$\gamma = -2\int_{0}^{\infty} Rg\xi^{2/\sigma-1}d\xi + \int_{0}^{\infty}\left(R_{\xi}^{2} - \frac{1}{\sigma+1}R^{2\sigma+2}\right)\xi^{2/\sigma-1}\ln\xi d\xi \quad (8.1.38)$$

can be evaluated explicitly as

$$\gamma = \frac{1}{2}\int R^{2}\xi^{2/\sigma-1}d\xi. \qquad (8.1.39)$$

Proof. The proof is based on the following identities derived from (8.1.9) and (8.1.24) satisfied by R and g respectively:

$$-2\sigma\int gR^{2\sigma+1}\xi^{\frac{2}{\sigma}-1}d\xi = \int R\frac{R_{\xi}}{\xi}\xi^{\frac{2}{\sigma}-1}d\xi \quad (8.1.40)$$

$$-2\int g\Delta R\,\xi^{\frac{2}{\sigma}-1}d\xi = \int R_{\xi}^{2}\xi^{\frac{2}{\sigma}-1}d\xi \quad (8.1.41)$$

$$-\int\left(R_{\xi}^{2} + R^{2} - R^{2\sigma+2}\right)\ln\xi\,\xi^{\frac{2}{\sigma}-1}d\xi = \int R\frac{R_{\xi}}{\xi}\xi^{\frac{2}{\sigma}-1}d\xi \quad (8.1.42)$$

$$\int\left((1-\sigma)R_{\xi}^{2} + R^{2} - \frac{R^{2\sigma+2}}{\sigma+1}\right)\ln\xi\,\xi^{\frac{2}{\sigma}-1}d\xi = \left(1 - \frac{\sigma}{2}\right)\int R^{2}\xi^{\frac{2}{\sigma}-1}d\xi.$$

$$(8.1.43)$$

To obtain (8.1.40), one multiplies (8.1.9) by g, (8.1.24) by R, subtracts the resulting equalities, and integrates on the whole space. To get (8.1.41), one combines (8.1.9) multiplied by ξg_{ξ} and (8.1.24) multiplied by ξR_{ξ}. To establish (8.1.42) and (8.1.43), one multiplies (8.1.9) by $R\ln\xi$ and $R_{\xi}\xi\ln\xi$, respectively, and integrates.

Using again (8.1.9), one gets from (8.1.40)–(8.1.41)

$$-2\int Rg\xi^{2/\sigma-1}d\xi = \int \left(R_\xi^2 + \frac{1}{\sigma}\frac{RR_\xi}{\xi}\right)\xi^{2/\sigma-1}d\xi. \tag{8.1.44}$$

Adding (8.1.42) and (8.1.43) leads to

$$\sigma\int \left(R_\xi^2 - \frac{1}{\sigma+1}R^{2\sigma+2}\right)\xi^{2/\sigma-1}\ln\xi d\xi = \int \left((\frac{\sigma}{2}-1)R^2 - \frac{RR_\xi}{\xi}\right)\xi^{2/\sigma-1}d\xi. \tag{8.1.45}$$

Finally, substituting these two last identities into (8.1.38) and using that

$$\int R_\xi^2\xi^{2/\sigma-1}d\xi = \frac{1}{\sigma+1}\int R^{2\sigma+2}\xi^{2/\sigma-1}d\xi = \frac{1}{\sigma}\int R^2\xi^{2/\sigma-1}d\xi, \tag{8.1.46}$$

one obtains (8.1.39).

8.1.3 A Nonuniform Limit

Proposition 8.4. *In the limit $a \to 0^+$, and $d(a) \downarrow 2/\sigma$, admissible solutions Q of (8.1.1)–(8.1.2) satisfy*

(i) *For $a\xi \ll 1$, $Q \approx R(\xi)e^{-ia\xi^2/4}$. Equivalently, the function P, solution of (8.1.7)–(8.1.8), approaches the ground state R.*
(ii) *For $a\xi \gg 1$, $Q \approx \mu\xi^{-1/\sigma-i/a}$, with $\mu^2 = \frac{\sigma d-2}{\sigma}N_c$.*
(iii) *The asymptotic behavior of $d(a)$ is given by*

$$\sigma d - 2 \approx \frac{2\sigma\nu_0^2}{N_c}\frac{1}{a}e^{-\pi/a}, \tag{8.1.47}$$

where $N_c = \int_0^\infty R^2\xi^{2/\sigma-1}d\xi$ is the critical mass for collapse and $\nu_0 = \lim_{\xi\to\infty}\xi^{\frac{d-1}{2}}e^\xi R(\xi)$.

Proof. It is useful to write the equation for Q in terms of amplitude and phase as in (8.1.3)–(8.1.4). Integration of the equation for the phase leads to

$$\theta_\xi + \frac{a}{2}\xi = -\frac{\sigma d-2}{\sigma\xi^{2/\sigma-1}W^2}\int_0^\xi \theta_\xi W^2\xi'^{2/\sigma-2}d\xi'. \tag{8.1.48}$$

(i) As $d \to 2/\sigma$, $a \to 0$, with fixed ξ, and W approaches the positive solution R of (8.1.9) and $\theta_\xi \approx -a\xi/2$, leading to a quadratic phase $\theta = -a\xi^2/4$.
(ii) At large distance ($a\xi \gg 1$), the right-hand side of (8.1.48) is, however, not negligible. From the asymptotic behavior of Q, we have in this limit

$$W \approx \mu\xi^{-1/\sigma} \quad \text{and} \quad \theta_\xi \approx -1/a\xi, \tag{8.1.49}$$

which implies that the integral $\int_0^\xi \theta_\xi W^2 \xi'^{2/\sigma - 2} d\xi'$ converges. The main contribution to this integral comes from $a\xi' \ll 1$, leading to

$$\int_0^\xi \theta_\xi W^2 \xi'^{2/\sigma - 2} d\xi' \approx -\frac{a}{2} N_c. \tag{8.1.50}$$

Since $\theta_\xi \approx -1/(a\xi)$ for $a\xi \gg 1$, equating the unbounded terms in (8.1.48) gives

$$\mu^2 = \frac{\sigma d - 2}{\sigma} N_c. \tag{8.1.51}$$

(iii) This point is obtained by matching the solution for $\xi \gg 2/a$ to the ground state R, which for ξ large behaves like

$$R(\xi) \approx \nu_0 \xi^{-(d-1)/2} e^{-\xi}. \tag{8.1.52}$$

Defining Z by

$$Q = e^{-ia\xi^2/4} \xi^{(1-d)/2} Z, \tag{8.1.53}$$

(7.1.2) is replaced by

$$-Z'' + \left(-\frac{a^2}{4}\xi^2 + 1 - ia\frac{d\sigma - 2}{2\sigma} - |\xi|^{(1-d)\sigma}|Z|^{2\sigma} + \frac{(d-1)(d-3)}{4\xi^2} \right) Z = 0. \tag{8.1.54}$$

The condition $Q \approx R e^{-ia\xi^2/4}$ valid for small or moderate ξ when $d \approx 2/\sigma$ becomes $Z \approx \xi^{(d-1)/2} R$, while for $\xi \gg 1$, (8.1.54) reduces near $d = 2/\sigma$ to

$$Z'' + \left(\frac{a^2 \xi^2}{4} - 1 \right) Z = 0. \tag{8.1.55}$$

Note that the nonlinearity of the Q-equation is relevant only for $\xi \ll 1$, when, up to a quadratic phase, Q is approximated by the ground state R. At larger distance, the equation can be viewed as essentially linear.

Equation (8.1.55) can be solved exactly as a linear combination of Weber parabolic functions (Koppel and Landman 1995). Alternatively, this equation can be analyzed in the limit $a \to 0$, using the WKB method (Sulem, and Sulem 1997). Writing $y = a\xi/2$ and $q(y) = 1 - y^2$, (8.1.55) reads

$$\frac{a^2}{4} Z_{yy} = q(y) Z. \tag{8.1.56}$$

For $y \ll 1$, the WKB approximation gives

$$Z^L_{WKB} \approx \frac{\nu_0}{(1 - y^2)^{1/4}} e^{-\frac{2}{a} \int_0^y \sqrt{1 - s^2} ds}. \tag{8.1.57}$$

For $y \ll 1$, $\int_0^y \sqrt{1 - s^2} ds \approx y$ and (8.1.57) reduces to

$$Z^L_{WKB} \approx e^{-\frac{2}{a} y} = C e^{-\xi}, \tag{8.1.58}$$

thus requiring $C = \nu_0$ to match.

When $y \to 1^-$, the integral in (8.1.57) can be approximated by $\frac{\pi}{4} - \sqrt{2}\frac{2}{3}(1-y)^{3/2}$, and in this limit

$$Z_{WKB}^{L} \approx \frac{C}{(1-y^2)^{1/4}} e^{-\frac{\pi}{2a} + \frac{4\sqrt{2}}{3a}(1-y)^{3/2}}. \qquad (8.1.59)$$

Near the turning point $y = 1$, we introduce the variable $\zeta = 2(1-y)a^{-2/3}$, and (8.1.55) can be replaced by the Airy equation

$$Z_{\zeta\zeta} = \zeta Z. \qquad (8.1.60)$$

The solution $Z = k_1 \mathrm{Ai}(\zeta) + k_2 \mathrm{Bi}(\zeta)$ is approximated in the limit $\zeta \to +\infty$ or equivalently $y \to 1^-$ with the condition $(1 - y) \gg a^{2/3}/2$, by

$$Z_{Airy} \approx \frac{1}{\sqrt{\pi}} \frac{a^{1/6}}{2^{1/4}} \frac{1}{(1-y)^{1/4}} \left(\frac{1}{2} k_1 e^{-\frac{4\sqrt{2}}{3a}(1-y)^{3/2}} + k_2 e^{\frac{4\sqrt{2}}{3a}(1-y)^{3/2}} \right). \qquad (8.1.61)$$

On the left side of the turning point, the exponential with the minus sign is negligible, and matching the remaining contribution with the WKB approximation (8.1.59) leads to

$$\frac{k_2}{\sqrt{\pi}} \frac{a^{1/6}}{2^{1/4}} = \frac{\nu_0}{2^{1/4}} e^{-\pi/2a}. \qquad (8.1.62)$$

To prevent oscillations in the amplitude of the solution on the right side of the turning point ($y \to 1^+$, or equivalently $\zeta \to -\infty$), we choose $k_1 = i k_2$, leading to

$$Z_{Airy} \approx \frac{k_2}{\sqrt{\pi}} \frac{1}{(-\zeta)^{1/4}} \exp i \left(\frac{\pi}{4} + \frac{2}{3}(-\zeta)^{3/2} \right)$$

$$\approx \frac{k_2}{\sqrt{\pi}} \frac{a^{1/6}}{2^{1/4}} \frac{1}{(y-1)^{1/4}} \exp i \left(\frac{\pi}{4} + \frac{4\sqrt{2}}{3a}(y-1)^{3/2} \right). \qquad (8.1.63)$$

This solution must be matched to the WKB approximation for $y > 1$,

$$Z_{WKB}^{R} \approx C_1 \frac{1}{(1-y^2)^{1/4}} \exp \left(\frac{2i}{a} \int_1^y \sqrt{s^2 - 1} \, ds \right). \qquad (8.1.64)$$

Note that $(1 - y^2)^{1/4} = e^{-i\pi/4}(y^2 - 1)^{1/4}$. Then as $y \to 1^+$,

$$Z_{WKB}^{R} \approx C_1 \frac{1}{2^{1/4}(y-1)^{1/4}} \exp \left(i \left(\frac{\pi}{4} + \frac{4\sqrt{2}}{3a}(y-1)^{3/2} \right) \right). \qquad (8.1.65)$$

The phases of (8.1.63) and (8.1.65) match exactly. To match the amplitudes, we choose

$$\frac{k_2}{\sqrt{\pi}} \frac{a^{1/6}}{2^{1/4}} = \frac{C_1}{2^{1/4}}. \qquad (8.1.66)$$

Finally, we match the WKB approximation (8.1.64) in the limit $y \gg 1$ with the large distance solution

$$Z_{\text{out}} = \mu e^{ia\xi^2/4 - (\ln \xi)/a} \xi^{-1/2}, \qquad (8.1.67)$$

prescribed by the asymptotic behavior of Q. Matching the amplitude gives

$$C_1 \frac{2^{1/2}}{a^{1/2}} = \mu. \tag{8.1.68}$$

We also check that the phases match correctly by computing the phase in (8.1.64) in the form

$$
\begin{aligned}
\text{Arg} Z_{WKB}^R &= \frac{\pi}{4} + \frac{2}{a} \left(\frac{y}{2}(y^2 - 1)^{1/2} - \frac{1}{2} \ln(y + (y^2 - 1)^{1/2}) \right) \\
&\approx \frac{a\xi^2}{4} - \frac{1}{a} \ln \xi. \tag{8.1.69}
\end{aligned}
$$

Combining the relations (8.1.62), (8.1.66), and (8.1.68), we finally obtain

$$\mu = \nu_0 2^{1/2} \frac{1}{a^{1/2}} e^{-\pi/2a}, \tag{8.1.70}$$

which, when substituted in (8.1.51), leads to the relation (8.1.47) between the dimension d and the coefficient a in (7.1.2) in the limit $a \to 0$, $d \to 2/\sigma$.

On the basis of this formal analysis and making use of geometrical methods for dynamical systems, Koppel and Landman (1995) established the existence of a unique admissible profile Q for sufficiently small a.

8.1.4   Remarks on the Critical Profile

We have shown that in order to give a sense to (8.1.1)–(8.1.2) for Q or (8.1.7)–(8.1.8) for P, we have to prescribe the coefficient

$$\nu(a) \equiv \frac{a(d\sigma - 2)}{2\sigma}, \tag{8.1.71}$$

which in the small a-limit is given by

$$\nu(a) \approx \frac{\nu_0^2}{N_c} e^{-\pi/a}, \tag{8.1.72}$$

and controls the distance to criticality. An alternative approach used by Dyachenko, Newell, Pushkarev, and Zakharov (1992) consists in regularizing (8.1.7) for P by adding a small corrective term in the form

$$\frac{d^2 P}{d\xi^2} + \frac{2/\sigma - 1}{\xi} \frac{dP}{d\xi} - P + \frac{a^2 \xi^2}{4} P - i\nu(a)P + |P|^{2\sigma} P = 0, \tag{8.1.73}$$

$$P_\xi(0) = 0; \quad P(\infty) = 0, \tag{8.1.74}$$

while keeping the dimension critical. The same analysis as that presented in the previous section then enables one to determine $\nu(a)$ in the form given by (8.1.72). The difference with the previous approach concerns only the dimension coefficient in the curvature term of the Laplacian (or equivalently the exponent of the power law nonlinearity), which is here taken exactly critical. This indicates that in the previous analysis, the deviation from criticality is important only in the term involving the complex coefficient

$i\nu(a) \equiv ia\frac{(d\sigma-2)}{2\sigma}$ and not in the Laplacian or the nonlinearity. This term $i\nu(a)P$ is indeed relevant at large distance, where it is responsible for the nonreal character of P or equivalently of the nonquadratic phase of Q. It may be seen as a complex correction to the eigenvalue $+1$ in (8.1.73) viewed as an eigenvalue problem. As stressed by Dyachenko, Newell, Pushkarev, and Zakharov (1992), this problem is not self-adjoint, and the eigenvalue has an imaginary part. In this framework, (8.1.73)–(8.1.74) considered in the limit $a \to 0$ can be interpreted as the continuation to critical dimension of (8.1.7) for the supercritical profile near criticality. It reduces to the ground state equation $\Delta R - R + R^{2\sigma+1} = 0$ for finite ξ but also includes corrective terms that are relevant for $a\xi \gg 1$ to correctly reproduce the large distance behavior of the profile. As discussed in Section 8.1.1, retaining only the term $\frac{a^2\xi^2}{4}P$ would lead to oscillations of $|P| = |Q|$. Those are prevented by including the imaginary coefficient term $i\nu(a)P$ with the adequate choice of the very small coefficient $\nu(a)$ given by (8.1.72).

8.2 Asymptotic Solutions at Critical Dimension

8.2.1 Construction of Asymptotic Solutions

At the critical dimension $d = 2/\sigma$, we write the NLS equation in the rescaled form

$$iu_\tau + u_{\xi\xi} + \frac{2/\sigma - 1}{\xi}u_\xi + ia(\tau)\left(\xi u_\xi + \frac{u}{\sigma}\right) + |u|^{2\sigma}u = 0, \qquad (8.2.1)$$

where

$$\psi(\mathbf{x}, t) = \frac{1}{L(t)^{1/\sigma}}u(\xi, \tau), \quad \xi = \frac{|\mathbf{x}|}{L(t)}, \quad \tau = \int_0^t \frac{ds}{L^2(s)}, \qquad (8.2.2)$$

and

$$a = -LL_t = -L_\tau/L. \qquad (8.2.3)$$

Making the change of variables

$$u = e^{i\tau - ia\xi^2/4}v, \qquad (8.2.4)$$

we get

$$iv_\tau + v_{\xi\xi} + \frac{2/\sigma - 1}{\xi}v_\xi - v + \frac{1}{4}b(\tau)\xi^2 v + |v|^{2\sigma}v = 0 \qquad (8.2.5)$$

with

$$b(\tau) = a^2 + a_\tau = -L^3 L_{tt}, \qquad (8.2.6)$$

where a_τ will a posteriori be checked to be negligible.

We now look for a solution v of (8.2.5) that is quasi-stationary at leading order in the limit $\tau \to \infty$. We write

$$v(\xi, \tau) = P(\xi; b(\tau)) + W(\xi, \tau), \tag{8.2.7}$$

where $W \ll P$ and the function P solves

$$P_{\xi\xi} + \frac{d-1}{\xi} P_\xi - P + \frac{b\xi^2}{4} P - i\nu(\sqrt{b})P + |P|^{2\sigma} P = 0 \tag{8.2.8}$$

with the boundary conditions

$$P_\xi(0) = 0, \qquad P(\xi) \to 0 \quad \text{as} \quad \xi \to \infty, \qquad P(0) \quad \text{real}, \tag{8.2.9}$$

and also satisfies the zero Hamiltonian condition

$$\int_0^\infty \left(|P_\xi|^2 - \frac{1}{\sigma+1} |P|^{2\sigma+2} + \sqrt{b}\Im(\xi P \bar{P}_\xi) + \frac{b}{4}\xi^2 |P|^2 \right) \xi^{d-1} d\xi = 0. \tag{8.2.10}$$

In (8.2.8),

$$\nu(\sqrt{b}) \approx \frac{\nu_0^2}{N_c} e^{-\pi/\sqrt{b}}, \tag{8.2.11}$$

and $d - 1 = \frac{2}{\sigma} - 1 + \frac{2}{\sqrt{b}}\nu(\sqrt{b})$, where the corrective term is neglected in the approach of Dyachenko, Newell, Pushkarev, and Zakharov (1992).

Our aim is now to determine b as a function of τ from the condition that (8.2.7) be an asymptotic solution of (8.2.5). One has the following propositions.

Proposition 8.5. *Collapsing solutions of the nonlinear Schrödinger equation at critical dimension $d\sigma = 2$ have the asymptotic form*

$$\psi(\mathbf{x}, t) \approx \frac{1}{L(t)} e^{i(\tau(t) - a(t)\frac{|\mathbf{X}|^2}{4L^2(t)})} P\left(\frac{|\mathbf{x}|}{L(t)}, b(t)\right) \tag{8.2.12}$$

near the singularity, where

$$\tau_t = L^{-2}, \quad -L_t L = a, \quad L^3 L_{tt} = -b, \tag{8.2.13}$$

and $b = a^2 + a_\tau \approx a^2$ obeys

$$b_\tau = -\frac{2N_c}{M}\nu(\sqrt{b}) \approx -\frac{2\nu_0^2}{M} e^{-\pi/\sqrt{b}} \tag{8.2.14}$$

with $N_c = \int_0^\infty R^2 \xi^{2/\sigma - 1} d\xi$ and $M = \frac{1}{4}\int_0^\infty R^2 \xi^2 \xi^{2/\sigma - 1} d\xi$.

Proof. Inserting (8.2.7) into (8.2.5) and using the fact that P satisfies (8.2.8), we obtain the following equation for W:

$$L_b W \equiv W_{\xi\xi} + \frac{d-1}{\xi} W_\xi - W + \frac{b\xi^2}{4} W - i\nu(\sqrt{b})W$$
$$+ (\sigma+1)|P|^{2\sigma} W + \sigma P^{\sigma+1} P^{*\sigma-1} \bar{W}$$

$$= -i\left(\frac{\partial P}{\partial b}b_\tau + W_\tau\right) + \frac{\nu(\sqrt{b})}{2\sqrt{b}}\frac{1}{\xi}(P_\xi + W_\xi)$$
$$-i\nu(\sqrt{b})(P + W) + o(W), \tag{8.2.15}$$

where $o(W)$ denotes higher order terms in W. As $b \to 0$, (8.2.15) reduces to leading order to

$$L_b W = -i\left(\frac{\partial P}{\partial b}b_\tau + \nu(\sqrt{b})P\right) + \frac{\nu(\sqrt{b})}{2\sqrt{b}}\frac{1}{\xi}P_\xi. \tag{8.2.16}$$

In addition,

$$\frac{\partial P}{\partial b}\Big|_{b=0} = \rho, \tag{8.2.17}$$

where ρ is defined by (8.1.15). Separating the real and imaginary parts of W in the form $W = S + iT$, we get to leading order in $d - 2/\sigma$,

$$S_{\xi\xi} + \frac{2/\sigma - 1}{\xi}S_\xi - S + (2\sigma + 1)R^{2\sigma}S = \frac{\nu(\sqrt{b})}{2\sqrt{b}}\frac{1}{\xi}R_\xi, \tag{8.2.18}$$

$$T_{\xi\xi} + \frac{2/\sigma - 1}{\xi}T_\xi - T + R^{2\sigma}T = -b_\tau\rho - \nu(\sqrt{b})R. \tag{8.2.19}$$

Equation (8.2.18) is always solvable, and consequently the presence of the right-hand side neglected in Dyachenko, Newell, Pushkarev, and Zakharov (1992) has no effect. In contrast, since the ground state R belongs to the null space of the operator acting on T in the left-hand side of (8.2.19), the solvability of this equation requires

$$\int_0^\infty \left(b_\tau\rho + \nu(\sqrt{b})R\right) R\xi^{2/\sigma-1}d\xi = 0, \tag{8.2.20}$$

or equivalently,

$$b_\tau = -\frac{2N_c}{M}\nu(\sqrt{b}), \tag{8.2.21}$$

where $N_c = \int_0^\infty R^2\xi^{2/\sigma-1}d\xi$, $M = 2\int_0^\infty R\rho\xi^{2/\sigma-1}d\xi$. Furthermore, we have $\nu(\sqrt{b}) \approx \frac{\nu_0^2}{N_c}e^{-\pi/\sqrt{b}}$. as $\tau \to \infty$. Using (8.1.9) and (8.1.15) satisfied by R and ρ, together with (8.1.11), one gets that $M = \frac{1}{4}\int_0^\infty \xi^2 R^2\xi^{2/\sigma-1}d\xi$ is strictly positive.

We thus obtain an asymptotic solution v of (8.2.5) as $\tau \to \infty$ in the form

$$v(\xi, \tau) = P(\xi, b(\tau)) - \frac{\nu(\sqrt{b})}{2\sqrt{b}}\left(g(\xi) - 2i\sqrt{b}h(\xi)\right) + \cdots, \tag{8.2.22}$$

where g satisfies (8.1.24), $d - 2/\sigma$ is given by (8.1.47), and h obeys

$$h_{\xi\xi} + \frac{2/\sigma - 1}{\xi}h_\xi - h + R^{2\sigma}h = 2\frac{N_c}{M}\rho(\xi) - R, \tag{8.2.23}$$

where ρ satisfies (8.1.15).

Proposition 8.6. *At leading order, as $\tau \to \infty$,*

$$a(\tau) \approx b^{1/2} \approx \frac{\pi}{\ln \tau} \tag{8.2.24}$$

and the corresponding scaling factor $L(t)$ in the primitive variables has the asymptotic form

$$L(t) \approx \left(\frac{2\pi(t_* - t)}{\ln \ln \frac{1}{t_* - t}} \right)^{1/2}. \tag{8.2.25}$$

In addition

$$\tau(t) \approx \frac{1}{2\pi} \ln \left(\frac{1}{t_* - t} \right) \ln \ln \left(\frac{1}{t_* - t} \right). \tag{8.2.26}$$

Proof.   We use that $b \approx a^2$ and the relation (8.1.47) between $d(a)$ and a in the limit $a \to 0$ to obtain

$$a_\tau \approx -\frac{\nu_0^2}{M} a^{-1} e^{-\pi/a}. \tag{8.2.27}$$

At leading order, as τ goes to infinity, (8.2.27) is solved as

$$a(\tau) \approx \frac{\pi}{\ln \tau}, \tag{8.2.28}$$

which validates the assumption $a_\tau \ll a^2$. When corrective terms are retained, one has

$$a(\tau) \approx \frac{\pi}{\ln \tau + 3 \ln \ln \tau}. \tag{8.2.29}$$

Numerical simulations of the two-dimensional NLS equation show a very good agreement with this law (see Fig.6 in Landman, Papanicolaou, Sulem, and Sulem 1988).

To estimate the scaling factor $L(t)$ in the limit $t \to t_*$, we write $L(t) = (t_* - t)^{1/2} q(t)$ where q is a slowly varying function. From $\frac{d\tau}{dt} = L^{-2}$, one gets to leading order

$$\tau \approx \frac{1}{q^2(t)} \ln \frac{1}{t_* - t}, \tag{8.2.30}$$

provided $q(t)$ varies more slowly that $\ln \frac{1}{t_* - t}$. From (8.2.28), it follows that $a \approx \pi/\ln \ln \left(\frac{1}{t_* - t} \right)$. Since $a = -LL_t$, the corrective factor is asymptotically given by $q(t) \approx \left(2\pi/\ln \ln \frac{1}{t_* - t} \right)^{1/2}$ in agreement with the ordering assumption. This leads to (8.2.25) and (8.2.26).

Including the next order term in the asymptotic form of $\tau(t)$, one gets (6.2.13), which, as mentioned in Chapter 6, is accurately fitted by the simulations of Akrivis, Dougalis, Karakashian, and McKinney (1997).

Proposition 8.7. *The asymptotic form of the solution given in Proposition 8.5 extends in the range $0 < r < r_{\text{out}}$ where*

$$r_{\text{out}} \approx 1/\sqrt{b} \sim \ln \ln \frac{1}{t_* - t}. \tag{8.2.31}$$

Proof. In order to evaluate the range of validity of the asymptotic behavior given in Proposition 8.5, we prescribe that the contribution of this range to the L^2-norm of the solution, remains finite. For this purpose, we estimate the L^2-norm $N = \int_0^\infty |u|^2 \xi^{(2/\sigma)-1} d\xi$ by dividing the integration domain into several ranges.

In the range $0 < \xi < \xi_0$ with $1 \ll \xi_0 \ll 2/a$, $|u|$ is approximated by the ground state R. In the range $\xi_0 < \xi < 2/a - \alpha$ (where α is taken small), we use the WKB approximation and write

$$\int_{\xi_0}^{2/a-\alpha} |u|^2 \xi^{\frac{2}{\sigma}-1} d\xi \approx \nu_0^2 \frac{2}{a} \int_{a\xi_0/2}^{1-\alpha'} \frac{1}{(1-y^2)^{1/2}} e^{-\frac{4}{a}\int_0^y \sqrt{1-s^2} ds} dy$$

$$\approx \nu_0^2 \frac{2}{a} \int_0^{\pi/2} e^{-\frac{2}{a}(\theta - \frac{1}{2}\sin 2\theta)} d\theta \approx \frac{1}{2} \nu_0^2 \tag{8.2.32}$$

by the Laplace method.

In the range $2/a - \alpha < \xi < 2/a + \alpha$, the solution is approximated in terms of the Airy function, and the contribution to N is negligible.

Finally, in the range $2/a + \alpha < \xi < \xi_{\text{out}}$, the integral reduces to

$$\frac{2}{a} C_1^2 \int_1^{a\xi_{\text{out}}/2} \frac{1}{(y^2-1)^{1/2}} dy \approx \frac{2C_1^2}{a} \ln(a\xi_{\text{out}}) \approx K \frac{1}{a} e^{-\pi/a} \ln(a\xi_{\text{out}}), \tag{8.2.33}$$

where K is a pure numerical constant. In order to keep this contribution finite, ξ_{out} should obey, up to numerical factors,

$$\ln(a\xi_{\text{out}}) \propto a e^{\pi/a} \approx \frac{\pi\tau}{\ln \tau}. \tag{8.2.34}$$

Noticing that

$$\ln \frac{1}{L} = \int^\tau \frac{\pi}{\ln \tau} d\tau \approx \frac{\pi\tau}{\ln \tau}, \tag{8.2.35}$$

we obtain that the condition (8.2.34) is fulfilled by taking $\xi_{\text{out}} \sim 1/(aL)$, or equivalently,

$$r_{\text{out}} \propto \frac{1}{a} \propto \ln \ln \frac{1}{t_* - t}, \tag{8.2.36}$$

which reproduces the result of Bergé and Pesme (1993b).

8.2.2 Effects of Mass and Hamiltonian Radiation

As noted by Malkin (1993), the lack of exact self-similarity of the solution near critical collapse, implies a slow variation of the mass N_s and of the

Hamiltonian H_s associated to the collapson. A simple but crude way to evaluate this variation is to consider the asymptotic form of the solution

$$\psi = \frac{1}{L(t)^{1/\sigma}} e^{i(\tau + \frac{L_t}{4L}r^2)} v, \qquad (8.2.37)$$

where

$$v \approx R + b\frac{\partial P}{\partial b}|_{b=0} = R + b\rho, \qquad (8.2.38)$$

and to compute N_s and H_s as integrals on the whole space. This yields

$$N_s \approx \int |v|^2 \xi^{d-1} d\xi \approx N_c + bM, \qquad (8.2.39)$$

where $M = 2 \int R\rho \xi^{d-1} d\xi = \frac{1}{4} \int \xi^2 R^2 \xi^{d-1} d\xi$, and

$$\frac{dN_s}{d\tau} \approx Mb_\tau \approx -2\nu_0^2 e^{-\pi/\sqrt{b}}. \qquad (8.2.40)$$

When a similar approach is used for the the Hamiltonian, one obtains

$$H_s L^2 = \int_0^\infty \left(|v_\xi|^2 - \frac{1}{\sigma+1}|v|^{2\sigma+2} + a\Im(\xi v\bar{v}_\xi) + \frac{a^2}{4}\xi^2|v|^2 \right) \xi^{\frac{2}{\sigma}-1} d\xi. \qquad (8.2.41)$$

Approximating v by means of (8.2.38), one gets

$$H_s \approx M\left((L_t)^2 - \frac{b}{L^2} \right) = -M\frac{a_\tau}{L^2}, \qquad (8.2.42)$$

when using $b = a^2 + a_\tau$ and $a = -L_t L$. However, v and thus $L^2 H_s$ have been computed with an error $o(b)$ and equation (8.2.42) does not provide any information.

A more accurate analysis is thus required. For this purpose, we define the mass N_s the Hamiltonian H_s in the collapson by

$$N_s = \int_0^{r_0} |\psi|^2 r^{d-1} dr \qquad (8.2.43)$$

$$H_s = \int_0^{r_0} \left(|\psi_r|^2 - \frac{1}{\sigma+1}|\psi|^{2\sigma+2} \right) r^{\frac{2}{\sigma}-1} dr, \qquad (8.2.44)$$

where $r_0 = L\xi_0$ and $\xi_0 \approx 2/\sqrt{b}$ is the turning point in eq. (8.1.55). A direct calculation gives

$$\frac{dN_s}{dt} = \xi_0^d L_t L^{d-1} |\psi(r_0, t)|^2 - 2L^{d-1}\xi_0^{d-1}\Im(\psi^*\psi_r(r_0, t)), \qquad (8.2.45)$$

$$\frac{dH_s}{dt} = L_t L^{d-1}\xi_0^d \left(|\psi_r|^2 - \frac{1}{\sigma+1}|\psi|^{2+4/d} \right)\Big|_{r=L\xi_0}$$

$$+ L^{d-1}\xi_0^{d-1}\Im(\psi_r\psi_{rr}^* + |\psi|^{4/d}\psi^*\psi_r)\Big|_{r=L\xi_0}. \qquad (8.2.46)$$

In terms of v defined in (8.2.37), one has have the form

$$\frac{dN_s}{dt} = -\frac{2}{L^2}\xi_0^{d-1}\Im(v^* v_\xi) \qquad (8.2.47)$$

$$\frac{dH_s}{dt} = L^{-4}a\xi_c^0|v_\xi|^2 - \frac{\sigma}{\sigma+1}L^{-4}a\xi_c^0|v|^{2+4/d} + 2L^{-4}\xi_0^{d-1}\Im(v_\xi v_{\xi\xi}^*)$$

$$-L^{-4}a\xi_0^{d-1}\Re(vv_{\xi\xi}^*) - \frac{1}{2}L^{-4}a^2\xi_0^{d+1}\Im(v_\xi v^*) + \frac{1}{2}L^{-4}a\xi_0^{d-1}\partial_\xi(|v|^2)$$

$$+2L^{-4}\xi_0^{d+1}\Im(|v|^{4/d}v^* v_\xi). \qquad (8.2.48)$$

where v and its derivatives are evaluated at ξ_0.

When the singularity time is approached, the function v is approximated by the solution P of (8.2.8), which, near the turning point $2/\sqrt{b}$, is written as $P = \xi^{(1-d)/2}Z$ where Z is approximated by (8.1.63). This gives

$$v(\xi) \approx \xi^{\frac{1-d}{2}}v_0 2^{-1/4}e^{-\pi/(2\sqrt{b})}e^{i\pi/4}\frac{1}{\left(\frac{\sqrt{b}\xi}{2}-1\right)^{1/4}}e^{\frac{4i\sqrt{2}}{3\sqrt{b}}\left(\frac{\sqrt{b}\xi_c}{2}-1\right)^{3/2}}, \qquad (8.2.49)$$

leading to

$$\frac{dN_s}{dt} \approx -2\frac{v_0^2}{L^2}e^{-\pi/\sqrt{b}}, \qquad (8.2.50)$$

or in terms of the rescaled time

$$\frac{dN_s}{d\tau} \approx -2v_0^2 e^{-\pi/\sqrt{b}} \approx M b_\tau. \qquad (8.2.51)$$

The mass contained in the collapson is thus radiated at an exponentially small rate, which reproduces the result obtained above by a cruder analysis.

We now consider the Hamiltonian. When we substitute (8.2.49) into the right-hand side of (8.2.48) the leading contribution originates from the term $-\frac{1}{2}L^{-4}a^2\xi_c^{d+1}\Im(v_\xi v^*)$. This gives

$$\frac{dH_s}{dt} \approx -2L^{-4}v_0^2 e^{-\pi/\sqrt{b}}, \qquad (8.2.52)$$

or in terms of the rescaled time

$$\frac{dH_s}{d\tau} \approx -2L^{-2}v_0^2 e^{-\pi/\sqrt{b}}. \qquad (8.2.53)$$

At this step, using that $-LL_t = a \approx b^{1/2}$, we rewrite (8.2.52) in the form

$$\frac{dH_s}{dt} \approx 2\frac{v_0^2}{\sqrt{b}}e^{-\pi/\sqrt{b}}\int^t \frac{L_t}{L^3}dt, \qquad (8.2.54)$$

where the variation of b is neglected compared to that of L. This finally leads to

$$L^2 H_s \approx -\frac{v_0^2}{\sqrt{b}}e^{-\pi/\sqrt{b}}. \qquad (8.2.55)$$

8.2.3   Adiabatic Approximation and Blowup Time Estimate

The evolution of the scaling factor $L(t)$ is given by the two equations

$$L_{tt} = -\frac{b}{L^3},$$ (8.2.56)

$$b_t = -\frac{\nu(b)}{L^2}.$$ (8.2.57)

Multiplying (8.2.56) by $2L_t$, one gets

$$((L_t)^2)_t = -b\frac{d}{dt}\left(\frac{1}{L^2}\right),$$ (8.2.58)

where the variation of b is very slow compared to that of L. This leads to

$$(L_t)^2 \approx \frac{b}{L^2},$$ (8.2.59)

or equivalently,

$$L(t) \sim \sqrt{2b^{1/2}(t_* - t)},$$ (8.2.60)

This is the "adiabatic law" given by Malkin (1993). Following Fibich (1996b), a more accurate description is obtained by retaining a constant in the integration of (8.2.58), thus replacing (8.2.59) by (8.2.42) in the form

$$(L_t)^2 \approx \frac{b}{L^2} + \frac{H_s}{M}.$$ (8.2.61)

Neglecting the variations of N_s and H_s compared to that of L, one gets, after multiplication by L^2 and integration,

$$(L^2)_t \approx \pm 2\left(b + \frac{H_s}{M}L^2\right)^{1/2}.$$ (8.2.62)

Integrating once more, one obtains

$$L^2(t) \approx L_0^2 \pm 2(b + \frac{H_s}{M}L_0^2)^{1/2}(t - t_0) + \frac{H_s}{M}(t - t_0)^2,$$ (8.2.63)

where t_0 is chosen sufficiently close to the singularity for the asymptotic regime to hold, and $L_0 = L(t_0)$. Observing that at the singularity time $L^2(t_*) = 0$, one has the estimate

$$t_* = \frac{L_0^2}{\sqrt{b} + \sqrt{b + H_s L_0^2/M}} \quad \text{when} \quad L_t(t_0) \leq 0$$ (8.2.64)

$$t_* = \frac{L_0^2}{\sqrt{b} - \sqrt{b + H_s L_0^2/M}} \quad \text{when} \quad L_t(t_0) > 0 \quad \text{and} \quad H_s < 0,$$ (8.2.65)

No blowup occurs if $L_t(0) > 0$ and $H_s > 0$. In (8.2.64)–(8.2.65), the parameter b is evaluated in terms of the mass by $N_s \approx N_c + Mb$. When the

initial conditions are sufficiently close to the asymptotic regime, this provides an explicit value of t_* which is supported by numerical simulations
with a satisfactory accuracy (Fibich 1996b).

8.2.4 Relation with Standing-Wave Instability

An interpretation of the deviation from self-similar collapse at critical dimension in terms of the marginal instability of the standing wave $e^{i\omega t}g_\omega$
associated to the ground state g_ω was proposed by Pelinovsky (1998). The
ground state satisfies

$$\Delta g_\omega - \omega g_\omega + g_\omega^{2\sigma+1} = 0, \tag{8.2.66}$$

and

$$\frac{d}{d\omega} \int g_\omega^2 d\mathbf{x} = 0. \tag{8.2.67}$$

A slower time $T = \epsilon t$ is then formally introduced, and the solution of the
NLS equation written as a modulation of the standing wave in the form

$$\psi = \varphi(x, T) \, e^{\frac{i}{\epsilon} \int_0^T \omega(s)ds} \tag{8.2.68}$$

with

$$\varphi(x, T) = g_\omega + \sum_k i\epsilon^{2k-1}\phi_{2k-1}(x, \omega, T) + \sum_k \epsilon^{2k}\phi_{2k}(x, \omega, T). \tag{8.2.69}$$

The parameter ω depends on the time variable T and the leading order
term can be interpreted as the usual ground state R, rescaled by the time-
dependent scaling factor $\omega(T)$, in the form $g_\omega(\mathbf{x}) = \omega^{1/(2\sigma)}R(\omega^{1/2}\mathbf{x})$.

A hierarchy of inhomogeneous linear equations is then derived for the
successive corrective terms. To leading order, one has

$$L_0\phi_1 = \frac{dg_\omega}{dT} \tag{8.2.70}$$

where L_0 is the linearized operator (4.1.5) and

$$\frac{d}{dT} = \frac{\partial}{\partial T} + \frac{d\omega}{dT}\frac{\partial}{\partial \omega}. \tag{8.2.71}$$

This equation is conveniently solved when L_0 is rewritten in the form
(4.1.8). Following the author, we consider for the sake of simplicity the
problem in one dimension (and thus a critical nonlinearity exponent $\sigma = 2$).
One has

$$\phi_1 = -g_\omega \int_0^x \frac{dx'}{g_\omega^2(x')} \int_0^{x'} g_\omega(x'')\frac{d}{dT}g_\omega(x'')dx'' \tag{8.2.72}$$

where the integrand is not singular because g_ω does not vanish. At large
distance $(x \to \infty)$, $g_\omega \sim A(\omega)e^{-\sqrt{\omega}|x|}$, and one obtains the asymptotic

approximation

$$\phi_1 \approx -\frac{1}{8}\frac{1}{\sqrt{\omega}A(\omega)}\frac{dN_s(\omega)}{dT}e^{\sqrt{\omega}|x|} - \frac{1}{8\omega}x^2e^{-\sqrt{\omega}|x|}, \qquad (8.2.73)$$

where $N_s = \int_{-\infty}^{+\infty} g_\omega^2 dx$ is the mass of the ground state. When substituting into (8.2.69) the expression of ϕ_1 given in (8.2.73), the second term of this expression can be viewed as resulting from the expansion of a complex phase, while the first term introduces an imaginary correction which diverges at large distance. One is then led to write for $x \to \infty$,

$$\varphi \approx A\left(\omega e^{-\sqrt{\omega}|x|} - \frac{i\epsilon}{8\sqrt{\omega}A(\omega)}\frac{dN_s}{dT}e^{\sqrt{\omega}|x|}\right)\exp\left(-\frac{i\epsilon}{8\omega}\frac{d\omega}{dT}x^2\right). \qquad (8.2.74)$$

The exponential contribution plays a role analogous to the complex term $i\nu P$ in (8.2.8), where it originates either from the deviation from criticality or from a complex correction to the eigenvalue problem of the linear operator associated to (8.2.8).

The analysis then proceeds in a way similar to that of Section 8.2, where the spatial domain is divided in three regions: the inner region where the profile φ is well described by the ground state with a quadratic phase, an intermediate region where φ satisfies a parabolic cylinder equation, and an outer region where ψ evolves according to the linear NLS equation $i\psi_t + \Delta\psi = 0$. Matching the boundary conditions in the various regions leads to the rate of variation of the wave energy in the form of (8.2.40). The relation with the analysis described in the previous sections is obtained by identifying $\frac{1}{\epsilon}\int_0^T \omega(s)ds$ with τ and $\frac{\epsilon}{2\omega^2}\frac{d\omega}{dT}$ with $a(\tau)$.

This method introduces a small parameter ϵ necessary to start the modulation analysis. However, as the singularity is approached, the scale separation is not preserved any more, and ϵ should be taken equal to 1 in order that the final results

$$\theta = \tau = \frac{1}{\epsilon}\int_0^T \omega(s)ds \qquad (8.2.75)$$

with

$$\omega \propto \frac{\ln|\ln(T_* - T)|}{T_* - T} \qquad (8.2.76)$$

reproduce the previous analyses of this problem. The time T reduces to the physical time and the initial assumption of a slow variation of the scaling factor ω compared to the phase θ, breaks down as the dynamics evolves from the slow modulation of a ground-state standing wave to the development of a finite-time blowup.

9
Perturbations of Focusing NLS

This chapter discusses various situations that can be viewed as perturbations of the elliptic cubic NLS equation. The present understanding results from numerical simulations and perturbative analysis near critical collapse.

9.1 Wave Dissipation at Critical Dimension

In this section we consider the NLS equation at critical dimension $\sigma d = 2$, with an additional dissipative term of the form $i\epsilon|\psi|^p\psi$ with $p = 2\sigma(1 + s)$ and $s > 0$, where ϵ is assumed to be small, in the form

$$i(\psi_t + \epsilon|\psi|^p\psi) + \Delta\psi + |\psi|^{2\sigma}\psi = 0 \qquad (9.1.1)$$

9.1.1 Effect of an Individual Collapse

As in the supercritical case, a balance argument can lead to an estimate of the amplitude of the wave and the size of the collapson when the dissipation enters into play. Neglecting the double log corrections, (7.2.4) indicates that the wave energy contained in the collapson and thus available for dissipation is finite and independent of ϵ, a property referred to as *strong* collapse. A more detailed analysis, taking into account the precise behavior of collapsing solutions, was performed by Dyachenko, Newell, Pushkarev, and Zakharov (1992) and leads to the following estimates.

Proposition 9.1. *In the limit $\epsilon \to 0$, the amount of dissipated energy ΔN in an individual collapse scales like*

$$\Delta N \propto \left(\ln\ln\frac{1}{\epsilon}\right)^{-2}. \tag{9.1.2}$$

Proof. When the dissipation coefficient ϵ is small, it plays a role only when the amplitude of the solution becomes large enough. At earlier times, the solution evolves as if it was approaching critical blowup. In order to estimate the dissipation, the solution is written

$$\psi(\mathbf{x}, t) = \frac{1}{L^{d/2}} e^{i\tau - ia\xi^2/4} v(\xi, \tau), \tag{9.1.3}$$

where v satisfies

$$i(v_\tau + \frac{\epsilon}{L(\tau)^{2s}}|v|^p v) + v_{\xi\xi} + \frac{2/\sigma - 1}{\xi}v_\xi - v + \frac{1}{4}b(\tau)\xi^2 v + |v|^{2\sigma}v = 0. \tag{9.1.4}$$

The dissipation ΔN is given by

$$\Delta N = C\epsilon \int_0^\infty \frac{1}{L(\tau)^{2s}}d\tau \tag{9.1.5}$$

with

$$C = \int |v|^{p+2}\xi^{d-1}d\xi \approx \int |R|^{p+2}\xi^{d-1}d\xi. \tag{9.1.6}$$

To evaluate the time integral, a modulation analysis is performed in the spirit of that developed in the context of critical collapse in order to obtain a dynamical equation for the function $a(\tau) = -L_\tau/L$. For this purpose, (9.1.4) is solved perturbatively by writing

$$v = P(\xi, b(\tau)) + W, \tag{9.1.7}$$

where P satisfies (8.2.8) with $\nu(\sqrt{b})$ given by (8.2.11), and W is a remainder. One proceeds as in Section 8.2. The equations for the real and imaginary parts of $W = S + iT$ reduce at leading order to

$$S_{\xi\xi} + \frac{2/\sigma - 1}{\xi}S_\xi - S + (2\sigma + 1)R^{2\sigma}S = \frac{\nu(\sqrt{b})}{2\sqrt{b}}\frac{1}{\xi}R_\xi, \tag{9.1.8}$$

$$T_{\xi\xi} + \frac{2/\sigma - 1}{\xi}T_\xi - T + R^{2\sigma}T = -b_\tau\rho - \nu(\sqrt{b})R - \frac{\epsilon}{L^{2s}}R^{1+p}, \tag{9.1.9}$$

where $\sqrt{b} \approx a$. Since the ground state R belongs to the null space of the differential operator arising in the left-hand side of (9.1.9), the solvability of this equation requires

$$aa_\tau + \alpha_1 e^{-\pi/a} + \alpha_2 \frac{\epsilon}{L^{2s}} = 0, \tag{9.1.10}$$

where $\alpha_1 = \nu_0^2/M$ and $\alpha_2 = \frac{1}{M}\int |R|^{p+2}\xi^{\frac{2}{\sigma}-1}d\xi$.

As long as dissipation is negligible, $a(\tau) \approx \pi/\ln\tau$. Denoting by τ_0 an estimate of the moment when dissipation comes into play, we rescale $a(\tau)$ by defining

$$A(\theta) = \mu a(\tau), \quad \text{with} \quad \mu = \frac{1}{\pi}\ln\tau_0 \gg 1, \tag{9.1.11}$$

where the new time scale θ is given by

$$\tau = \tau_0 + \mu\theta. \tag{9.1.12}$$

Noting that $L(\tau) = L_0 e^{-\int_0^\tau a(\tau')d\tau'}$, the modulation equation (9.1.10) becomes

$$\frac{1}{\mu^3}AA_\theta + \alpha_1 e^{-\pi\mu/A} + \alpha_2 L_0^{-2s}\epsilon e^{2s\int_0^{\tau_0} a(\tau')d\tau'} e^{2s\int_0^\theta A(\theta')d\theta'} = 0. \tag{9.1.13}$$

The time τ_0 is estimated by writing

$$\epsilon e^{2s\int_0^{\tau_0} a(\tau')d\tau'} \propto \frac{1}{\mu^3}. \tag{9.1.14}$$

Using that $\mu = \frac{1}{\pi}\ln\tau_0$ and that a is slowly varying, one gets that $\tau_0 \propto \ln\frac{1}{\epsilon}$ and $\mu \propto \ln\ln\frac{1}{\epsilon}$.

One can now estimate ΔN. The dissipation being negligible for $\tau < \tau_0(\epsilon)$, one has

$$\Delta N \approx C\epsilon \int_{\tau_0}^\infty \frac{1}{L^{2s}}d\tau = C\frac{\epsilon}{L_0^{2s}}e^{-2s\int_0^{\tau_0} a(\tau')d\tau'} \int_{\tau_0}^\infty e^{-2s\int_{\tau_0}^\tau a(\tau')d\tau'}d\tau$$

$$\propto \frac{1}{\mu^2}\int_0^\infty e^{-2s\int_0^\theta A(\theta')d\theta'}d\theta \propto \left(\ln\ln\frac{1}{\epsilon}\right)^{-2}. \tag{9.1.15}$$

Linear dissipation is often considered to model Landau damping in plasmas, in the form $\int \gamma(k)\hat{\psi}(\mathbf{k},t)e^{i\mathbf{k}\cdot\mathbf{x}}d\mathbf{k}$ where $\gamma(k) = k^2 h(k/k_D)$ and h is a cutoff function that makes a smooth transition from zero to a constant value as k increases through k_D. In this case, the amount of the burnout energy varies as $(\ln\ln k_D)^{-1}$. More precisely, numerical simulations in one space dimension and quintic nonlinearity show that the fraction of the energy of the Townes soliton absorbed in the collapse is given by Dyachenko, Newell, Pushkarev, and Zakharov (1992),

$$r = \frac{\Delta N}{N_c} \approx \frac{1}{9.5\ln\ln k_D - 8.7}, \tag{9.1.16}$$

where $N_c = \int R^2\xi^{(2/\sigma)-1}d\xi$ is the energy of the Townes soliton. In the considered range $200 < k_D < 900$, r varies from 15 to a few percent. A fraction of 15% to 25% is reported in the case of a nonlinear dissipation by Shvets and Zakharov (1990). The multifocus structure of a light beam developing in a medium with a weak nonlinear dissipation is described by Dyshko, Lugovoĭ, and Prokhorov (1972).

9.1.2 The Turbulent Regime

The effect of a large number of collapses is considered by Newell, Rand, and Russel (1988). In such a turbulent regime, the energy dissipation results not only from the sudden energy transfer from large to small scales due to the collapses, but also from the resonant wave interactions, whose contribution can be estimated in the framework of the wave turbulence theory (Zakharov, L'vov , and Falkovich 1992, Dyachenko, Newell, Pushkarev, and Zakharov 1992). Numerical simulations performed in the fully turbulent regime show that the contributions from the two effects are comparable, while in situations of low amplitude waves, the dissipation due to wave turbulence is negligible compared to that due to collapses.

The mean dissipation rate $\langle \gamma \rangle$ due to collapses is given by

$$\langle \gamma \rangle = \omega_c \langle r \rangle N_c, \tag{9.1.17}$$

where ω_c denotes the occurrence frequency of collapses and $\langle r \rangle$ the energy fraction burned out on average in a collapse event. Numerical simulations show that the product $\langle \gamma \rangle \omega_c^{-1}$ is almost independent of the mean turbulent energy level. On the other hand, $\omega_c \langle r \rangle$ appears to be independent of the mechanism of dissipation and of the scale at which it takes place, since, for example, as k_d is increased, the factor $\langle r \rangle$ is slowly reduced, but the frequency ω_c of the collapses increases accordingly, so that the average dissipation rate remains essentially unchanged. Estimates of these quantities in the context of the solar corona and of the interstellar medium are presented in Champeaux, Gazol, Passot, and Sulem (1997).

9.2 Non-Elliptic Schrödinger Equation

We consider here the non-elliptic NLS equation of the form

$$i\psi_t + \sum_{i=1}^{d} \varepsilon_i \frac{\partial^2 \psi}{\partial x_i^2} + f(|\psi|^2)\psi = 0, \tag{9.2.1}$$

where all the constant coefficients ε_i do not have the same sign and f denotes a polynomial having no constant or linear terms. In some instances, it is useful to rescale the various coordinates and take $\varepsilon_i = \pm 1$ and $\varepsilon_j \neq \varepsilon_k$ for some (j, k). The existence of a local solution for this equation is a consequence of the decay estimates and the Strichartz inequalities, which have the same form for the elliptic or non-elliptic linear Schrödinger equations (Ghidaglia and Saut 1990, see also Remark 1 at the end of Section 3.1). It is still unknown whether initially smooth solutions can blow up in a finite time.

An application to plasma physics concerns, for example, the dynamics of electrostatic wave packets propagating in the direction of an ambient magnetic field (Nishinari, Abe, and Satsuma 1994). Another important example arises in the context of the propagation of an ultra-short laser pulse in a medium with normal time dispersion (see Chapter 1).

It is important to notice that the non-elliptic NLS equation does not admit nontrivial solutions of the form of standing waves $\psi(\mathbf{x}, t) = e^{i\omega t} S(\mathbf{x})$ with a localized profile S and a frequency $\omega \in \mathbf{R}$ (see, e.g., Ghidaglia and Saut 1996). Such a profile would indeed satisfy

$$- \omega S + \sum_{i=1}^{d} \varepsilon_i \frac{\partial^2 S}{\partial x_i^2} + f(|S|^2)S = 0. \qquad (9.2.2)$$

Multiplying (9.2.2) by $x_j \frac{\partial S^*}{\partial x_j} - x_k \frac{\partial S^*}{\partial x_k}$, taking the real part, and integrating over $\mathbf{R}^d$ leads, after simple calculations, to

$$- \varepsilon_j \int \left| \frac{\partial S}{\partial x_j} \right|^2 d\mathbf{x} + \varepsilon_k \int \left| \frac{\partial S}{\partial x_k} \right|^2 d\mathbf{x} = 0. \qquad (9.2.3)$$

Using this equality for various couples $(\varepsilon_j, \varepsilon_k)$ of opposite signs, it follows that ∇S and thus S vanish identically.

9.2.1 Estimates for the Partial Variances

We restrict ourselves to the non-elliptic NLS equation with a power law nonlinearity $|\psi|^{2\sigma}\psi$. Following Bergé, Kuznetsov, and Rasmussen (1996) and Bergé and Rasmussen (1996), we establish estimates concerning the evolution of the partial variances. Although they do not lead to a definite conclusion on the existence of blowup solutions, they can shed some light on the dynamics.

Using the notations of Section 2.4.2, we first estimate $|\nabla_\pm \psi|^2_{L^2}$. For this purpose, defining $\psi = Ae^{i\varphi}$ with A and φ real, we write

$$|\nabla_\pm \psi|^2_{L^2} = |\nabla_\pm A|^2_{L^2} + |A\nabla_\pm \varphi|^2_{L^2}, \qquad (9.2.4)$$

where the two terms in the right-hand side are estimated separately. Using the uncertainty principle (5.1.7)

$$\int |A|^2 d\mathbf{x} \le \frac{2}{d_\pm} |\mathbf{x}_\pm A|_{L^2} |\nabla_\pm A|_{L^2}, \qquad (9.2.5)$$

it follows that

$$|\nabla_\pm A|^2_{L^2} \ge \frac{d_\pm^2}{4} \frac{N^2}{V_\pm}. \qquad (9.2.6)$$

On the other hand, from the equation

$$\frac{1}{2} \partial_t(A^2) + \nabla_+ \cdot (A^2 \nabla_+ \varphi) - \nabla_- \cdot (A^2 \nabla_- \varphi) = 0 \qquad (9.2.7)$$

one gets

$$\dot{V}_{\pm} = \pm 4 \int \mathbf{x}_{\pm} \cdot A^2 \nabla_{\pm} \varphi \, d\mathbf{x} \le 4 |A\nabla\varphi|_{L^2} |\mathbf{x}_{\pm} A|_{L^2}, \qquad (9.2.8)$$

where the dot indicates time derivative. It follows that

$$|A\nabla\varphi|_{L^2}^2 \ge \frac{\dot{V}_{\pm}^2}{16V_{\pm}}. \qquad (9.2.9)$$

By substituting in (9.2.4), one gets

$$|\nabla_{\pm}\psi|_{L^2}^2 \ge \frac{d_{\pm}^2}{4} \frac{N^2}{V_{\pm}} + \frac{\dot{V}_{\pm}^2}{16V_{\pm}}. \qquad (9.2.10)$$

Consider first the partial variance V_-. It results from the above estimates that

$$\ddot{V}_- \ge 8|\nabla_-\psi|_{L^2}^2 \ge 8\left(\frac{d_-^2}{4}\frac{N^2}{V_-} + \frac{\dot{V}_-^2}{16V_-}\right). \qquad (9.2.11)$$

One easily checks that this implies

$$\frac{d^2}{dt^2}(V_-^{1/2}) \ge \frac{d_-^2 N^2}{V_-^{3/2}}. \qquad (9.2.12)$$

Assume now that the beam focuses in the directions associated to the components of $\mathbf{x}_-$, in the sense that $\dot{V}_- < 0$. After multiplying the above inequality by $\dot{V}_-$, one gets

$$\frac{d}{dt}\left(\left(\frac{d}{dt}V_-^{1/2}\right)^2 + \frac{d_-^2 N^2}{V_-}\right) \le 0. \qquad (9.2.13)$$

It follows that if V_- decreases, so does $(\frac{d}{dt}V_-^{1/2})^2 + \frac{d_-^2 N^2}{V_-}$. As a consequence, V_- cannot vanish. It is thus possible to estimate

$$\frac{d^2}{dt^2}(V_-^2) \ge 2V_-\ddot{V}_- \ge 16|\nabla_-\psi|_{L^2}^2 V_- \ge 4d_-^2 N^2, \qquad (9.2.14)$$

indicating that if the solution exists for long time, $\dot{V}_-$ should become positive after some time.

When $\dot{V}_- > 0$, one has

$$\frac{d}{dt}\left((\frac{d}{dt}V_-^{1/2})^2 + \frac{d_-^2 N^2}{V_-}\right) \ge 0. \qquad (9.2.15)$$

V_- increases, and one concludes that it grows at least like t^2. It follows that if the solution remains smooth, the bulk of the wave amplitude will tend to move away in the direction associated to the components of $\mathbf{x}_-$, either by spreading or by displacing the amplitude maximum.

One now turns to the partial variance V_+. The equation for $\ddot{V}_+$ is rewritten

$$\frac{d^2}{dt^2}V_+ = 4\left[H + |\nabla_+\psi|_{L^2}^2 + |\nabla_-\psi|_{L^2}^2 + \frac{1 - \sigma d_+}{\sigma + 1}\int |\psi|^{2(\sigma+1)}d\mathbf{x}\right]$$

$$\geq 4\left[H + |\nabla_+\psi|_{L^2}^2 + \frac{1 - \sigma d_+}{\sigma + 1}\int |\psi|^{2(\sigma+1)}d\mathbf{x}\right], \qquad (9.2.16)$$

which leads to partial conclusions in two important special cases.

The Case $d_+ = d_- = 1/\sigma$

This corresponds to the two-dimensional "hyperbolic" NLS equation which arises for example in the water-wave problem for deep water (see Section 11.1). One has in this case

$$\frac{d^2}{dt^2}V_+ \geq 4(H + |\nabla_+\psi|_{L^2}^2) \geq 4H + d_+^2\frac{N^2}{V_+} + \frac{1}{4V_+}\left(\frac{dV_+}{dt}\right)^2. \qquad (9.2.17)$$

Defining $X_+ = V_+^{3/4}$, this inequality becomes

$$\frac{d^2}{dt^2}X_+ \geq 3HX_+^{-1/3} + \frac{3}{4}d_+^2\frac{N^2}{X_+^{5/3}}. \qquad (9.2.18)$$

In the case of a solution focusing in the $\mathbf{x}_+$-directions ($\dot{V}_+ < 0$), then by multiplying the two sides of the above inequality by $\dot{X}_+$, one gets

$$\frac{d}{dt}\left[\left(\frac{dX_+}{dt}\right)^2 - 9HX_+^{2/3} + \frac{9}{4}d_+^2N^2X_+^{-2/3}\right] \leq 0, \qquad (9.2.19)$$

which implies that X_+ and thus V_+ cannot vanish. Furthermore, since

$$\frac{d^2V}{dt^2} \equiv \frac{d^2V_+}{dt^2} - \frac{d^2V_-}{dt^2} = 8H, \qquad (9.2.20)$$

then for large t, $V_+ \approx V_- + 4Ht^2$. Since H is assumed negative, one concludes that V_- should grow faster than $4|H|t^2$ in this limit.

The Case $d_+ = 2/\sigma$, $d_- = 1/\sigma$

For $\sigma = 1$, this situation arises in the context of nonlinear optics for a beam focusing in the transverse directions in the presence of a normal group velocity dispersion.

The variance identities become

$$\frac{d^2V_+}{dt^2} = 8(H + |\nabla_-\psi|_{L^2}^2) \geq 8H + \frac{2d_-^2N^2}{V_-}, \qquad (9.2.21)$$

$$\frac{d^2V_-}{dt^2} = -4H + 4|\nabla_-\psi|_{L^2}^2 + 4|\nabla_+\psi|_{L^2}^2 \geq -4H + \frac{d_+^2N^2}{V_+}. \qquad (9.2.22)$$

When assuming focusing in all the directions ($\dot{V}_\pm < 0$), one easily gets

$$\frac{d}{dt}\left(\frac{dV_+}{dt}\frac{dV_-}{dt} - 8HV_- + 4HV_+ - 2d_-^2 N^2 \ln V_- - d_+^2 N^2 \ln V_+ \right) \leq 0.$$

(9.2.23)

Since the beam is assumed to focus in the directions associated to the components of $\mathbf{x}_-$, the partial variance V_- is bounded from above, and as previously shown, also from below. It follows that V_+ cannot vanish in this case. Vanishing of V_+ is, however, not excluded if the beam spreads in the $\mathbf{x}_-$ directions ($\dot{V}_- > 0$). As previously shown, such a spreading will take place after some time if the solution remains smooth.

9.2.2 Numerical Observations in Two Dimensions

As already mentioned, an important example of the two-dimensional "hyperbolic" Schrödinger equation,

$$i\partial_t\psi + \partial_{xx}\psi - \partial_{yy}\psi + |\psi|^2\psi = 0,$$

(9.2.24)

comes from the envelope dynamics of weakly-nonlinear surface gravity-wave trains in deep water (Zakharov 1968b) (see Section 11.1). It also arises in various problems of plasma physics (Pereira, Sen, and Bers 1978, Litvak, Petrova, Sergeev, and Yunakovskii 1983).

Numerical integrations in spatially periodic domains show no tendency to collapse, and Fermi–Pasta–Ulam recurrence was reported (Yuen and Ferguson 1978b). However, in contrast with the focusing one-dimensional problem (Yuen and Ferguson 1978a, Hafizi 1981), this recurrence could be only approximate, since a small fraction of the energy is pumped to higher frequencies during each cycle because the region of unstable wave vectors is unbounded (Martin and Yuen 1980, Yuen and Lake 1980, Litvak, Petrova, Sergeev, and Yunakovskii 1983).

The nonlinear development of the instability of a one-dimensional soliton subject to a transverse harmonic perturbation with a wavelength large compared to the width of the soliton (Zakharov and Rubenchik 1974, see also Section 1.6) is described by Pereira, Sen, and Bers (1978), who observed the breaking of the soliton in smaller pieces that cannot keep themselves together but spread out throughout the domain. It is indeed expected that any localized wave packet disperses outward asymptotically in time or in other words, if a solution exists in the large in time, it is not localized. This observation comes from the variance identity that for two-dimensional hyperbolic NLS equation reduces to (Myra and Liu 1980)

$$\frac{d^2}{dt^2}\int |\mathbf{x}|^2|\psi|^2 d\mathbf{x} = 8\int |\nabla\psi|^2 d\mathbf{x}.$$

(9.2.25)

If $\int |\nabla\psi|^2 d\mathbf{x}$ is bounded from below in the form $\int |\nabla\psi|^2 d\mathbf{x} > \epsilon$, it follows that for large t, $\int |\mathbf{x}|^2|\psi|^2 d\mathbf{x} > 8\epsilon t^2$, contradicting the hypothesis of local-

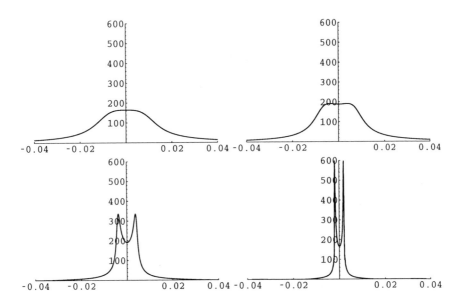

FIGURE 9.1. Snapshots at the propagation distances $z = 0.04216,\ 0.04221,$ $0.04231,\ 0.04234$ (from left to right and top to bottom), of the amplitude $|\psi(0, t, z)|$ plotted versus t for the solution of the 3D non-elliptic NLS equation with $\epsilon = 1$ and initial condition $\psi_0(r, t) = 6e^{-\frac{r^2}{2} - \frac{t^2}{2}}$.

ized solutions. Numerical simulations using Gaussian initial conditions with various degrees of anisotropy are presented in Litvak, Petrova, Sergeev, and Yunakovskii (1988). For short time, a monotonic expansion of the distribution is seen along the y-axis, accompanied with complicated changes in the width along the x-direction. This leads to a breakup of the structure and formation of additional maxima. There is no beam collapse, and for large t, the dynamics always convert into defocusing.

9.2.3   Numerical Simulations in Three Dimensions

We consider the NLS equation

$$i\partial_z\psi + \Delta_\perp\psi - \epsilon\partial_{tt}\psi + |\psi|^2\psi = 0, \qquad (9.2.26)$$

in the form used in optics where it is viewed as an initial value problem in the propagation coordinate. As established in Section 1.2.4, it corresponds to the case of a normal (i.e., negative) time dispersion.

When there is a range in the t variable such that the L^2-norm in the transverse directions (the beam power) of the incident wave is several times critical, numerical simulations show that transverse focusing leads to the splitting of the original pulse in the t variable (Zharova, Litvak, Petrova, Sageev, and Yunakovskii 1986, Chernev and Petrov 1992, Rothenberg 1992). Experimental observations of this phenomenon including multiple splitting when the power beam is sufficient are reported by Ranka, Schirmer, and Gaeta (1996) and more recently by Diddams, Eaton, Zozulya, and Clement (1998). Simulations were performed by Bergé, Rasmussen, Kuznetsov, Shapiro, and Turitsyn (1996) with a Gaussian initial condition $\psi_0(r,t) = Ae^{-\frac{r^2}{2} - \frac{t^2}{32}}$ and $\epsilon = 1$. They observed different evolutions according to the wave amplitude. For $A < 1.5$, the pulse just spreads out and disperses for large z. For $1.5 < A < 1.7$, a single splitting is observed: The maximum of the pulse first slightly increases at the origin, then splits in t and eventually spreads out. For $1.7 < A < 2$, a multisplitting is reported, but this regime is at the limit of the accuracy of the simulation performed with a uniform grid in the $(\sqrt{r}, z)$-variables using a 251×256 resolution. For $A > 2$, the limitations of the numerical scheme prevent the authors from discriminating between a finite-distance collapse and a strong growth of the peak amplitude preceding a possible splitting.

Simulations using the dynamic rescaling method (with 300×400 grids points uniformly distributes in the rescaled variables) are displayed in Fig. 9.1 for initial conditions $\psi_0(r,t) = 6e^{-\frac{r^2}{2} - \frac{t^2}{2}}$. One splitting is clearly seen, but we cannot conclude about the further dynamics, due to the loss of accuracy resulting from the development of strong gradients even in the rescaled variables. A simulation with a larger amplitude (8 instead of 6) also displays a unique splitting, although amplitudes of several thousands are reached before accuracy is lost.

Finite-distance blowup cannot be excluded in the case of high-energy beams, and the existence of multiple splitting could require the presence of saturating processes. After the first splitting, the solution is fully three-dimensional, and the dynamics are not constrained to repeat themselves in the form of recursive splitting until the power of the individual structures become subcritical.

The effect of a nonuniform phase quadratically dependent on the radial coordinate for the incident pulse ("chirped pulse") was considered by Cao, Agrawal, and McKinstrie (1994) and Bergé, Rasmussen, Kuznetsov, Shapiro, and Turitsyn (1996). They conclude that with an appropriate choice of the sign of the chirp parameter, the self-focusing of the pulse can be enhanced. Another way of accelerating self-contraction of the beam consists in assuming that it propagates in an inhomogeneous medium characterized by a parabolic density profile (Bergé 1997 and references therein).

9.2.4 Effect of a Small Time Dispersion on Critical Collapse

Even when small, time dispersion can become significant after the beam starts to focus. Gradients in the t variable are indeed enhanced, because focusing takes place at different rate in the various transverse planes. A partial understanding of the effect of dispersion on critical collapse, is provided by the perturbative analysis presented in this section, where the parameter ϵ in (9.2.26) is small. It is based on the works of Luther, Newell, and Moloney (1994) and Fibich, Malkin, and Papanicolaou (1995), who established modulation equations based on the assumption that the solution remains close to the critical profile at fixed t. Luther, Newell, and Moloney (1994) assumed that the solution depends on t appears only through the location of the collapse point $z_0 = z_0(t)$. If the incident wave packet is maximum at $t = 0$, the first collapse also occurs at $t = 0$, where $z_0'(0) = 0$ and $z_0''(0) > 0$. Under this hypothesis,

$$\partial_t \psi = -\partial_t z_0 \partial_z \psi \qquad (9.2.27)$$

and

$$\partial_{tt} \psi = -\partial_{tt} z_0 \partial_z \psi + (\partial_t z_0)^2 \partial_{zz} \psi. \qquad (9.2.28)$$

Near $t = 0$, the last term is negligible, and (9.2.26) is rewritten as

$$i\psi_z + \Delta_\perp \psi + |\psi|^2 \psi + i\gamma(1 - i\gamma)^{-1}(\Delta_\perp \psi + |\psi|^2 \psi) = 0 \qquad (9.2.29)$$

with $\gamma = \epsilon \partial_{tt} z_0(0)$. In the above equation, one can take $1 - i\gamma \approx 1$.

Assuming radial symmetry in the (x, y) plane, the solution $\psi = \psi(\rho, t, z)$, with $\rho^2 = x^2 + y^2$, is rescaled as

$$\psi(\rho, t, z) \approx \frac{1}{L} v(\xi, \zeta) e^{i\zeta - ia\xi^2/4}, \qquad (9.2.30)$$

with

$$\xi = \frac{\rho}{L}, \quad \zeta = \int_0^z L(s)^{-2} ds, \ a = -LL_z. \qquad (9.2.31)$$

Substituting into (9.2.29), one gets

$$i\partial_\zeta v - v + \Delta_\xi v + \frac{b}{4}\xi^2 v + |v|^2 v = -\gamma\left(i\Delta_\xi v + a\xi\partial_\xi v + av - i\frac{a^2}{4}\xi^2 v + i|v|^2 v\right). \qquad (9.2.32)$$

At lowest order, one replaces $\Delta_\xi v + |v|^2 v$ in the right-hand side of the previous equation by $-i\partial_\zeta v + v - \frac{b}{4}\xi^2 v$ and gets

$$i\partial_\zeta v - v + \Delta_\xi v + \frac{b}{4}\xi^2 v + |v|^2 v = -\gamma\left(\partial_\zeta v + a\xi\partial_\xi v + av + iv - i\frac{b + a^2}{4}\xi^2 v\right). \qquad (9.2.33)$$

As for critical NLS, one develops v in the form $v = \chi_0 + \chi_1$, with $\chi_1 \ll \chi_0$, where χ_0 satisfies

$$\Delta\chi_0 - \chi_0 + \frac{b}{4}\xi^2\chi_0 + |\chi_0|^2\chi_0 - i\nu(\sqrt{b})\chi_0 = 0 \qquad (9.2.34)$$

and $\nu(\sqrt{b}) \approx e^{-\pi/\sqrt{b}}$. Following the proof of Proposition 8.5, one writes the equation satisfied by χ_1 at lowest order, and gets a solvability condition in the form

$$b_\zeta = -\frac{2N_c}{M}\nu(\sqrt{b}) - 2\gamma\left(\frac{N_c}{M} - \frac{a^2+b}{4}\right) \tag{9.2.35}$$

with $M = \frac{1}{4}\int \xi^2 R^2 \xi \, d\xi$ and $N_c = \int R^2 \xi \, d\xi$. Together with $b = a_\zeta + a^2$ and $a = -\frac{L_\zeta}{L}$, this provides the modulation equations associated to a weak time dispersion. Luther, Newell, and Moloney (1994) integrated numerically these reduced equations and compared them with direct integration of the NLS equation. They observed that the solution predicted by the modulation equations agrees quantitatively with the direct numerical simulations, and displays an arrest of the collapse and a decrease of the intensity at the center of the beam. At larger propagation distances, the solutions of the two problems do not coincide any longer due to the development of strong three-dimensional effects.

Fibich, Malkin, and Papanicolaou (1995) derived modulation equations without assuming that the solution depends on the variable t only through the location of the blowup point $z_0(t)$. The scaling factor L and thus ξ, ζ and a depend on t. This leads to partial differential equations for the modulation. They also suppose that the solution takes the self-similar form (9.2.30) and develop $\chi = \chi_0 + \chi_1$ with χ_0 satisfying (9.2.34). The solvability condition for the equation for χ_1 is

$$Mb_\zeta + 2N_c\nu(\sqrt{b}) = 2\epsilon L^3 \int \partial_{tt}\left(\frac{R}{L}e^{i\zeta - ia\xi^2/4}\right) Re^{-i\zeta + ia\xi^2/4}\xi \, d\xi. \tag{9.2.36}$$

At lowest order, the right-hand side of (9.2.36) is equal to $2\epsilon L^2 \zeta_{tt} \int R^2 \xi \, d\xi$, and (9.2.36) reduces to

$$b_z + \frac{2N_c}{ML^2}\nu(\sqrt{b}) = \frac{2\epsilon N_c}{M}\zeta_{tt}, \tag{9.2.37}$$

where

$$b = -L^3 L_{zz} \quad \text{and} \quad \zeta_z = L^{-2}. \tag{9.2.38}$$

Since the factor $\nu(\sqrt{b})$ is exponentially small in b, it can be neglected.

Numerical integration of this reduced system shows that as z increases, $1/L(z,t)$ as a function of t splits into two peaks, in good agreement with the numerical simulations performed on the primitive NLS equation. However, the validity of the system breaks down when the parameter b reaches large values.

Special solutions of the system (9.2.37)–(9.2.38) can be obtained under the assumption that there is a unique singularity curve $z_0(t)$ for the solution and that near this curve the solution depends mostly on the distance to this curve. This is expressed by writing

$$L(z,t) = L(z_0(t) - z), \quad \beta(z,t) = \beta(z_0(t) - z), \quad \zeta(z,t) = \zeta(z_0(t) - z). \tag{9.2.39}$$

Equation (9.2.37) then reduces to

$$b_z = \epsilon'(-z_0''\zeta_z + z_0'^2\zeta_{zz}), \tag{9.2.40}$$

where $\epsilon' = \frac{2N_c}{M}\epsilon$. Integrating once with respect to z and defining the variable $A = 1/L$, one has $b = A_{\zeta\zeta}/A)$, which leads to an Airy type equation for A in the form

$$A_{\zeta\zeta}(b_0 - \epsilon'z_0''\zeta)A + \epsilon'z_0'^2 A^3 = 0, \tag{9.2.41}$$

with $b_0 = b_0(t) = b(0,t) - \epsilon'z_0'^2 L(0,t)^{-2}$. A discussion of the solution and of its validity is given in Fibich, Malkin, and Papanicolaou (1995).

An understanding of the splitting phenomenon was proposed by Bergé and Rasmussen (1996) and Bergé, Rasmussen, Kuznetsov, Shapiro, and Turitsyn (1996). They constructed a quasi-self-similar solution where the scaling factors are prescribed in such a way that the following constraints are satisfied: The time derivatives are negligible compared to the transverse ones; the phase of the solution includes a quadratic term in the rescaled time variable τ, with a slowly varying coefficient whose effect in the equation for the rescaled profile is comparable to that originating from the usual quadratic phase (in the transverse variable). This additional phase affects the critical modulation equation $b_\zeta \approx -e^{-\pi\sqrt{b}}$ by replacing the factor π by a function of the rescaled time τ. In this context, they defined a value τ_* such that the beam contracts for $\tau < \tau_*$ and disperses for $\tau > \tau_*$. The authors conclude that the transverse scale reaches a minimum at τ_* and that the amplitude becomes maximum at this time. In a system that is symmetric in the time variable, this scenario leads to a splitting of the pulse into two temporally separated structures.

To conclude, we stress that in spite of the achievements described in this section, the effect of a normal time dispersion on critical collapse is still not completely understood. Several questions remain open. It is still unknown whether a finite-distance collapse can occur. The conditions for the existence of multisplitting are also to be specified. Furthermore, the analytical approaches are based on perturbations on the two-dimensional NLS equation and thus break down as the pulse tends to split.

9.3 Saturated Nonlinearity

In situations where the paraxial approximation remains valid while focusing has proceeded, the problem can still be described in terms of a nonlinear Schrödinger equation,

$$i\psi_t + \Delta\psi + f(|\psi|^2)\psi = 0, \tag{9.3.1}$$

but the nonlinear coupling $f(|\psi|^2)\psi$ no longer reduces to a cubic nonlinearity. The latter should indeed be viewed as the leading-order term in a small-amplitude expansion. We are interested here in saturated nonlinearities, several examples of which are considered in the literature. Algebraic nonlinearities of the form $f(s) = s/(1 + \gamma s)$ with positive γ were found to accurately describe the variation of the dielectric constant of gas vapors where a laser beam propagates (Tikhonenko, Christou, and Luther-Davies 1996). Exponential nonlinearities $f(s) = 1 - e^{-\gamma s}$ were used in the context of laser beams in plasmas (Cohen, Lasinski, Langdon, and Cummings 1991, Johnston, Vidal, and Fréchette 1997). In both cases, the nonlinearity saturates when the wave amplitude reaches large values.

The NLS equation with saturated nonlinearity is also relevant in the description of Bose superfluids at zero temperature, in the Hartree approximation (Gross–Pitaevskii equation). In this context, the wave amplitude keeps a finite value at infinity (see, e.g., Barashenkov, Gocheva, Makhankov, and Puzynin 1989, Barashenkov and Panova 1993, Josserand, Pomeau, and Rica 1995, Josserand and Rica 1997).

A simplified model resulting from the expansion of the previous ones is provided by the polynomial nonlinearity

$$f(s) = s - \gamma s^2 \tag{9.3.2}$$

with $\gamma \ll 1$, which causes the beam to defocus when the amplitude becomes large. Conservation of the Hamiltonian

$$H = |\nabla\psi|^2_{L^2} - \frac{1}{2}|\psi|^4_{L^4} + \frac{\gamma}{3}|\psi|^6_{L^6} \tag{9.3.3}$$

implies the inequality

$$|\nabla\psi|^2_{L^2} + \frac{\gamma}{3}|\psi|^6_{L^6} \le H + \frac{1}{2}|\psi|^2_{L^2}|\psi|^2_{L^6} \le H + \frac{1}{4\gamma}|\varphi|^2_{L^2} + \frac{\gamma}{6}|\psi|^6_{L^6}. \tag{9.3.4}$$

This ensures that for an initial condition $\psi(\mathbf{x}, 0) = \varphi \in H^1(\mathbf{R}^d)$ in dimensions $d = 2$ or 3, the norms $|\nabla\psi|_{L^2}$ and $|\psi|_{L^6}$ remain uniformly bounded. An equivalent result is given in Cazenave (1979) in the case of an exponential nonlinearity.

An important question is the limit of solutions when $\gamma \to 0$. A rigorous result presented in Merle (1992b) concerns nonlinearities that are critical when $\gamma = 0$.

Theorem 9.2. *Let ψ_γ be the solution of*

$$i\psi_t + \Delta\psi + |\psi|^{\frac{4}{d}}\psi - \gamma|\psi|^{2q}\psi = 0 \tag{9.3.5}$$

for $\mathbf{x} \in \mathbf{R}^d$ and $\frac{2}{d} < q < \frac{2}{d-2}$, with initial condition $\psi_\gamma(\mathbf{x}, 0) = \varphi(\mathbf{x}) \in H^1$ and a finite variance $\int |\mathbf{x}|^2|\varphi|^2 d\mathbf{x}$. Suppose that the solution of (9.3.5) with $\gamma = 0$ blows up in a finite time T_*. For $T_0 > T_*$, one has the following alternative: Either the variance $\int |\mathbf{x}|^2\psi_\gamma(\mathbf{x}, T_0)d\mathbf{x} \to \infty$ as $\gamma \to 0$, or there is a constant C such that $\int |\mathbf{x}|^2\psi_\gamma(\mathbf{x}, T_0)d\mathbf{x} \le C$. In the latter case, one has the following properties.

(i) Compactness outside the origin in L^2 for $t < T_0$. *There is an application* $t \mapsto \psi^*$ *defined for* $t < T_0$ *such that for any* $K > 0$,

$$\psi^* \in C([0, T_0), L^2(|\mathbf{x}| > K)) \tag{9.3.6}$$

$$\psi_\gamma \to \psi^* \quad in \quad C([0, T_0), L^2(|\mathbf{x}| > K)) \quad as \quad \gamma \to 0. \tag{9.3.7}$$

(ii) Concentration at the origin. *For* $t < T_0$, *there is* $m(t) \geq 0$ *such that*

$$|\psi_\gamma(\mathbf{x}, t)|^2 \to m(t)\delta_{\mathbf{x}=0} + |\psi^*(\mathbf{x}, t)|^2 \tag{9.3.8}$$

in the sense of distributions. If $m(t) \neq 0$, *then* $|\psi_\gamma(t)|_{H^1} \to \infty$ *and* $m(t) > |R|_{L^2}^2$ *where* R *is the ground state solution of* $\Delta R - R + R^{\frac{4}{d}+1} = 0$. *Alternatively, if* $m(t) = 0$, *there exists a constant* c *such that for all* γ, $|\psi_\gamma(t)|_{H^1} < c$ *and* $\psi_\gamma(t) \to \psi^*(t)$ *in* L^2.

(iii) Conservation of mass. *For all* $t < T_0$, $m(t) + |\psi^*(t)|_{L^2}^2 = |\varphi|_{L^2}^2$.

9.3.1 Standing-Wave Solutions

We look for solutions in the form of standing waves

$$\psi(\mathbf{x}, t) = e^{i\lambda^2 t} S(\mathbf{x}) \tag{9.3.9}$$

where S satisfies

$$\Delta S - \lambda^2 S + f(|S|^2)S = 0, \tag{9.3.10}$$

where (as assumed in this paragraph), one can take $\lambda^2 = 1$ by rescaling S and the space variables. The coefficient γ is then replaced by $\gamma\lambda^2$. The general theory for equations of the form $-\Delta u = g(u)$ gives sufficient (and "almost" necessary) conditions for existence of ground states, i.e., positive, radial solutions that decrease exponentially at infinity (Berestycki and Lions 1983). One of these conditions is $\int ug(u(\mathbf{x})) d\mathbf{x} > 0$. Since the ground-state is decreasing monotonically with the radial variable, this condition is satisfied if there exists $s > 0$ such that $\int_0^s g(s')ds' > 0$. In the case of algebraic nonlinearity, $f(s) = s/(1 + \gamma s)$, this requires $\gamma < 1$, while for the polynomial case $f(s) = s - \gamma s^2$, this condition reduces to $0 < \gamma < 3/16$ (Anderson 1971).

Two-dimensional standing waves displaying a phase singularity at the center, of the form

$$S(x, y) = A(r)e^{il\theta} \tag{9.3.11}$$

with $r = \sqrt{x^2 + y^2}$, θ the polar angle, $A(r)$ real, and l a nonzero integer, are considered in Skryabin and Firth (1998) and references therein.

The stability of three-dimensional standing waves in the case of an exponential nonlinearity $f(s) = 1 - e^{-s}$ is addressed by Laedke and Spatschek (1984), who conclude to the existence of a critical value for λ, below which the ground-state standing-wave is unstable and above which it is (orbitally) stable. Shatah (1985b) proved that for polynomial saturation, the three-dimensional ground-state standing waves are orbitally stable for γ close to $3/16$ and unstable for γ close to 0.

Numerical simulations in two dimensions in the case of the algebraic saturated nonlinearity with $\gamma = 1$, indicate that when varying the parameter λ^2 in the range $0 \leq \lambda^2 \leq 1$ of existence, the L^2-norm of the ground-state standing waves grows monotonically with λ^2. As expected, these solutions are then found to be are stable. In contrast, standing waves associated to other bound states display instabilities that break the azimuthal symmetry of the system (Soto-Crespo, Wright, and Akhmediev 1991). A similar instability is reported in three dimensions by Edmundson (1997). Furthermore, the two-dimensional ground-state standing wave appears to be unstable relative to three-dimensional perturbations (Akhmediev, Korneev, and Nabiev 1992).

Ground-state standing waves in one space dimension for a polynomial nonlinearity of the form $(|\psi|^6 - \gamma|\psi|^{12})\psi$ (supercritical for $\gamma = 0$) were considered by Grikurov (1997) who showed numerically that the L^2-norm of the ground state associated to $\lambda^2 = 1$, decreases with γ until it reaches a (flat) minimum for a critical value γ_* and then increases. As a consequence, the ground-state standing wave will be stable or unstable depending on $\gamma > \gamma_*$ or $\gamma < \gamma_*$.

9.3.2 Numerical Observations

A conspicuous effect observed numerically with saturated nonlinearities is the development of multiple foci as the variable t (viewed as time or propagation distance according to the context) increases. This effect is associated to an oscillatory behavior of the solution due to a competition between the focusing of the wave and its spreading when the nonlinearity saturates (Dawes and Marburger 1969, Cohen, Lasinski, Langdon, and Cummings 1991, Zakharov and Synakh 1976 and references therein). Furthermore, both laboratory experiments of laser beam propagation in a Kerr medium (Campillo, Shapiro, and Sydam 1973, 1974) and numerical simulations of the NLS equation with a saturated nonlinearity in two space dimensions without special symmetry (Soto-Crespo, Wright, and Akhmediev 1992) show the development of transverse instabilities when the initial L^2-norm is several times critical. The dynamics lead to a splitting of the beam into several filaments with an isotropic shape that, according to the initial conditions appear to remain located at a fixed distance from the center of the emitted beam or to tend to separate. Complicated dynamics can then result from filament interactions, an important effect in periodic geometry (Konno and Suzuki 1979, Johnston, Vidal, and Fréchette 1997).

Neglecting this effect, many numerical simulations deal with radially symmetric solutions. Simulations using the dynamic rescaling method are presented by LeMesurier, Papanicolaou, Sulem, and Sulem (1988a) in the case of saturated algebraic and polynomial nonlinearities with small γ, both in two and three space dimensions.

Simulations of the two-dimensional NLS equation with a nonlinearity $(|\psi|^2 - \gamma|\psi|^4)\psi$ and fixed initial conditions $\psi_0(\mathbf{x}) = 4e^{-|\mathbf{x}|^2}$ are presented Fig. 9.2 and show different regimes when the parameter γ is decreased. For $\gamma = 10^{-2}$, one observes a relatively quick convergence of the solution towards a function of a ground-state standing wave. For smaller values of γ, the solution displays oscillations with in some instances a modulation of the amplitude or even more complex oscillations. The amplitude of the oscillations seems however to decrease and although the numerical integration was not carried out on sufficiently long times to reach the asymptotic regime, one may expect the eventual convergence of the solution to a stable standing wave, at least for some ranges of parameters. A qualitatively similar behavior is observed with an algebraic nonlinearity, except that an example of persistence of purely periodic oscillations is reported for a moderately small value $\gamma = 10^{-2}$ of the control parameter while modulated oscillations with a decreasing amplitude are observed with $\gamma = 10^{-4}$.We note that in all the cases, the profile of the solution near the maximum is monotonic, while at larger distances moderate oscillations associated to the emission of radiation of moderate amplitude that escapes to infinity are visible. Detailed simulations recently performed by Vidal and Johnston (1997) in the case of the algebraic saturated nonlinearity show that

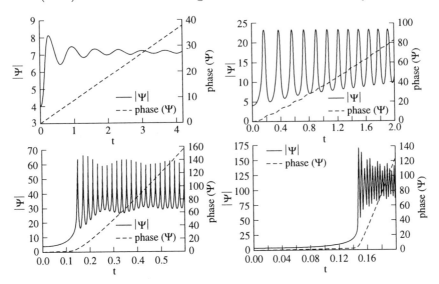

FIGURE 9.2. Time evolution of the amplitude and (asymptotically linear) phase at the origin of the solution of the two-dimensional NLS equation with polynomial potential $|\psi|^2 - \gamma|\psi|^4$ with, from left to right and top to bottom, $\gamma = 10^{-2}, 10^{-3}, 10^{-4}, 10^{-5}$ and initial conditions $\psi(\mathbf{x}) = 4e^{-|\mathbf{x}|^2}$ (from LeMesurier et al. 1988a).

for initial conditions far from the ground state, the system can display complicated oscillations, while for initial conditions close to it, three main types of oscillations are identified. They are characterized by one or more well-defined frequencies and by the presence or the absence of a significant damping associated to the energy radiation.

The simulations of LeMesurier, Papanicolaou, Sulem, and Sulem (1988a) in three dimensions with an initial condition $\psi_0(\mathbf{x}) = 6\sqrt{2}e^{-|\mathbf{x}|^2}$ show that for very small γ, the profile of the solution is more complex than in two dimensions, and may display a flat maximum at the origin, secondary peaks (ring focusing) with an amplitude sometimes comparable to that of the central peak, and a ripple pattern. During the oscillations, the origin may even become a local minimum.

The situation is simpler when taking as the initial condition an unstable ground state subject to a weak perturbation. Simulations were performed in this context by Grikurov (1997), in one space dimension, with a nonlinearity of the form $(|\psi|^6 - \gamma|\psi|^{12})\psi$. The solution is seen to oscillate with radiation of energy toward large distances, and to eventually approach a stable standing wave of the form $\frac{1}{L_0}S(\frac{x}{L_0})e^{\frac{it}{L_0^2}}$ where S is the ground-state solution of

$$\frac{d^2S}{dx^2} - S + S^7 - \epsilon S^{13} = 0, \tag{9.3.12}$$

where the scaling factor L_0 is sufficiently small to make the parameter $\epsilon = \frac{\gamma}{L_0^2}$ within the range of existence and stability for the corresponding standing-wave solution.

Simulations in three dimensions with an algebraic nonlinearity and $\psi_0(\mathbf{x}) = 6e^{-|\mathbf{x}|^2}$ show a similar behavior (Papanicolaou, Sulem, and Sulem 1988a). The profile remains monotonic near the origin, although an important radiation takes place, associated with the development of ripples at large distance. For $\gamma = 10^{-2}$, the solution oscillates with an amplitude which decreases monotonically in time. For $\gamma = 10^{-4}$, a more complex evolution is however visible, and longer integrations would be necessary to characterize the asymptotic regime.

The development of transverse instabilities in the three-dimensional case was analyzed by Akhamediev and Sote-Crespo (1993) for axisymmetric solutions. The description is presented in the context of nonlinear optics, where the problem corresponds to the propagation of a beam in a medium with anomalous group velocity dispersion and algebraic saturated nonlinearity. Numerical simulations show that in situations corresponding to a moderate coefficient γ, an initially homogeneous self-focusing beam breaks into "light clumps" that through oscillations of the peak intensity and energy radiation converges to a train of three-dimensional standing waves referred to as "optical bullets."

9.3.3 Saturation of Critical Collapse

In contrast with the supercritical case where no analytical approach is presently available, some of the phenomena observed in numerical simulations of radially symmetric solutions at critical dimension can be interpreted in the context of an asymptotic analysis in the limit of small γ. For this purpose, we consider an initial condition (or a beam when using the language of optics) with an L^2-norm (a power) which is in slight excess to critical, and concentrate on the case of a polynomial nonlinearity $|\psi|^2 - \gamma|\psi|^4$ with $\gamma \ll 1$. This nonlinearity can indeed be viewed as the expansion of more general saturated nonlinearities that leads to similar dynamics. Following Malkin (1993) (see also Fibich and Papanicolaou 1997b), we present an analysis based on the variation of the mass and of the Hamiltonian of the collapson, resulting from radiation effects.

Proceeding as in Chapter 8 (and keeping the variable t to denote the evolution variable), we make the transformation

$$\psi(\mathbf{x}, t) = \frac{1}{L(t)} v(\xi, \tau) e^{i\tau - ia\xi^2/4}, \tag{9.3.13}$$

where

$$\xi = \frac{|\mathbf{x}|}{L(t)}, \quad \tau = \int_0^t \frac{ds}{L^2(s)}, \tag{9.3.14}$$

and

$$a(\tau) = -LL_t = -L_\tau/L. \tag{9.3.15}$$

The function v satisfies

$$iv_\tau + \Delta v - v + \frac{1}{4}b(\tau)\xi^2 v + |v|^2 v - \epsilon(\tau)|v|^4 v = 0 \tag{9.3.16}$$

with

$$b(\tau) = a^2 + a_\tau = -L^3 L_{tt} \quad \text{and} \quad \epsilon(\tau) = \gamma/L^2. \tag{9.3.17}$$

We look for a solution v that within the collapson takes the form

$$v(\xi, \tau) \approx P(\xi, b(\tau), \epsilon(\tau)) \tag{9.3.18}$$

where P satisfies

$$\Delta P - P + b\frac{\xi^2}{4}P + |P|^2 P - \epsilon|P|^4 P - i\nu(\sqrt{b})P = 0. \tag{9.3.19}$$

In the limit of small b, the function $\nu(b)$ required for existence of admissible solutions keeps the same functional form

$$\nu(\sqrt{b}) \approx Ce^{-\pi/\sqrt{b}} \tag{9.3.20}$$

as for the focusing cubic NLS equation. At leading order, P can be approximated in the form

$$P \approx R(\xi) + b(\tau)\rho(\xi) + \epsilon(\tau)h(\xi), \tag{9.3.21}$$

where R is, as in the previous chapters, the positive solution (ground state) of $\Delta R - R + R^3 = 0$, while ρ and h are solutions of

$$\Delta\rho - \rho + 3R^2\rho = -\frac{\xi^2}{4}R \qquad (9.3.22)$$

and

$$\Delta h - h + 3R^2 h = R^5 \qquad (9.3.23)$$

respectively. Using (9.3.18) and (9.3.21) to compute the mass N_s of the collapson, one has

$$N_s = N_c + 2b(\tau)\int R\rho\,\xi d\xi + 2\epsilon(\tau)\int Rh\,\xi d\xi, \qquad (9.3.24)$$

where

$$N_c = \int R^2\xi d\xi \qquad (9.3.25)$$

is the critical mass for the cubic problem,

$$2\int R\rho\xi d\xi = \frac{1}{4}\int \xi^2 R^2\xi d\xi = M > 0 \qquad (9.3.26)$$

and

$$\int Rh\xi d\xi = \frac{1}{3}\int R^6\xi d\xi = C_1 > 0. \qquad (9.3.27)$$

The excess of mass above critical

$$\widetilde{N} \equiv N_s - N_c \qquad (9.3.28)$$

is thus given by

$$\widetilde{N} = Mb(\tau) + 2C_1\epsilon(\tau). \qquad (9.3.29)$$

Similarly, the Hamiltonian associated to the collapson H_s satisfies

$$H_s L^2 = -C_1\epsilon(\tau) - Mb(\tau) + Ma^2(\tau). \qquad (9.3.30)$$

Using (9.3.29) and the definition of a, the above equation is rewritten as

$$\dot{L}^2 = \frac{H_s}{M} - \frac{\gamma C_1}{ML^4} + \frac{\widetilde{N}}{ML^2}, \qquad (9.3.31)$$

where we denote by a dot the derivative with respect to the physical time t. Introducing $y = L^2$, we have

$$\dot{y}^2 = \frac{4H_s}{My}\left(y^2 + \frac{\widetilde{N}}{H_s}y - \frac{\gamma C_1}{H_s}\right). \qquad (9.3.32)$$

Since H_s is assumed negative, then, provided that $\gamma \leq \frac{\widetilde{N}^2}{4|H_s||C_1|}$, it follows that y and thus the scaling factor L display an oscillatory behavior, as observed in the numerical simulations. Indeed, (9.3.32) becomes

$$\dot{y}^2 = -\frac{4|H_s|}{My}(y - y_{\mathrm{m}})(y - y_{\mathrm{M}}),$$ (9.3.33)

or equivalently,

$$L^4\left(\frac{dL}{dt}\right)^2 = -\frac{|H_s|}{M}(L^2 - L_{\mathrm{m}}^2)(L^2 - L_{\mathrm{M}}^2).$$ (9.3.34)

Here L_{M} and L_{m} denote the maximum and minimum of the scaling factor and

$$y_{\mathrm{m}} = L_{\mathrm{m}}^2 = \frac{\widetilde{N}}{2|H_s|}\left(1 - \left(1 - \frac{4\gamma C_1|H_s|}{\widetilde{N}^2}\right)^{1/2}\right),$$ (9.3.35)

$$y_{\mathrm{M}} = L_{\mathrm{M}}^2 = \frac{\widetilde{N}}{2|H_s|}\left(1 + \left(1 - \frac{4\gamma C_1|H_s|}{\widetilde{N}^2}\right)^{1/2}\right),$$ (9.3.36)

where the variations of the mass and of the Hamiltonian of the collapson have been neglected, leading to a local period

$$T = 2\int_{y_{\mathrm{m}}}^{y_{\mathrm{M}}}\frac{1}{L|\dot{L}|}dy = 2\int_{y_{\mathrm{m}}}^{y_{\mathrm{M}}}\frac{1}{\sqrt{y}}\frac{1}{\sqrt{\frac{|H_s|}{M}(y_{\mathrm{M}} - y)(y - y_{\mathrm{m}})}}dy.$$ (9.3.37)

Using the change of variables

$$\cos^2 u = (y - y_{\mathrm{m}})/(y_{\mathrm{M}} - y_{\mathrm{m}}),$$ (9.3.38)

the integral becomes

$$T = 2\sqrt{\frac{M}{|H_s|}}\int_0^{\pi/2}\frac{1}{\sqrt{y_{\mathrm{M}} - (y_{\mathrm{M}} - y_{\mathrm{m}})\sin^2 u}}du$$

$$= \sqrt{\frac{My_{\mathrm{M}}}{|H_s|}}E\left(1 - \frac{y_{\mathrm{m}}}{y_{\mathrm{M}}}\right),$$ (9.3.39)

where $E(x) = \int_0^{\pi/2}\sqrt{1 - x\sin^2 u}\,du$ is the elliptic integral of second kind (Abramovitz and Stegun 1965).

A slow variation of the period is nevertheless possible as the result of the energy radiation that takes place on a longer time scale and leads to an evolution of L_{m} and L_{M}. A complete theory is still missing, but some information can be obtained by writing the rate of variation of the mass and of the Hamiltonian trapped in the collapson in the form

$$\frac{dN_s}{d\tau} = -J_N, \qquad \frac{dH_s}{d\tau} = -J_H.$$ (9.3.40)

In the absence of saturation ($\gamma = 0$), as shown in Section 8.2.2,

$$J_N = 2\nu_0^2 e^{-\pi/\sqrt{b}},$$ (9.3.41)

$$J_H = \frac{2v_0^2}{L^2} e^{-\pi/\sqrt{b}}. \tag{9.3.42}$$

The same functional form for J_N and J_H will be used in the case of a saturated nonlinearity but with a function b now given by

$$b = \frac{\widetilde{N}}{M} \left(1 - \frac{2\gamma C_1}{\widetilde{N} L^2} \right), \tag{9.3.43}$$

obtained from (9.3.29)–(9.3.30). In the case where $L_m \ll L_M$, or equivalently, $\frac{4\gamma C_1 |H_s|}{N^2} \ll 1$, one gets

$$L_m^2 \approx \gamma C_1 / \widetilde{N}, \tag{9.3.44}$$

$$L_M^2 \approx \widetilde{N} / |H_s|, \tag{9.3.45}$$

$$b \approx \frac{\widetilde{N}}{M} \left(1 - 2\frac{L_m^2}{L^2} \right). \tag{9.3.46}$$

Thus,

$$\frac{d\widetilde{N}}{d\tau} \approx -C e^{-\Lambda} e^{-\Lambda \frac{N_c}{C_1} \frac{L_m^2}{L^2}}, \tag{9.3.47}$$

$$\frac{dH_s}{d\tau} \approx -\frac{C}{L^2} e^{-\Lambda} e^{-\Lambda \frac{N_c}{C_1} \frac{L_m^2}{L^2}} \tag{9.3.48}$$

with $\Lambda = \pi\sqrt{M/\widetilde{N}}$, indicating that in this context, $\widetilde{N}$ and H_s decay monotonically in time. Concerning the extrema of L, one sees from (9.3.44)–(9.3.45) that L_m increases, while since

$$\frac{dL_M^2}{d\tau} = \frac{J_N}{H_s} \left(1 - \frac{\widetilde{N}}{L^2 H_s} \right) < 0, \tag{9.3.49}$$

L_M decreases. The effect of the radiation thus appears to slowly reduce the amplitude of the oscillations.

The (small) lost of mass during one oscillation of duration Δt is given by

$$\Delta N = N_s(t + \Delta t) - N_s(t) = M \int_t^{t+\Delta t} \frac{v(\sqrt{b})}{L^2} dt. \tag{9.3.50}$$

Following Fibich and Papanicolaou (1997b), the above integral can be estimated using the Laplace method, since $v(\sqrt{b})$ rapidly decays away from its maximum $v(\sqrt{b_M})$, which occurs at the instant $t = t_M$, when the scale factor reaches its maximum L_M, and the amplitude is thus minimum. One gets

$$\Delta N \approx M L_M^{-2} v(\sqrt{b_M}) \int_t^{t+\Delta t} e^{-A(t-t_M)^2} dt$$

$$\approx (2\pi)^{1/2} M A^{-1/2} L_M^{-2} v(\sqrt{b_M}) \tag{9.3.51}$$

with

$$A = -\frac{\pi}{4}\frac{1}{b_M^{3/2}}\partial_{tt}b(t_M).$$ (9.3.52)

One now evaluates

$$\partial_{tt}b(t_M) = \partial_L b \, \partial_{tt}L(t_M).$$ (9.3.53)

From (9.3.43) and (9.3.34) one has

$$\partial_L b(t_M) = \frac{4\gamma C_1}{ML_M^3}$$ (9.3.54)

and

$$\partial_{tt}L(t_M) = -\frac{|H_s|}{ML_M^3}(L_M^2 - L_m^2).$$ (9.3.55)

Thus,

$$A = \gamma\frac{\pi C_1}{M^2}\frac{|H_s|}{b_M^{3/2}}\frac{L_M^2 - L_m^2}{L_M^6}$$ (9.3.56)

and

$$\Delta N \approx \frac{M^2}{\sqrt{C_1}}\nu(\sqrt{b_M})\,b_M^{3/4}\frac{1}{(\gamma|H_s|)^{1/2}}\left(1 - \frac{L_m^2}{L_M^2}\right)^{-1/2}.$$ (9.3.57)

Using (9.3.44)-(9.3.45) and $\widetilde{N} \approx Mb_M$, one gets to leading order in the limit $L_m \ll L_M$

$$\Delta N \approx M\frac{1}{b_M^{1/4}}\nu(\sqrt{b_M})\frac{L_M}{L_m}$$ (9.3.58)

where $b_M \approx |H_s|L_M^2/M$. This analysis which assumes large amplitude oscillation cannot in principle describe the relation of the solution to a standing wave. Since it assumes that the loss $|\Delta N|$ is small compared to the excess of mass above critical $\widetilde{N}$, it is also limited to moderately small values of the parameter γ since ΔN scales like $\gamma^{-1/2}$. Furthermore, the analysis seems too local to reproduce the persistent multi-frequency oscillatory regimes observed numerically by Vidal and Johnston (1997).

9.4 Other Pertubations

9.4.1 Weak Nonparaxiality

We have seen that when the canonical NLS equation at critical dimension is weakly perturbed by saturated nonlinearities, the collapse is arrested, and the reduced system displays slowly decaying focusing–defocusing oscillations. It turns out that this phenomenon is not specific to saturated nonlinearities. A weak departure from paraxiality, which among other effects becomes also relevant near collapse, has a similar influence. It was observed by Feit and Fleck (1988) by comparison of numerical simulations of the Helmholtz wave equation with cubic refractive index with the paraxial approximation. A perturbative analysis was given by Fibich (1996c), starting from the equation

$$\epsilon\psi_{zz} + i\psi_z + \Delta\psi + |\psi|^2\psi = 0, \qquad \mathbf{x} \in \mathbf{R}^2, \tag{9.4.1}$$

as it results from the Helmholtz equation.

One writes the balance of the wave energy in the form

$$\frac{d}{dz} \int |\psi|^2 d\mathbf{x} = -2\epsilon\Im \int \psi_{zz}\psi^* d\mathbf{x}. \tag{9.4.2}$$

Using a change of variables equivalent to (9.3.13), (9.3.14), and (9.3.15) and the approximation

$$v \approx R + 2b\rho, \tag{9.4.3}$$

gives

$$Mb_z = -2\epsilon N_c \left(\frac{1}{L^2}\right)_z, \tag{9.4.4}$$

or after integration,

$$b = b_0 - \frac{2\epsilon N_c}{ML^2}, \tag{9.4.5}$$

where b_0 is a constant. The Hamiltonian is given by

$$H = \int \left(|\nabla\psi|^2 - \frac{1}{2}|\psi|^4 - \epsilon|\psi_z|^2\right) d\mathbf{x}, \tag{9.4.6}$$

and its restriction H_s to the collapson satisfies

$$H_s L^2 = Ma^2 - Mb - \epsilon\frac{N_c}{L^2}. \tag{9.4.7}$$

Combining (9.4.5) and (9.4.7), one gets

$$H_s L^2 = Ma^2 - Mb_0 + \epsilon\frac{N_c}{L^2}. \tag{9.4.8}$$

Introducing $y = L^2$, (9.4.8) becomes

$$\dot{y}^2 = \frac{4H_s}{My}\left(y^2 + \frac{Mb_0}{H_s}y - \frac{\epsilon N_c}{H_s}\right). \tag{9.4.9}$$

Note that this equation identifies with (9.3.32) for saturated nonlinearities, where the constant C_1 is replaced by N_c and $\tilde{N}/M$ replaced by b_0. When H_s is negative and $b_0 > 0$, (9.4.9) is rewritten in the form

$$\dot{y}^2 = -\frac{4|H_s|}{My}(y - y_\mathrm{m})(y - y_\mathrm{M}) \tag{9.4.10}$$

with

$$y_m = L_m^2 = \frac{Mb_0}{2|H_s|}\left(1 - \left(1 - \frac{4\epsilon N_c|H_s|}{M^2 b_0^2}\right)^{1/2}\right), \qquad (9.4.11)$$

$$y_M = L_M^2 = \frac{Mb_0}{2|H_s|}\left(1 + \left(1 - \frac{4\epsilon N_c|H_s|}{M^2 b_0^2}\right)^{1/2}\right) \qquad (9.4.12)$$

and displays an oscillatory behavior. The calculation is pursued as in the previous section, leading to a period of the oscillations given by (9.3.37).

9.4.2 General Formalism

A general approach of perturbed critical NLS equation is presented by Fibich and Papanicolaou (1997b, 1998) where the effect on self-focusing of additional small terms of different kinds is studied. In addition to the two problems discussed above, it is applied to various examples including fiber arrays, Debye relaxation, effect of random impureties, random potentials, and also to situations where two different perturbations are retained. Other examples of NLS equations with additional corrective terms are reviewed in Ghidaglia and Saut (1993).

The combined effect of small time dispersion and nonparaxiality is addressed in Fibich and Papanicolaou (1997a), where additional terms corresponding to "shock term", "group velocity nonparaxiality", etc. are also retained and shown to be possibly dominant. A related discussion of "higher-order effects" on critical collapse is also presented in Malkin (1997), where the perturbation approach initiated in Malkin (1993) is extended to more general settings.

Let us consider a general perturbed NLS equation at critical dimension in the form

$$i\psi_t + \Delta\psi + |\psi|^2\psi + \epsilon F(\psi) = 0, \qquad \mathbf{x} \in \mathbf{R}^2. \qquad (9.4.13)$$

where ϵF holds for a small corrective term.

In this section we briefly review the derivation of the modulation equations in a general context. Using the change of variables (9.3.13), (9.3.14), and (9.3.15), equation (9.4.1) becomes

$$iv_\tau + \Delta v - v + \frac{1}{4}b(\tau)\xi^2 v + |v|^2 v + \epsilon L^3 e^{-iS} F\left(\frac{v}{L}e^{iS}\right) = 0 \qquad (9.4.14)$$

with

$$b(\tau) = a^2 + a_\tau = -L^3 L_{tt} \qquad (9.4.15)$$

and

$$S = \tau - a\xi^2/4. \qquad (9.4.16)$$

As for the unperturbed NLS equation, the function v is approximated by

$$v \approx V_0 + V_1 + \cdots, \qquad (9.4.17)$$

where V_0 is a solution of

$$\Delta V_0 - V_0 + \frac{b}{4}\xi^2 V_0 + |V_0|^2 V_0 - i\nu(\sqrt{b})V_0 + \epsilon w(V_0) = 0, \qquad (9.4.18)$$

and

$$w(V_0) = L^3 \Re\left[e^{-iS} F\left(\frac{V_0}{L} e^{iS}\right)\right]. \qquad (9.4.19)$$

Then V_0 is expanded in the two small parameters b and ϵ in the form

$$V_0 \approx R(\xi) + b\rho(\xi) + \epsilon h(\xi, \tau), \qquad (9.4.20)$$

where ρ satisfies (9.3.22) and h is a solution of

$$\Delta h - h + 3R^2 h = -w(R). \qquad (9.4.21)$$

At lowest order in b, the equation for V_1 equation reduces to

$$\Delta V_1 - V_1 + 2|V_0|^2 V_1 + V_0^2 \bar{V}_1$$
$$= -i(b_\tau \rho + \epsilon h_\tau + \nu(\sqrt{b})R) - i\epsilon L^3 \Im\left[F\left(\frac{R}{L} e^{iS}\right)e^{-iS}\right], \qquad (9.4.22)$$

where we have used (9.4.20) in the right-hand side of (9.4.22).

The equation for the real part of V_1 is always solvable, but when solving the equation for its imaginary part, one needs to satisfy the solvability condition

$$\int \left(b_\tau \rho + \epsilon h_\tau + \nu(\sqrt{b})R + \epsilon L^3 \Im\left[F\left(\frac{R}{L} e^{iS}\right)e^{-iS}\right]\right) R\xi d\xi = 0. \qquad (9.4.23)$$

This differential equation involving a, b, and L is to be coupled with (9.4.15) and (9.4.16) with $a = -L_\tau/L$. The resulting system provides the modulation equations associated to weak perturbations of critical collapse.

Among the various perturbations amenable to this formalism, the effects of a random potential describing the effect of impurities on the propagation of a laser beam is of special interest. An important issue noted by Fibich and Papanicolaou (1997a) concerns the probability of escape (corresponding to $L \to \infty$) due to the random inhomogeneities.

Part IV

Coupling to a Mean Field

10

Mean Field Generation

This chapter is devoted to the modulation of a weakly nonlinear wave train in the case where a mean field is driven due to quadratic nonlinearities. The analysis is presented in the context of a scalar problem, first in a general framework and then exemplified on a few simple wave equations. An example of degenerate case is also presented.

10.1 General Formalism

Consider a wave equation, assumed for the sake of simplicity to be scalar, written in the symbolic form

$$L(\partial_t, \boldsymbol{\partial})u + G\{u\} = 0. \qquad (10.1.1)$$

Here L denotes a dispersive linear operator with real constant coefficients where $\boldsymbol{\partial}$ holds for the spatial gradient, and G refers to a nonlinear term depending on the field u and on its derivatives. For a plane wave $e^{i(\mathbf{k}\cdot\mathbf{x}-\omega t)}$, the dispersion relation associated to L reads

$$L(-i\omega, i\mathbf{k}) = 0. \qquad (10.1.2)$$

In order to describe the weakly nonlinear regime, we introduce the slow variables $T = \epsilon t$ and $\mathbf{X} = \epsilon \mathbf{x}$, where ϵ measures the magnitude of the wave amplitude. In a multiple-scale analysis, ∂_t is replaced by $\partial_t + \epsilon \partial_T$ and $\boldsymbol{\partial}$ by $\boldsymbol{\partial} + \epsilon \nabla$, where ∇ denotes the gradient with respect to the large-scale variables. We also expand the solution in the form

$$u = \epsilon(u_0 + \epsilon u_1 + \epsilon^2 u_2 + \cdots). \qquad (10.1.3)$$

The linear operator L becomes

$$L(\partial_t + \epsilon \partial_T, \boldsymbol{\partial} + \epsilon \nabla) = L(\partial_t, \boldsymbol{\partial}) + \epsilon [L_0(\partial_t, \boldsymbol{\partial}) \partial_T + \sum_i L_i(\partial_t, \boldsymbol{\partial}) \partial_{X_i}]$$

$$+ \frac{\epsilon^2}{2} [L_{00}(\partial_t, \boldsymbol{\partial}) \partial_{TT} + 2 \sum_i L_{0i}(\partial_t, \boldsymbol{\partial}) \partial_{TX_i}$$

$$+ \sum_{i,j} L_{ij}(\partial_t, \boldsymbol{\partial}) \partial_{X_i X_j}] + \cdots, \qquad (10.1.4)$$

where L_μ and $L_{\mu\nu}$ (with $\mu, \nu = 0, 1, \ldots, d$) hold for the formal first and second derivatives of the operator L with respect to ∂_t (index 0) or to ∂_i (index $i = 1, \ldots, d$).

At successive orders of the expansion, we have

$$L(\partial_t, \boldsymbol{\partial})u_0 = 0, \qquad (10.1.5)$$

$$L(\partial_t, \boldsymbol{\partial})u_1 = -[L_0(\partial_t, \boldsymbol{\partial}) \partial_T + \sum_i L_i(\partial_t, \boldsymbol{\partial}) \partial_{X_i}]u_0 - G_2\{u_0^2\}, \qquad (10.1.6)$$

$$L(\partial_t, \boldsymbol{\partial})u_2 = -[L_0(\partial_t, \boldsymbol{\partial}) \partial_T + \sum_i L_i(\partial_t, \boldsymbol{\partial}) \partial_{X_i}]u_1$$

$$- \frac{1}{2}[L_{00}(\partial_t, \boldsymbol{\partial}) \partial_{TT} + 2 \sum_i L_{0i}(\partial_t, \boldsymbol{\partial}) \partial_{TX_i}$$

$$+ \sum_{i,j} L_{ij}(\partial_t, \boldsymbol{\partial}) \partial_{X_i X_j}]u_0 - G_3\{u_0^3, u_0 u_1, u_0 \nabla u_0\}, \qquad (10.1.7)$$

where G_2 and G_3 refer to the nonlinearities arising at order ϵ^2 and ϵ^3, respectively. The quadratic nonlinearity is assumed to include derivatives that at this order are taken with respect to the fast variables.

Equation (10.1.5) reproduces the linearized problem, which is solved as

$$u_0 = \psi_0(\mathbf{X}, T)e^{i(\mathbf{k} \cdot \mathbf{x} - \omega t)} + \text{c.c.} + \phi_0(\mathbf{X}, T), \qquad (10.1.8)$$

where the wave amplitude now depends on the slow variables. The notation c.c. stands for complex conjugate. In the case where the equation includes a quadratic nonlinearity, the "mean field" ϕ_0, independent of the fast variables, is also retained to balance nonoscillating resonant terms arising at higher order in the ϵ-expansion.

Equation (10.1.6) is not always solvable. The solvability condition that cancels out the oscillating resonant terms reads

$$\left(L_0(-i\omega, i\mathbf{k}) \partial_T + \sum_j L_j(-i\omega, i\mathbf{k}) \partial_{X_j}\right) \psi_0 = 0. \qquad (10.1.9)$$

By differentiating the dispersion relation, one gets

$$- L_0(-i\omega, i\mathbf{k})\omega_j' + L_j(-i\omega, i\mathbf{k}) = 0, \qquad (10.1.10)$$

where $\mathbf{v}_g = (\omega_1', \cdots, \omega_d')$ is the group velocity associated to the carrier wave vector $\mathbf{k}$. The solvability condition becomes

$$L_0(-i\omega, i\mathbf{k})\Big(\partial_T + \mathbf{v}_g \cdot \nabla\Big)\psi_0 = 0. \qquad (10.1.11)$$

This condition is always satisfied when $L_0(-i\omega, i\mathbf{k}) = 0$, i.e., when $\omega(k)$ is a double root of the dispersion relation. We concentrate here on the generic case $L_0(-i\omega, i\mathbf{k}) \neq 0$. The solvability condition then prescribes that to leading order, on spatial and temporal scales of the order of the inverse wave amplitude ϵ^{-1}, the modulated wave train is only transported at the group velocity. Equation (10.1.6) is then solved in the form

$$u_1 = \psi_1(\mathbf{X}, T)e^{i(\mathbf{k}\cdot\mathbf{x}-\omega t)} + \gamma\psi_0^2(\mathbf{X}, T)e^{2i(\mathbf{k}\cdot\mathbf{x}-\omega t)} + \text{c.c.} + \phi_1(\mathbf{X}, T), \qquad (10.1.12)$$

where γ is a numerical constant and where we have included elements of the null space of L_0 that will be useful to prevent resonances at the next order. The term ϕ_1 is introduced when quadratic nonlinearities are present, to cancel nonoscillating resonant terms that cannot be eliminated by a mean field ϕ_0 arising at the leading order.

Several terms may contribute to the nonlinearity G_3. Cubic nonlinearities in the primitive equations lead to a contribution of the form u_0^3 on which derivatives with respect to fast variables may act. When the primitive equation includes quadratic nonlinearities, contributions proportional to $\partial_{x_i}(u_0 u_1)$ or $u_0\partial_{X_i}u_0$ are also possible.

The solvability condition that removes oscillating resonant terms reads

$$L_0(-i\omega, i\mathbf{k})\Big(\partial_T + \mathbf{v}_g \cdot \nabla\Big)\psi_1 + \frac{1}{2}\Big[L_{00}(-i\omega, i\mathbf{k})\partial_{TT}$$
$$+ 2\sum_j L_{0j}(-i\omega, i\mathbf{k})\partial_{TX_j} + \sum_{j,l} L_{jl}(-i\omega, i\mathbf{k})\partial_{X_j X_l}\Big]\psi_0$$
$$+ \Big(g_1|\psi_0|^2 + g_2\phi_1 + \sum_j g_{3,j}\partial_{X_j}\phi_0\Big)\psi_0 = 0, \qquad (10.1.13)$$

where g_1, g_2, and $g_{3,i}$ are coefficients. Only the first nonlinear coupling occurs when the leading nonlinearity in the primitive equation, is cubic. In the presence of quadratic nonlinearities, terms proportional to $\phi_1\psi_0$ or to $(\partial_{X_i}\phi_0)\psi_0$ are also present, depending, as seen below, on the linear operator in the primitive equation. Note that contributions of the form $\phi_0\partial_{X_i}\psi_0$ are not permitted, since they would be associated at previous order to a term proportional to $\phi_0\psi_0$ (since at that order, the derivative is taken with respect to the fast variable). Since we assume that no mean field is present in the absence of modulation, such a coupling is not possible.

When looking at the modulation as an initial value problem in time, and assuming $L_0(-i\omega, i\mathbf{k})$ nonzero [1] we use (10.1.11) to rewrite (10.1.13) as

$$L_0(-i\omega, i\mathbf{k})\left(\partial_T + \mathbf{v}_g \cdot \nabla\right)\psi_1$$
$$+ \frac{1}{2}\sum_{j,l}\left(L_{00}(-i\omega, i\mathbf{k})\omega_j'\omega_l' - 2L_{0j}(-i\omega, i\mathbf{k})\omega_l' + L_{jl}(-i\omega, i\mathbf{k})\right)\partial_{X_j X_l}\psi_0$$
$$+ \left(g_1|\psi_0|^2 + g_2\phi_1 + \sum_j g_{3,j}\partial_{X_j}\phi_0\right)\psi_0 = 0. \tag{10.1.14}$$

Differentiating the dispersion relation twice, one gets

$$i\omega_{jl}''L_0(-i\omega, i\mathbf{k}) + \omega_j'\omega_l'L_{00}(-i\omega, i\mathbf{k}) - \omega_l'L_{0j}(-i\omega, i\mathbf{k})$$
$$-\omega_j'L_{0l}(-i\omega, i\mathbf{k}) + L_{jl}(-i\omega, i\mathbf{k}) = 0 \tag{10.1.15}$$

where $\omega_{ij}'' = \frac{\partial^2\omega}{\partial k_i \partial k_j}$ is the second-order derivative of the carrier frequency with respect to the corresponding components of the wave vector. Equation (10.1.14) thus becomes

$$L_0(-i\omega, i\mathbf{k})\left((\partial_T + \mathbf{v}_g \cdot \nabla)\psi_1 - \frac{i}{2}\sum_{j,l}\omega_{jl}''\partial_{X_j X_l}\psi_0\right)$$
$$+ \left(g_1|\psi_0|^2 + g_2\phi_1 + \sum_j g_{3,j}\partial_{X_j}\phi_0\right)\psi_0 = 0. \tag{10.1.16}$$

Combining (10.1.11) and (10.1.14), and defining $\psi = \psi_0 + \epsilon\psi_1$, we obtain up to subdominant contributions[2]

$$L_0(-i\omega, i\mathbf{k})\left((\partial_T + \mathbf{v}_g \cdot \nabla)\psi - i\epsilon\nabla \cdot (\mathcal{D}\nabla\psi)\right)$$
$$+ \epsilon\left(g_1|\psi|^2 + g_2\phi_1 + \sum_j g_{3,j}\partial_{X_j}\phi_0\right)\psi = 0 \tag{10.1.17}$$

where $\mathcal{D}$ denotes the matrix of elements $\frac{1}{2}\omega_{ij}''$. Since the coefficient $L_0(-i\omega, i\mathbf{k})$ is assumed to be non zero, (10.1.14) reduces to

$$i(\partial_T\psi + \mathbf{v}_g \cdot \nabla\psi) + \epsilon\nabla \cdot (\mathcal{D}\nabla\psi) + \epsilon\left(q_1|\psi|^2 + q_2\phi_1 + \sum_j q_{3,j}\partial_{X_j}\phi_0\right)\psi = 0,$$
$$\tag{10.1.18}$$

where q_i q_2 and $q_{3,j}$ are real coupling coefficients because of the conservative character of the problem.

---

[1] When $L_0(-i\omega, i\mathbf{k}) = 0$, a nonlinear Klein–Gordon equation is obtained. A example is given in Lange and Newell (1971) in the context of the postbuckling problem of a thin elastic shell. Another situation is provided by the Kelvin–Helmholtz instability near onset (Weissman 1979, see also Craik 1985).

[2] Alternatively, as in Section 1.1, one can take $\psi_1 = 0$ and using (10.1.11), write ψ_0 as a function of $\xi = \mathbf{X} - \mathbf{v}_g T$ and of a longer time scale $\tau = \epsilon^2 t$.

In (10.1.7), the non oscillating terms present in the right-hand side are also resonant. The corresponding solvability condition reads

$$L_0(0,0)\partial_T\phi_1 + \sum_i L_i(0,0)\partial_{X_i}\phi_1 + \frac{1}{2}L_{00}(0,0)\partial_{TT}\phi_0 + \sum_i L_{0i}(0,0)\partial_{X_iT}\phi_0$$

$$+ \frac{1}{2}\sum_{i,j} L_{ij}(0,0)\partial_{X_iX_j}\phi_0 + \sum_i \lambda_i\partial_{X_i}|\psi_0|^2 = 0, \qquad (10.1.19)$$

where the λ_i's (which at least some of them are non-zero) are real coefficients originating from the nonlinearity G_3 assumed to include derivatives.

If the linear operator of the primitive equation includes first order but not second order derivatives, as illustrated by the Korteveg–de Vries equation discussed in Section 10.2.1, we take $\phi_0 = 0$ and the solvability condition reduces to

$$L_0(0,0)\partial_T\phi_1 + \sum_i \left(L_i(0,0)\partial_{X_i}\phi_1 + \lambda_i\partial_{X_i}|\psi_0|^2\right) = 0. \qquad (10.1.20)$$

If differently the lowest order derivatives are of second order, the solvability condition reads

$$\frac{1}{2}L_{00}(0,0)\partial_{TT}\phi_0 + \sum_i L_{0i}(0,0)\partial_{X_iT}\phi_0$$

$$+ \frac{1}{2}\sum_{i,j} L_{ij}(0,0)\partial_{X_iX_j}\phi_0 + \sum_i \lambda_i\partial_{X_i}|\psi_0|^2 = 0, \qquad (10.1.21)$$

where again some of the coefficients may vanish. This case is illustrated in Section 10.2.2 by the Boussinesq equation.

In contrast, for the Kadomtsev–Petviashvili equation considered in Section 10.2.3, the amplitude modulation does not drive mean fields at the level of (10.1.19) since $\lambda_i = 0$. In this case, $\phi_0 = 0$ and the mean field equation arises at the next order in the form

$$\frac{1}{2}L_{00}(0,0)\partial_{TT}\phi_1 + \sum_i L_{0i}(0,0)\partial_{X_iT}\phi_1$$

$$+ \frac{1}{2}\sum_{i,j} L_{ij}(0,0)\partial_{X_iX_j}\phi_1 + \sum_{i,j} \lambda_{ij}\partial_{X_iX_j}|\psi_0|^2 = 0. \qquad (10.1.22)$$

Writing the equations in the frame moving at the group velocity of the wave packet, we define $\boldsymbol{\xi} = \mathbf{X} - \mathbf{v}_g T$, and introduce the slower time $\tau = \epsilon T$. One is then led to replace ∂_T by $-\mathbf{v}_g \cdot \nabla + \epsilon\partial_\tau$, and to leading order, the dynamics of the mean fields are slaved to those of the amplitude carrier. We obtain the system

$$i\partial_\tau\psi + \frac{1}{2}\sum_{j,l} \omega''_{jl}\partial_{\xi_j\xi_l}\psi + \left(q_1|\psi|^2 + q_2\phi_1 + \sum_j q_{3,j}\partial_{\xi_j}\phi_0\right)\psi = 0, \quad (10.1.23)$$

where, according to the form of the linear operator in the primitive equation, the mean field is given by

$$\sum_i \partial_{\xi_i}\left(-L_0(0,0)\omega_i' + L_i(0,0))\phi_1 + \lambda_i|\psi|^2\right) = 0, \qquad (10.1.24)$$

or

$$\sum_{i,j}\left(\frac{1}{2}L_{00}(0,0)\omega_i'\omega_j' - L_{0i}(0,0)\omega_j' + \frac{1}{2}L_{ij}(0,0)\right)\partial_{\xi_i\xi_j}\phi_0 + \sum_j \lambda_j\partial_{\xi_j}|\psi|^2 = 0,$$

$$(10.1.25)$$

or

$$\sum_{i,j}\left(\frac{1}{2}L_{00}(0,0)\omega_i'\omega_j' - L_{0i}(0,0)\omega_j' + \frac{1}{2}L_{ij}(0,0)\right)\partial_{\xi_i\xi_j}\phi_1 + \sum_{i,j}\lambda_{i,j}\partial_{\xi_i\xi_j}|\psi|^2 = 0.$$

$$(10.1.26)$$

These mean field equations are solvable in one space dimension, and in this case, ϕ_1 or $\partial_\xi\phi_0$ is proportional to $|\psi|^2$. The equation for the wave amplitude then reduces to the canonical one-dimensional cubic Schrödinger equation where the nonlinear coupling is renormalized by the interaction with the mean field.

10.2   A few simple examples

We consider here a few typical examples of scalar wave equations with quadratic nonlinearity. Following Newell (1985), we first consider the Korteweg–de Vries equation (see also Zakharov and Kuznetsov 1986a). On the time scale of the envelope dynamics, the mean field is slaved and proportional to the square amplitude of the envelope. Differently, for the Boussinesq equation, it is the mean field gradient that is proportional to this quantity. In both cases, the cubic nonlinear Schrödinger equation is recovered with a coupling coefficient renormalized by the interaction with the mean field. This is not the case in higher dimensions, where the non-linear Schrödinger equation is coupled to an equation for the mean flow (Davey–Stewartson system). This situation is illustrated in the case of the Kadomtsev–Petviashvili equation, also considered by Ablowitz, Manakov and Schultz(1990) and Ablowitz and Clarkson (1991, p. 264).

10.2.1   The Korteweg–de Vries equation

The Korteweg–de Vries (KdV) equation

$$u_t + \lambda u_{xxx} + uu_x = 0 \qquad (10.2.1)$$

is classically obtained as the canonical description of the long-wave dynamics (in a frame moving at the phase velocity) of a one-dimensional weakly nonlinear wave with a linear dispersion relation given by $\omega = \alpha k + \beta k^3 + \cdots$ in the limit of small wave numbers (Segur 1978). It was first derived in the context of shallow water-waves (see, e.g., Ablowitz and Segur 1981, Newell 1985). A simple example concerning plasma waves is given in Washimi and Taniuti (1966). It is a main paradigm of a soliton equation integrable by inverse scattering transform. A historical survey of the discovery of KdV solitons is found in Allen (1998).

Since the nonlinear term includes a first order derivative, a resonant term $\partial_X |\psi_0|^2$ arises in the equation obtained at the third order of the ϵ-expansion. This term drives a mean field that because of the form of the linear operator in the KdV equation (first order in time and third order in space) occurs in the correction term u_1.

Proceeding as in the previous section, we obtain at the successive orders of expansion

$$\partial_t u_0 + \lambda \partial_x^3 u_0 = 0, \tag{10.2.2}$$

$$\partial_t u_1 + \lambda \partial_x^3 u_1, = -\partial_T u_0 - 3\lambda \partial_x^2 \partial_X u_0 - \frac{1}{2}\partial_x u_0^2 \tag{10.2.3}$$

$$\partial_t u_2 + \lambda \partial_x^3 u_2, = -\partial_T u_1 - 3\lambda \partial_x^2 \partial_X u_1 - 3\lambda \partial_x \partial_X^2 u_0$$
$$- \partial_x (u_0 u_1) - \frac{1}{2}\partial_X u_0^2. \tag{10.2.4}$$

At leading order, we have

$$u_0 = \psi_0 e^{i(kx-\omega t)} + \text{c.c.} \tag{10.2.5}$$

with $\omega = -\lambda k^3$. As expected, at the next order, the solvability condition reads

$$\partial_T \psi_0 + \omega' \partial_X \psi_0 = 0, \tag{10.2.6}$$

and we solve

$$u_1 = \psi_1 e^{i(kx-\omega t)} + \frac{1}{6\lambda k^2}\psi_0^2 e^{2i(kx-\omega t)} + \text{c.c.} + \phi, \tag{10.2.7}$$

where the mean flow arises at a subdominant order. At the third order, solvability conditions are written in order to eliminate both resonant terms proportional to $e^{i(kx-\omega t)}$ (and its complex conjugate) and those independent of the fast variables. Defining $\psi = \psi_0 + \epsilon\psi_1$ and using (10.2.6), we get

$$\partial_T \psi - 3\lambda k^2 \partial_X \psi + i\epsilon \{3\lambda k \partial_X^2 \psi + \frac{1}{6\lambda k}|\psi|^2 \psi + k\phi\psi\} = 0 \tag{10.2.8}$$

$$\partial_T \phi + \partial_X |\psi|^2 = 0. \tag{10.2.9}$$

Since ϕ arises in the $O(\epsilon)$-term in (10.2.8), it is enough to compute it at leading order, and thus to replace (10.2.9) by

$$\partial_T \phi + \frac{1}{3\lambda k^2}\partial_T |\psi|^2 = 0 \tag{10.2.10}$$

or

$$\phi = -\frac{1}{3\lambda k^2}|\psi|^2. \tag{10.2.11}$$

Equivalently, one can argue that since the mean field ϕ evolves on a time scale of order ϵ^{-1}, faster than the typical time of order ϵ^{-2} of the carrier modulation, its dynamics on the latter time scale adiabatically follows that of the carrier.

The equation for the wave amplitude becomes

$$\partial_T\psi - 3\lambda k^2\partial_X\psi + i\epsilon\{3\lambda k\partial_X^2\psi - \frac{1}{6\lambda k}|\psi|^2\psi\} = 0. \tag{10.2.12}$$

Using the reference frame moving at the group velocity and rescaling time, we define $\xi = x + 3\lambda k^2 t$ and $\tau = \epsilon T$. We finally get

$$i\partial_\tau\psi - 3\lambda k\psi_{\xi\xi} + \frac{1}{6\lambda k}|\psi|^2\psi = 0. \tag{10.2.13}$$

Note that the Benjamin–Feir instability is prevented by the effect of the mean field, which has a stabilizing effect.

10.2.2 The Boussinesq equation

A simple example where the mean flow arises at the leading order is provided by the Boussinesq equation

$$u_{tt} - c^2 u_{xx} + \lambda u_{xxxx} = \beta u_x u_{xx}, \tag{10.2.14}$$

which describes the dynamics of two dispersive counterpropagating waves. In the long-wave, weakly nonlinear regime $\lambda \sim \beta \ll 1$, these two waves decouple, and the dynamics of each of them are governed by the KdV equation (Ablowitz and Segur 1981). Here we assume the dispersion finite and the nonlinearity weak.

When introducing the slow variables, the linear operator arising in the left-hand side of the equation can be rewritten in the form

$$L = L^{(0)} + \epsilon L^{(1)} + \epsilon^2 L^{(2)} + \ldots, \tag{10.2.15}$$

where

$$L^{(0)} = \partial_{tt} - c^2\partial_{xx} + \lambda\partial_{xxxx}, \tag{10.2.16}$$

$$L^{(1)} = 2(\partial_t\partial_T - c^2\partial_x\partial_X + 2\lambda\partial_x^3\partial_X), \tag{10.2.17}$$

$$L^{(2)} = \partial_{TT} - c^2\partial_X^2 + 6\lambda\partial_x^2\partial_X^2. \tag{10.2.18}$$

Taking λ and β of order unity, one expands $u = \epsilon(u_0 + \epsilon u_1 + \cdots)$. At the successive orders, the ϵ-expansion leads to the system

$$L^{(0)}u_0 = 0, \tag{10.2.19}$$

$$L^{(0)}u_1 = -L^{(1)}u_0 + \beta(\partial_x u_0)(\partial_x^2 u_0), \tag{10.2.20}$$

$$L^{(0)}u_2 = -L^{(2)}u_0 - L^{(1)}u_1 + \beta G_2, \tag{10.2.21}$$

where

$$G_2 = (2\partial_x\partial_X u_0 + \partial_x^2 u_1)\partial_x u_0 + (\partial_X u_0 + \partial_x u_1)\partial_{xx} u_0. \qquad (10.2.22)$$

At leading order, we solve

$$u_0 = \psi_0 e^{i(kx-\omega t)} + \text{c.c.} + \phi_0, \qquad (10.2.23)$$

with the dispersion relation

$$-\omega^2 + c^2 k^2 + \lambda k^4 = 0. \qquad (10.2.24)$$

In (10.2.20), the linear term of the right-hand side is resonant and thus leads to the solvability condition

$$(\partial_t\partial_T - c^2\partial_x\partial_X + 2\lambda\partial_x^3\partial_X)\psi_0 = 0, \qquad (10.2.25)$$

or

$$i\omega(\partial_T + \omega'\partial_X)\psi_0 = 0. \qquad (10.2.26)$$

Equation (10.2.20) is then solved as

$$u_1 = -\frac{i\beta}{12\lambda k}\psi_0^2 e^{2i(kx-\omega t)} + \psi_1 e^{i(kx-\omega t)} + \text{c.c.} \qquad (10.2.27)$$

The solvability condition that cancels out the resonant terms proportional to $e^{i(kx-\omega t)}$ in (10.2.21), is

$$-2i\omega(\partial_T + \omega'\partial_X)\psi_1 - \omega\omega''\partial_X^2\psi_0 - \frac{\beta^2 k^2}{6\lambda}|\psi_0|^2\psi_0 + \beta k^2\psi_0\partial_X\phi = 0. \qquad (10.2.28)$$

The solvability condition that eliminates the nonoscillating terms in (10.2.21) is

$$(\partial_T^2 - c^2\partial_X^2)\phi = \beta k^2\partial_X|\psi_0|^2. \qquad (10.2.29)$$

Defining $\psi = \psi_0 + \epsilon\psi_1$ and $\rho = \partial_X\phi$ and combining the successive solvability conditions originating from the oscillating terms, we obtain the system

$$2i(\partial_T + \omega'\partial_X)\psi + \epsilon\{\omega''\partial_X^2\psi + \frac{\beta^2 k^2}{6\lambda\omega}|\psi|^2\psi - \frac{k^2\beta}{\omega}\rho\psi\} = 0, \qquad (10.2.30)$$

$$(\partial_T^2 - c^2\partial_X^2)\rho - \beta k^2\partial_X^2|\psi|^2 = 0. \qquad (10.2.31)$$

At leading order

$$(\partial_T^2 - c^2\partial_X^2)|\psi|^2 = (\omega'^2 - c^2)\partial_X^2|\psi|^2, \qquad (10.2.32)$$

and (10.2.31) becomes

$$(\partial_T^2 - c^2\partial_X^2)\left(\rho - \frac{\beta k^2}{\omega'^2 - c^2}|\psi|^2\right) = 0. \qquad (10.2.33)$$

Excluding the long-wave short-wave resonance $\omega' = c$ where the group velocity of the carrier equals the phase velocity of the mean field, we get

$$\rho = \frac{\beta k^2}{\omega'^2 - c^2}|\psi|^2. \qquad (10.2.34)$$

In other words, the low frequency field ρ which on the time scale ϵ^{-1} obeys a wave equation driven by the amplitude modulation of the carrier, becomes slaved on the time scale ϵ^{-2}.

After elimination of the mean field, one finally obtains

$$2i(\partial_T + \omega'\partial_X)\psi + \epsilon\left(\omega''\partial_X^2\psi - \frac{k^2\beta^2\omega''}{6\lambda(\omega'^2 - c^2)}|\psi|^2\psi\right) = 0, \qquad (10.2.35)$$

which after the change of frame $\xi = X - \omega'T$ and the introduction of a slower time scale $\tau = \epsilon T = \epsilon^2 t$, reduces to the canonical nonlinear Schrödinger equation

$$2i\partial_\tau\psi + \omega''\partial_{\xi\xi}\psi - \frac{k^2\beta^2\omega''}{6\lambda(\omega'^2 - c^2)}|\psi|^2\psi = 0. \qquad (10.2.36)$$

Near the long-wave short-wave resonance $\omega' = c$, we rewrite the system (10.2.30)–(10.2.31) in the frame moving at the group velocity of the wave packet. Using the variables $\xi = X - \omega'T$ and $\tau = \epsilon T$, and neglecting $O(\epsilon^2)$-corrections, we get,

$$2i\partial_\tau\psi + \omega''\psi_{\xi\xi} + \left(\frac{\beta^2 k^2}{6\lambda\omega}|\psi|^2 - \frac{k^2\beta}{\omega}\rho\right)\psi = 0 \qquad (10.2.37)$$

$$-2\omega'\epsilon\partial_\tau\rho + (\omega'^2 - c^2)\partial_\xi\rho - \beta k^2\partial_\xi|\psi|^2 = 0. \qquad (10.2.38)$$

Near the resonance $(c - \omega' = \nu\epsilon)$, the amplitude is rescaled by writing $\psi = \epsilon^{1/2}\widetilde{\psi}$. Equivalently, the wave amplitude can be kept of order ϵ and the distance to the resonance taken as $c - \omega' = \nu\epsilon^{2/3}$. The other variables are also differently rescaled in the form $\xi = \epsilon^{2/3}(x - \omega't)$ and $\tau = \epsilon^{4/3}t$, while $\rho = \partial_x\phi$ is obtained by scaling the physical variable by $\epsilon^{4/3}$.

Neglecting subdominant terms and dropping the tildes, we get

$$2i\partial_\tau\psi + \omega''\psi_{\xi\xi} - \frac{k^2\beta}{\omega}\rho\psi = 0 \qquad (10.2.39)$$

$$2\omega'\partial_\tau\rho + \nu\partial_x\rho + \frac{\beta k^2}{2\omega'}\partial_\xi|\psi|^2 = 0. \qquad (10.2.40)$$

By simple rescaling, this system takes the canonical form (Benney 1977)

$$i\psi_\tau + \lambda\psi_{\xi\xi} - \rho\psi = 0 \qquad (10.2.41)$$

$$\rho_\tau + \nu\rho_\xi + \alpha(|\psi|^2)_\xi = 0, \qquad (10.2.42)$$

where λ and α are positive constants. Note that at the resonance, the mean field is no longer slaved to the carrier modulation. System (10.2.41)-(10.2.42) was solved by inverse scattering technique in the case $\lambda = 1/2$ and $\alpha = 1$ by Yajima and Oikawa (1976), and for $\nu = 0$ and arbitrary positive λ and α by Ma (1978). The well-posedness of the system when the (subdominant) cubic nonlinearity is retained in the carrier amplitude equation is studied by Bekiranov, Ogawa, and Ponce (1996) (see also references therein).

10.2.3   The Kadomtsev–Petviashvili equation

The equation, named after Kadomtsev and Petviashvili (1970) can be viewed as the multidimensional generalization of the KdV equation when slow variations in a direction transverse to the propagation are retained. A systematic derivation of this equation in the context of surface water waves is found in Freeman and Davey (1975). It reads

$$(u_t + u_{xxx} + uu_x)_x + \lambda u_{yy} = 0, \tag{10.2.43}$$

where $\lambda = \pm 1$. The case $\lambda = -1$ is called KPI, while $\lambda = +1$ is known as KPII. Both are integrable by inverse scattering transform (Ablowitz and Clarkson 1991).

The carrying wave being assumed to propagate in the x direction, we introduce the slow variables $X = \epsilon x$, $Y = \epsilon y$, and $T = \epsilon t$ and expand in powers of ϵ.

At order ϵ we have

$$\partial_x(\partial_t + \partial_x^3)u_0 = 0, \tag{10.2.44}$$

leading to

$$u_0 = \psi_0(X, Y, T)e^{i(kx - \omega t)} + \text{c.c.} \tag{10.2.45}$$

with $\omega = -k^3$.

At order ϵ^2, we have

$$\partial_x(\partial_t + \partial_x^3)u_1 = -(\partial_x\partial_T + \partial_t\partial_X + 4\partial_X\partial_x^3)u_0 - \frac{1}{2}\partial_{xx}u_0^2. \tag{10.2.46}$$

The solvability condition reads

$$\partial_T\psi_0 + \omega'\partial_X\psi_0 = 0, \tag{10.2.47}$$

with $\omega' = -3k^2$. The equation for u_1 becomes

$$\partial_x(\partial_t + \partial_x^3)u_1 = -\frac{1}{2}\partial_{xx}(\psi_0^2 e^{2i(kx - \omega t)} + \text{c.c.}), \tag{10.2.48}$$

and is solved as

$$u_1 = \frac{1}{6k^2}\psi_0^2 e^{2i(kx - \omega t)} + \psi_1(X, Y, T)e^{i(kx - \omega t)} + \text{c.c.} + \phi_1(X, Y, T), \tag{10.2.49}$$

where elements spanning the null space have been included.

At third order, we have

$$\partial_x(\partial_t + \partial_x^3)u_2 = -(\partial_x\partial_T + \partial_X\partial_t + 4\partial_X\partial_x^3)u_1$$
$$-(\partial_X\partial_T + 6\partial_X^2\partial_x^2)u_0 - \partial_x\partial_X u_0^2 - \partial_x^2(u_0u_1) - \lambda\partial_{YY}u_0. \tag{10.2.50}$$

The solvability condition reads

$$(-ik\partial_T + 3ik^3\partial_X)\psi_1 - \partial_{XT}\psi_0 + 6k^2\partial_{XX}\psi_0$$
$$- \lambda\partial_{YY}\psi_0 + \frac{1}{6}|\psi_0|^2\psi_0 + k^2\phi_1\psi_0 = 0. \tag{10.2.51}$$

Defining, as before, $\psi = \psi_0 + \epsilon\psi_1$, and using that $\partial_T\psi = 3k^2\partial_X\psi + O(\epsilon)$, we obtain

$$- ik(\partial_T - 3k^2\partial_X)\psi + \epsilon\{3k^2\partial_{XX}\psi - \lambda\partial_{YY}\psi + (\frac{1}{6}|\psi|^2 + k^2\phi_1)\psi\} = 0,$$
(10.2.52)

or

$$i(\partial_T - 3k^2\partial_X)\psi + \epsilon\{-3k\partial_X^2\psi + \frac{\lambda}{k}\partial_Y^2\psi - (\frac{1}{6k}|\psi|^2 + k\phi_1)\psi\} = 0. \quad (10.2.53)$$

The mean field ϕ_1 is obtained by considering the order ϵ^4, which reads

$$\partial_x(\partial_t + \partial_x^3)u_3 = -(\partial_x\partial_T + \partial_X\partial_t + 4\partial_X\partial_x^3)u_2 -$$
$$-(\partial_X\partial_T + 6\partial_X^2\partial_x^2)u_1 - \lambda\partial_Y^2 u_1 - 4ik\partial_X^3 u_0$$
$$-\frac{1}{2}\partial_X^2 u_0^2 - \partial_x\partial_X(u_0 u_1) - \frac{1}{2}\partial_x^2(u_1^2 + 2u_0 u_2).$$
(10.2.54)

Elimination of the non oscillating resonant terms requires the solvability condition, which to leading order, reads

$$\partial_T\partial_X\phi + \lambda\partial_{YY}\phi + \partial_{XX}|\psi|^2 = 0. \tag{10.2.55}$$

Using the reference frame moving at the group velocity, we define

$$\xi = X + 3k^2 T, \tag{10.2.56}$$

and introducing $\tau = \epsilon^2 t$, we have

$$i\partial_\tau\psi - 3k\partial_{\xi\xi}\psi + \frac{\lambda}{k}\partial_{YY}\psi - (\frac{1}{6k}|\psi|^2 + k\phi)\psi = 0, \tag{10.2.57}$$
$$\epsilon\partial_\tau\partial_\xi\phi + 3k^2\partial_{\xi\xi}\phi + \lambda\partial_{YY}\phi + \partial_{\xi\xi}|\psi|^2 = 0. \tag{10.2.58}$$

In the limit $\epsilon \to 0$, that is to say on the time scale $O(\epsilon^{-2})$, the mean flow obeys

$$3k^2\partial_{\xi\xi}\phi + \lambda\partial_{YY}\phi = -\partial_{\xi\xi}|\psi|^2 \tag{10.2.59}$$

and thus follows adiabatically the amplitude dynamics. The resulting "Davey–Stewartson (DS) system" can be written in a canonical form by rescaling the dependent and independent variables in the form $\psi = 6\tilde{\psi}$, $\phi = \frac{6}{k^2}\tilde{\phi}$, $\xi = k\tilde{\xi}$, $Y = 3^{-\frac{1}{2}}\tilde{Y}$, and $\tau = \frac{k}{3}\tilde{\tau}$. After dropping the tildes, we get

$$i\partial_\tau\psi - \partial_{\xi\xi}\psi + \lambda\partial_{YY}\psi - 2(|\psi|^2 + \phi)\psi = 0 \tag{10.2.60}$$
$$\partial_{\xi\xi}\phi + \lambda\partial_{YY}\phi + 2\partial_{\xi\xi}|\psi|^2 = 0, \tag{10.2.61}$$

where $\lambda = \pm 1$. This system is discussed in details in Chapter 12.

10.3   A Degenerate Case

As exemplified above, the cubic NLS equation (or in several dimensions, the DS system) is the canonical equation for the envelope dynamics of a weakly nonlinear dispersive wave train. In some instances, however, the interactions of the carrying wave with the mean field and with the second harmonics exactly cancel out in the solvability condition that generically leads to the NLS equation. At the usual temporal and spatial scales for the envelope evolution, the dynamics are then purely dispersive. Nonlinear effects become relevant only in considering a wave of larger amplitude, or equivalently a modulation at larger scales. This situation arises in the context of the two-dimensional water-wave problem in the special case where the product kh of the carrier wave number k and the undisturbed fluid depth h is close to a critical value (about 1.363 in the absence of surface tension), for which the coefficient in front of the nonlinear term in the (one-dimensional) NLS equation for the wave amplitude vanishes (Johnson 1977). Another example not restricted to a specific choice of the carrier wave number, is provided by the Benjamin–Ono equation

$$\frac{\partial u}{\partial t} + \frac{\partial^2}{\partial x^2} \int_{-\infty}^{+\infty} \frac{u(x')}{x - x'} dx' + u \frac{\partial u}{\partial x} = 0, \qquad (10.3.1)$$

where the integral is taken in the sense of principal value. This equation arises in the description of long internal waves propagating in a deep stratified fluid (Benjamin 1967b, Davies and Acrivos 1967, Ono 1975).

Using that the Hilbert transform

$$Hu(x) = \int_{-\infty}^{+\infty} \frac{u(x')}{x - x'} dx'$$

reads in Fourier space $\widehat{Hu}(k) = i\,\mathrm{sgn}\,(k)\,\hat{u}(k)$, (10.3.1) can be rewritten in the form

$$\frac{\partial u}{\partial t} - i|D_x|D_x u + iuD_x u = 0, \qquad (10.3.2)$$

where $D_x = -i\frac{\partial}{\partial x}$ and $|D_x|$ corresponds to the Fourier multiplier $|k|$.

In introducing the slow variable $X = \epsilon x$, $|D_x|$ is replaced by $|D_x + \epsilon D_X|$, which we expand in powers of ϵ. This can be done formally as in the case of a usual Taylor expansion (Craig, Sulem, and Sulem 1992). It should be noticed that the operator $|D_x + \epsilon D_X|$ is given by $|D_x|^{-1}D_x(D_x + \epsilon D_X)$ when it acts on a function depending on both the slow and fast variables, while it reduces to $\epsilon|D_X|$ when applied to functions of the slow variable only. It follows that the linear operator appearing on the left-hand side of (10.3.2) is expanded as $L = L^{(0)} + \epsilon L^{(1)} + \epsilon^2 L^{(2)} + \cdots$, where

$$L^{(0)} = \partial_t - i|D_x|D_x, \qquad (10.3.3)$$
$$L^{(1)} = \partial_T - 2i|D_x|D_X, \qquad (10.3.4)$$

and

$$L^{(2)} = -iD_x|D_x|^{-1}D_{XX},$$ (10.3.5)

when acting on functions of both the slow and fast variables, or

$$L^{(2)} = -i|D_X|D_X$$ (10.3.6)

when acting on functions of the sole slow variables.

In order that in the envelope equation the dispersion be comparable to the nonlinear terms, expected of the form $|\psi|^4\psi$ or $\partial_X(|\psi|^2)\psi$, we are led to expand $u = \epsilon^{1/2}u_0 + \epsilon u_1 + \epsilon^{3/2}u_2 + \cdots$. This leads to the hierarchy

$$L^{(0)}u_0 = 0,$$ (10.3.7)

$$L^{(0)}u_1 = -\frac{i}{2}D_x(u_0^2),$$ (10.3.8)

$$L^{(0)}u_2 = -L^{(1)}u_0 - iD_x(u_0 u_1),$$ (10.3.9)

$$L^{(0)}u_3 = -L^{(1)}u_1 - iD_x(\frac{1}{2}u_1^2 + u_0 u_2) - \frac{i}{2}D_X(u_0^2),$$ (10.3.10)

$$L^{(0)}u_4 = -L^{(1)}u_2 - L^{(2)}u_0 - iD_x(u_1 u_2 + u_0 u_3) - iD_X(u_0 u_1),$$ (10.3.11)

$$L^{(0)}u_5 = -L^{(1)}u_3 - L^{(2)}u_1 - iD_x(u_0 u_4 + u_1 u_3 + \frac{1}{2}u_2^2) - iD_X(u_0 u_2 + \frac{1}{2}u_1^2).$$ (10.3.12)

Equation (10.3.7) is solved as

$$u_0 = \psi_0 e^{i\theta} + \text{c.c.}$$ (10.3.13)

with $\theta = kx - \omega t$ and the dispersion relation

$$\omega + |k|k = 0.$$ (10.3.14)

Equation (10.3.8) is solved as

$$u_1 = \frac{\psi_0^2}{2|k|}e^{2i\theta} + \psi_1 e^{i\theta} + \text{c.c.} + \phi_1.$$ (10.3.15)

Equation (10.3.9) requires the solvability condition

$$(\partial_T - 2|k|\partial_X)\psi_0 + ik\left(\frac{|\psi_0|^2}{2|k|} + \phi_1\right)\psi_0 = 0,$$ (10.3.16)

where $-2|k|$ is identified with the group velocity ω'. It then gives

$$u_2 = \frac{\psi_0^3}{4k^2}e^{3i\theta} + \frac{\psi_0\psi_1}{|k|}e^{2i\theta} + \psi_2 e^{i\theta} + \text{c.c.} + \phi_2.$$ (10.3.17)

The solvability conditions required for (10.3.10) read

$$(\partial_T - 2|k|\partial_X)\psi_1 + i\,k(\psi_1\phi_1 + \psi_0\phi_2) + i\,\text{sgn}\,k\,\left(\frac{1}{2}\psi_0^2\psi_1^* + |\psi_0|^2\psi_1\right) = 0$$ (10.3.18)

and

$$\partial_T \phi_1 + D_X |\psi_0|^2 = 0. \tag{10.3.19}$$

One then solves (10.3.10) in the form

$$u_3 = \alpha e^{4i\theta} + \beta e^{3i\theta} + \gamma e^{2i\theta} + \psi_3 e^{i\theta} + \text{c.c.} + \phi_3 \tag{10.3.20}$$

where α, β, and γ are slowly varying functions that are easily obtained by identification in (10.3.10). In fact, only γ is to be computed explicitely, in the form

$$\gamma = \frac{i}{4|k|k}\left(-\frac{1}{|k|}\partial_T\psi_0^2 + 3\partial_X\psi_0^2\right)$$

$$+ \frac{1}{|k|}\left(\psi_0\psi_2 + \frac{1}{2}\psi_1^2 + \frac{1}{2|k|}\phi_1\psi_0^2 + \frac{1}{4|k|^2}|\psi_0|^2\psi_0^2\right). \tag{10.3.21}$$

The solvability of (10.3.11) requires

$$(\partial_T - 2|k|\partial_X)\psi_2 + i\,\text{sgn}k\partial_{XX}\psi_0 + i\,\text{sgn}k(\frac{1}{2}\psi_0^2\psi_2^* + |\psi_1|^2\psi_0 + \frac{1}{8|k|^2}|\psi_0|^4\psi_0)$$

$$+ ik(\phi_1\psi_2 + \phi_2\psi_1 + \phi_3\psi_0 + \gamma\psi_0^*) + \partial_X(\psi_0\phi_1 + \frac{1}{2|k|}|\psi_0|^2\psi_0) = 0, \tag{10.3.22}$$

and

$$\partial_T \phi_2 + \partial_X(\psi_0\psi_1^* + \psi_0^*\psi_1) = 0. \tag{10.3.23}$$

The system formed by the solvability conditions (10.3.16), (10.3.18), (10.3.19), (10.3.22), and (10.3.23) will be closed by expressing ϕ_3 by means of the solvability condition suppressing the nonoscillating resonant terms from the right-hand side of (10.3.12). It reads

$$\partial_T \phi_3 - |D_X|\partial_X\phi_1 + \partial_X\left(\frac{1}{4k^2}|\psi_0|^4 + \frac{1}{2}\phi_1^2 + |\psi_1|^2 + \psi_0\psi_2^* + \psi_0^*\psi_2\right) = 0. \tag{10.3.24}$$

One then defines $\psi = \psi_0 + \epsilon^{1/2}\psi_1 + \epsilon\psi_2$ and $\phi = \phi_1 + \epsilon^{1/2}\phi_2 + \epsilon\phi_3$. Using the frame moving with the group velocity by defining $\xi = X - \omega'T$, and introducing $\tau = \epsilon T$, one obtains

$$\epsilon\partial_\tau\psi + i\,\epsilon\,\text{sgn}\,k\partial_{\xi\xi}\psi + i\,k\left(\phi + \frac{|\psi|^2}{2|k|}\right)\psi + i\,\epsilon\text{sgn}\,k\frac{1}{2|k|}|\psi|^2\psi\left(\phi + \frac{|\psi|^2}{2|k|}\right)$$

$$+ i\,\epsilon\,\text{sgn}\,k\frac{1}{8|k|^2}|\psi|^4\psi - \frac{\epsilon}{4|k|}\psi^*\partial_\xi\psi^2 + \epsilon\partial_\xi((\phi + \frac{|\psi|^2}{2|k|})\psi) = O(\epsilon^{3/2}), \tag{10.3.25}$$

$$\epsilon\partial_\tau\phi + \partial_\xi\left(2|k|\phi + |\psi|^2 + \epsilon\frac{\phi^2}{2} + \epsilon\frac{|\psi|^4}{4|k|^2}\right) + \epsilon|D_\xi|\partial_\xi\phi = O(\epsilon^{3/2}). \tag{10.3.26}$$

Equation (10.3.26) is solved at order $O(\epsilon^{3/2})$ in the form

$$2|k|\phi + |\psi|^2 = \epsilon \left(\frac{-3}{8k^2}|\psi|^4 - \frac{1}{2|k|}|D_\xi|(|\psi|^2) + \partial_\tau \int^\xi \frac{|\psi|^2}{2|k|}d\xi \right) + O(\epsilon^{3/2}).$$

(10.3.27)

Replacing this expression in (10.3.25), we have

$$\partial_\tau \psi + i \operatorname{sgn} k \, \partial_{\xi\xi}\psi - \frac{i}{2}\operatorname{sgn} k \left(\frac{1}{8k^2}|\psi|^4 + \frac{1}{2|k|}|D_\xi|(|\psi|^2) + \partial_\tau \int^\xi \frac{|\psi|^2}{2|k|}d\xi \right)\psi$$

$$- \frac{1}{4|k|}\psi^*\partial_\xi\psi^2 = 0.$$

(10.3.28)

The term $\partial_\tau \int^\xi \frac{|\psi|^2}{2|k|}d\xi$ in (10.3.28) is calculated by multiplying this equation by ψ^*, integrating over $(-\infty,\xi]$, and taking the imaginary part. This gives

$$\partial_\tau \int^\xi \frac{|\psi|^2}{2|k|}d\xi = \frac{1}{2i|k|}\operatorname{sgn} k(\psi^*\partial_\xi\psi - \psi\partial_\xi\psi^*) + \frac{1}{8|k|^2}|\psi|^4.$$

(10.3.29)

Substituting in (10.3.28), we get

$$\partial_\tau \psi + i \operatorname{sgn} k \, \partial_{\xi\xi}\psi - \frac{i}{2}\operatorname{sgn} k \left(\frac{1}{2|k|}|D_\xi|(|\psi|^2) + \frac{i}{2k}(\psi^*\partial_\xi\psi - \psi\partial_\xi\psi^*) \right)\psi$$

$$- \frac{1}{4|k|}\psi^*\partial_\xi\psi^2 = 0.$$

(10.3.30)

Equivalently, the equation rewrites

$$\partial_\tau \psi + i \operatorname{sgn} k \, \partial_{\xi\xi}\psi - \frac{i}{4k}|D_\xi|(|\psi|^2)\psi - \frac{1}{4|k|}\psi\partial_\xi(|\psi|^2) = 0$$

(10.3.31)

or

$$2i\partial_\tau\psi + \omega'' \partial_{\xi\xi}\psi - \frac{1}{2|k|}\partial_\xi\left((\operatorname{sgn} k \, H + i)|\psi|^2\right)\psi = 0,$$

(10.3.32)

where $\omega'' = -2\operatorname{sgn} k$ for $k \neq 0$, and H is the Hilbert transform. This equation was derived by Pelinovsky (1995) (see also Pelinovsky and Grimshaw 1996) who revisited a previous analysis by Tanaka (1982). The integrability by inverse scattering transform was established in Pelinovsky and Grimshaw (1995).

Remark: Another example of degeneracy concerns the vicinity of a caustic of the dispersive wave where the group velocity is stationary and the dispersive effects thus relatively weak. The NLS equation must then be modified by the inclusion of third-order derivatives. The existence of wave envelopes of permanent form in this regime is discussed by Akylas and Kung (1990).

and

11

Gravity-Capillary Surface Waves

We consider in this chapter the weakly nonlinear dynamics of a wave train propagating at the surface of a liquid, a question also referred to as the water-wave problem. When the wave is modulated in the direction of propagation only, the long-time large-scale dynamics of the envelope is governed by the cubic Schrödinger equation in one space dimension (see Zakharov 1968b for the case of infinitely deep water, Hasimoto and Ono 1972 for the extension to a finite depth, and Kawahara 1975 for the inclusion of the surface tension). We review here the more general case where perturbations in the transverse directions are also permitted. In this case, a mean flow and a mean elevation of the free surface are induced by the amplitude modulation of the wave. The original derivations of the envelope equations in the case of pure gravity waves are due to Benney and Roskes (1969), who retained the dynamics of the mean fields, and Davey and Stewartson (1974), who consider these fields on the long-time scale where they are slaved to the amplitude modulation (see also Johnson 1997). It is noticeable that the former system is identical with that arising in situations where the carrying wave modulation drives low-frequency acoustic waves (Zakharov and Rubenchik 1972, Zakharov and Schulman 1991), thus providing another example of the canonical character of the amplitude equations. The effect of surface tension was included by Djordjevic and Redekopp (1977) and Ablowitz and Segur (1979).

11.1 The Water-Wave Problem

11.1.1 Equations Governing the Interface Motion

In this section a formal derivation of the Davey–Stewartson (DS) system including the surface tension is presented. The modulation analysis is performed, starting from a reduced formulation of the water-wave problem in terms of the free surface elevation and the trace of the velocity potential on this surface. We consider the motion of the free surface $\mathbf{r} = (x, y, z = h(x, y))$ of a three-dimensional fluid with surface tension (normalized by the density) β, under the influence of gravity g. The domain is infinite in the horizontal directions and has a fixed bottom at $z = -H_0$. The fluid is assumed to be incompressible, inviscid, and irrotational, so that the motion is described by a velocity potential Φ that satisfies (Whitham 1974, Karpman 1975)

$$\Delta \Phi = 0 \quad \text{for} \quad -H_0 < z < h(x, y, t), \tag{11.1.1}$$

with the boundary conditions

$$\partial_z \Phi = 0 \quad \text{on} \quad z = -H_0, \tag{11.1.2}$$

and

$$\partial_t \Phi + \frac{1}{2}(\nabla \Phi)^2 + gh - \beta \Gamma(h) = 0, \tag{11.1.3}$$

$$\partial_t h + \nabla_\perp \Phi \cdot \nabla_\perp h - \partial_z \Phi = 0, \tag{11.1.4}$$

on the free surface $z = h(x, y, t)$ over the fluid domain. Here $\nabla_\perp = (\partial_x, \partial_y)$ is the horizontal gradient, and

$$\Gamma(h) = \frac{h_{xx}(1 + h_y^2) + h_{yy}(1 + h_x^2) - 2h_x h_y h_{xy}}{(1 + |\nabla_\perp h|^2)^{3/2}} \tag{11.1.5}$$

is the mean curvature of the free surface. Proceeding as in Craig and Sulem (1993), we first reduce (11.1.1)–(11.1.4) to a system where all the functions are evaluated at the free surface only by introducing the trace $\phi(x, y, t) = \Phi(x, y, h(x, y, t), t)$ of the velocity potential Φ at the surface and defining the Dirichlet–Neumann operator G that relates ϕ to the normal derivative $\partial_n \Phi$ of the potential by

$$G(h)\phi = \sqrt{1 + |\nabla_\perp h|^2} \, \partial_n \Phi \Big|_{z=h} = \left(-\nabla_\perp \Phi \cdot \nabla_\perp h + \partial_z \Phi \right) \Big|_{z=h}. \tag{11.1.6}$$

On the free surface $z = h(x, y, t)$, we additionally have

$$\partial_t \Phi = \partial_t \phi - \partial_t h \partial_z \Phi, \qquad \partial_{x_i} \Phi = \partial_{x_i} \phi - \partial_{x_i} h \partial_z \Phi, \tag{11.1.7}$$

where x_i, with $i = (1, 2)$, stands for x or y.

The water-wave problem is a Hamiltonian system in terms of the canonical variables (h, ϕ), where the Hamiltonian identifies with the total energy of the fluid (see Zakharov 1998 for a recent review).

Proposition 11.1. *The free surface elevation* $h(x, y, t)$ *and the trace* $\phi(x, y, t)$ *of the velocity potential on this surface satisfy the system*

$$\partial_t h - G(h)\phi = 0, \qquad (11.1.8)$$

$$\partial_t \phi + gh + \frac{1}{2(1 + |\nabla_\perp h|^2)} \Big(|\nabla_\perp \phi|^2 - (G(h)\phi)^2 - 2(\nabla_\perp h \cdot \nabla_\perp \phi)G(h)\phi$$

$$+ |\nabla_\perp h|^2 |\nabla_\perp \phi|^2 - (\nabla_\perp h \cdot \nabla_\perp \phi)^2 \Big) - \beta\Gamma(h) = 0. \qquad (11.1.9)$$

When linearizing the water-wave equations about the static solution $h = H_0$, $\phi = 0$, one gets harmonic waves obeying the dispersion relation

$$\omega^2 = k(g + \beta k^2)\tanh(H_0 k). \qquad (11.1.10)$$

This leads to the definition of a critical wavelength $\lambda_m = 2\pi(\beta/g)^{1/2}$ (approximately equal to 1.73 cm in the case of water), for the relative importance of gravity and surface tension (see, e.g., Whitham 1974, Chap. 12). For perturbations whose wavelength is large compared to λ_m, surface tension is negligible, and such perturbations are referred to as *gravity waves*. In the opposite regime where the wavelength of the perturbation is small compared to λ_m, gravity is negligible and the waves are *capillary waves*. The regime where both effects are comparable corresponds to *gravity–capillary waves*.

In the *shallow water limit* $kH_0 \to 0$, the dispersion relation is expanded as

$$\omega = \pm(gH_0)^{1/2} k \left[1 - \frac{1}{2}\left(\frac{1}{3} - \frac{\beta}{gH_0^2}\right) k^2 H_0^2 + \cdots \right]. \qquad (11.1.11)$$

The Korteweg–de Vries and, in several dimensions, the Kadomtsev–Petviashvili equations are obtained in this limit by including nonlinear effects with a strength comparable to the (weak) dispersion (see, e.g., Ablowitz and Segur 1981, Section 4.1).

In contrast, in the *deep water limit*, $kH_0 \to \infty$,

$$\omega^2 = gk\left(1 + \frac{\beta k^2}{g}\right), \qquad (11.1.12)$$

and the gravity waves are strongly dispersive.

11.1.2 Formal Modulation Analysis

In the following, we need the Taylor expansion of the operator G in powers of the surface elevation h in the form

$$G(h) = \sum_{j=0}^{\infty} G_j(h), \qquad (11.1.13)$$

where $G_j(h)$ is a pseudo-differential operator homogeneous in h of degree j. In order to compute explicitly these operators, we consider the particular family of functions

$$f_p(x, y, z) = e^{i(p_1 x + p_2 y)} \cosh(|p|(z + H_0)), \tag{11.1.14}$$

where $|p| = \sqrt{p_1^2 + p_2^2}$. These functions are harmonic in the domain $\{z > -H_0\}$ and satisfy $\partial_z f_p = 0$ on the bottom boundary $z = -H_0$. By definition,

$$G(h)f_p = \partial_z f_p - \nabla_\perp f_p \cdot \nabla_\perp h \Big|_{z=h}. \tag{11.1.15}$$

We substitute (11.1.14) into (11.1.15), expand the hyperbolic functions near $z = 0$, and replace the left-hand side of (11.1.15) by its expansion $\sum_{j=0}^{\infty} G_j(z)f_p$. This leads to an identity where by equating the terms of degree j in h we get a recursion formula for the G_j's (Craig and Groves 1994). Using the notation $|D| = (-\Delta)^{1/2}$, the first three terms of the expansion read

$$G_0 = |D| \tanh(H_0|D|), \tag{11.1.16}$$
$$G_1(h) = D \cdot hD - G_0 h G_0, \tag{11.1.17}$$
$$G_2(h) = -\frac{1}{2}\left(G_0 h^2 |D|^2 + |D|^2 h^2 G_0 - 2G_0 h G_0 h G_0\right). \tag{11.1.18}$$

In a weakly nonlinear analysis, we expand the solution in the form

$$h = \epsilon h^{(1)} + \epsilon^2 h^{(2)} + \cdots, \tag{11.1.19}$$
$$\phi = \epsilon \phi^{(1)} + \epsilon^2 \phi^{(2)} + \cdots, \tag{11.1.20}$$
$$G = G^{(0)} + \epsilon G^{(1)} + \cdots. \tag{11.1.21}$$

We also introduce a large-scale spatial variable $\mathbf{X} = (X, Y) = \epsilon(x, y)$ and a slow time $T = \epsilon t$. At leading order, we have the linearized system

$$L\begin{pmatrix} h \\ \phi \end{pmatrix} = 0 \tag{11.1.22}$$

with $(\Delta_\perp = \partial_{xx} + \partial_{yy})$

$$L = \begin{pmatrix} \partial_t & -G_0 \\ g - \beta \Delta_\perp & \partial_t \end{pmatrix}. \tag{11.1.23}$$

It admits solutions of the form

$$h^{(1)} = \frac{i\omega}{g + \beta k^2} \psi_1(X, T) e^{i(kx - \omega t)} + \text{c.c.}, \tag{11.1.24}$$
$$\phi^{(1)} = \psi_1(X, T) e^{i(kx - \omega t)} + \text{c.c.} + \varphi(X, T), \tag{11.1.25}$$

where we include a mean potential φ, which will be useful at the next order of the expansion. The wave number k is related to ω by the dispersion relation $\omega^2 = (g + \beta k^2) k \tanh(H_0 k)$. For concise notation, we write

$$\chi = H_0 k, \quad \theta = \tanh \chi. \tag{11.1.26}$$

We also use $D_x = -i\partial_x$, $D_X = -i\partial_X$, and $D_Y = -i\partial_Y$ and define $|D_x| = (-\partial_x^2)^{1/2}$. We expand $|D| = (-\Delta)^{1/2}$ and $\tanh(h|D|)$ in the form

$$|D| = |D_x| + \epsilon D_x |D_x|^{-1} D_X + \frac{\epsilon^2}{2} |D_x|^{-1} D_Y^2 + O(\epsilon^3), \qquad (11.1.27)$$

and

$$\tanh(H_0|D|) = \tanh(H_0|D_x|) + \epsilon H_0 D_x |D_x|^{-1} \big(1 - \tanh^2(H_0 D_x)\big) D_X$$
$$+ \epsilon^2 \Big(\frac{H_0}{2} |D_x|^{-1} \big(1 - \tanh^2(H_0 D_x)\big) D_Y^2$$
$$- H_0^2 \big(1 - \tanh^2(H_0 D_x)\big) \tanh(H_0|D_x|) \Big) D_X^2 + O(\epsilon^3).$$
$$(11.1.28)$$

To obtain the coefficients $G^{(n)}$, we first expand G in powers of h and then use the expansion of h in terms of ϵ, together with (11.1.27)–(11.1.28). At leading order, we recover

$$G^{(0)} = D_x \tanh(H_0 D_x). \qquad (11.1.29)$$

The terms of order ϵ and ϵ^2 are respectively

$$G^{(1)} = D_x h^{(1)} D_x - G^{(0)} h^{(1)} G^{(0)} + \tanh(H_0 D_x) D_X$$
$$+ H_0 D_x \big(1 - \tanh^2(H_0 D_x)\big) D_X \qquad (11.1.30)$$

and

$$G^{(2)} = \frac{1}{2} |D_x|^{-1} \tanh(H_0|D_x|) D_Y^2 + H_0 \big(1 - \tanh^2(H_0 D_x)\big) D_X^2$$
$$+ \frac{H_0}{2} \big(1 - \tanh^2(H_0 D_x)\big) D_Y^2$$
$$- H_0^2 D_x \big(1 - \tanh^2(H_0 D_x)\big) \tanh(H_0 D_x) D_X^2$$
$$+ D_x h^{(2)} D_x - G^{(0)} h^{(2)} G^{(0)} + D_x h^{(1)} D_X + D_X h^{(1)} D_x$$
$$- G^{(0)} h^{(1)} \big(\tanh(H_0 D_x) D_X + H_0 D_x \big(1 - \tanh^2(H_0 D_x)\big) D_X \big)$$
$$- \big(\tanh(H_0 D_x) D_X + H_0 D_x \big(1 - \tanh^2(H_0 D_x)\big) D_X \big) h^{(1)} G^{(0)}$$
$$- \frac{1}{2} \big(G^{(0)} h^{(1)2} D_x^2 + D_x^2 h^{(1)2} G^{(0)} - 2 G^{(0)} h^{(1)} G^{(0)} h^{(1)} G^{(0)} \big).$$
$$(11.1.31)$$

We now expand equations (11.1.8)–(11.1.9) in powers of ϵ, which at order $n > 1$ gives the inhomogeneous linear system

$$L \begin{pmatrix} h^{(n)} \\ \phi^{(n)} \end{pmatrix} = \begin{pmatrix} A_n \\ B_n \end{pmatrix}. \qquad (11.1.32)$$

The solvability of (11.1.32) requires that its right-hand side be orthogonal to the kernel of the adjoint operator

$$L^\dagger = \begin{pmatrix} -\partial_t & g + \beta D_x^2 \\ -G^{(0)} & -\partial_t \end{pmatrix} \qquad (11.1.33)$$

of L. Since the kernel of $L^\dagger$ is spanned by

$$\begin{pmatrix} 1 \\ \frac{-i\omega}{g+\beta k^2} \end{pmatrix} e^{i(kx-\omega t)} \quad \text{and} \quad \begin{pmatrix} 1 \\ 0 \end{pmatrix}, \qquad (11.1.34)$$

the solvability conditions are given by two conditions:

(S1) A_n does not contain terms independent of $(kx - \omega t)$.

(S2) The coefficients P_n and Q_n of $e^{i(kx-\omega t)}$ in A_n and B_n, respectively, satisfy

$$P_n + \frac{i\omega}{g + \beta k^2} Q_n = 0 . \qquad (11.1.35)$$

At order $n = 2$, we have

$$A_2 = -h_T^{(1)} + G^{(1)}\phi^{(1)}, \qquad (11.1.36)$$

$$B_2 = -\phi_T^{(1)} - \frac{1}{2}(\phi_x^{(1)})^2 + \frac{1}{2}\left(G^{(0)}\phi^{(1)}\right)^2 + 2\beta h_{xX}^{(1)}. \qquad (11.1.37)$$

Using the expressions of $h^{(1)}$, $\phi^{(1)}$, and $G^{(1)}$ given in (11.1.24), (11.1.25), and (11.1.30), we get

$$A_2 = i\left(\frac{-\omega}{g+\beta k^2}\psi_{1T} - (\theta + \chi(1 - \theta^2))\psi_{1X}\right) e^{i(kx-\omega t)}$$
$$+ \frac{2i\omega k^2}{g+\beta k^2}\psi_1^2\left(1 - \theta\tanh(2\chi)\right)e^{2i(kx-\omega t)} + \text{c.c.}, \qquad (11.1.38)$$

$$B_2 = -\varphi_T + k^2(\theta^2 - 1)|\psi_1|^2 - \left(\psi_{1T} + \frac{2\omega k\beta}{g+\beta k^2}\psi_{1X}\right)e^{i(kx-\omega t)}$$
$$+ \frac{1}{2}k^2\psi_1^2(1 + \theta^2)e^{2i(kx-\omega t)} + \text{c.c.} \qquad (11.1.39)$$

At this order, the solvability condition thus reads

$$\frac{2\omega}{g+\beta k^2}\psi_{1T} + \left(\theta + \chi(1 - \theta^2) + \frac{2\beta k^2\theta}{g+\beta k^2}\right)\psi_{1X} = 0, \qquad (11.1.40)$$

or equivalently, in terms of the group velocity ω',

$$\psi_{1T} + \omega'\psi_{1X} = 0, \qquad (11.1.41)$$

Equation (11.1.32) at order $n = 2$ is then solved in the form

$$h^{(2)} = a_1 e^{i(kx-\omega t)} + a_2 e^{2i(kx-\omega t)} + \text{c.c.} + \overline{h}, \qquad (11.1.42)$$

$$\phi^{(2)} = b_1 e^{i(kx-\omega t)} + b_2 e^{2i(kx-\omega t)} + \text{c.c.} \qquad (11.1.43)$$

where the coefficients are given by

$$a_1 = \frac{1}{g + \beta k^2}\psi_{1T} + \frac{\theta + \chi(1 - \theta^2)}{\omega}\psi_{1X} + \frac{i\omega}{g + \beta k^2}\psi_2, \quad (11.1.44)$$

$$a_2 = \frac{(\theta^2 - 3)k^2}{2(\beta k^2(\theta^2 - 3) + g\theta^2)}\,\psi_1^2, \quad (11.1.45)$$

$$\overline{h} = \frac{1}{g}\big(k^2(\theta^2 - 1)|\psi_1|^2 - \varphi_T\big), \quad (11.1.46)$$

$$b_1 = \psi_2, \quad (11.1.47)$$

$$b_2 = \frac{i\omega k\left(2\frac{g + 4\beta k^2}{g + \beta k^2}(1 - \theta^2) + \frac{1}{2}(1 + \theta^2)^2\right)}{4\theta(\beta k^2(\theta^2 - 3) + g\theta^2)}\,\psi_1^2. \quad (11.1.48)$$

A mean velocity potential φ was retained at leading order, and it is not necessary to introduce a correction at the present order. The quantities ψ_1, ψ_2, and φ are slowly varying functions that are still undetermined. As stressed by Djordjevic and Redekoff (1977), the amplitude of the second harmonics becomes infinite when the carrier wave number k is such that $\frac{\beta k^2}{g} = \frac{\theta^2}{3 - \theta^2}$. Wave numbers satisfying this condition have the property that the phase velocities $\omega(k)/k$ and $\omega(2k)/(2k)$ of the carrier and of the second harmonics are equal, resulting in the phenomenon of "second harmonic resonance." The usual asymptotics break down near such wave numbers, and a special analysis is required (see McGoldrick 1972 and references therein). In the following, the system is assumed to be far from this resonance.

At order ϵ^3, we have

$$A_3 = -h_T^{(2)} + G^{(2)}\phi^{(1)} + G^{(1)}\phi^{(2)}, \quad (11.1.49)$$

$$\begin{aligned}B_3 = &-\phi_T^{(2)} - \phi_x^{(1)}\phi_X^{(1)} - \phi_x^{(1)}\phi^{(2)} \\ &+ G^{(0)}\phi^{(1)}\big(G^{(0)}\phi^{(2)} + G^{(1)}h^{(1)}\big) + h_x^{(1)}\phi_x^{(1)}G^{(0)}\phi^{(1)} \\ &+ \beta\big(h_{XX}^{(1)} + h_{YY}^{(1)} + 2h_{xX}^{(2)} - \tfrac{3}{2}h_{xx}^{(1)}h_x^{(1)2}\big).\end{aligned} \quad (11.1.50)$$

Elimination of the non-oscillating resonant terms in A_3 leads to

$$\overline{h}_T + H_0(\varphi_{XX} + \varphi_{YY}) + \frac{2\omega k}{g + \beta k^2}|\psi_1|_X^2 = 0. \quad (11.1.51)$$

To express the solvability condition (S2), we write P_3 and Q_3 in the form

$$\begin{aligned}P_3 = &-a_{1T} - i\big(\theta + \chi(1 - \theta^2)\big)b_{1X} - \frac{1}{2k}\big(\theta + \chi(1 - \theta^2)\big)\psi_{1YY} \\ &- H_0(1 - \theta^2)(1 - \chi\theta)\psi_{1XX} + k^2(1 - \theta^2)\overline{h}\psi_1 \\ &- \frac{2i\omega k^2}{g + \beta k^2}\big(1 - \theta\tanh(2\chi)\big)\psi_1^* b_2 - k^2(1 + \theta^2)\psi_1^* a_2 \\ &+ \frac{\omega k}{g + \beta k^2}\psi_1\varphi_X + \frac{k^3\theta\omega^2}{(g + \beta k^2)^2}\big(-1 + 2\theta\tanh(2\chi)|\psi_1|^2\psi_1\big), \quad (11.1.52)\end{aligned}$$

$$Q_3 = -b_{1T} - ik\psi_1\varphi_X - 2k^2\left(1 - \theta\tanh(2\chi)\right)\psi_1^*b_2$$

$$+ \frac{i\omega\beta}{g + \beta k^2}(\psi_{1XX} + \psi_{1YY}) + 2ik\beta a_{1X} + \frac{3}{2}i\beta k^4\left(\frac{\omega}{g + \beta k^2}\right)^3|\psi_1|^2\psi_1$$

$$+ \frac{ik^3\omega\theta}{g + \beta k^2}\left(1 - 2\theta\tanh(2\chi)\right)|\psi_1|^2\psi_1. \tag{11.1.53}$$

Solvability condition (S2) reads

$$2i(\psi_{2T} + \omega'\psi_{2X}) + \omega''\psi_{1XX} + \frac{\omega'}{k}\psi_{1YY}$$

$$= \chi_1|\psi_1|^2\psi_1 - \frac{g + \beta k^2}{\omega}\left(\frac{k^2(1 - \theta^2)}{g}\varphi_T - \frac{2\omega k}{g + \beta k^2}\varphi_X\right)\psi_1, \tag{11.1.54}$$

where

$$\chi_1 = \frac{k^4}{2\omega}\left(\frac{(1 - \theta^2)(9 - \theta^2) + \widetilde{\beta}(3 - \theta^2)(7 - \theta^2)}{\theta^2 - \widetilde{\beta}(3 - \theta^2)}\right.$$

$$\left. + 8\theta^2 - 2(1 - \theta^2)^2(1 + \widetilde{\beta}) - \frac{3\theta^2\widetilde{\beta}}{1 + \widetilde{\beta}}\right) \tag{11.1.55}$$

with $\widetilde{\beta} = k^2\beta/g$. Defining $\psi = \psi_1 + \epsilon\psi_2$ and combining (11.1.41) and (11.1.54), one gets

$$i(\psi_T + \omega'\psi_X) + \epsilon\left(\frac{\omega''}{2}\psi_{XX} + \frac{\omega'}{2k}\psi_{YY}\right)$$

$$= \epsilon\left(\frac{\chi_1}{2}|\psi|^2 - \left(\frac{k^2(1 - \theta^2)(1 + \widetilde{\beta})}{2\omega}\varphi_T - k\varphi_X\right)\right)\psi. \tag{11.1.56}$$

Equations (11.1.46) and (11.1.51) govern the dynamics of the quantities $\bar{h}$ and φ associated to the mean elevation of the free surface and the mean velocity potential on this surface (whose magnitudes are of order ϵ and ϵ^2, respectively). They can be rewritten as

$$\varphi_T + g\bar{h} + k^2(1 - \theta^2)|\psi|^2 = 0, \tag{11.1.57}$$

$$\bar{h}_T + H_0(\varphi_{XX} + \varphi_{YY}) + \frac{2\omega k}{g(1 + \widetilde{\beta})}(|\psi|^2)_X = 0. \tag{11.1.58}$$

The system of equations (11.1.56)–(11.1.58) extends the equations derived by Benney and Roskes (1969) in the absence of surface tension. It describes the interaction of high-frequency gravity–capillary waves with low-frequency gravity waves. Note that the envelope equation is written for the amplitude of the free surface displacement in Benney and Roskes (1969), while we use here the hydrodynamic potential at the surface. Furthermore, when replacing φ_T by its expression in terms of $\bar{h}$ in the amplitude equation for the carrying wave, the system is identified with that given by Zakharov and Rubenchik (1972) and Zakharov and Schulman (1991) on the basis of a Hamiltonian formalism for the interaction of a high-frequency wave with low-frequency waves of acoustic type, driven by the modulation of the carrier (see Chapter 15).

Eliminating the surface elevation from (11.1.46) and (11.1.51), one gets to leading order

$$\varphi_{TT} - gH_0(\varphi_{XX} + \varphi_{YY}) = \alpha_1(|\psi|^2)_X \qquad (11.1.59)$$

with

$$\alpha_1 = \frac{2\omega k}{1 + \widetilde{\beta}} + \omega'(1 - \theta^2)k^2. \qquad (11.1.60)$$

One now uses a reference frame moving at the group velocity ω' by introducing $\xi = X - \omega'T$ and also defines the longer time $\tau = \epsilon T$. Retaining the leading order in (11.1.59), the mean potential φ follows adiabatically the amplitude modulation of the carrying wave. To leading order, (11.1.59) and (11.1.56) lead to the so-called Davey–Stewartson system

$$2i\psi_\tau + \omega''\psi_{\xi\xi} + \frac{\omega'}{k}\psi_{YY} = \chi_1|\psi|^2\psi + \chi_2\psi\varphi_\xi, \qquad (11.1.61)$$

$$\alpha\varphi_{\xi\xi} + \varphi_{YY} = -\gamma|\psi|^2_\xi, \qquad (11.1.62)$$

where χ_1 is given in (11.1.55) and the other constants are given by

$$\chi_2 = k\left(2 + \frac{\omega'k}{\omega}(1 - \theta^2)(1 + \widetilde{\beta})\right),$$

$$\alpha = 1 - \frac{\omega'^2}{gH_0},$$

$$\gamma = \frac{k}{gH_0}\left(\frac{2\omega}{1 + \widetilde{\beta}} + \omega'(1 - \theta^2)k\right)$$

in terms of $\widetilde{\beta} = k^2\beta/g$, $\theta = \tanh(H_0 k)$, and $\omega^2 = (g + \beta k^2)k\tanh(H_0 k)$.

The adiabatic approximation for the mean flow is not valid near the long-wave short-wave resonance where α vanishes. Near the resonance, a different scaling is required and the mean flow is no longer slaved to the amplitude modulation. Retaining the leading order deviation from adiabaticity, (11.1.59) rewrites in the moving frame

$$2\omega'\epsilon\varphi_{\xi\tau} + (gH_0 - \omega'^2)\varphi_{\xi\xi} + gH_0\varphi_{YY} = -\alpha_1(|\psi|^2)_\xi. \qquad (11.1.63)$$

When keeping the wave amplitude fixed, the longitudinal scale for the modulation near the resonance is smaller than in the usual setting (see Section 10.2.2). Equivalently, it is simpler in the present formalism to keep this scale fixed and reduce the carrying-wave amplitude by defining $\psi = \epsilon^{1/2}\widetilde{\psi}$. One then considers a neighborhood of the resonance such that $\sqrt{gH_0} - \omega' = \epsilon\nu$, with finite ν. One also assumes a larger scale for the transverse modulation by defining $\widetilde{Y} = \epsilon^{1/2}Y$. In this regime, (11.1.63) becomes

$$\left(\varphi_\tau + v\varphi_\xi + \frac{\alpha_1}{2\omega'}|\psi|^2\right)_\xi + \frac{\omega'}{2}\varphi_{\widetilde{Y}\widetilde{Y}} = 0. \tag{11.1.64}$$

In the amplitude equation (11.1.56) for the carrier, the terms ψ_{YY} and $|\psi|^2\psi$ associated to diffraction and the self-interaction of the carrier are negligible. Furthermore, φ_τ is to leading order replaced by $-\omega'\varphi_\xi$. The amplitude equation thus becomes

$$i\partial_\tau\psi + \frac{\omega''}{2}\psi_{\xi\xi} = \delta\varphi_\xi\psi, \tag{11.1.65}$$

with

$$\delta = k\left(1 + \frac{k\omega'}{2\omega}(1-\theta^2)(1+\widetilde{\beta})\right). \tag{11.1.66}$$

The longitudinal mean-field velocity φ_ξ obeys the dynamic equation (11.1.64) and is relatively larger than in the non-resonant regime. In the absence of transverse modulation, one recovers equations (4.12a)–(4.12b) of Djordjevic and Redekopp (1977) (see also Benney 1977).

Assuming that the system is far from the long-wave short-wave resonance, it is convenient to rewrite the DS system using the nondimensional variables $\widetilde{\xi} = k\xi$, $\widetilde{Y} = kY$, $\widetilde{\tau} = \omega_0\tau$, $\widetilde{\psi} = (k^2/\omega_0)\psi$, and $\widetilde{\varphi} = (k^2/\omega_0)\varphi$ with $\omega_0^2 = gk$. After dropping the tildes, one gets

$$i\psi_\tau + \lambda\psi_{\xi\xi} + \mu\psi_{YY} = (v_1|\psi|^2 + v_2\varphi_\xi)\psi, \tag{11.1.67}$$

$$\alpha\varphi_{\xi\xi} + \varphi_{YY} = -\delta|\psi|_\xi^2, \tag{11.1.68}$$

where

$$\lambda = \frac{k^2\omega''}{2\omega_0}, \quad \mu = \frac{k\omega'}{2\omega_0} \geq 0,$$

$$v_1 = \frac{\omega_0}{4\omega}\left(\frac{(1-\theta^2)(9-\theta^2) + \widetilde{\beta}(3-\theta^2)(7-\theta^2)}{\theta^2 - \widetilde{\beta}(3-\theta^2)}\right.$$

$$\left. + 8\theta^2 - 2(1-\theta^2)^2(1+\widetilde{\beta}) - \frac{3\theta^2\widetilde{\beta}}{1+\widetilde{\beta}}\right),$$

$$v_2 = 1 + \frac{k\omega'}{2\omega}(1-\theta^2)(1+\widetilde{\beta}) \geq 0,$$

$$\alpha = 1 - \frac{\omega'^2}{gH_0},$$

$$\delta = \frac{\omega}{\omega_0 kH_0}\left(\frac{k\omega'}{\omega}(1-\theta^2) + \frac{2}{1+\widetilde{\beta}}\right) \geq 0.$$

Except for wave numbers close to the two resonance conditions $\omega' = (gH_0)^{1/2}$ and $\widetilde{\beta} = \frac{\theta^2}{3-\theta^2}$, the Davey–Stewartson system (11.1.61)–(11.1.62) describes, to leading order, the evolution of a nearly monochromatic gravity–capillary wave. Different boundary conditions are, however, required for the mean flow according to the sign of the parameter $\alpha = 1 - M$, where the "Mach number" $M = \omega'/\sqrt{gH_0}$ is defined as the ratio of the wave-packet group velocity (which increases with the relative importance of the superficial tension) to the phase velocity of the mean field (which plays the role of low-frequency acoustic waves). The motion of the wave packet is thus said to be "subsonic" when $\alpha > 0$ and "supersonic" when $\alpha < 0$ (Ablowitz and Segur 1981). In this context, an important point concerns the boundary conditions, which in the case of a localized wave packet depend on the sign of α. For $\alpha > 0$, they reduce to $\psi \to 0$, and $\varphi \to 0$ when $\xi^2 + Y^2 \to \infty$. For $\alpha < 0$, in contrast, the boundary conditions are of radiation type, that is, $\psi \to 0$ when $\xi^2 + Y^2 \to \infty$ and $\varphi \to 0$ when the characteristic variables $\xi \pm \sqrt{-\alpha}Y \to -\infty$ (or $+\infty$), with no condition when $\xi \pm \sqrt{-\alpha}Y \to +\infty$ (repectively $-\infty$) (Ablowitz, Manakov, and Schultz 1990, Ablowitz and Clarkson 1991 Section 5.5.8). In this case, we also assume that $\delta > 0$, since a situation with $\alpha < 0$ and $\delta < 0$ never occurs in the context of the water-wave problem.

In fact, some parameters are reducible in the above equations, and by a further rescaling of the dependent and independent variables, the DS system finally takes the form

$$i\psi_\tau + \sigma_1 \psi_{\xi\xi} + \psi_{YY} = (\sigma_2 |\psi|^2 + \varphi_\xi)\psi, \qquad (11.1.69)$$

$$a\varphi_{\xi\xi} + \varphi_{YY} = -b|\psi|_\xi^2 \qquad (11.1.70)$$

with $\sigma_1 = \operatorname{sgn}\lambda$, $\sigma_2 = \operatorname{sgn}\nu_1$, $a = \frac{\alpha\mu}{|\lambda|}$, $b = \frac{\delta\mu\nu_2}{|\lambda||\nu_1|}$. Two important limit cases are to be considered.

In the deep water limit $kH_0 \to \infty$, the parameter δ vanishes, and no mean flow is driven by the wave modulation. The amplitude then obeys the NLS equation

$$i\psi_\tau + \lambda_\infty \psi_{\xi\xi} + \psi_{YY} = \nu_\infty |\psi|^2 \psi, \qquad (11.1.71)$$

where

$$\lambda_\infty = -\frac{\omega_0}{8\omega} \frac{1 - 6\tilde{\beta} - 3\tilde{\beta}^2}{1 + \tilde{\beta}}$$

$$\mu_\infty = \frac{\omega_0(1 + 3\tilde{\beta})}{4\omega}$$

$$\nu_\infty = \frac{\omega_0}{4\omega} \frac{8 + \tilde{\beta} + 2\tilde{\beta}^2}{(1 - 2\tilde{\beta})(1 + \tilde{\beta})},$$

where the factor $1 - 2\tilde{\beta}$ in the denominator of ν_∞ reflects the second harmonic resonance, which is still present. We also mention that the inclusion of higher-order corrective terms was proposed to reproduce the dynamics of waves with a steepness larger than permitted by the NLS equation (Hogan 1985 and references therein).

In the long-wave or shallow-water limit $kH_0 \to 0$, one has

$$i\psi_\tau - \sigma\psi_{\xi\xi} + \psi_{YY} = (\sigma|\psi|^2 + \varphi_\xi)\psi, \qquad (11.1.72)$$

$$\sigma\varphi_{\xi\xi} + \varphi_{YY} = -2|\psi|^2_\xi, \qquad (11.1.73)$$

where $\sigma = \text{sgn } (\frac{1}{3} - \widetilde{\beta})$. The case $\sigma = -1$ is usually referred to as DSI, and $\sigma = 1$ as DSII. These two limiting cases are integrable by inverse scattering and admit explicit solutions that generalize the concept of soliton to multidimensional problems (Ablowitz and Clarkson 1991, Section 5.5, and references therein). The double limit $\epsilon \to 0$, $kH_0 \to 0$ was shown to be uniform by Freeman and Davey (1975), who checked that it can be identified with the procedure consisting in first deriving a KP equation for the long-wavelength dynamics by taking $kH_0 \to 0$ with $\epsilon/(kH_0)^2$ fixed and then performing a modulational analysis in the weak nonlinearity limit $\epsilon/(kH_0)^2 \to 0$.

11.2   Error Bounds

In this section we justify the modulation approximation by proving that the approximate wave packet constructed formally by means of the modulational analysis satisfies approximatively the water-wave equations during a time interval of order ϵ^{-2}, where ϵ estimates the wave amplitude. Details of this analysis can be found in Craig, Sulem, and Sulem (1992) for the two-dimensional water-wave problem and in Craig, Schanz, and Sulem (1997) for the extension to three dimensions.

11.2.1   Preliminaries

We denote by $\mathbf{W}$ the water-wave operator such that

$$\mathbf{W}(h, \phi) = \begin{pmatrix} W_1(h, \phi) \\ W_2(h, \phi) \end{pmatrix}, \qquad (11.2.1)$$

where W_1 and W_2 are defined as the operators arising in the left-hand side of (11.1.8) and (11.1.9), respectively. A solution of the water-wave problem satisfies $\mathbf{W}(h, \phi) = 0$.

In the previous section we constructed a formal approximation of the solution of the water-wave problem (11.1.8)–(11.1.9) based on the functions $\psi(\xi, \eta, \tau)$ and $\varphi(\xi, \eta, \tau)$ which satisfy the Davey–Stewartson system (11.1.61)–(11.1.62), in the form given by

$$\widetilde{h} = \epsilon h^{(1)} + \epsilon^2 h^{(2)} + \epsilon^3 h^{(3)},$$

$$\widetilde{\phi} = \epsilon\phi^{(1)} + \epsilon^2\phi^{(2)} + \epsilon^3\phi^{(3)}, \qquad (11.2.2)$$

with $(h^{(1)}, \phi^{(1)})$ defined by (11.1.24)–(11.1.25) and $(h^{(2)}, \phi^{(2)})$ by (11.1.42)–(11.1.43). The next-order term $(h^{(3)}, \phi^{(3)})$ can also be computed and is given in Craig, Schanz, and Sulem (1997). The question that we address here is to what extent the above formal solution deviates from being a true solution of the full Euler equations, that is, a solution of the system $\mathbf{W}(h, \phi) = 0$. More precisely, consider the initial condition $(h_0, \phi_0) = (\widetilde{h}, \widetilde{\phi})|_{t=0}$ of small magnitude ϵ, and let (h, ϕ) be the solution of the initial value problem $\mathbf{W}(h, \phi) = 0$ with the initial condition (h_0, ϕ_0). An optimal result would be to compare the exact solution (h, ϕ) with the modulation approximation $(\widetilde{h}, \widetilde{\phi})$ in an appropriate norm, with an estimate of the form $\sup_{t \in I} |(\widetilde{h}, \widetilde{\phi}) - (h, \phi)| \leq o(\epsilon^3)$, over an interval of time $I = [0, t_0]$ with $t_0 = O(\epsilon^{-2})$. Such a result would require an existence theory for the water-wave problem in three space dimensions over long time intervals, given initial conditions of size $O(\epsilon)$, which currently remains an open problem. We thus limit ourselves to a weaker result, evaluating the accuracy to which the modulational solution $(\widetilde{h}, \widetilde{\phi})$ satisfies the water-wave problem. Equivalently, we give an estimate of $\mathbf{W}(\widetilde{h}, \widetilde{\phi})$ in a Sobolev norm and show that it is $o(\epsilon^3)$ uniformly in the time interval I.

Rigorous justifications of modulational analysis were established in the context of other physical problems. For example, Collet and Eckmann (1990) give a comparison of solutions of the Swift–Hohenberg equation with approximations based on solutions of a Ginzburg–Landau equation for the envelope. Kirrmann, Shneider, and Mielke (1992) have results for the Sine–Gordon problem. Also, Pierce and Wayne (1995) studied the modulational regime for a one-dimensional wave equation that involves the interaction of left- and right-moving periodic wave trains.

11.2.2   Modulation of the Water-Wave Operator

The analysis of the water-wave operator requires several steps: (i) Proof of the analyticity of the Dirichlet–Neumann operator $G(h)$, (ii) analysis of pseudo-differential operators in a multiple-scale regime, (iii) well-posedness of the initial value problem for the Davey–Stewartson system, (iv) estimates for $\mathbf{W}(\widetilde{h}, \widetilde{\phi})$.

The Dirichlet–Neumann Operator $G(h)$

In the previous section, we computed the Taylor expansion of the Dirichlet–Neumann operator in terms of the surface elevation h. Indeed, each G_j is homogeneous of degree j in h. It is a combination of powers of h and of its derivatives with explicit Fourier multipliers, and can be obtained by a simple recursive formula. The following theorem justifies this derivation by proving the analyticity of G as a function of h for small enough h in the C^1-norm and provides estimates of the Taylor remainders. The proof is based on the analysis of singular integral operators of potential theory. Details can be found in Coifman and Meyer (1985) for the two-dimensional case and in Craig, Schanz, and Sulem (1997) for the three-dimensional problem. Here, the spaces $W^{s,p}$ where s is a nonnegative integer and $1 \leq p < \infty$ are the Sobolev spaces equipped with the norm $|u|_{s,p} = \left(\sum_{0 \leq |\alpha| \leq s} |\partial^\alpha u|_{L^p}^p \right)^{1/p}$, where α is a multi-index.

Theorem 11.2. *Let $1 < q < +\infty$. There is a constant δ such that the operator $G(h)$ is analytic in h in the neighborhood $\{h : |h|_{C^1} < \delta, |h|_{C^{s+1}} < \infty\}$, as a mapping $G(h) : W^{s+1,q} \to W^{s,q}$. The Taylor remainders $R_j(h)$ from the expansion of $G(h)$ $(j = 1, 2, 3)$,*

$$G(h) = \Sigma_{i=1}^j G_i(h) + R_j(h), \tag{11.2.3}$$

satisfy

$$|R_j(h)\xi|_{s,q} \leq C(s)|h|_{C^1}^{j-1} \left(|h|_{C^1}|\xi|_{s+1,q} + |h|_{C^{s+1}}|\xi|_{1,q} \right). \tag{11.2.4}$$

Action of Pseudo-Differential Operators on Multiple-Scale Functions

Under multiple-scale expansion, the terms G_j in the expansion of G are approximated by differential operators whose action on multiple-scale functions is to be analyzed. The modulational regime of the water-wave problem involves classes of functions whose space dependence is of the form $u(\mathbf{x}, \mathbf{X})|_{\mathbf{X}=\epsilon\mathbf{x}} = \exp(i\mathbf{k} \cdot \mathbf{x})c(\mathbf{X})|_{\mathbf{X}=\epsilon\mathbf{x}}$, for $\mathbf{x} \in [0, 2\pi]^2$ and $\mathbf{X} \in \mathbf{R}^2$.

When a function c depends upon the slow variables alone, that is, $c = c(\epsilon\mathbf{x})$, its Sobolev norm in that variable, $\|c\|_{s,q} = \sum_{|\alpha| \leq s} \left(\int |\partial_{\mathbf{X}}^\alpha c|^q d\mathbf{X} \right)^{1/q}$, is related to the usual Sobolev norm $|c|_{s,q}$ viewed as a function of $\mathbf{x}$ by $|c|_{s,q} \leq \epsilon^{-d/q} \|c\|_{s,q}$ for any $\epsilon \leq 1$. For a multiple-scale function $u = u(\mathbf{x}, \epsilon\mathbf{x})$ we define

$$|u|_{(a,b),q} = \sum_{\substack{|\alpha| \leq a \\ |\beta| \leq b}} \left(\int_{\mathbf{R}^d} \int_{[0,2\pi]^d} |\partial_{\mathbf{x}}^\alpha \partial_{\mathbf{X}}^\beta u(\mathbf{x}, \mathbf{X})|^q d\mathbf{x} d\mathbf{X} \right)^{1/q}. \tag{11.2.5}$$

A Fourier multiplier operator is defined as

$$M(D)u(\mathbf{x}) = \frac{1}{(2\pi)^d} \int e^{i\mathbf{k}\cdot\mathbf{x}} M(\mathbf{k})\hat{u}(\mathbf{k})d\mathbf{k}, \tag{11.2.6}$$

where $\hat{u}$ is the Fourier transform of u. For u a multiple-scale function of $\mathbf{x} \in [0, 2\pi]^d$, $\mathbf{X} \in \mathbf{R}^d$, $M(D)u$ depends upon ϵ and $\mathbf{x}$, but is not in general a multiple-scale function. However, it does have an asymptotic expansion in terms of such functions. This is the object of the next theorem.

Theorem 11.3. *Assume that the Fourier multiplier $M(D)$ has the property that $|\partial_{\mathbf{k}}^j M(\mathbf{k})| \leq c_j(1 + |\mathbf{k}|^2)^{(m-j)/2}$, $0 \leq j \leq m$. Then its action on a multiple-scale function has the asymptotic expansion*

$$\big(M(D)u\big)(\mathbf{x}; \epsilon) = \left(\sum_{j=0}^N \frac{\epsilon^j}{j!} \partial_{\mathbf{k}}^j M(D_{\mathbf{x}}) \left(\frac{1}{i}\partial_{\mathbf{X}} \right)^j u \right)(\mathbf{x}, \epsilon\mathbf{x}) + R_{N+1}u. \tag{11.2.7}$$

The remainder obeys the estimate

$$|R_{N+1}u|_{s,q} \leq C_{sq}\epsilon^{N+1}|u|_{(\ell_1,\ell_2),q}, \tag{11.2.8}$$

where the subscripts are $\ell_1 = m-(N+1)+s+\sigma$, $\ell_2 = s+\sigma+\max(m,N+1)$ for any $\sigma > n$ and $1 < q < \infty$.

Initial value problem for the Davey–Stewartson System

The main properties of the Davey–Stewartson system (DS) are detailed in the Chapter 12. Here, we need only the local existence of solutions of the DS system (11.1.61)–(11.1.62).

We first recall that the boundary conditions for the mean field φ depend on the sign of α. When $\alpha > 0$, $\varphi \to 0$ as $\xi^2+Y^2 \to +\infty$. When $\alpha < 0$, $\varphi \to 0$ when the characteristic variables $\xi \pm \sqrt{-\alpha}Y \to -\infty$, with no condition when $\xi \pm \sqrt{-\alpha}Y \to +\infty$.

The following results were established by Ghidaglia and Saut (1990) in the case where the equation for the mean field is elliptic ($\alpha > 0$), and by Hayashi (1997) in the case where it is hyperbolic ($\alpha < 0$).

Theorem 11.4. *Given an initial condition ψ_0 in $H^r(\mathbf{R}^2)$, $r \geq 2$:*
(i) For $\alpha > 0$, there exists a unique solution (ψ,φ) of the DS system during a finite interval of time $[0,\tau_1)$, such that $\psi \in C([0,\tau_1), H^r(\mathbf{R}^2))$ and $\varphi \in C([0,\tau_1), H^{r+1}(\mathbf{R}^2))$.
(ii) For $\alpha < 0$ and $\delta > 0$, there exists a unique solution (ψ,φ) during a finite interval of time $[0,\tau_2)$ with $\psi \in C([0,\tau_2), H^{r-1}(\mathbf{R}^2)) \cup L^\infty([0,\tau_2), H^r(\mathbf{R}^2))$ and $\varphi \in L^\infty([0,\tau_2), W^{r-1,\infty}(\mathbf{R}^2))$.

Accuracy of the Approximation: Estimates of $\mathbf{W}(\widetilde{h},\widetilde{\phi})$

It is convenient to consider separately the cases where the equation for the mean field is elliptic and where it is hyperbolic.

Case $\alpha > 0$: Starting with an initial condition $\psi_0 \in H^r(\mathbf{R}^2)$, there exists a unique solution (ψ,φ) of the Davey-Stewartson system, with ψ and φ being continuous functions of $\tau \in [0,\tau_1]$ with values in $H^r(\mathbf{R}^2)$ and $H^{r+1}(\mathbf{R}^2)$ respectively (τ_1 may be finite or infinite). By Sobolev embedding, these functions are also in $W^{s,q}$ and $W^{s-1,q}$ for $s = r - \frac{1}{2}(\frac{1}{2}-\frac{1}{q})$. From this solution of the DS system, proceeding as described in Section 11.2.1, one first constructs the *formal* perturbative approximation, as the functional $\widetilde{h} = \widetilde{h}(\psi,\varphi)$ and $\widetilde{\phi} = \widetilde{\phi}(\psi,\varphi)$, in the time interval $I = [0,\epsilon^{-2}\tau_1]$.

Case $\alpha < 0$: Again, one starts with an initial condition $\psi_0 \in H^r(\mathbf{R}^2)$. There exists a unique solution (ψ,φ) such that ψ is a continuous function of τ in $[0,\tau_2]$ with values in $H^r(\mathbf{R}^2)$ and thus, by Sobolev embedding, in $W^{s,q}(\mathbf{R}^2)$, and φ a continuous function of τ with values only in $W^{s-1,\infty}$ and thus not integrable. To overcome this difficulty, it is convenient to introduce a cutoff function χ infinitely differentiable with compact support and equal to 1 in the ball B_R of center 0 and radius R. One then constructs the formal approximation $\bar{h} = \widetilde{h}(\psi, \chi\varphi)$ and $\bar{\phi} = \widetilde{\phi}(\psi, \chi\varphi)$ where the mean field has been truncated at large distance.

The next step is to show that these formal approximations do satisfy approximatively the water-wave problem, by establishing an upper bound for Sobolev norms of $\mathbf{W}(\widetilde{h}, \widetilde{\phi})$ when $\alpha > 0$ and of $\mathbf{W}(\bar{h}, \bar{\phi})$ when $\alpha < 0$, in the interval I in terms of powers of ϵ and of Sobolev norms of ψ and φ. In both cases, this bound involves high order derivatives of ψ and φ, resulting from the expansion of integral operators. To quantify the amount of derivatives required, one introduces the notion of estimating factors. These are functions $F(s_1, s_2)$ that are nonnegative, continuous, and satisfy $F(0,0) = 0$, with polynomial growth in (s_1, s_2), and coefficients depending on k and ω. This leads to the following result.

Theorem 11.5. *Let* (ψ, φ) *denote a solution of the Davey–Stewartson system (11.1.61)–(11.1.62). It defines through the modulation expansion an approximate solution* $(\widetilde{h}, \widetilde{\phi})$ *of the water-wave problem that during a time interval* $I = [0, \epsilon^{-2}\tau_1]$ *with* τ_1 *finite and, depending on* α, *satisfies the error estimate:*
(i) *For* $\alpha > 0$,

$$\sup_{t \in I} |\mathbf{W}(\widetilde{h}, \widetilde{\phi})|_{s,q}$$
$$\leq \epsilon^{4-2/q} F(\sup_{t \in I} |\psi|_{C^{s+3}}, \sup_{t \in I} |\varphi|_{C^{s+2}})(\|\psi\|_{s+6,q} + \|\varphi\|_{s+6,q}) \quad (11.2.9)$$

(ii) *For* $\alpha < 0$,

$$\sup_{t \in I} |\chi_1 \mathbf{W}(\bar{\eta}, \bar{\xi})|_{s,q}$$
$$\leq \epsilon^{4-2/q} F(\sup_{t \in I} |\psi|_{C^{s+3}}, \sup_{t \in I} |\varphi|_{C^{s+2}})(\|\psi\|_{s+6,q} + \|\chi\varphi\|_{s+6,q}) \quad (11.2.10)$$

for all $\chi_1 \in C_0^\infty(\mathbf{R}^2)$ *with* $\mathrm{supp}(\chi_1) \subseteq \mathrm{supp}(\chi)$, *where* χ *and* χ_1 *denote cutoff functions centered at the origin.*

This result expresses that the approximation constructed from the solution of the Davey–Stewartson equations, which is defined on a time scale $O(\epsilon^{-2})$, solves the water-wave equations up to an error that for large q, scales almost like ϵ^4. This does not however demonstrate that, on a comparable time scale, it provides an accurate approximation of the solution of the water-wave problem because the latter problem is proved to be well-posed only on a time scale of order ϵ^{-1}, a time during which the variation of the envelope is weak and only the simple translational motion of the solution at the group velocity is significant. In other words, the constructed approximation is accurate on a time scale $O(\epsilon^{-2})$, provided the primitive problem does not loose regularity at an earlier time.

12
The Davey–Stewartson System

12.1   General Setting

12.1.1   Boundary Conditions

The Davey–Stewartson system (DS) that provides a canonical description of the amplitude dynamics of a weakly nonlinear two-dimensional wave packet when a mean field is driven by the modulation, is written in the form

$$i\partial_t\psi + \sigma_1\partial_{xx}\psi + \partial_{yy}\psi = (\sigma_2|\psi|^2 + \partial_x\varphi)\psi, \qquad (12.1.1)$$

$$\alpha\partial_{xx}\varphi + \partial_{yy}\varphi = -\gamma\partial_x|\psi|^2, \qquad (12.1.2)$$

where $\mathbf{x} = (x, y) \in \mathbf{R}^2$, $t \in \mathbf{R}$, $\sigma_1 = \pm 1$, $\sigma_2 = \pm 1$. Furthermore, γ and α are constant parameters with $\gamma > 0$. A derivation of this system in the context of the surface water waves is given in Chapter 11.

The wave amplitude ψ is supposed to vanish at infinity,

$$\psi(x, y, t) \to 0 \quad \text{as} \quad x^2 + y^2 \to \infty. \qquad (12.1.3)$$

In contrast, the boundary conditions on the mean field φ depend on the sign of α (Ablowitz, Manakov, and Schultz 1990, Ablowitz and Clarkson 1991). If $\alpha > 0$ ("subsonic" wave packet), it is of Dirichlet type

$$\varphi(x, y, t) \to 0 \quad \text{as} \quad x^2 + y^2 \to \infty. \qquad (12.1.4)$$

If $\alpha < 0$ ("supersonic" wave packet), it is of radiation type. Defining the characteristic coordinates

$$\xi = \frac{1}{\sqrt{2}}(x + |\alpha|^{\frac{1}{2}}y), \quad \eta = \frac{1}{\sqrt{2}}(-x + |\alpha|^{\frac{1}{2}}y), \qquad (12.1.5)$$

the boundary conditions on φ should be of the form

$$\lim_{\xi \to -\infty} \varphi(\xi, \eta, t) = \varphi_1(\eta, t), \quad \lim_{\eta \to -\infty} \varphi(\xi, \eta) = \varphi_2(\xi, t), \qquad (12.1.6)$$

where φ_1 and φ_2 can be chosen at will. They are respectively assumed to decay for large values of η and ξ, and their prescription results in imposing an external mean flow far upstream in the (ξ, η)-plane.

It is useful to classify the DS system as elliptic–elliptic, hyperbolic–elliptic, elliptic–hyperbolic and hyperbolic–hyperbolic according to the signs of (σ_1, α): $(+,+)$, $(+,-)$, $(-,+)$, and $(-,-)$.

When the parameters are such that $(\sigma_1, \sigma_2, \alpha, \gamma) = (1, \pm 1, -1, \mp 2)$, the system is usually referred to as DSI. When $(\sigma_1, \sigma_2, \alpha, \gamma) = (-1, 1, 1, 2)$ and $(\sigma_1, \sigma_2, \alpha, \gamma) = (-1, -1, 1, -2)$, the system is called DSII defocusing and DSII focusing, respectively. The DSI and DSII systems have the remarkable property of being integrable by inverse scattering transform. Schul'man (1983) showed that they are the only integrable cases. A detailed study of these two systems can be found in Ablowitz and Clarkson (1991) and references therein.

Extension of the DS system to higher dimensions was considered by Zakharov and Schulman (1991) and their mathematical properties studied by Ghidaglia and Saut (1992), and Kenig, Ponce, and Vega (1995). An example of physical application is given by Nishinari, Abe, and Satsuma (1994) in the context of electrostatic ion wave packets propagating in arbitrary direction in a magnetized plasma.

12.1.2   Expression of the Mean Flow

"Subsonic" Wave Packet

For $\alpha > 0$, the mean field φ obeys an elliptic equation and is solved in terms of ψ. One has $\varphi_x = \gamma(-\Delta_\alpha)^{-1}|\psi|^2_{xx}$, with $\Delta_\alpha = \alpha\partial_{xx} + \partial_{yy}$, or $\varphi_x = \mathcal{B}(|\psi|^2)$, where the operator $\mathcal{B}$ is defined in Fourier variables by

$$\widehat{\mathcal{B}(f)}(k_1, k_2) = -\frac{\gamma k_1^2}{\alpha k_1^2 + k_2^2} \widehat{f}(k_1, k_2). \qquad (12.1.7)$$

The operator $\mathcal{B}$ is of order zero. One has $|\varphi_x|_{L^2} \leq \frac{\gamma}{\alpha}|\psi|^2_{L^4}$, and $|\varphi_x|_{L^p} \leq C|\psi|^2_{L^{2p}}$, for $1 < p < \infty$. The DS system can thus be written in the form of a nonlinear Schrödinger equation with a nonlocal cubic nonlinearity

$$i\partial_t\psi + \sigma_1\partial_{xx}\psi + \partial_{yy}\psi + \mathcal{L}(|\psi|^2)\psi = 0 \qquad (12.1.8)$$

with $\mathcal{L}(|\psi|^2) = -\sigma_2|\psi|^2 - \mathcal{B}(|\psi|^2)$. As seen in the following, the properties of the DS system are in this case very similar to those of the cubic NLS equation in two dimensions.

"Supersonic" Wave Packet

For $\alpha < 0$, it is convenient to rewrite the DS system using the characteristic variables (ξ, η). The equation for the mean field reads

$$\varphi_{\xi\eta} = \frac{\gamma}{2\sqrt{2\alpha}}(|\psi|_\xi^2 - |\psi|_\eta^2). \tag{12.1.9}$$

Integrating this equation and substituting in the equation for the envelope, one obtains

$$i\psi_t + L\psi = S, \tag{12.1.10}$$

where, renaming for convenience ξ and η by x and y,

$$S = \left(\sigma_2 - \frac{\gamma}{2\alpha}\right)|\psi|^2\psi$$
$$+\frac{\gamma}{4\alpha}\left(\int_{-\infty}^{y}\partial_x|\psi|^2\,dy' + \int_{-\infty}^{x}\partial_y|\psi|^2\,dx' + \varphi_1(\xi) + \varphi_2(\eta)\right) \tag{12.1.11}$$

and where the operator

$$L = \frac{1}{2}(\sigma_1 - \alpha)(\partial_{xx} + \partial_{yy}) - (\sigma_1 + \alpha)\partial_{xy} \tag{12.1.12}$$

is elliptic or hyperbolic according to $\sigma_1 = 1$ or $\sigma_1 = -1$. Furthermore, the initial and boundary conditions read

$$\psi(x, y, 0) = \psi_0(x, y), \tag{12.1.13}$$
$$\psi(x, y, t) \to 0 \quad \text{as} \quad x^2 + y^2 \to \infty. \tag{12.1.14}$$

12.1.3 Conservation Properties

It is convenient in this section to rewrite the DS system in the form

$$i\psi_t + \nabla \cdot (\mathcal{D}\nabla\psi) = \sigma_2|\psi|^2\psi \tag{12.1.15}$$
$$\nabla \cdot (\mathcal{A}\nabla\varphi) = -\gamma\partial_x|\psi|^2 \tag{12.1.16}$$

where

$$\mathcal{D} = \begin{pmatrix} \sigma_1 & 0 \\ 0 & 1 \end{pmatrix} \quad \text{and} \quad \mathcal{A} = \begin{pmatrix} \alpha & 0 \\ 0 & 1 \end{pmatrix}.$$

Lemma 12.1. *The solutions of the DS system satisfy:*

$$\frac{\partial}{\partial t}|\psi|^2 + \nabla \cdot \mathcal{D}\mathbf{P} = 0, \tag{12.1.17}$$

where $\boldsymbol{P} = i(\psi\nabla\psi^* - \psi^*\nabla\psi)$,

$$\partial_t\mathcal{H} + \nabla\cdot\mathcal{S} = 0, \tag{12.1.18}$$

with

$$\mathcal{H} = \nabla\psi\cdot\mathcal{D}\nabla\psi^* + \frac{\sigma_2}{2}|\psi|^4 - \frac{1}{2\gamma}\mathcal{A}\nabla\phi\cdot\nabla\phi \tag{12.1.19}$$

$$\mathcal{S} = -2\Re[\psi_t^*\mathcal{D}\nabla\psi] + \frac{1}{\gamma}\phi\mathcal{A}\nabla\phi_t + \phi|\psi|_t^2\mathbf{e}_x, \tag{12.1.20}$$

and

$$\frac{\partial\mathcal{P}_j}{\partial t} + \partial_i\mathcal{T}_{ij} = 0 \tag{12.1.21}$$

where

$$\mathcal{T} = 2\Re[2\mathcal{D}\nabla\psi\otimes\nabla\psi^* - (\nabla\psi\cdot\mathcal{D}\nabla\psi^*)I - ((\psi\nabla\cdot(\mathcal{D}\nabla\psi^*)I]$$
$$+\sigma_2|\psi|^4 I + 2|\psi|^2\mathbf{e}_x\otimes\nabla\varphi + \frac{2}{\gamma}\mathcal{A}\nabla\varphi\otimes\nabla\varphi - \frac{1}{\gamma}(\nabla\varphi\cdot\mathcal{A}\nabla\varphi)I. \tag{12.1.22}$$

Proof. As in the case of the NLS equation, the above conservation properties can formally be viewed as a consequence of invariance properties of the associated action, through the Noether theorem. In the case where the equation for the mean field is hyperbolic ($\alpha < 0$), some of the integrals involving this field may however diverge. We thus establish the lemma directly from the DS system.

To obtain (12.1.17), one multiplies (12.1.15) by ψ^* and takes the imaginary part. To get (12.1.18), one multiplies (12.1.15) by ψ_t^* and takes the real part. One also differentiates (12.1.16) and multiplies the result by φ. Direct combination of the obtained equations then leads to (12.1.18).

The identity (12.1.21) is obtained as follows. From (12.1.15), one gets

$$\frac{1}{2}\partial_t P + \Re[\nabla\psi^*\nabla\cdot(\mathcal{D}\nabla\psi) - \psi\nabla(\nabla\cdot\mathcal{D}\nabla\psi^*)]$$
$$= -\frac{\sigma_2}{2}\nabla|\psi|^4 + \partial_x(|\psi|^2)\nabla\varphi - \partial_x(|\psi|^2\nabla\varphi). \tag{12.1.23}$$

The second term in the right-hand side of (12.1.23) is evaluated from (12.1.16). One gets

$$\partial_x(|\psi|^2)\partial_i\varphi = -\frac{1}{\gamma}\nabla\cdot(\partial_i\varphi\mathcal{A}\nabla\varphi) + \frac{1}{2\gamma}\partial_i(\nabla\varphi\cdot\mathcal{A}\nabla\varphi). \tag{12.1.24}$$

Writing

$$\Re[\partial_i\psi^*\nabla\cdot(\mathcal{D}\nabla\psi) - \psi\partial_i(\nabla\cdot\mathcal{D}\nabla\psi^*)]$$
$$= \Re[2\nabla\cdot(\partial_i\psi^*\mathcal{D}\nabla\psi) - \partial_i(\nabla\psi\cdot\mathcal{D}\nabla\psi^*) - \partial_i(\psi\nabla\cdot\mathcal{D}\nabla\psi^*)], \tag{12.1.25}$$

one obtains (12.1.21).

Proposition 12.2. *In the case $\alpha > 0$, the DS system (12.1.1)–(12.1.2) preserves:*
the wave energy

$$N = \int |\psi(\mathbf{x}, t)|^2 d\mathbf{x}, \qquad (12.1.26)$$

the linear momentum

$$\mathbf{P} = \int i(\psi^* \nabla \psi - \psi \nabla \psi^*) \, d\mathbf{x}, \qquad (12.1.27)$$

and the Hamiltonian

$$H = \int \left(\sigma_1 |\psi_x|^2 + |\psi_y|^2 + \frac{1}{2}\left(\sigma_2 |\psi|^4 - \frac{\alpha}{\gamma}|\varphi_x|^2 - \frac{1}{\gamma}|\varphi_y|^2 \right) \right) d\mathbf{x}. \quad (12.1.28)$$

Furthermore, the "variance"

$$\mathcal{V}(t) = \int (\sigma_1 x^2 + y^2)|\psi|^2 d\mathbf{x} \qquad (12.1.29)$$

obeys

$$\frac{d^2}{dt^2}\mathcal{V}(t) = 2\sum_{i=1}^{2} \int T_{ii} d\mathbf{x} = 8H. \qquad (12.1.30)$$

Proof. The conservation of the wave energy, of the momentum, and of the Hamiltonian follows directly from the integration of (12.1.17), (12.1.18), and (12.1.21) over $\mathbf{R}^2$, provided that all functions decay fast enough at infinity, which restricts the results to $\alpha > 0$.

To get the variance identity, we note that from (12.1.17) and (12.1.21), one has

$$\partial_{tt}|\psi|^2 - \mathcal{D}_{ij}\partial_{ik}T_{kj} = 0. \qquad (12.1.31)$$

This implies that

$$\frac{d^2}{dt^2}\mathcal{V}(t) = 2\sum_{i=1}^{2} \int T_{ii} \, dx$$

$$= \int \left(8(\sigma_1 |\psi_x|^2 + |\psi_y|^2) + 4\sigma_2 |\psi|^4 + 4|\psi|^2 \partial_x \varphi \right) d\mathbf{x}. \qquad (12.1.32)$$

Again, from (12.1.16),

$$\int |\psi|^2 \partial_x \varphi d\mathbf{x} = -\frac{1}{\gamma}(\alpha\varphi_x^2 + \varphi_y^2), \qquad (12.1.33)$$

and the right-hand side of (12.1.32) reduces to $8H$.

Remark. As in the case of the critical NLS equation, the variance identity (12.1.30) can be viewed as the consequence of the additional invariance of the action by the pseudo-conformal transformation (Kuznetsov and Turitsyn 1985, Ozawa 1992):

$$\mathbf{x} \mapsto \mathbf{x}' = \frac{\mathbf{x}}{t_* - t},$$

$$t \mapsto t' = \frac{t}{t_*(t_* - t)},$$

$$\psi \mapsto \psi'(\mathbf{x}', t') = (t_* - t)\psi(\mathbf{x}, t)e^{\frac{i(\sigma_1 x^2 + y^2)}{4(t_* - t)}},$$

$$\varphi \mapsto \varphi'(\mathbf{x}', t') = (t_* - t)\varphi(\mathbf{x}, t), \tag{12.1.34}$$

which leads to the conservation of

$$C = \int \Big(\sigma_1 |(x + 2i\sigma_1 t \partial_x)\psi|^2 + |(y + 2it\partial_y)\psi|^2 + 2t^2(\sigma_2|\psi|^4$$

$$- \frac{\gamma}{\alpha}|\partial_x\varphi|^2 - \gamma|\partial_y\varphi|^2)\Big)d\mathbf{x}. \tag{12.1.35}$$

12.2   Standing-Wave Solutions

12.2.1   The Elliptic–Elliptic Case

We look for standing-wave solutions of (12.1.8) in the form

$$\psi(\mathbf{x}, t) = e^{i\omega t}\Phi(\mathbf{x}), \tag{12.2.1}$$

where Φ is real-valued and belongs to $H^1(\mathbf{R}^2)$. The function Φ must solve

$$\Delta\Phi - \omega\Phi + \mathcal{L}(\Phi^2)\Phi = 0, \tag{12.2.2}$$

where $\mathcal{L}(\Phi^2) = -\sigma_2\Phi^2 - \mathcal{B}(\Phi^2)$ and $\mathcal{B}$ is the operator defined in (12.1.7). It should obey

$$V(\Phi) \equiv \int \Big(\frac{1}{2}|\Phi|^2\mathcal{B}(|\Phi|^2) - \frac{\sigma_2}{2}|\Phi|^4 - \omega|\Phi|^2 \Big)d\mathbf{x} = 0, \tag{12.2.3}$$

which requires the condition $\gamma/\alpha - \sigma_2 > 0$.

Furthermore, the function Φ is a solution of (12.2.2) if and only if it is a critical point of the action $\mathcal{S}$ defined by

$$\mathcal{S}(f) = \int \Big(\frac{1}{2}|\nabla f|^2 - \frac{1}{4}|f|^2\mathcal{L}(|f|^2) + \frac{\omega}{2}|f|^2 \Big)d\mathbf{x}. \tag{12.2.4}$$

By extending the analysis developed for the standing-wave solutions of the NLS equation, Cipolatti (1992,1993) proved the existence of a ground state for (12.2.2):

Theorem 12.3. *Assume $\gamma/\alpha - \sigma_2 > 0$. Let $\mathcal{X}$ be the set of functions in H^1 satisfying (12.2.2) and $\mathcal{G}$ the set of the elements of $\mathcal{X}$ that minimize $\mathcal{S}$.*
(i) $\mathcal{G}$ contains a real positive function (ground state),

(ii) Φ *belongs to* $\mathcal{G}$ *if and only if it realizes the minimum of the functional*

$$I(f) = \int |\nabla f|^2 d\mathbf{x} \qquad (12.2.5)$$

over all $f \in H^1(\mathbf{R}^2)$ *such that* $V(f) = 0$.

(iii) *Furthermore, if* Φ *is a solution of (12.2.2), then* $\Phi \in C^2$, *and there exist positive constants* C *and* ν *such that*

$$|\Phi(\mathbf{x})| + |\nabla\Phi(\mathbf{x})| \le Ce^{-\nu|\mathbf{x}|} \qquad (12.2.6)$$

for all $\mathbf{x} \in \mathbf{R}^2$.

Theorem 12.4. *Ground-state standing waves are unstable.*

Cipolatti (1992, 1993) established the above results for more general power law nonlinearities both in dimension 2 and 3. Further extensions are given by Ohta (1995).

An alternative way of characterizing the solutions of (12.2.2) with $\sigma_2 = -1$, is presented in Papanicolaou, Sulem, Sulem, and Wang (1994) by generalizing the approach developed by Weinstein (1983) for the NLS equation.

Theorem 12.5. *The functional*

$$J(u) = \frac{|u|_{L^2}^2 |\nabla u|_{L^2}^2}{\langle F(|u|^2), u^2 \rangle}, \qquad (12.2.7)$$

where $\langle ., . \rangle$ *is the* L^2*-scalar product, attains its infimum at a function* $\Phi \in H^1(\mathbf{R}^2)$.

Corollary 12.6. *The optimal constant for the inequality*

$$|\langle \mathcal{L}(|u|^2), u^2 \rangle| \le C_{\text{opt}} \, |u|_{L^2}^2 \, |\nabla u|_{L^2}^2, \qquad (12.2.8)$$

is $C_{\text{opt}} = 2/|S|_{L^2}^2$, *where* S *is the positive solution of*

$$\Delta S - S + \mathcal{L}(S^2)S = 0. \qquad (12.2.9)$$

Proof. Equation (12.2.8) follows immediately from Theorem 12.5 with $C_{\text{opt}} = 1/I$, where $I = \inf J(u)$. Furthermore, the minimizing function Φ satisfies $|\Phi|_{L^2} = |\nabla\Phi|_{L^2} = 1$ and solves the Euler–Lagrange equation

$$\Delta\Phi - \Phi + 2I\mathcal{L}(\Phi^2)\Phi = 0. \qquad (12.2.10)$$

The function $S = \sqrt{2I}\phi$, satisfies (12.2.9). Since $|\Phi|_{L^2}^2 = 1$, we have $I = |S|_{L^2}^2/2$. Therefore,

$$C_{\text{opt}} = \frac{1}{I} = \frac{2}{|S|_{L^2}^2}. \qquad (12.2.11)$$

This estimate enables us to give a sharp condition on the initial amplitude of the wave for global existence of solutions for the Davey–Stewartson equations.

Remark. Consider the system for $\sigma_2 = -1$,

$$\Delta S - S + S^3 - SX = 0,$$
$$\alpha X_{\xi_1\xi_1} + X_{\xi_2\xi_2} + \gamma |S|^2_{\xi_1\xi_1} = 0. \qquad (12.2.12)$$

For $\gamma = 0$, we have $X = 0$, and S reduces to the ground state $S_0 = R$, which is the positive solution of $\Delta R - R + R^3 = 0$. This solution is radially symmetric and decreasing. Up to a space translation, it is unique. For small γ, the solution (S, X) with $S > 0$ can be computed by an expansion of the form

$$S = S_0(|\xi|) + \gamma S_1 + \gamma^2 S_2 + \cdots,$$
$$X = \gamma X_1 + \gamma^2 X_2 + \cdots. \qquad (12.2.13)$$

This leads to a hierarchy of inhomogeneous linear equations for the perturbations (S_n, X_n) at the successive orders $n > 0$. The solvability of these systems is established in Papanicolaou, Sulem, Sulem, and Wang (1994), which validates the above expansion of S and X in powers of γ.

12.2.2   The Hyperbolic–Elliptic case

We rewrite the hyperbolic–elliptic DS system in the form

$$i\partial_t \psi - \partial_{xx} \psi + \partial_{yy} \psi = \sigma_2 |\psi|^2 \psi + \beta \varphi_x \psi, \qquad (12.2.14)$$
$$\Delta \varphi = \partial_x(|\psi|^2). \qquad (12.2.15)$$

We look for solutions $(\psi, \varphi) = (u(x, y)e^{i\omega t}, q(x, y))$ with $u \in H^1(\mathbf{R}^2)$ with real values and $\nabla \varphi \in L^2$. The pair (u, φ) satisfies

$$-\partial_{xx} u + \partial_{yy} u = \sigma_2 |u|^2 u + \beta\left(q_x + \frac{\omega}{\beta}\right)u, \qquad (12.2.16)$$

$$\Delta q = \partial_x(|u|^2). \qquad (12.2.17)$$

It is convenient to make the change of variables $p = q + \frac{\omega}{\beta}x$, which leads to the system

$$-\partial_{xx} u + \partial_{yy} u = \sigma_2 |u|^2 u + \beta p_x u, \qquad (12.2.18)$$
$$\Delta p = \partial_x(|u|^2). \qquad (12.2.19)$$

Necessary conditions for existence of wave-guides are given in Ghidaglia and Saut (1996):

Proposition 12.7. *The system (12.2.18)–(12.2.19) has solutions only if* $\sigma_2 = -1$ *and* $\beta > 1$.

Proof.   The following calculations suppose that $\nabla p \in L^2$. The condition that $\nabla \varphi \in L^2$ thus implies that $\omega = 0$. Multiplying (12.2.18) by u and integrating over $\mathbf{R}^2$, one gets

$$\int \left(u_x^2 - u_y^2 - \sigma_2 u^4 - \beta|\nabla p|^2\right) dx = 0. \qquad (12.2.20)$$

Multiplying (12.2.18) by xu_x and yu_y, one obtains respectively

$$\int \left(u_x^2 + u_y^2 + \frac{\sigma_2}{2} u^4 + \frac{\beta}{2} (p_x^2 - p_y^2) \right) d\mathbf{x} = 0 \qquad (12.2.21)$$

$$\int \left(u_x^2 + u_y^2 - \frac{\sigma_2}{2} u^4 - \frac{\beta}{2} (p_x^2 + 3p_y^2) \right) d\mathbf{x} = 0. \qquad (12.2.22)$$

Adding eqs. (12.2.21)–(12.2.22) yields

$$\int \left(|\nabla u|^2 - \beta p_y^2 \right) d\mathbf{x} = 0. \qquad (12.2.23)$$

This last equality implies that $\beta > 0$. Subtracting (12.2.20) from (12.2.21), one obtains

$$\int \left(-2u_y^2 - \frac{3}{2}\sigma_2 u^4 - \frac{\beta}{2} (3p_x^2 + p_y^2) \right) d\mathbf{x} = 0. \qquad (12.2.24)$$

This equality implies that σ_2 should be strictly negative and thus equal to -1. Finally, subtracting (12.2.21) from (12.2.22) leads to

$$\int \left(\sigma_2 u^4 + \beta |\nabla p|^2 \right) d\mathbf{x} = 0. \qquad (12.2.25)$$

Let us define $v = u^2$. By the Parseval's identity, we have

$$\int |\nabla p|^2 d\mathbf{x} = \int |\mathbf{k}|^2 |\hat{p}|^2 d\mathbf{k} = \int \frac{k_1^2}{k_1^2 + k_2^2} |\hat{v}|^2 d\mathbf{k}, \qquad (12.2.26)$$

where a hat denotes the Fourier transform. Rewriting (12.2.25) in Fourier space, we get

$$\int \left(\sigma_2 |\hat{v}|^2 + \beta \frac{k_1^2}{k_1^2 + k_2^2} |\hat{v}|^2 \right) d\mathbf{k} = 0, \qquad (12.2.27)$$

which implies that $\beta > |\sigma_2| = 1$.

Proposition 12.8. *The system (12.2.18)–(12.2.19) has radial solutions only if $\sigma_2 = -1$ and $\beta = 2$, which corresponds to focusing DSII.*

Proof. Since v is radial, this is also the case for its Fourier transform $\hat{v}$, and we have

$$\int \frac{k_1^2}{k_1^2 + k_2^2} |\hat{v}|^2 \, d\mathbf{k} = \frac{k_2^2}{k_1^2 + k_2^2} |\hat{v}|^2 \, d\mathbf{k} = \int \frac{1}{2} |\hat{v}|^2 \, d\mathbf{k}. \qquad (12.2.28)$$

Equation (12.2.27) now becomes

$$\left(-1 + \frac{\beta}{2} \right) \int |\hat{v}|^2 \, d\mathbf{k} = 0, \qquad (12.2.29)$$

which implies that $\beta = 2$.

By analogy with the lump solitons of the Kadomtsev–Petviashvili equation, one looks for radial solutions in the form of $u = C/(x^2 + y^2 + a^2)$ and find

$$u = \frac{2\sqrt{2}a}{x^2 + y^2 + a^2}, \quad q = \frac{4x}{x^2 + y^2 + a^2}. \tag{12.2.30}$$

A more general family of solutions was derived by Arkadiev, Pogrebkov, and Polivanov(1989) using the inverse scattering transform (see Ablowitz and Clarkson 1991, Section 5.5.3).

Using the invariance of the system under the pseudo-conformal transformation (12.1.34) and the existence of the lump soliton (u, q) given in (12.2.30), one constructs a family of exact self-similar solutions in the hyperbolic–elliptic case that blow up in a finite time. They have the form

$$\psi(\mathbf{x}, t) = \frac{1}{t_* - t} u\left(\frac{\mathbf{x}}{t_* - t}\right) e^{\frac{i(x^2 - y^2)}{4(t_* - t)}}, \quad \varphi(\mathbf{x}) = \frac{1}{t_* - t} q\left(\frac{\mathbf{x}}{t_* - t}\right). \tag{12.2.31}$$

When $t \to t_*$, the "mass density" $|\psi(\mathbf{x}, t)|^2$ converges to a Dirac measure at the origin.

12.3 The Initial Value Problem

12.3.1 Subsonic Wave Packet

In this section, we state results of existence, uniqueness, regularity, and continuity with respect to initial conditions for the elliptic–elliptic or hyperbolic–elliptic DS system (Ghidaglia and Saut 1990).

Theorem 12.9. (Existence and uniqueness in L^2)
(i) *For ψ_0 in $L^2(\mathbf{R}^2)$, there exists a unique maximal solution (ψ, φ) of (12.1.1)–(12.1.2) during a finite time $(0, T^*)$, $T^* > 0$, such that for $0 \le t < T^*$,*

$$\psi \in C((0, T^*), L^2(\mathbf{R}^2)) \cap L^4((0, t) \times \mathbf{R}^2), \quad \nabla\varphi \in L^2((0, t), L^2(\mathbf{R}^2)),$$

and it satisfies the conservation of mass (12.1.26).
(ii) *Furthermore, if ψ_0 is sufficiently small in $L^2(\mathbf{R}^2)$, then $T^* = \infty$.*

Theorem 12.10. (Regularity)
(i) *If $\psi_0 \in H^1(\mathbf{R}^2)$, the above solution satisfies*

$$\psi \in C((0, T^*), H^1(\mathbf{R}^2)) \cap C^1((0, T^*), H^{-1}(\mathbf{R}^2)), \quad \nabla\psi \in L^4((0, t) \times \mathbf{R}^2)$$
$$\nabla\varphi \in C((0, t), L^p(\mathbf{R}^2)), \quad \nabla^2\varphi \in L^4((0, t), L^q(\mathbf{R}^2)),$$

for all $0 \le t < T^*$, $2 \le p < \infty$, $2 \le q \le 4$, together with the conservation of the Hamiltonian. In addition, the map $\psi_0 \mapsto (\psi, \nabla\varphi)$ is continuous from $H^1(\mathbf{R}^2)$ into $C([0, t), H^1(\mathbf{R}^2)) \times C([0, t), L^p(\mathbf{R}^2))$, $p > 2$.
(ii) *If $\psi_0 \in H^2(\mathbf{R}^2)$, the above solution satisfies*

$$\psi \in C((0, T^*), H^2(\mathbf{R}^2)) \cap C^1((0, T^*), L^2(\mathbf{R}^2)),$$
$$\nabla\varphi \in C((0, T^*), H^2(\mathbf{R}^2)).$$

There is also a gain of regularity in the form

$$\psi \in L^2((0,t), H_{\text{loc}}^{5/2}(\mathbf{R}^2)), \ \nabla\varphi \in L^1((0,t), H_{\text{loc}}^{3/2}(\mathbf{R}^2)).$$

(iii) *If $\psi \in \Sigma$ (i.e., has a finite variance), the above solution satisfies*

$$\psi \in C((0,T^*),\Sigma),$$

and the identity (12.1.30) holds in the sense of distributions on $(0,T^*)$.

Long-time behavior of global solutions similar to that obtained for the NLS equation and existence of the scattering operators are presented in Guzmán-Gómez (1994).

Theorem 12.11. (Global solution in the elliptic–elliptic case) *Let the initial amplitude be $\psi_0 \in H^1(\mathbf{R}^2)$. If $\sigma_2 = 1 > \frac{\gamma}{\alpha}$ or if $\sigma_2 = -1$ with the condition $|\psi_0|_{L^2} < |\Phi|_{L^2}$ where Φ denotes the ground state solution of (12.2.2), the maximal solution obtained in Theorem 12.4 exists for all time.*

Proof. It suffices to prove the existence of a uniform bound for $|\nabla\psi|_{L^2}$. When $\sigma_2 = 1$ and $\gamma < \alpha$, this is a direct consequence of the Hamiltonian identity together with estimate $|\varphi_x|_{L^2} \leq \frac{\gamma}{\alpha}|\psi|_{L^4}^2$. When $\sigma_2 = -1$, denoting by $\langle\cdot,\cdot\rangle$ the L^2-scalar product, we have from (12.1.28),

$$H = |\nabla\psi|_{L^2}^2 - \frac{1}{2}\langle\mathcal{L}\psi^2, \psi^2\rangle, \tag{12.3.1}$$

and from Corollary 12.6,

$$\langle\mathcal{L}\psi^2, \psi^2\rangle \leq \frac{2}{|\Phi|_{L^2}}|\nabla\psi|_{L^2}^2|\psi|_{L^2}^2. \tag{12.3.2}$$

Thus,

$$|\nabla\psi|_{L^2}^2 \leq |H| + \frac{1}{2}\langle\mathcal{L}\psi^2, \psi^2\rangle \leq |H| + \frac{|\psi_0|_{L^2}^2}{|\Phi|_{L^2}^2}|\nabla\psi|_{L^2}^2, \tag{12.3.3}$$

and $|\nabla\psi|_{L^2}^2 \leq C$, for some constant C if $|\psi_0|_{L^2}^2 < |\Phi|_{L^2}^2$. This ensures global existence.

Remark. A cruder but more explicit condition for global existence when $\sigma_2 = -1$ is $|\psi_0|_{L^2}^2 < \frac{1}{1+\gamma/\alpha}|R|_{L^2}^2$, where R denotes the positive solution of $\Delta R - R + R^3 = 0$ and satisfies $|R|_{L^2}^2 = 1.86\dots$.

When the coefficients σ_2 and β in the hyperbolic–elliptic DS system (12.2.14)–(12.2.15) have the special values $(-1,2)$ or $(1,-2)$ (which corresponds to focusing and defocusing DSII, respectively), stronger results of existence and long-time behavior were obtained by Sung (1994, 1995) using the inverse scattering method. He proved in particular the following result.

Theorem 12.12. (DSII system) *Under the assumption that the initial condition $\psi_0 \in L^p$ for some $1 \leq p < 2$, and that its Fourier transform $\hat{\psi}_0$ belongs to $L^1 \cap L^\infty$ with the smallness condition $|\hat{\psi}_0|_{L^1}|\hat{\psi}_0|_{L^\infty} < \frac{\pi^3}{2}(\frac{\sqrt{5}-1}{2})^2$ in the focusing case (no condition in the defocusing case), there exists a* *unique global solution $\psi \in C(\mathbf{R}, C^0 \cup L^2)$, $\varphi_x \in C(\mathbf{R}, L^q)$ to DSII system for any $2 \leq q < \infty$. The solution ψ satisfies also the decay estimate $|\psi(t)|_{L^\infty} < C/t$.*

Furthermore, if the initial condition ψ belongs to the Schwartz space $\mathcal{S}(\mathbf{R}^2)$, there is an infinite number of conserved quantities.

We now turn to the existence of solutions that blow up in finite time in the elliptic–elliptic case. As for the NLS equation, the proof results from the variance identity and holds for initial conditions with negative Hamiltonian and finite variance. We thus start by establishing a necessary and sufficient conditions for the existence of initial conditions with negative Hamiltonian.

Proposition 12.13. *There exist initial conditions $\psi_0 \in H^1(\mathbf{R}^2)$ with negative Hamiltonian $H(\psi_0)$ if and only if either $\sigma_2 = -1$ or $\sigma_2 = 1 > \frac{\gamma}{\alpha}$.*

Proof. Defining $f = |\psi|^2$, one writes the Hamiltonian in terms of the Fourier transform of ψ in the form

$$H = \int |\mathbf{k}|^2 |\widehat{\psi}|^2 d\mathbf{k} + \frac{1}{2} \int \left(\sigma_2 - \frac{\gamma}{\alpha} \right) \frac{\alpha k_1^2}{k_1^2 + k_2^2} |\widehat{f}|^2 d\mathbf{k}. \qquad (12.3.4)$$

Theorem 12.14. (Finite-time blowup in the elliptic–elliptic case)
Let $\psi_0 \in \Sigma$ such that $H(\psi_0) < 0$. The maximal solution constructed in Theorem 12.4 satisfies $\lim_{t \to T^*} |\nabla \psi|_{L^2} = \infty$.

Proof. As for the NLS equation, we integrate twice in time the variance identity (12.1.30) and get $\mathcal{V}(t) = 4Ht^2 + \mathcal{V}'(0)t + \mathcal{V}(0)$. Thus, if $H < 0$, $\lim_{t \to t_0} \mathcal{V}(t) = 0$. The uncertainty principle $|f|_{L^2} \leq \frac{2}{d} |\nabla f|_{L^2} |xf|_{L^2}$ shows that there exists a time $T^* \leq t_0$ such that $\lim_{t \to T^*} |\nabla \psi|_{L^2} = \infty$.

12.3.2   Supersonic Wave Packets

A first result concerning the initial value problem was obtained by Fokas and Sung (1992) for the DSI system:

Theorem 12.15. *If $\psi \in \mathcal{S}(\mathbf{R}^2)$, where $\mathcal{S}(\mathbf{R}^2)$ is the Schwartz space, and $\varphi_1, \varphi_2 \in C(\mathbf{R}, \mathcal{S}(\mathbf{R}))$, then DSI has a unique solution $\psi \in C(\mathbf{R}, \mathcal{S}(\mathbf{R}^2))$.*

For general DS systems, results of existence of classical solutions make use of the weighted Sobolev spaces

$$H^{m,l} = \{ f \in L^2; |(1 - \partial_x^2 - \partial_y^2)^{m/2}(1 + x^2 + y^2)^{l/2} f|_{L^2} < \infty \}. \qquad (12.3.5)$$

Linares and Ponce (1993) proved the following theorems.

Theorem 12.16.
(i) hyperbolic–hyperbolic case: *For a sufficiently small initial condition ψ_0 in the space $Y_s = H^s(\mathbf{R}^2) \cap H^{3,2}(\mathbf{R}^2)$ with $s \geq 6$ and boundary conditions $\varphi_1 = \varphi_2 = 0$ for the mean flow, there exists a unique classical solution $\psi \in C([0, T], Y_s)$, where $T > 0$ depends on the initial conditions.*

(ii) elliptic–hyperbolic case: *For a sufficiently small initial condition ψ_0 in the space $W_s = H^s(\mathbf{R}^2) \cap H^{6,6}(\mathbf{R}^2)$ with $s \geq 12$ and $\varphi_1 = \varphi_2 = 0$, there exists a unique classical solution $\psi \in C([0,T], W_s)$, where $T > 0$ depends on the initial conditions.*

The order of the Sobolev spaces was reduced by Hayashi (1996) in the following form.

Theorem 12.17. (elliptic–hyperbolic and hyperbolic–hyperbolic cases) *For a sufficiently small initial condition ψ_0 in the space $Z_\delta = H^{\delta,0}(\mathbf{R}^2) \cap H^{0,\delta}(\mathbf{R}^2)$, $\partial_x\varphi_1 \in C(\mathbf{R}, H_x^{\delta,0})$, and $\partial_y\varphi_2 \in C(\mathbf{R}, H_y^{\delta,0})$, there exists a unique classical solution $\psi \in C([0,T], Z_\delta)$, and*

$$\int_0^T \left| (1 + |\mathbf{x}|^2)^{-\delta/4}(1 - \Delta)^{\frac{\delta+1/2}{2}} \psi(t) \right|_{L^2} dt < \infty, \qquad (12.3.6)$$

where $\delta = 1 + \epsilon$, with ϵ small enough.

Theorem 12.18. (elliptic–hyperbolic case) *For an initial condition ψ_0 such that $(1+|\mathbf{x}|^2)^{1/2}\psi_0$ is small enough in the space $X_\delta = H^{\delta-1/2,0}(\mathbf{R}^2) \cap H^{0,\delta-1/2}(\mathbf{R}^2)$, with $\partial_x\varphi_1 \in C(\mathbf{R}, H_x^{\delta-1/2,0})$ and $\partial_y\varphi_2 \in C(\mathbf{R}, H_y^{\delta-1/2,0})$, there exists a unique classical solution $\psi \in C([0,T], X_\delta) \cap C([0,T], H_{\mathrm{loc}}^{\delta+1/2,0})$, where $\delta = 1 + \epsilon$, ϵ small.*

The constraint of small initial conditions was removed by Hayashi and Saut (1995) by working in spaces of analytic functions decreasing exponentially at infinity. Let $B = (B_1, \ldots, B_j)$ be a vector field of derivations, which in the present context will be one of the following operators

$$\partial = (\partial_x, \partial_y), \quad R(t) = (\partial, J_1(t), J_2(t)), \quad \widetilde{R}(t) = (\partial, J_x(t), J_y(t)), \quad (12.3.7)$$

with

$$J_x(t) = x + 2it\partial_x, \quad J_y(t) = y + 2it\partial_y, \quad J_1(t) = y + it\partial_x, \quad J_2(t) = x + it\partial_y. \tag{12.3.8}$$

Let X denote a Banach space equipped with a norm $\|.\|_X$. They define the generalized analytic function space

$$G^A(B; X) = \{ f \in X; \|f\|_{G^A(B;X)} = \sum_\beta \frac{A^{|\beta|}}{|\beta|!} \|B^\beta f\|_X < \infty \} \qquad (12.3.9)$$

where A is a positive constant, $\beta = (\beta_1, \ldots, \beta_j)$, with $\beta_k \in \mathbf{N}$, $|\beta| = \sum_{1 \leq k \leq j} \beta_k$ and $B^\beta = B_1^{\beta_1} \ldots B_j^{\beta_j}$.

They also define the generalized Sobolev spaces

$$B^{m,p} = \{ f \in L^p; \|f\|_{B^{m,p}} = \sum_{|\beta| \leq m} \|B^\beta f\|_{L^p} < \infty \}. \qquad (12.3.10)$$

In the following, the Banach space X will be identified with such a space. This leads to define spaces $G^A(B'(t); B^{m,p}(t))$ where $B(t)$ and $B'(t)$ are time-dependent vector fields of derivation.

In the case of zero boundary conditions $\varphi_1 = \varphi_2 = 0$, they prove the following results.

Theorem 12.19. (hyperbolic–hyperbolic or elliptic–hyperbolic system) *For an initial condition $\psi_0 \in G^A(\partial; W^{m,2})$ where $A > 0$ and $m \geq 3$, there exists a unique solution ψ and a positive constant T such that $\psi \in C([-T, T]; G^{A_1}(\partial; W^{m,2}))$.*

Theorem 12.20. (hyperbolic–hyperbolic case) *For an initial condition $\psi_0 \in G^A(R(0); R^{m,2}(0))$ with $\|\psi_0\|_{G^A(R(0);R^{m,2}(0))}$ small enough, where $A > 0$ and $m \geq 3$, there exists a unique solution $\psi \in G^{A_1}(R(t); R^{m,2}(t))$ for any t, where $A_1 < A$.*

Theorem 12.21. (elliptic–hyperbolic case) *For $\psi_0 \in G^A(R(0); \widetilde{R}^{m,2}(0))$ with $\|\psi_0\|_{G^A(R(0);\widetilde{R}^{m,2}(0))}$ small enough, where $A > 0$ and $m \geq 3$, there exists a unique solution $\psi \in G^{A_2}(\widetilde{R}(t); \widetilde{R}^{m,2}(t))$ for any t, where $A_2 < A$.*

In the elliptic–hyperbolic case, it is possible to remove the hypotheses of analyticity and exponential decay and work in the usual Sobolev spaces. Hayashi and Hirata (1996) proved the following result.

Theorem 12.22. *For $\psi_0 \in H^s(\mathbf{R}^2)$, $s > \frac{5}{2}$, $\partial_x \varphi_1 \in C(\mathbf{R}, H_x^s)$, $\partial_y \varphi_2 \in C(\mathbf{R}, H_y^s)$, and $|\psi_0|_{L^2}$ sufficiently small, there exists a unique solution locally in time in $C([0, T], H^s)$.*

Theorem 12.23. *For small enough initial conditions in the weighted Sobolev space $H^{3,0} \cap H^{0,3}$ and $\partial_x^{j+1} \varphi_1 \in C(\mathbf{R}, L_x^\infty)$ and $\partial_y^{j+1} \varphi_2 \in C(\mathbf{R}, L_y^\infty)$, $0 \leq j \leq 2$, decreasing fast enough at infinity, there exists a unique solution, globally in time, such that*

$$\psi \in L_{\text{loc}}^\infty(\mathbf{R}, H^{3,0} \cap H^{0,3}) \cap C(\mathbf{R}, H^{2,0} \cap H^{0,2}). \qquad (12.3.11)$$

Furthermore, $|\psi(t)|_{L^\infty} \leq C/(1 + |t|)$, and the solution becomes asymptotically free as $|t| \to \infty$.

Recently, Hayashi (1997) obtained local existence of solutions in H^2 without smallness conditions on initial conditions:

Theorem 12.24. *Let $\psi_0 \in H^s(\mathbf{R}^2)$, $\varphi_1 \in C(\mathbf{R}, H_x^s)$, $\varphi_2 \in C(\mathbf{R}, H_y^s)$, $s \geq 2$. Then there exists a unique solution locally in time in $C([-T, T], H^{s-1}) \cap L^\infty([-T, T], H^s)$.*

12.4 Rate of Blowup for Elliptic–Elliptic DS

The dynamics of blowup solutions of the Davey–Stewartson system was studied numerically in the elliptic–elliptic case ($\sigma_1 = 1$ and $\alpha > 0$) by Papanicolaou, Sulem, Sulem, and Wang (1994) using the anisotropic dynamic rescaling method. The main observation is that the blowup is very similar to that of critical NLS, except that the profile of the wave amplitude displays a moderate anisotropy, the ratio of the typical scales along the two coordinate axes converging to a finite value.

In order to present an asymptotic construction of such self-focusing solutions, it is convenient to rewrite the DS system in terms of the amplitude ψ and the longitudinal velocity $u_1 = \varphi_x$ in the form

$$i\psi_\tau + \Delta\psi - \sigma_2|\psi|^2\psi = u_1\psi, \tag{12.4.1}$$

$$\Delta_\alpha u_1 = -\gamma|\psi|^2_{xx}, \tag{12.4.2}$$

where Δ_α denotes the operator $\alpha\dfrac{\partial^2}{\partial x^2} + \dfrac{\partial^2}{\partial y^2}$. Since the numerical simulations suggest that the typical scales remain comparable in the x and y directions, the same scaling factor is used in both directions. We define

$$\xi_1 = \frac{x}{L(t)}, \quad \xi_2 = \frac{y}{L(t)}, \quad \tau = \int_0^t \frac{1}{L^2(s)}\,ds \tag{12.4.3}$$

and

$$U(\xi_1, \xi_2, \tau) = L(t)\psi(x, y, t), \quad W(\xi_1, \xi_2, \tau) = L^2(t)u_1(x, y, t). \tag{12.4.4}$$

Equations (12.4.1)–(12.4.2) become

$$iU_\tau + ia(U + \xi \cdot \nabla U) + \Delta U - \sigma_2|U|^2U = WU \tag{12.4.5}$$

$$\Delta_\alpha W = -\gamma|U|^2_{\xi_1\xi_1}, \tag{12.4.6}$$

with $a = -L_t L$ and $\xi = (\xi_1, \xi_2)$. Writing

$$U = e^{i\tau - ia\frac{\xi^2}{4}}\, V, \quad b(\tau) = a_\tau + a^2, \tag{12.4.7}$$

we see that V and W satisfy

$$iV_\tau - V + \Delta V + \frac{b(\tau)}{4}|\xi|^2V - \sigma_2|V|^2V = VW, \tag{12.4.8}$$

$$\Delta_\alpha W = -\gamma|V|^2_{\xi_1\xi_1}. \tag{12.4.9}$$

The asymptotic behavior of the scaling factors cannot be obtained by directly taking $a = 0$ and neglecting time derivatives in (12.4.9). As in the case of critical NLS, a more refined analysis is required.

The construction of singular isotropic solutions for critical NLS presented in Chapter 8 was based on the modulation of the self-similar solutions that develop in dimension $d > 2$. This dimension was varied in time, tending to 2 as the singularity is approached. For DS equations, due to the anisotropy of the problem, we keep the dimension equal to 2 and formally vary the degree of the nonlinearity. The two approaches are obviously equivalent for isotropic NLS. We thus introduce the steady system

$$\Delta P - P + \frac{b}{4}|\xi|^2P + i\left(\frac{1}{p} - 1\right)\sqrt{b}P - \sigma_2|P|^{2p}P = PZ, \tag{12.4.10}$$

$$\Delta_\alpha Z = -\gamma|P|^{2p}_{\xi_1\xi_1}, \tag{12.4.11}$$

where $p > 1$ and $b > 0$. As in the context of the NLS equation, this nonlinear eigenvalue problem is expected to have no nontrivial solutions with monotonic decreasing profiles except when a specific relation $b = b(p)$, or equivalently $p = p(b)$, holds. Since we are interested in situations where a varies in time, we are led to consider (12.4.10)–(12.4.11) with $b = b(\tau)$ and $p = p(b(\tau)) > 1$, and the conditions $b \to 0$ and $p \to 1$ as $\tau \to \infty$. We denote by $\big(P(b(\tau), \xi_1, \xi_2), Z(b(\tau), \xi_1, \xi_2)\big)$ the corresponding solution, supposed to exist. Note that in the limit $b \to 0$, $p \to 1$, (P, Z) tends to (S, X), a solution of

$$\Delta S - S - \sigma_2 S^3 - SX = 0, \tag{12.4.12}$$

$$\Delta_\alpha X + \gamma |S|^2_{\xi_1 \xi_1} = 0. \tag{12.4.13}$$

For the unsteady problem (12.4.8)–(12.4.9), we make the ansatz

$$V(\xi_1, \xi_2, \tau) = P(b(\tau), \xi_1, \xi_2) + v(\xi_1, \xi_2, \tau), \tag{12.4.14}$$

$$W(\xi_1, \xi_2, \tau) = Z(\xi_1, \xi_2) + w(\xi_1, \xi_2, \tau). \tag{12.4.15}$$

We assume that as $\tau \to \infty$, we have $|v| \ll |P|$, $|w| \ll |Z|$, and that b tends slowly to zero. Using (12.4.10)–(12.4.11), we obtain a system for the remainders (v, w) that, to leading order as $\tau \to \infty$ reads

$$\Delta v - v + \frac{b|\xi|^2}{4} v - \sigma_2 (2|P|^2 v + P^2 \bar{v}) - vZ - Pw$$
$$= -b_\tau \frac{\partial P}{\partial b} - \sigma_2 (p-1)|P|^2 P \ln |P|^2 + i\sqrt{b}\Big(\frac{1}{p} - 1\Big)P, \tag{12.4.16}$$

$$\Delta_\alpha w = -\gamma (\bar{P}v + P\bar{v})_{\xi_1 \xi_1} - (p-1)(|P|^2 \ln |P|^2)_{\xi_1 \xi_1}. \tag{12.4.17}$$

Proposition 12.25. *The system (12.4.16)–(12.4.17) is solvable if the function $b(\tau)$ satisfies the differential equation*

$$b_\tau = \frac{2N_0}{M}\Big(\frac{1}{p} - 1\Big)\sqrt{b} \tag{12.4.18}$$

with $N_0 = \frac{1}{2\pi} \int |S|^2 \, d\xi$ and $M = \frac{1}{8\pi} \int |S|^2 |\xi|^2 d\xi$.

Proof. The pair $(\rho, \beta) = \Big(\dfrac{\partial P}{\partial b}, \dfrac{\partial Z}{\partial b}\Big)\Big|_{b=0}$ satisfies the linear system

$$\Delta \rho - \rho - 3\sigma_2 S^2 \rho - S\beta - X\rho = -\frac{1}{4}\xi^2 S, \tag{12.4.19}$$

$$\Delta_\alpha \beta + \gamma(2S\rho)_{\xi_1 \xi_1} = 0. \tag{12.4.20}$$

with zero boundary conditions at infinity. Existence of solutions requires that the right-hand side of the above system be orthogonal to the kernel of the adjoint of the operator arising in the left-hand side, which is defined by the solutions (λ, μ) of

$$\Delta \lambda - \lambda - 3\sigma_2 S^2 \lambda + 2\gamma S\mu_{\xi_1 \xi_1} - X\lambda = 0, \tag{12.4.21}$$

$$\Delta_\alpha \mu - S\lambda = 0. \tag{12.4.22}$$

The set of solutions is spanned by

$$\left\{ \begin{pmatrix} S_{\xi_1} \\ -\frac{1}{2\gamma} \int^{\xi_1} X d\xi_1 \end{pmatrix}, \begin{pmatrix} S_{\xi_2} \\ \frac{-1}{2\gamma} \int^{\xi_1} d\xi_1' (\int^{\xi_1'} X_{\xi_2}(z, \xi_2) dz) \end{pmatrix} \right\}.$$

The solvability condition thus reads

$$\int \xi^2 S\, S_{\xi_i}\, d\xi_1 d\xi_2 = 0, \tag{12.4.23}$$

or equivalently, $\int \xi_i S^2\, d\xi_1 d\xi_2 = 0$. This condition is satisfied when, as confirmed by the numerical simulations, S is even with respect to the variables ξ_1 and ξ_2. As a consequence, (ρ, β) is well-defined.

We now consider (12.4.16)–(12.4.17) in the limit $b \to 0$ and write $v = v_1 + iv_2$. To lowest order, we have

$$\Delta v_1 - v_1 - 3\sigma_2 S^2 v_1 - v_1 X - Sw = -\sigma_2 (p-1) S^3 \ln(S^2), \tag{12.4.24}$$

$$\Delta v_2 - v_2 - \sigma_2 S^2 v_2 - v_2 X = -b_\tau \rho + (\frac{1}{p} - 1)\sqrt{b} S, \tag{12.4.25}$$

$$\alpha w_{\xi_1 \xi_1} + w_{\xi_2 \xi_2} - 2\gamma (Sv_1)_{\xi_1 \xi_1} = -(p-1)(S^2 \ln(S^2))_{\xi_1 \xi_1}. \tag{12.4.26}$$

Note that (12.4.25) is decoupled from (12.4.24) and (12.4.26). The solvability conditions for (12.4.24) and (12.4.26) are again satisfied, provided that S is an even function of ξ_1 and ξ_2. Since S satisfies (12.4.12), the solvability condition for (12.4.25) is

$$- b_\tau \int S\rho\, d\xi + \left(\frac{1}{p} - 1 \right) \sqrt{b} \int S^2\, d\xi = 0, \tag{12.4.27}$$

or

$$b_\tau = C_1 \left(\frac{1}{p} - 1 \right) \sqrt{b}, \tag{12.4.28}$$

with $C_1 = \frac{\int |S|^2 d\xi}{\int S\rho\, d\xi}$. Note that $b_\tau < 0$ and $p > 1$. It follows that, for consistency, the constant C_1 must be positive. This property results from the identity $2 \int S\rho\, d\xi = \frac{1}{4} \int S^2 |\xi|^2 d\xi$, proved in Papanicolaou, Sulem, Sulem, and Wang (1994). Equation (12.4.18) follows. It provides a differential equation for $b(\tau)$ if a relation between p and b is known. The latter relation, given in the next proposition is obtained by analyzing (12.4.10)–(12.4.11) in the limit $b \to 0$, $p \to 1$.

Proposition 12.26. *In the limit $b \to 0^+$, $p \to 1^+$, the solution (P, Z) of (12.4.10)–(12.4.11) satisfies:*
(i) In the limit $\sqrt{b}|\xi| \ll 1$, $(P, Z) \approx (S, X)$, where (S, X) are solutions of (12.4.12)–(12.4.13).
(ii) For $\sqrt{b}|\xi| \gg 1$, $P \approx \mu |\xi|^{-1/p - i/\sqrt{b}} e^{i\sqrt{b}\xi^2/4}$, with $\mu^2 = 2(1 - \frac{1}{p})N_0$.

(iii) *The asymptotic behavior of p when b tends to 0 is*

$$1 - \frac{1}{p} \approx \frac{\nu_1^2}{N_0} \frac{1}{\sqrt{b}} e^{-\pi/\sqrt{b}}, \qquad (12.4.29)$$

where $N_0 = \frac{1}{2\pi} \int S^2 d\xi$, *and* ν_1 *is a numerical constant.*

Proof. Writing $P = Be^{i\theta}$, the system (12.4.10)–(12.4.11) becomes

$$\Delta B - B - \sigma_2 B^{2p+1} - (|\nabla \theta|^2 - \frac{b\xi^2}{4})B - ZB = 0 \qquad (12.4.30)$$

$$\nabla B^2 \cdot \nabla \theta + B^2 \left(\Delta \theta - \sqrt{b} + \frac{\sqrt{b}}{p}\right) = 0, \qquad (12.4.31)$$

$$\alpha Z_{\xi_1 \xi_1} + Z_{\xi_2 \xi_2} + \gamma (B^{2p})_{\xi_1 \xi_1} = 0. \qquad (12.4.32)$$

Letting $\chi = \nabla \theta$, (12.4.31) becomes

$$\operatorname{div} \left\{ B^2 \chi |\xi|^{\frac{2}{p}-2} \right\} - 2B^2 \left(\frac{1}{p} - 1\right) |\xi|^{2(\frac{1}{p}-2)} \xi \cdot \left(\chi - \frac{\sqrt{b}\xi}{2}\right) = 0. \quad (12.4.33)$$

Let U satisfy

$$\operatorname{div} U = 2B^2 |\xi|^{2(\frac{1}{p}-2)} \xi \cdot \left(\chi - \frac{\sqrt{b}\xi}{2}\right). \qquad (12.4.34)$$

Then

$$\chi = \frac{1}{B^2 |\xi|^{\frac{2}{p}-2}} \left(\frac{1}{p} - 1\right) U. \qquad (12.4.35)$$

The limit $p \to 1$, $b \to 0$, $\xi \to \infty$ is nonuniform. When $p \to 1$, with fixed ξ, B approaches S. On the other hand, for large $|\xi|$ and fixed b, the nonlinear terms are negligible in (12.4.30) and so are the azimuthal derivatives in the Laplacian. We can therefore repeat the analysis of the isotropic case given in Section 8.2. For $|\xi| \to \infty$ we have

$$B \sim \mu |\xi|^{-1/p}, \quad \theta \sim \frac{\sqrt{b}\xi^2}{4} - \frac{1}{\sqrt{b}} \ln |\xi|, \quad \chi \sim \frac{\sqrt{b}\xi}{2} - \frac{1}{\sqrt{b}} \frac{\xi}{|\xi|}. \quad (12.4.36)$$

In order to characterize the behavior of U, we write

$$U = \nabla q_1 + \nabla \times \mathbf{q}_2 \qquad (12.4.37)$$

with

$$\Delta q_1 = 2|B|^2 |\xi|^{2(\frac{1}{p}-2)} \xi \cdot \left(\chi - \frac{\sqrt{b}\xi}{2}\right). \qquad (12.4.38)$$

Up to a divergence-free term we thus have

$$U(\xi) = \frac{1}{\pi} \int_{\mathbf{R}^2} \frac{\xi - \xi'}{|\xi - \xi'|^2} B^2(\xi') |\xi'|^{2(\frac{1}{p}-2)} \xi' \cdot \left(\chi - \frac{\sqrt{b}\xi'}{2}\right) d\xi'. \quad (12.4.39)$$

It follows that

$$U \approx \frac{1}{\pi} \frac{\xi}{|\xi|^2} \int_{\mathbf{R}^2} B^2(\xi')|\xi'|^{2(\frac{1}{p}-2)}\xi' \cdot \left(\chi - \frac{\sqrt{b}\xi'}{2}\right)d\xi'. \qquad (12.4.40)$$

The main contribution in this integral is from $|\xi'|$ of order one. Since we can approximate B by S defined by (12.4.12) and neglect χ, which is proportional to $\frac{1}{p} - 1 \ll \sqrt{b}$, we get

$$U \approx -\frac{1}{2\pi} \frac{\sqrt{b}\xi}{|\xi|^2} \int S^2(\xi')d\xi'. \qquad (12.4.41)$$

Therefore, equating the unbounded terms in (12.4.35), we have

$$\mu^2 \approx 2\left(1 - \frac{1}{p}\right)\frac{1}{2\pi} \int S^2(\xi')d\xi' = 2\left(1 - \frac{1}{p}\right)N_0. \qquad (12.4.42)$$

Finally, we match the solution for $|\xi| \gg 2/\sqrt{b}$ to the ground state S, solution of (12.4.10). When $|\xi| \to \infty$, the nonlinear terms are negligible, and

$$S \approx \nu_1 |\xi|^{-1/2} e^{-\xi}. \qquad (12.4.43)$$

We define $Z = B|\xi|^{1/2}$. In the range of interest, (12.4.30) reduces to

$$Z'' + \left(\frac{b|\xi|^2}{4} - 1\right)Z = 0. \qquad (12.4.44)$$

From here, we follow the WKB argument given in Section 8.2. Using that $a \approx \sqrt{b}$, we get

$$\mu = \nu_1 \sqrt{2} b^{-1/4} e^{-\pi/(2\sqrt{b})} \qquad (12.4.45)$$

which, when substituted into (12.4.42), gives

$$1 - \frac{1}{p} \approx \frac{1}{N_0} \nu_1^2 b^{-1/2} e^{-\pi/\sqrt{b}}. \qquad (12.4.46)$$

Proposition 12.27. *Blowup solutions of the elliptic–elliptic DS system (12.4.1)–(12.4.2) have the following asymptotic form near the singularity:*

$$\psi(\mathbf{x},t) \approx \frac{1}{L(t)} e^{i(\tau(t)-a(t)\frac{|\mathbf{X}|^2}{4}L(t)^2)} P\left(\frac{|\mathbf{x}|}{L(t)}, b(t)\right), \qquad (12.4.47)$$

$$u_1(\mathbf{x},t) \approx \frac{\gamma}{L(t)^2}(-\Delta_\alpha)^{-1}(|P|^2)_{\xi_1\xi_1}, \qquad (12.4.48)$$

where $\tau_t = L^{-2}$, $-L_t L = a$, $L^3 L_{tt} = -b$ and $b = a^2 + a_\tau \approx a^2$ is given by

$$b_\tau = -2\frac{\nu_1^2}{M} e^{-\pi/\sqrt{b}} \qquad (12.4.49)$$

with $N_0 = \frac{1}{2\pi} \int_{\mathbf{R}^2} S^2 d\xi$ and $M = \frac{1}{4}\frac{1}{2\pi}\int S^2 |\xi|^2 d\xi$.

Proof. Putting (12.4.18) and (12.4.29) together, we get (12.4.49). As for critical NLS, $a^2 \approx b$, and thus

$$a_\tau \approx -\frac{\nu_1^2}{M} \frac{1}{a} e^{-\pi/a}. \tag{12.4.50}$$

To leading order in the limit $\tau \to \infty$, this gives $a \sim \dfrac{C}{\ln \tau}$. Returning to the primitive variables, this corresponds to a blowup rate of the form

$$L(t) \propto (t_* - t)^{1/2} \left(\ln \ln \frac{1}{t_* - t} \right)^{-1/2}. \tag{12.4.51}$$

Remarks.
1. When the coefficient γ in (12.4.2) is small, the system (12.4.1)–(12.4.2) can be viewed as a perturbation of the cubic two-dimensional focusing equation and is amenable to a perturbation analysis that confirms that the additional coupling does not modify the rate of blowup (Fibich and Papanicolaou 1997b).
2. Another example of a nonlocal equation displaying the same form of blowup is the so-called Chern–Simons gauged planar NLS equation arising in quantum physics, in the form

$$i(\partial_t + iA^0)\psi = -\frac{1}{2}(\nabla - i\mathbf{A})^2\psi - g|\psi|^2\psi \tag{12.4.52}$$

with

$$\mathbf{A}(\mathbf{x}, t) = \frac{1}{\kappa} \int G(\mathbf{x} - \mathbf{x}') |\psi(\mathbf{x}', t)|^2 d\mathbf{x}',$$

$$A^0(\mathbf{x}, t) = \frac{1}{\kappa} \int G(\mathbf{x} - \mathbf{x}') \cdot \Im(\psi^*(\nabla - i\mathbf{A})\psi)(\mathbf{x}', t)d\mathbf{x}', \tag{12.4.53}$$

where

$$G(\mathbf{x}) = \frac{1}{2\pi} \nabla \times (\ln |\mathbf{x}| \, \mathbf{e}_\perp)$$

and κ and $g > 0$ denote numerical constants (Bergé, de Bouard, and Saut 1995 and references therein).

12.5 Solutions of Elliptic–Hyperbolic DS

Special attention has been devoted to the DSI system

$$i\partial_t\psi + \partial_{xx}\psi + \partial_{yy}\psi = -\sigma|\psi|^2\psi + \varphi_x\psi, \tag{12.5.54}$$

$$\partial_{xx}\varphi - \partial_{yy}\varphi = 2\sigma\partial_x|\psi|^2. \tag{12.5.55}$$

For $\sigma = -1$, Boiti, Leon, Martina, and Pempinelli (1988) showed that DSI admits "two-dimensional breather solutions" corresponding to localized wave packets decaying exponentially in both space coordinates. These solutions were generalized to arbitrary time-dependent boundary conditions φ_1 and φ_2 by Fokas and Santini (1990), who named these solutions "dromions" (see also Hietarinta and Hirota 1990 and Gilson and Nimmo 1991 for further generalizations). In contrast with the one-dimensional solitons, dromions do not preserve their form after interaction (Gilson and Nimmo 1991). Note that when $\varphi_1 = \varphi_2 = 0$, arbitrary data disperse as $t \to \infty$ (Schultz and Ablowitz 1989). A discussion of these solutions is made by Ablowitz and Clarkson (1991) who conjecture that any arbitrary initial condition will decompose as $t \to \infty$ into either a number of breather solutions for time-independent boundary conditions or a number of dromions for time-dependent boundary conditions.

Besse and Bruneau (1997) performed numerical simulations on the general elliptic–hyperbolic DS system

$$i\partial_t \psi + \partial_{xx}\psi + \partial_{yy}\psi = \chi|\psi|^2\psi + b(\partial_x\varphi + \partial_y\varphi)\psi, \qquad (12.5.56)$$

$$\partial_{xy}\varphi = \frac{\sigma}{4}(\partial_x|\psi|^2 + \partial_y|\psi|^2). \qquad (12.5.57)$$

For the time stepping, they used a "relaxed" Crank–Nicholson scheme

$$i\frac{\psi^{n+1} - \psi^n}{\delta t} + \Delta\frac{\psi^{n+1} + \psi^n}{2} = \frac{\chi}{4}(|\psi^{n+1}|^2 + |\psi^n|^2)(\psi^{n+1} + \psi^n)$$

$$+ b\frac{\psi^{n+1} + \psi^n}{2}(\varphi_x^{n+1/2} + \varphi_y^{n+1/2}), \qquad (12.5.58)$$

$$\partial_{xy}\frac{\varphi^{n+1/2} + \varphi^{n-1/2}}{2} = \frac{\sigma}{4}\left((|\psi^n|^2)_x + (|\psi^n|^2)_y\right), \qquad (12.5.59)$$

where the first step is computed by identifying $\varphi^{-1/2}$ and $\varphi^{1/2}$. This scheme can be viewed as an extension of the classical fully implicit Crank–Nicholson scheme that conserves exactly the mass and when the mean field vanishes at infinity, the Hamiltonian of the wave packet. It enables a decoupling of the two equations. An extension consisting in replacing the cubic term in the amplitude equation by a term of the form $\frac{1}{2}u^{n+1/2}(\psi^{n+1} + \psi^n)$ with $\frac{1}{2}(u^{n+1/2} + u^{n-1/2}) = |\psi^n|^2$ enables one to avoid the use of an iterative resolution (Besse 1997). For the space discretization, they used a second-order finite difference. The scheme was first shown to be efficient in calculating dromions solutions, and was then applied to the case of Gaussian initial conditions $\psi_0 = 4e^{-(x^2+y^2)}$ and $\varphi_1 = \varphi_2 = 0$.

For $\sigma = 0$ and $\chi = -1$ in (12.5.56)–(12.5.57), one recovers the usual critical NLS equation that blows up in a finite time. Besse and Bruneau (1997) integrated (12.5.58)–(12.5.59) with $b = 1$, $\chi = -1$, and various values of σ ranging from -0.3 to 0.2. They observed that the solutions blow up in a finite time and that the singularity time increases with σ. They also integrated the system with $\chi = -1$, $\sigma = -2$ and various values of the parameter b and concluded that blowup occurs for $b > 0$ and for small negative values of b, while the solution remains bounded if b is sufficiently negative. This study demonstrates a qualitative sensitivity of the elliptic–

hyperbolic DS system to the values of the parameters. Further study is needed to understand in detail the structure of the blowup.

One can interpret the numerical observations for fixed σ and small $b = \epsilon$ (or equivalently, fixed b and small σ) by looking at the system (12.5.56)–(12.5.57) as a perturbation of the two-dimensional focusing cubic NLS equation. One can then perform a perturbation analysis similar to that done by Fibich and Papanicolaou (1997b) for the elliptic–elliptic DS system and predict that the blowup is similar to that of the critical NLS. Introducing the scaled variables

$$\xi_1 = \frac{x}{L(t)}, \quad \xi_2 = \frac{y}{L(t)}, \quad \tau = \int_0^t \frac{1}{L^2(s)} \, ds \qquad (12.5.60)$$

and

$$v(\xi_1, \xi_2, \tau) = L(t)\psi(x, y, t)e^{-i\tau + ia\xi^2/4}, \quad w(\xi_1, \xi_2, \tau) = L^2(t)\varphi(x, y, t), \qquad (12.5.61)$$

the system (12.5.56)–(12.5.57) becomes

$$iv_\tau + \Delta v + |v|^2 v + \frac{b(\tau)}{4}\xi^2 v = \epsilon(w_{\xi_1} + w_{\xi_2})v, \qquad (12.5.62)$$

$$w_{\xi_1\xi_2} = -\frac{\sigma}{4}(\partial_{\xi_1}|v|^2 + \partial_{\xi_2}|v|^2), \qquad (12.5.63)$$

with $a = -L_t L$ and $\xi = (\xi_1, \xi_2)$. Proceeding as in Chapter 9, one gets the reduced equation

$$\int \left(b_\tau \rho + \epsilon h_\tau + \nu(\sqrt{b})R\right) R d\xi_1 d\xi_2 = 0 \qquad (12.5.64)$$

where $b = a^2 + a_\tau$,

$$\nu(\sqrt{b}) = \frac{\nu_0^2}{N_c} e^{-\pi/\sqrt{b}}, \qquad (12.5.65)$$

and h satisfies

$$\Delta h - h + 3R^2 h = (\widetilde{w}_{\xi_1} + \widetilde{w}_{\xi_2})R, \qquad (12.5.66)$$

with $\widetilde{w}$ solution of

$$\widetilde{w}_{\xi_1\xi_2} = -\frac{\sigma}{4}\left((R^2)_{\xi_1} + (R^2)_{\xi_2}\right). \qquad (12.5.67)$$

As before, R is the positive solution of $\Delta R - R + R^3 = 0$, and ρ is the solution of $\Delta\rho - \rho + 3R^2\rho = -\frac{1}{4}|\xi|^2 R^2$. Since $h_\tau = 0$, the modulation equation is the same as for the critical NLS equation and leads to the same rate of blowup.

13
Langmuir Oscillations

Langmuir oscillations, (also called Langmuir waves or electron plasma waves), constitute an important phenomenon taking place in a nonmagnetized or weakly magnetized plasma, and have applications to both astrophysical situations such as solar radio emission (ter Haar and Tsytovich 1975, Gurnett, Maggs, Gallagher, Kurth, and Scarf 1981, Goldman1983, 1984) and laboratory experiments (Wong and Cheung 1984, Cheung and Wong 1985). Application to the ionosphere is considered by Rose, Dubois, Russel, and Bezzerides (1991). A review is found in Robinson (1997). Modulational instability leads to the formation of regions of lower plasma density (cavities) where high-frequency oscillations of the electric field are trapped. The nonlinear evolution leads to the collapse of the cavities and to a strong amplification of the oscillation amplitude. This provides an effective mechanism for the dissipation of long-wavelength plasma waves. Collapse thus plays a central role in the dynamics of Langmuir turbulence, and strong attention has been paid to this problem after Zakharov (1972) derived the equations governing the coupled dynamics of the electric-field amplitude and of the low-frequency density fluctuations of the ions (see also Zakharov 1984, Rubenchik, Sagdeev, and Zakharov 1985). The energy absorption that could occur in the postcollapse evolution of the cavities as the result of the drawing of new Langmuir waves into them is considered by Malkin (1991).

13.1 Derivation of the Zakharov Equations

13.1.1 The Two-Fluid Model

We consider a plasma as two interpenetrating fluids, an electron fluid and an ion fluid. We denote by m_e, $n_e(\mathbf{x}, t)$, and $\mathbf{v}_e(\mathbf{x}, t)$ (respectively m_i, $n_i(\mathbf{x}, t)$ and $v_i(\mathbf{x}, t)$) the mass, number density (number of particles per unit volume) and velocity of the electrons (respectively of the ions). The continuity equations for each of the fluid species read

$$\partial_t n_e + \nabla \cdot (n_e \mathbf{v}_e) = 0, \qquad (13.1.1)$$

$$\partial_t n_i + \nabla \cdot (n_i \mathbf{v}_i) = 0. \qquad (13.1.2)$$

Denoting by $-e$ the charge of the electron and by e that of the ions assumed to reduce to protons (extension to ions with a larger charge is straightforward), the momentum equations read

$$m_e n_e (\partial_t \mathbf{v}_e + \mathbf{v}_e \cdot \nabla \mathbf{v}_e) = -\nabla p_e - e n_e (\mathbf{E} + \frac{1}{c} \mathbf{v}_e \times \mathbf{B}), \quad (13.1.3)$$

$$m_i n_i (\partial_t \mathbf{v}_i + \mathbf{v}_i \cdot \nabla \mathbf{v}_i) = -\nabla p_i + e n_i (\mathbf{E} + \frac{1}{c} \mathbf{v}_i \times \mathbf{B}), \quad (13.1.4)$$

where the right-hand sides include the pressure gradient force and the Lorentz force per unit volume.

The (macroscopic) electric and magnetic fields $\mathbf{E}$ and $\mathbf{B}$ are provided by the Maxwell equations

$$\nabla \cdot \mathbf{E} = 4\pi \rho, \qquad (13.1.5)$$

$$\nabla \cdot \mathbf{B} = 0, \qquad (13.1.6)$$

$$\nabla \times \mathbf{E} = -\frac{1}{c} \partial_t \mathbf{B}, \qquad (13.1.7)$$

$$\nabla \times \mathbf{B} = \frac{4\pi}{c} \mathbf{j} + \frac{1}{c} \partial_t \mathbf{E}, \qquad (13.1.8)$$

where $\rho = -e(n_e - n_i)$ and $\mathbf{j} = -e(n_e \mathbf{v}_e - n_i \mathbf{v}_i)$ are the densities of total charge and total current respectively.

Equations (13.1.7) and (13.1.8) yield

$$\frac{1}{c^2} \partial_{tt} \mathbf{E} + \nabla \times (\nabla \times \mathbf{E}) + \frac{4\pi}{c^2} \partial_t \mathbf{j} = 0, \qquad (13.1.9)$$

where using the equations for the densities and momenta of the two species, we have

$$\partial_t \mathbf{j} = e \left(\nabla \cdot (n_e \mathbf{v}_e) \mathbf{v}_e + n_e \mathbf{v}_e \cdot \nabla \mathbf{v}_e + \frac{1}{m_e} \nabla p_e + \frac{e n_e}{m_e} (\mathbf{E} + \frac{1}{c} \mathbf{v}_e \times \mathbf{B}) \right)$$
$$- e \left(\nabla \cdot (n_i \mathbf{v}_i) \mathbf{v}_i + n_i \mathbf{v}_i \cdot \nabla \mathbf{v}_i + \frac{1}{m_i} \nabla p_i - \frac{e n_i}{m_i} (\mathbf{E} + \frac{1}{c} \mathbf{v}_i \times \mathbf{B}) \right).$$
$$(13.1.10)$$

13.1.2   The Vector Zakharov Equations

A heuristic derivation of the equations governing the envelope dynamics of Langmuir waves in the presence of density fluctuations is presented in Robinson (1997) and Bergé (1998). Here we use a more formal approach based on a multiple-scale modulation analysis. We thus consider a long-wavelength small-amplitude Langmuir oscillation of the form

$$\mathbf{E} = \frac{\epsilon}{2}\left(\mathcal{E}(\mathbf{X},T)e^{-i\omega_e t} + \text{c.c.}\right) + \epsilon^2\overline{\mathcal{E}}(\mathbf{X},T) + \cdots, \qquad (13.1.11)$$

where $\mathbf{X} = \epsilon\mathbf{x}$ and $T = \epsilon^2 t$. It induces fluctuations for the density and velocity of the electrons and of the ions whose dynamical time will be seen to be $\tau = \epsilon t$, thus shorter than T. We write

$$n_e = n_0 + \frac{\epsilon^2}{2}\left(\tilde{n}_e(\mathbf{X},\tau)e^{-i\omega_e t} + \text{c.c.}\right) + \epsilon^2\overline{n}_e(\mathbf{X},\tau) + \cdots, \, (13.1.12)$$

$$n_i = n_0 + \frac{\epsilon^2}{2}\left(\tilde{n}_i(\mathbf{X},\tau)e^{-i\omega_e t} + \text{c.c.}\right) + \epsilon^2\overline{n}_i(\mathbf{X},\tau) + \cdots, \, (13.1.13)$$

$$\mathbf{v}_e = \frac{\epsilon}{2}\left(\tilde{\mathbf{v}}_e(\mathbf{X},\tau)e^{-i\omega_e t} + \text{c.c.}\right) + \epsilon^2\overline{\mathbf{v}}_e(\mathbf{X},\tau) + \cdots, \qquad (13.1.14)$$

$$\mathbf{v}_i = \frac{\epsilon}{2}\left(\tilde{\mathbf{v}}_i(\mathbf{X},\tau)e^{-i\omega_e t} + \text{c.c.}\right) + \epsilon^2\overline{\mathbf{v}}_i(\mathbf{X},\tau) + \cdots, \qquad (13.1.15)$$

where n_0 is the unperturbed plasma density.

In the long-wavelength limit where the particles travel only a fraction of the wavelength in one wave period, the compression of the wave is adiabatic. For small fluctuations, we write $\nabla p_e = \gamma_e T_e \nabla n_e$ and $\nabla p_i = \gamma_i T_i \nabla n_i$, where γ_e and γ_i denote the specific heat ratios of the electrons and the ions and T_e and T_i their respective temperatures in energy units.

It is easily seen from (13.1.3) considered to leading order that the amplitude of the electron velocity oscillations is given by

$$\tilde{\mathbf{v}}_e = -\frac{ie}{m_e\omega_e}\mathcal{E}, \qquad (13.1.16)$$

while the velocity oscillations of the ions is negligible because of their large mass. We thus take

$$\tilde{\mathbf{v}}_i = 0. \qquad (13.1.17)$$

It follows that the density fluctuations are given by

$$\tilde{n}_e = i\frac{n_0}{\omega_e}\nabla\cdot\tilde{\mathbf{v}}_e = \frac{en_0}{m_e\omega_e^2}\nabla\cdot\mathcal{E}, \qquad (13.1.18)$$

$$\tilde{n}_i = 0. \qquad (13.1.19)$$

On the other hand, as seen from (13.1.7), the magnetic field $\mathbf{B}$ is of order ϵ^2 and will not contribute to (13.1.16).

To leading order, the equation for the electric field reduces to

$$-\frac{\omega_e^2}{c^2}\mathcal{E} + i\omega_e\frac{4\pi}{c^2}\frac{ie^2 n_0}{m_e\omega_e}\mathcal{E} = 0, \qquad (13.1.20)$$

from which we get the "electron plasma frequency"

$$\omega_e = \sqrt{\frac{4\pi e^2 n_0}{m_e}}. \tag{13.1.21}$$

At the next order, provided that no large-scale magnetic field is generated, we obtain the amplitude equation in the form

$$-2i\frac{\omega_e}{c^2}\partial_T \mathcal{E} + \nabla \times (\nabla \times \mathcal{E}) - \frac{\gamma_e T_e}{m_e c^2}\nabla(\nabla \cdot \mathcal{E}) + \frac{4\pi e^2}{c^2 m_e}\overline{n}_e \mathcal{E} = 0, \tag{13.1.22}$$

where the contributions of $\mathbf{j}$ originate only from the pressure gradient force of the electron fluid and from the electric force acting on these particles. Due to their large mass, the contribution of the ions is negligible.

We then define the electron thermal velocity as

$$v_e = \sqrt{\frac{T_e}{m_e}} \tag{13.1.23}$$

and note that if we assume a small collision frequency, the change in temperature of the electrons during the compression along the direction of propagation will not be transmitted to the other two directions. It follows that in the formula for the ratio of specific heats $\gamma_e = \frac{2+D}{D}$, the number D of dimensions involved in the increase in temperature is $D = 1$, leading to $\gamma_e = 3$ (Nicholson 1983, Thornhill and ter Haar 1978). The amplitude equation becomes

$$i\partial_T \mathcal{E} - \frac{c^2}{2\omega_e}\nabla \times (\nabla \times \mathcal{E}) + \frac{3v_e^2}{2\omega_e}\nabla \cdot (\nabla \cdot \mathcal{E}) = \frac{\omega_e}{2}\frac{\overline{n}_e}{n_0}\mathcal{E}. \tag{13.1.24}$$

Furthermore, we have for the mean electron density fluctuations $\overline{n}_e$

$$\partial_\tau \overline{n}_e + n_0 \nabla \cdot \overline{\mathbf{v}}_e = 0. \tag{13.1.25}$$

The mean electron velocity $\overline{\mathbf{v}}_e$ obeys

$$m_e\left(\partial_\tau \overline{\mathbf{v}}_e + \frac{1}{4}(\widetilde{\mathbf{v}}_e \cdot \nabla \widetilde{\mathbf{v}}_e^* + \widetilde{\mathbf{v}}_e^* \cdot \nabla \widetilde{\mathbf{v}}_e)\right) = -\frac{\gamma_e T_e}{n_0}\nabla \overline{n}_e - e\overline{\mathbf{E}}, \tag{13.1.26}$$

where

$$\frac{1}{4}(\widetilde{\mathbf{v}}_e \cdot \nabla \widetilde{\mathbf{v}}_e^* + \widetilde{\mathbf{v}}_e^* \cdot \nabla \widetilde{\mathbf{v}}_e) = \frac{e^2}{4m_e^2 \omega_e^2}\nabla |\mathcal{E}|^2, \tag{13.1.27}$$

and the time derivative on the left-hand side is negligible due to the small mass of the electron. Furthermore, $\overline{\mathbf{E}}$ denotes the leading contribution (of order ϵ^3) of the mean electric field. We thus get

$$\frac{e^2}{4m_e \omega_e^2}\nabla |\mathcal{E}|^2 = -\frac{\gamma_e T_e}{n_0}\nabla \overline{n}_e - e\overline{\mathbf{E}}. \tag{13.1.28}$$

In order to express the mean electric field $\overline{\mathbf{E}}$, we consider the equation for the mean ion velocity $\overline{\mathbf{v}}_i$ and the mean ion density fluctuations $\overline{n}_i$,

$$m_i \partial_\tau \overline{\mathbf{v}}_i = -\frac{\gamma_i T_i}{n_0} \nabla \overline{n}_i + e\overline{\mathbf{E}}. \tag{13.1.29}$$

The system is closed by using the quasi-neutrality of the plasma in the form

$$\overline{n}_e = \overline{n}_i, \tag{13.1.30}$$

$$\overline{\mathbf{v}}_e = \overline{\mathbf{v}}_i, \tag{13.1.31}$$

which we denote by $\overline{n}$ and $\overline{\mathbf{v}}$, respectively. Adding (13.1.28) and (13.1.29), we get

$$\partial_\tau \overline{\mathbf{v}} = -\frac{c_s^2}{n_0} \nabla \overline{n} - \frac{1}{4\pi m_i n_0} \nabla |\mathcal{E}|^2, \tag{13.1.32}$$

where the speed of sound c_s is defined by

$$c_s^2 = \eta \frac{T_e}{m_i}, \tag{13.1.33}$$

with

$$\eta = \frac{\gamma_e T_e + \gamma_i T_i}{T_e}. \tag{13.1.34}$$

Equations (13.1.24), (13.1.25), and (13.1.32) yield the vector Zakharov equations (Kuznetsov 1974, Thornhill and ter Haar 1978, Nicholson 1983, Zakharov 1984)

$$i\partial_T \mathcal{E} - \frac{c^2}{2\omega_e} \nabla \times (\nabla \times \mathcal{E}) + \frac{3v_e^2}{2\omega_e} \nabla(\nabla \cdot \mathcal{E}) = \frac{\omega_e}{2} \frac{\overline{n}}{n_0} \mathcal{E}, \tag{13.1.35}$$

$$\epsilon^2 \partial_{TT} \overline{n} - c_s^2 \Delta \, \overline{n} = \frac{1}{16\pi m_i} \Delta |\mathcal{E}|^2. \tag{13.1.36}$$

These equations describe the dynamics of the complex envelope of the electric field oscillations near the electron plasma frequency and the slow variations of the density perturbations. These variations, which affect the effective refractive index felt by the electric field, are driven by the ponderomotive force resulting from the spatial variation of the modulated electric-field amplitude. At least as long as the ion inertia term $\epsilon^2 \partial_{TT} \overline{n}$ is negligible, the ponderomotive force leads to a local reduction of the density and an increase of the refractive index and thus to a focusing of the wave. Whether this effect can be balanced by the wave dispersion depends on the space dimension.

If the temperature of the ions is neglected compared to that of the electrons and, arguing that the motion is slow, the specific ratio γ_e is taken equal to unity (Thornhill and ter Haar 1978), then $\eta = 1$ and c_s can be identified with the ion-sound velocity $(T_e/m_i)^{1/2}$.

The above equations are conveniently written in a nondimensional form. Denoting by

$$\lambda_d = \sqrt{\frac{T_e}{4\pi e^2 n_0}} \qquad (13.1.37)$$

the Debye length and by $\mu = m_e/m_i$ the ratio of the electron to the ion masses, we define the normalized variables

$$T' = \frac{2\eta}{3}\mu\omega_e T, \quad \mathbf{X}' = \frac{2}{3}(\eta\mu)^{1/2}\frac{\mathbf{X}}{\lambda_d}, \qquad (13.1.38)$$

$$\overline{n}' = \frac{3}{4\eta}\frac{1}{\mu}\frac{\overline{n}}{n_0}, \quad \mathcal{E}' = \frac{1}{\eta}\frac{1}{\mu^{1/2}}\left(\frac{3}{64\pi n_0 T_e}\right)^{1/2}\mathcal{E}. \qquad (13.1.39)$$

Defining

$$\alpha = \frac{c^2}{3v_e^2} = \frac{c^2}{3\omega_e^2\lambda_d^2} \qquad (13.1.40)$$

and dropping the primes, we finally get the vector Zakharov equations in a nondimensional form:

$$i\partial_T\mathcal{E} - \alpha\nabla\times(\nabla\times\mathcal{E}) + \nabla(\nabla\cdot\mathcal{E}) = \overline{n}\mathcal{E}, \qquad (13.1.41)$$

$$\epsilon^2\partial_{TT}\overline{n} - \Delta\,\overline{n} = \Delta|\mathcal{E}|^2. \qquad (13.1.42)$$

Typical values of the parameters for astrophysical and laboratory plasmas are given in Thornhill and ter Haar (1978, table 1, page 47). It is noticeable that the parameter α is large in realistic situations, ranging from 20 in laboratory plasmas to 2×10^5 in the interstellar gas. Such extreme regimes are amenable to an asymptotic approximation, as described in Section 13.1.3.

Note that in the "subsonic limit", where the density fluctuations are assumed to follow adiabatically the modulation of the Langmuir wave, we obtain the vector NLS equation

$$i\partial_T\mathcal{E} - \alpha\nabla\times(\nabla\times\mathcal{E}) + \nabla(\nabla\cdot\mathcal{E}) + |\mathcal{E}|^2\mathcal{E} = 0. \qquad (13.1.43)$$

13.1.3 The Electrostatic Limit

A simpler form of (13.1.41)–(13.1.42) is obtained in the limit $v_e^2/c^2 \ll 1$ or $\alpha \to \infty$, which corresponds to a plasma that is not too hot. In this limit, the electric field is close to being potential. A weak solenoidal component is, however, produced by the coupling to the density fluctuations. One thus expands

$$\mathcal{E} = \nabla\psi + \frac{1}{\alpha}\mathcal{E}_1 + \cdots. \qquad (13.1.44)$$

The leading order term $\nabla\psi$ is usually called the "longitudinal component" in the sense that when ψ is monochromatic, this contribution to the electric field is parallel to the associated wave vector. The correction

$\frac{1}{\alpha}\mathcal{E}_1$ contributes to leading order to $\alpha\nabla \times (\nabla \times \mathcal{E})$ and should thus be retained. Nevertheless, a close equation for the electric potential ψ is obtained by taking the divergence of the resulting amplitude equation, in the form (Zakharov 1972, Zakharov, Mastryukov, and Synakh 1975)

$$\Delta(i\partial_T\psi + \Delta\psi) = \nabla \cdot (n\nabla\psi), \qquad (13.1.45)$$

$$\epsilon^2 \partial_{TT} n - \Delta n = \Delta(|\nabla\psi|^2), \qquad (13.1.46)$$

which describes the interaction of the electrostatic potential with the plasma density in the limit of large α. The above system is usually called the Zakharov equations.

Equation (13.1.45) can be rewritten as

$$i\partial_t\nabla\psi + \Delta(\nabla\psi) = \mathbf{H}(n\nabla\psi), \qquad (13.1.47)$$

where $\mathbf{H}$ denotes the zeroth-order operator $\Delta^{-1}\text{grad div}$. In one space dimension, where the operator $\mathbf{H}$ reduces to the identity, the Zakharov equations admit solitary waves that are, however, not proper solitons, since they are not stable relative to collisions (Gibbons, Thornhill, Wardrop, and ter Haar 1977). As in the case of the NLS equation, these structures are unstable relatively to large-scale transverse perturbations both in the general context and the electrostatic limit (Degtyarev, Zakharov, and Rudakov 1975, Schmidt 1975, Pereira, Sudan, and Denavit 1977, Laedke and Spatschek 1978, 1979, Wardrop and ter Haar 1979, Zakharov 1984, Kuznetsov, Rubenchik, and Zakharov 1986).

Both in laboratory and space plasmas, an ambient magnetic field is usually present and affects the Langmuir waves (see, e.g., Pelletier, Sol, and Asseo 1988 and Newman, Goldman, and Ergun 1994). It is shown that in the case where this field is weak, (13.1.45) is replaced by

$$\Delta(i\partial_T\psi + \Delta\psi) - \sigma\Delta_\perp\psi + \nabla \cdot (n\nabla\psi) = 0, \qquad (13.1.48)$$

where σ is a nondimensional parameter proportional to the magnitude of the external magnetic field. The transverse Laplacian is associated to the directions perpendicular to this field. The ion dynamics are governed by (13.1.46). In the adiabatic limit, one has

$$\Delta(i\partial_T\psi + \Delta\psi) - \sigma\Delta_\perp\psi + \nabla \cdot (|\nabla\psi|^2\nabla\psi) = 0. \qquad (13.1.49)$$

Several studies have been devoted to this problem, characterized by a strong anisotropy, among them Krasnosel'skikh and Sotnolov (1987), Rolland and Tagare (1982), Assalauov and Zakharov (1985), Kuznetsov and Turitsyn (1990), Hadžievski, Škorić, Rubenchik, Shapiro, and Turitsin (1990), Hadžievski, and Škorić (1993). Note that related envelope equations also arise in other physical contexts that we do not review here, like upper-hybrid and lower-hybrid waves in a weakly magnetized plasma (see, e.g., Musher and Sturman 1975, Sturman 1976, Kuznetsov and Škorić 1988, Shapiro, Shevshenko, Solov'vev, Kalimin, Bingham, Sagdeev, Ashour-Abdalla, Dawson, and Su 1993).

13.1.4   Generation of a Large-Scale Magnetic Field

As noticed by several authors (Bel'kov and Tsytovich 1979, Kono, Škorić, and ter Haar 1981, ter Haar , and Tsytovich 1981, Li 1993), the above derivation overlooked the effect of a magnetic field $\epsilon^2 \overline{\mathbf{B}}$ that develops at large scale under the effect of the modulation. Before estimating such a field, we note that it will affect the contribution of $\partial_t \mathbf{j}$ to the amplitude equation, which when written in the dimensional form (13.1.22) will include on the left-hand side an additional term,

$$S = \frac{4\pi e^2 n_e}{c^3 m_e} \widetilde{\mathbf{v}}_e \times \overline{\mathbf{B}} = -i \frac{4\pi n_e e^3}{c^3 m_e^2 \omega_e} \mathcal{E} \times \overline{\mathbf{B}} = -i \frac{e\omega_e}{c^3 m_e} \mathcal{E} \times \overline{\mathbf{B}}. \quad (13.1.50)$$

The mean magnetic field is given by

$$\nabla \times \overline{\mathbf{B}} = \frac{4\pi}{c} \overline{\mathbf{j}} + \frac{1}{c} \frac{\partial \overline{\mathbf{E}}}{\partial \tau}. \quad (13.1.51)$$

By the Ampère law, the magnetic field obeys

$$\frac{\epsilon^2}{c^2} \partial_{TT} \overline{\mathbf{B}} + \nabla \times (\nabla \times \overline{\mathbf{B}}) = \frac{4\pi}{c} \nabla \times \overline{\mathbf{j}}. \quad (13.1.52)$$

Proceeding as in Bel'kov and Tsytovich (1979) and Kono, Škorić, and ter Harr (1981), we neglect the displacement current $\frac{1}{c} \partial_\tau \overline{\mathbf{E}}$, thus assuming adiabatic dynamics of the magnetic field. Making this assumption while keeping the full dynamics of the density fluctuations is justified by the smallness of the ion acoustic velocity compared to the speed of light ($c_s \ll c$). We thus write

$$\nabla \times \overline{\mathbf{B}} = \frac{4\pi}{c} \overline{\mathbf{j}}. \quad (13.1.53)$$

Denoting by $\overline{\mathbf{v}}_i^{(3)}$ and $\overline{\mathbf{v}}_e^{(3)}$ the third-order contributions to the mean velocity of the ions and of the electrons, respectively, the mean current is given by

$$\overline{\mathbf{j}} = -\frac{e}{4} (\widetilde{n}_e \widetilde{\mathbf{v}}_e^* + \text{c.c.}) + e n_0 (\overline{\mathbf{v}}_i^{(3)} - \overline{\mathbf{v}}_e^{(3)}), \quad (13.1.54)$$

since the oscillating velocity of the ions is negligible and the mean velocity of the ions and that of the electrons are equal to leading order. This last point contrasts with the assertion of Kono, Škorić, and ter Haar (1981), who consider the mean velocity of the ions negligible compared to that of the electrons. Furthermore, the contribution of the mean velocity difference of the ions and the electrons is omitted in Li (1993) who, however, retained subdominant nonlinearities.

It is easily seen from (13.1.16) and (13.1.18) that

$$\widetilde{n}_e \widetilde{\mathbf{v}}_e^* + \text{c.c.} = \frac{-i}{4\pi m_e \omega_e} \left((\nabla \cdot \mathcal{E}) \mathcal{E}^* - \text{c.c.} \right). \quad (13.1.55)$$

The contribution of the mean velocities is delicate to compute precisely because it involves high-order terms in the expansion, but it is needed to ensure the solenoidal character of $\bar{\jmath}$. Clearly, this (nonoscillating) contribution arising at order ϵ^3 should be linear with respect to $\mathcal{E}$, $\mathcal{E}^*$, and ∇. This suggests that

$$en_0(\overline{\mathbf{v}}_i^{(3)} - \overline{\mathbf{v}}_e^{(3)}) = \frac{-ie}{16\pi m_e \omega_e}((\mathcal{E} \cdot \nabla)\mathcal{E}^* - (\mathcal{E}^* \cdot \nabla)\mathcal{E}) \qquad (13.1.56)$$

in such a way that

$$\bar{\jmath} = \frac{-ie}{16\pi m_e \omega_e}\nabla \times (\mathcal{E} \times \mathcal{E}^*). \qquad (13.1.57)$$

A derivation of this expression for the current $\bar{\jmath}$ from the Vlasov equation, is presented in Bel'kov and Tsytovich (1979) and Kono, Škorić, and ter Haar (1980, 1981).

It follows that (see also Relke and Rubenchik 1988)

$$\mathbf{B} = \frac{-ie}{4cm_e \omega_e}(\mathcal{E} \times \mathcal{E}^*). \qquad (13.1.58)$$

As a consequence, the amplitude equation (13.1.22) includes an additional term

$$S = \frac{-e^2}{4c^4 m_e^2}\mathcal{E} \times (\mathcal{E} \times \mathcal{E}^*). \qquad (13.1.59)$$

In this context, the system (13.1.41)–(13.1.42) for electromagnetic oscillations is replaced by

$$i\partial_T \mathcal{E} - \alpha \nabla \times (\nabla \times \mathcal{E}) + \nabla(\nabla \cdot \mathcal{E}) = \bar{n}\mathcal{E} + \beta \mathcal{E} \times (\mathcal{E}^* \times \mathcal{E}),$$
$$(13.1.60)$$

$$\epsilon^2 \partial_{TT} \bar{n} - \Delta \bar{n} = \Delta |\mathcal{E}|^2, \qquad (13.1.61)$$

where $\beta = \eta v_e^2/c^2 = \eta/(3\alpha)$. As expected, the additional term is thus negligible in the electrostatic limit.

The subsonic limit thus takes the form of the complete vector NLS equation

$$i\partial_T \mathcal{E} + \nabla(\nabla \cdot \mathcal{E}) - \alpha \nabla \times (\nabla \times \mathcal{E}) + |\mathcal{E}|^2 \mathcal{E} - \beta \, \mathcal{E} \times (\mathcal{E}^* \times \mathcal{E}) = 0, \quad (13.1.62)$$

which appears to be the generic amplitude equation for a weakly nonlinear vector wave train, as expected from symmetry arguments.

13.2   Rigorous Results

13.2.1   Existence Theory

Neglecting the possible effect of the self-generated magnetic field, we write the vector Zakharov equations in the equivalent forms

$$i\partial_t \mathbf{E} - \alpha \nabla \times (\nabla \times \mathbf{E}) + \nabla(\nabla \cdot \mathbf{E}) = n\mathbf{E}, \tag{13.2.1}$$

$$\partial_{tt} n - \Delta n = \Delta |\mathbf{E}|^2 \tag{13.2.2}$$

or

$$i\partial_t \mathbf{E} - \alpha \nabla \times (\nabla \times \mathbf{E}) + \nabla(\nabla \cdot \mathbf{E}) = n\mathbf{E}, \tag{13.2.3}$$

$$\partial_t n + \nabla \cdot \mathbf{v} = 0, \tag{13.2.4}$$

$$\partial_t \mathbf{v} + \nabla(n + |\mathbf{E}|^2) = 0. \tag{13.2.5}$$

Since in the present asymptotics the velocity is irrotational, we can also introduce the hydrodynamic potential U such that $\mathbf{v} = -\nabla U$. We then have

$$\partial_t U = n + |\mathbf{E}|^2 \tag{13.2.6}$$

and

$$\partial_t n = \Delta U. \tag{13.2.7}$$

Furthermore, we denote by $\mathbf{E}_0(\mathbf{x})$, $n_0(\mathbf{x})$, and $n_1(\mathbf{x})$ the initial values of $\mathbf{E}$, n, and $\partial_t n$, respectively. The Zakharov equations derive from a Lagrangian density (Gibbons, Thornhill, Wardrop , and ter Haar 1977, ter Haar 1979)

$$\mathcal{L} = \frac{i}{2}\{(\mathbf{E}^* \cdot \partial_t \mathbf{E} - \partial_t \mathbf{E}^* \cdot \mathbf{E})\} - \alpha|\nabla \times \mathbf{E}|^2 - |\nabla \cdot \mathbf{E}|^2 + \frac{1}{2}\{(\partial_t U - |\mathbf{E}|^2)^2 - |\nabla U|^2\}. \tag{13.2.8}$$

Several conserved quantities result from the associated invariance properties. In particular, if $(\mathbf{E}, n)$ is a smooth solution of (13.2.1)–(13.2.2), the wave energy (also called the number of Langmuir quanta) $N = |\mathbf{E}|_{L^2}^2$ and the Hamiltonian

$$H = \alpha|\nabla \times \mathbf{E}|_{L^2}^2 + |\nabla \cdot \mathbf{E}|_{L^2}^2 + \frac{1}{2}|n|_{L^2}^2 + \frac{1}{2}|\nabla U|_{L^2}^2 + \int n|\mathbf{E}|^2 d\mathbf{x} \tag{13.2.9}$$

are conserved. Other invariants are the linear and angular momenta $\mathbf{P}$ and $\mathbf{M}$ defined by

$$\mathbf{P} = \int \left(\frac{i}{2}\sum_j (E_j \nabla E_j^* - E_j^* \nabla E_j) + n\mathbf{v}\right) d\mathbf{x} \tag{13.2.10}$$

and

$$\mathbf{M} = \int (i\mathbf{E} \times \mathbf{E}^* + \mathbf{x} \times \mathbf{P}) d\mathbf{x}. \tag{13.2.11}$$

Denoting by $\nabla \mathbf{E}$ the tensor of elements $\frac{\partial E_i}{\partial x_j}$, we have the estimate, valid in dimension $d \leq 3$,

$$|\nabla \mathbf{E}|^2_{L^2} + \frac{1}{4}|n|^2_{L^2} + \frac{1}{2}|\mathbf{v}|^2_{L^2} \leq |H| + c_d\, N^{(4-d)/2}|\nabla \mathbf{E}|^d_{L^2} \qquad (13.2.12)$$

with $c_1 = 4$, $c_3 = 24$. In dimension 2, the best constant c_2 is obtained using Weinstein's estimate (see Section 3.2.1) in the form $c_2 = 1/|R|_{L^2}$, where R denotes the positive localized solution (ground state) of $\Delta R - R + R^3 = 0$. This leads to the following theorem.

Theorem 13.1. (Sulem and Sulem 1979) *For initial conditions $\mathbf{E}_0 \in (H^1(\mathbf{R}^d))^d$, $n_0 \in L^2(\mathbf{R}^d)$, and $n_1 \in H^{-1}(\mathbf{R}^d)$, arbitrary in dimension 1 and obeying $|\mathbf{E}_0|_{L^2} < |R|_{L^2}$ in dimension 2 or $N|H| < 4/(27c_3^2)$ and $|\nabla \mathbf{E}_0|_{L^2} \leq |H|$ in dimension 3, there exists a weak solution $\mathbf{E} \in L^\infty(\mathbf{R}^+, (H^1(\mathbf{R}^d))^d)$, $n \in L^\infty(\mathbf{R}^+, (L^2(\mathbf{R}^d))$, to (13.2.1)–(13.2.2).*

Remark. As shown by Added and Added (1984), this result can be improved in two dimensions in the special case $\alpha = 1$, where the linear operator in (13.2.1) reduces to a Laplacian (a case relevant for the scalar model discussed in Chapter 14), by use of the inequality (Brezis and Gallouët 1980) $|u|_{L^\infty} \leq c(1+|u|_{H^1}(\ln(1+|\Delta u|_{L^2})))$, valid if $u \in H^2(\mathbf{R}^2)$. In this case, if $|\mathbf{E}_0|_{L^2} < |R|_{L^2}$, there exists a unique solution $\mathbf{E} \in L^\infty_{loc}(\mathbf{R}^+, (H^m(\mathbf{R}^2))^2)$, $n \in L^\infty_{loc}(\mathbf{R}^+, H^{m-1}(\mathbf{R}^2))$ for (13.2.1)–(13.2.2).

Theorem 13.2. *In one dimension, for initial conditions, $\mathbf{E}_0 \in H^m(\mathbf{R})$, $n_0 \in H^{m-1}(\mathbf{R})$, and $n_1 \in H^{m-2}(\mathbf{R})$ with $m \leq 3$, there exists a unique solution $\mathbf{E} \in L^\infty(\mathbf{R}^+, H^m(\mathbf{R}))$, $n \in L^\infty(\mathbf{R}^+, H^{m-1}(\mathbf{R}))$ for (13.2.1)–(13.2.2).*

Theorem 13.3. *In dimensions 2 and 3, for initial conditions $\mathbf{E}_0 \in (H^m(\mathbf{R}^d))^d$, $n_0 \in H^{m-1}(\mathbf{R}^d)$, and $n_1 \in H^{m-2}(\mathbf{R}^d)$ with $m \leq 3$, there exists a unique solution $\mathbf{E} \in L^\infty([0, T^*), (H^m(\mathbf{R}^d))^d)$, $n \in L^\infty([0, T^*), H^{m-1}(\mathbf{R}^d))$ for (13.2.1)–(13.2.2), where the time T^* depends on the initial conditions.*

Remark. These results have been extended by Laurey (1995) to the case where the Zakharov system is coupled to a driven magnetic field $\mathbf{B}$ (see Section 13.1.4) by an additional term $-i\mathbf{E} \times \mathbf{B}$ in the right-hand side of (13.2.1) and where $\mathbf{B}$ obeys

$$\Delta \mathbf{B} - i\eta \nabla \times (\mathbf{E} \times \mathbf{E}^*) + \mathbf{A} = 0,$$

with the additional vector valued function $\mathbf{A}$ defined as either $\mathbf{A} = \beta \mathbf{B}$, and β a nonpositive constant, or $\mathbf{A} = -\gamma \partial_t \int \mathbf{B}(\mathbf{y}, t)|\mathbf{x} - \mathbf{y}|^{-2}d\mathbf{y}$.

13.2.2   The Subsonic Limit

In the electrostatic approximation ($\alpha \gg 1$), the subsonic limit (adiabatic dynamics of the ions) of the Zakharov equations reads

$$i\partial_T(\nabla\psi) + \Delta(\nabla\psi) + \Delta^{-1}\text{grad div}(|\nabla\psi|^2\nabla\psi) = 0, \qquad (13.2.13)$$

or

$$\Delta(i\partial_T\psi + \Delta\psi) + \nabla \cdot (|\nabla\psi|^2\nabla\psi) = 0. \qquad (13.2.14)$$

Equation (13.2.14) was considered by Colin (1993a), who proved the existence locally in time of a unique solution ψ in $C([0, T[, \mathcal{H})$ where T depends on the initial condition ψ_0 and $\mathcal{H} = \{\psi \in L^6 \cap C_0(\mathbf{R}^3)), \nabla\psi \in H^1)\}$ equipped with the norm $|\psi|_{\mathcal{H}} = |\nabla\psi|_{H^1}$. Furthermore, if $\psi_0 \in \mathcal{H}$ with $\nabla\psi_0 \in H^2$, the solution satisfies $\nabla\psi \in C([0, T), H^2)$. If the initial norm $|\psi_0|_{\mathcal{H}} = |\nabla\psi_0|_{H^1}$ is small enough, this solution is global in time. As in the case of the usual scalar NLS equation, (13.2.14) admits standing wave solutions that are unstable in the physical case of three space dimensions and cubic nonlinearity (Colin 1993b). Furthermore, (13.2.13) simplifies in the case of radially symmetric solutions. Since in this geometry any radially symmetric vector field is a gradient, the nonlocal operator $\Delta^{-1}\text{grad div}$ then disappears. A variance estimate and the existence of solutions blowing up in a finite time are then easily obtained (Degtyarev, Zakharov, and Rudakov 1975, Colin 1993a).

The validity of the electrostatic limit in the subsonic regime was recently addressed by Galusinski (1998), who establishes a convergence result for solutions of (13.1.43) to solutions of (13.2.13) as $\alpha \to \infty$, during their existence time.

13.2.3   The Vector NLS Equation

A Modified Variance Identity

Omitting for the sake of simplicity the skew nonlinear term associated to the magnetic field generation, we establish for the vector NLS equation (13.1.43) a "modified variance identity" that simplifies in the special case $\alpha = 1$, where it leads to a blowup result (Goldman and Nicholson 1978, Goldman, Rypdal, and Hafizi 1980). The existence of blowup solutions for other values of α is not proved.

Proposition 13.4. *Solutions of the vector NLS equation (13.1.43) obey the modified variance identity*

$$\frac{d^2}{dt^2}\left(\int |\mathbf{x}|^2|\mathbf{E}|^2 d\mathbf{x} - (1-\alpha)\int_0^t \int x_j K_j d\mathbf{x}\,dt\right) = -2\alpha \int \text{tr}\,\mathbf{T}\,d\mathbf{x}$$
$$(13.2.15)$$

with $\mathcal{K}_j = 2 \, \Im(E_j^* \nabla \cdot \mathbf{E})$ and

$$\int \operatorname{tr} \mathbf{T} \, dx = -2dH - 2\alpha(2-d)\left(\alpha|\nabla \mathbf{E}|_{L^2}^2 - \frac{1-\alpha}{2\alpha}|\nabla \cdot \mathbf{E}|_{L^2}^2\right), \quad (13.2.16)$$

where $H = \alpha|\nabla \mathbf{E}|_{L^2}^2 + (1-\alpha)|\nabla \cdot \mathbf{E}|_{L^2}^2 - \frac{1}{2}|\mathbf{E}|_{L^4}^4$ is the Hamiltonian.

Proof. It is convenient to rewrite (13.1.43) in the form

$$i\partial_t \mathbf{E} + \alpha\Delta \mathbf{E} + (1-\alpha)\nabla(\nabla \cdot \mathbf{E}) + |\mathbf{E}|^2 \mathbf{E} = 0. \quad (13.2.17)$$

The continuity equation for the "mass" reads

$$\frac{\partial}{\partial t}|\mathbf{E}|^2 + \nabla \cdot \mathcal{J} = 0. \quad (13.2.18)$$

Here $\mathcal{J} = \alpha\mathcal{P} + (1-\alpha)\mathcal{K}$, where $\mathcal{P}_j = 2 \, \Im(\mathbf{E}^* \cdot \partial_j \mathbf{E})$ are the components of the momentum density and $\mathcal{K}_j = 2 \, \Im(E_j^* \nabla \cdot \mathbf{E})$. The equation for the momentum is

$$\frac{\partial \mathcal{P}_j}{\partial t} = \partial_i T_{ij} \quad (13.2.19)$$

with

$$T_{ij} = -2\alpha \, \Re(\partial_i \mathbf{E}^* \, \partial_j \mathbf{E}) + 2\alpha \, \Re(\mathbf{E}^* \, \partial_{ij} \mathbf{E}) - 2(1-\alpha) \, \Re(\nabla \cdot \mathbf{E}^* \, \partial_j E_i)$$
$$+ 2(1-\alpha) \, \Re(\partial_j(\nabla \cdot \mathbf{E}) \, E_i^*) + \delta_{ij} \, |\mathbf{E}|^4. \quad (13.2.20)$$

To obtain (13.2.15), one writes

$$\int |\mathbf{x}|^2 \partial_j(\partial_t K_j) dx = -2\partial_{tt}\left(\int_0^t \int x_j K_j dx \, dt\right) \quad (13.2.21)$$

and $\int |\mathbf{x}|^2 \partial_{ij} T_{ij} \, dx = 2 \int \operatorname{tr} \mathbf{T} dx$ with

$$\int \operatorname{tr} \mathbf{T} \, d\mathbf{x} = -4\alpha \int |\nabla \mathbf{E}|^2 dx - 4(1-\alpha)\int |\nabla \cdot \mathbf{E}|^2 dx + d\int |\mathbf{E}|^4 \, dx. \quad (13.2.22)$$

Equation (13.2.15) follows.

Remark. Existence of blowup solutions is not known when $\alpha \neq 1$ because the modified variance does not have a fixed sign. In contrast, when $\alpha = 1$, (13.2.15) reduces to

$$\frac{d^2}{dt^2}\int |\mathbf{x}|^2 |\mathbf{E}|^2 dx = 4dH + 4(2-d)\int |\nabla \mathbf{E}|^2 dx, \quad (13.2.23)$$

and the usual blowup analysis is easily done. This case can actually be viewed as a system of nonlinearly coupled scalar NLS equations, a problem also addressed in a different context by McKinstrie and Russel (1988). The case where the additional skew nonlinear term $-\beta \mathbf{E} \times (\mathbf{E} \times \mathbf{E}^*)$ is included in the vector NLS equation as in (13.1.62) is considered by Kono, Škorić, and ter Haar (1980, 1981) in a setting corresponding, in fact, to $\alpha = 1$. This leads to an equation for the variance itself and again to the existence of solutions blowing up in finite time.

Existence of Standing Waves

The existence of standing waves for the vector NLS equation (13.1.43) was addressed by Colin and Weinstein (1996) and discussed in terms of the parameter α. For $\alpha \neq 1$, they showed that in dimension 2 and 3, there exists a number α_* such that if $\alpha < \alpha_*$, the ground state is not the gradient of a radial function.

Bound states of the vector NLS equation (13.1.62) with $\alpha = 1$ but including the skew nonlinear term originating from the magnetic field generated at large scale were considered by Stubbe and Vázquez (1989). They conclude that in dimensions $d = 2$ and 3, the ground state, which displays a generalized rotational symmetry, is unstable. Ground states for the vector NLS equation with saturating nonlinearity were considered by Laedke and Spatschek (1984).

13.3 Evidence of Collapse

13.3.1 Heuristic Arguments

No rigorous results concerning the existence of a finite-time singularity are presently available for the vector Zakharov equations. In this section, following Zakharov, Mastryukov, and Synakh (1975), we argue that a blowup is expected in dimension $d \geq 2$.

We first note that the Hamiltonian H can be negative only if the density perturbation n is sufficiently negative. In this case, $\int n|\mathbf{E}|^2 d\mathbf{x} < H < 0$. Replacing n by its minimal value $n_{\min}$, one has $n_{\min} \int |\mathbf{E}|^2 d\mathbf{x} < H < 0$ or $|n_{min}| > \frac{|H|}{N}$. Since H and N are conserved, it follows that in the case $H < 0$, the density well produced by the plasma wave is maintained whatever the evolution of the wave packet trapped in it.

In situations where the plasma wave packet tends to a standing wave $n = -|\mathbf{E}|^2$, $\mathbf{E}_t = i\lambda\mathbf{E}$ after emitting part of its energy, one has

$$-\lambda^2 \mathbf{E} + \nabla(\nabla \cdot \mathbf{E}) - \alpha\nabla \times (\nabla \times \mathbf{E}) + |\mathbf{E}|^2\mathbf{E} = 0. \tag{13.3.1}$$

Multiplying by $\mathbf{E}^*$ or by $(\mathbf{x} \cdot \nabla)\mathbf{E}^*$, combining the results with the complex conjugates, and integrating, one obtains

$$\lambda^2 N + \int (|\nabla \cdot \mathbf{E}|^2 + \alpha|\nabla \times \mathbf{E}|^2 - |\mathbf{E}|^4)d\mathbf{x} = 0 \tag{13.3.2}$$

and

$$d\lambda^2 N + (d-2)\int(|\nabla \cdot \mathbf{E}|^2 + \alpha|\nabla \times \mathbf{E}|^2)d\mathbf{x} - \frac{1}{2}N\int|\mathbf{E}|^4 d\mathbf{x} = 0. \tag{13.3.3}$$

Since for a standing wave $H = \int(|\nabla \cdot \mathbf{E}|^2 + \alpha|\nabla \times \mathbf{E}|^2 - \frac{1}{2}|\mathbf{E}|^4)d\mathbf{x}$, it follows that in this case $H = \frac{d-2}{4-d}\lambda^2 N$, indicating that for $d = 1$, one has $H = -\frac{\lambda^2}{3}N$, while for $d = 2$, $H = 0$ holds and for $d = 3$, $H = \lambda^2 N$.

During the evolution of the plasma wave packet, the Hamiltonian in the density well can only decrease as the result of sound emission. If it was initially negative, it remains so. As a consequence, in two and three dimensions, a wave packet with a negative Hamiltonian cannot evolve to a standing wave. Such a regime can be established only in one-dimension, where a solitary wave can form (Gibbons, Thornhill, Wardrop, and ter Haar 1977).

Furthermore, an oscillatory behavior is also not expected, since in this case the presence of a localized oscillating source in the wave equation would lead to an unlimited radiation of acoustic energy from the density well and again to a decrease of the Hamiltonian. The existence of a finite-time singularity in dimension $d \geq 2$ is thus suspected in the case of a negative Hamiltonian.

13.3.2 Simulations in the Electrostatic Approximation

Extensive efforts both analytical and numerical have been devoted to the understanding of Langmuir collapse in the electrostatic approximation (Zakharov 1972) governed by (13.1.45)–(13.1.46). A review is presented in Rudakov and Tsytovich (1978), Zakharov (1984), Rubenchik, Sagdeev, and Zakharov 1985, Robinson (1997). If the potential is assumed to be radially symmetric, the electric field vanishes at the origin, a configuration that is not realistic. The wave packet takes the form of a spherical layer collapsing to the origin, a configuration unstable with respect to perturbations breaking the spherical symmetry (Degtyarev, Zakharov, and Rudakov 1975).

Collapse was also simulated in a fully two-dimensional geometry and in three dimensions assuming an axial symmetry with the axis directed along the initial electric field (Degtyarev, Zakharov, and Rudakov 1976). These calculations show the convergence of the collapsing solutions to the self-similar form

$$\psi = \frac{1}{(t_* - t)^{1-2/d}} g(\boldsymbol{\xi}) e^{i \int^t \lambda(t')dt'}, \quad n = \frac{V(\boldsymbol{\xi})}{(t_* - t)^{4/d}}, \qquad (13.3.4)$$

with

$$\lambda(t) = \frac{\lambda_0}{(t_* - t)^{4/d}}, \quad \boldsymbol{\xi} = \frac{\mathbf{x}}{(t_* - t)^{2/d}}, \qquad (13.3.5)$$

the profiles taking a dipolar form: The collapsing cavity has an asymmetrical oblate shape with the electric field at the cavity center directed along the short size (see Dyachenko, Pushkarev, Rubenchik, Sagdeev, Shvets, and Zakharov 1991 for a detailed description).

Three-dimensional simulations are reported by Robinson, Newman and Goldman (1988), Newman, Robinson, and Goldman (1989, 1991), Robinson and Newman (1990), who conclude to the formation of collapsing structures that are predominantly pancake-shaped. Such solutions were also analyzed by Malkin, Khudik, and Fedoruk (1990).

Experimental observations of Langmuir collapse with transverse contraction rates of the wave packet following the self-similar scaling are reported by Wong and Cheung (1984) (see ter Haar and Tsytovich 1981 for references to previous experimental work). Predictions of the Zakharov equations were also favorably compared against direct simulations of the Vlasov equation until the moment where the self-focusing of the Langmuir oscillation takes the system out of the region of validity of the asymptotic description (Sigov and Zakharov 1979, Rowland, Lyon, and Papadopoulos 1981, Anisimov, Berezovskiĭ, Ivanov, Petrov, Rubenchik, and Zakharov 1982). Comparison of numerical simulations of the two-fluid equations with the Zakharov model are reported in one space dimension by Nicholson, Payne, Sheerin, and Mei-Mei Shen (1991) who observe the same qualitative behavior when the electric field magnitude is in the range of validity of the model, while for larger fields, they conclude that Langmuir wave breaking can possibly occur.

13.3.3 Simulations of the Vector Equations

Numerical integrations of (13.1.41)–(13.1.42) and of (13.1.43) in three space dimensions are presented in Papanicolaou, Sulem, Sulem, and Wang (1991) for various initial conditions associated to a negative Hamiltonian. In dealing with the Zakharov system, the initial density fluctuations were taken as $n_0 = -|\mathbf{E}_0|^2$ with $\partial_t n_0 = 0$. In all the computations, $\alpha = 0.5$. The development of the collapse is accurately followed by using the dynamic rescaling method where a change of the dependent and independent variables is introduced in order that the system actually solved numerically be nonsingular.

Let $D(t)$ be a 3×3 matrix function of time, $\mathbf{x}_0(t)$ a vector function of time and $\Omega(t)$ a positive scalar function. One writes

$$\xi = D^{-1}(t)(\mathbf{x} - \mathbf{x}_0), \quad \tau = \int_0^t \frac{1}{\Omega^2(s)} ds, \qquad (13.3.6)$$

$$\mathbf{E}(\mathbf{x}, t) = \frac{O(t)^T}{L(t)} \mathbf{u}(\boldsymbol{\xi}, \tau), \qquad (13.3.7)$$

$$n(\mathbf{x}, t) = \frac{1}{M(t)} v(\boldsymbol{\xi}, \tau), \qquad (13.3.8)$$

$$n_t(\mathbf{x}, t) = \frac{1}{N(t)} w(\boldsymbol{\xi}, \tau), \qquad (13.3.9)$$

where the matrix $D(t)$ has the form

$$D(t) = O^T(t)\Lambda(t) \qquad (13.3.10)$$

with $O(t)$ an orthogonal matrix and $\Lambda(t)$ a diagonal matrix whose diagonal elements are $\lambda_i(i = 1, \ldots, 3)$. One takes

$$x_0^i = \frac{\int x_i |\mathbf{E}|^{2p} d\mathbf{x}}{\int |\mathbf{E}|^{2p} d\mathbf{x}} \qquad (13.3.11)$$

to be the centroid defined by the $2p$ power ($p = 3$ was used) of $|\mathbf{E}|$, which for large τ will be located near the blowup point. The matrix $D(t)$ is chosen such that

$$\frac{\int \xi_i \xi_j |\mathbf{u}|^{2p} d\boldsymbol{\xi}}{\int |\mathbf{u}|^{2p} d\boldsymbol{\xi}} = \delta_{ij}. \qquad (13.3.12)$$

With this condition, the second moment of $|\mathbf{u}|^{2p}$ is the identity matrix or, using (13.3.7),

$$D^{-1} S (D^{-1})^T = I, \qquad (13.3.13)$$

where $S = (s_{ij})$ and

$$s_{ij} = \frac{\int (x^i - x_0^i)(x^j - x_0^j) |\mathbf{E}|^{2p} d\mathbf{x}}{\int |\mathbf{E}|^{2p} d\mathbf{x}}. \qquad (13.3.14)$$

This leads to a closed system for the rescaled fields $\mathbf{u}$, v, and w which is given in the original paper.

The scaling functions are chosen such that the various terms of the rescaled equation remain finite. One sets $M(t) = \Omega^2(t)$, $N(t) = \Omega^4(t)$, and $\Omega^2(t) = 3/\sum(1/\lambda_i^2(t))$. The relation, between L and Ω depends on the regime that is considered. When ϵ is of order unity, the condition $L(t) = \Omega^{3/2}(t)$ was prescribed, while for the vector NLS equation $L(t) = \Omega(t)$ was used. The near subsonic limit is more delicate. When ϵ is very small, the correct scaling at early times is $L(t) = \Omega(t)$, which is consistent with the scaling of the vector NLS equation. As seen in the following, at later times, the system evolves to the supersonic regime, and the scaling must be changed to $L(t) = \Omega^{3/2}(t)$.

Simulations were performed with an initial electric field $\mathbf{E}_0$ whose three components are equal, with an intensity isotropically distributed around the origin, and $\epsilon = 1$. For both the Zakharov and the vector NLS equations, the amplitude of the solutions develop a mild anisotropy when the collapse is approached, an effect also observed by Robinson, Newman, and Goldman (1988) and Newman, Robinson, and Goldman (1989) in the electrostatic case. The difference between the Zakharov and the vector NLS equations appears in considering the evolution of the coefficients $a_{ii} = -\frac{d}{d\tau}(\ln \lambda_i)$, which remain equal but tend to zero like $2/\tau$ for the Zakharov equations, while they converge to a constant for the vector NLS equation. This reflects the different rates of blowup of the solution in the two regimes: the coefficients λ_i behave like $(t_* - t)^{2/3}$ for the Zakharov equation, while for the vector NLS equation they behave like $(t_* - t)^{1/2}$, as in the scalar problem. For both equations, the contours of the wave amplitude are ellipsoids pointing in the direction of the initial electric field. The distribution of the electric field intensity is thus weakly anisotropic in both problems, although the anisotropy is weaker for vector NLS.

When the Zakharov equations are solved for a very small value of the parameter ϵ in front of the density time derivatives $\partial_{tt}n$, using the same initial conditions as above, the dynamics are subsonic at early times. The λ_i's are then equal and behave like $(t_* - t)^{1/2}$. However, sufficiently close to the singularity, Δn becomes negligible compared to $\partial_{tt}n$, and the λ_i's, which are still equal, behave like $(t_* - t)^{2/3}$.

For initial conditions corresponding to an anisotropic intensity distribution, the solution of the Zakharov equations displays a long anisotropic transient before reaching a stable profile where the contours are again ellipsoidal. The rates of collapse are the same in all directions, the λ_i's all behaving like $(t_* - t)^{2/3}$. In this case, the vector NLS equation develops a very strong anisotropy at early times and then breaks apart into two peaks. A similar result is observed with the Zakharov equation with a very small ϵ. The dynamic rescaling method in its present formulation is unable to follow such a solution further. However, the computations do give information about the behavior of the solution up to the breakup. In particular, one of the normalized scaling factors λ_i converges to zero much faster than the others.

It appears that increasing the initial anisotropy makes the dynamics more violent. Furthermore, the process of collapse is somewhat slower and less violent for the Zakharov equations than for the vector NLS equation. This can be seen directly from the equations. Indeed, as the solution tends to blowup, it weakens the ponderomotive force, especially in the early stage, when the solution has not yet stabilized, i.e., when the self-similar regime is not yet reached. For the vector NLS equation, the nonlinear term is always enhanced as the solution blows up. It was also noted that in the case of the vector NLS equation and the near subsonic limit of the Zakharov equations, when the initial electric field is sufficiently anisotropic, the solution contracts much faster in one direction than in the two others, before breaking apart into two peaks.

Remark. The vector NLS equation (13.1.62) previously derived as the subsonic limit of the Zakharov equations for Langmuir oscillations, including the effect of the magnetic field generated at large scale, also occurs in two space dimensions in the special case $\alpha = 0$, in the context of the transverse collapse (filamentation) of Alfvén wave trains in the small dispersion limit (see Section 15.3.1). Numerical simulations show that small scales are still formed in this case, but their structures do not reduce to foci. The wave amplitude remains moderate, but strong gradients develop on elongated structures (Champeaux, Passot, and Sulem 1998a). Whether a blowup of the gradient arises in a finite time remains an open problem.

14
The Scalar Model

In the special case $\alpha = 1$, the vector Zakharov equations discussed in Chapter 13 can be viewed as a system of three scalar Schrödinger equations for the components of the electric field envelope, coupled through the equation for the ion density. Assuming that only one of these components, denoted by ψ, is initially nonzero, one obtains the system

$$i\psi_t + \Delta\psi = n\psi, \qquad (14.0.1)$$

$$n_{tt} - \Delta n = \Delta|\psi|^2 \qquad (14.0.2)$$

with initial conditions $\psi(\mathbf{x}, 0) = \psi_0(\mathbf{x})$, $n(\mathbf{x}, 0) = n_0(\mathbf{x})$, $n_t(\mathbf{x}, 0) = n_1(\mathbf{x})$, usually called the "scalar Zakharov equations." It is often used as a simplified description of Langmuir waves where the vector character of the electric field is neglected. It furthermore admits meaningful radially symmetric solutions (Budneva, Zakharov, and Synakh 1975). Other applications of these equations are mentioned in Fraiman (1979), who noted that this system describes the electron-phonon coupling in a solid-state plasma and the strictive self-focusing of three-dimensional clusters of electromagnetic oscillations. In the subsonic limit where n_{tt} is negligible, the scalar model reduces to the cubic NLS equation. Furthermore, (14.0.2) can be rewritten in the form

$$n_t + \boldsymbol{\nabla} \cdot \mathbf{v} = 0, \qquad (14.0.3)$$

$$\mathbf{v}_t + \boldsymbol{\nabla} n = -\boldsymbol{\nabla}|\psi|^2. \qquad (14.0.4)$$

The scalar Zakharov system derives from a Lagrangian. The associated action is invariant under time and space translations, rotation, and phase shift, which leads to the existence of conserved quantities: the wave energy

$$N = |\psi|_{L^2}^2, \tag{14.0.5}$$

the linear momentum

$$\mathbf{P} = \int \left(\frac{i}{2}(\psi \nabla \psi^* - \psi^* \nabla \psi) + n\mathbf{v} \right) d\mathbf{x}, \tag{14.0.6}$$

the angular momentum

$$\mathbf{M} = \int \mathbf{x} \times \mathbf{P} \, d\mathbf{x}, \tag{14.0.7}$$

and the Hamiltonian

$$H = \int \left(|\nabla \psi|^2 + n|\psi|^2 + \frac{1}{2}|\mathbf{v}|^2 + \frac{1}{2}n^2 \right) d\mathbf{x}. \tag{14.0.8}$$

As already noted in Section 13.1.3, in one space dimension there exist localized solutions in the form of solitary waves, although the system is probably not integrable by the inverse scattering method. As in the case of the NLS equation, these solutions are unstable with respect to long-wavelength transverse perturbations.

14.1   Self-Similar Solutions

14.1.1   Formal Construction

In the radially symmetric case, self-similar and asymptotically self-similar singular solutions have been predicted in two and three dimensions, respectively.

In three dimensions, self-similar solutions can exist only asymptotically close to the collapse when the regime is strongly supersonic with the pressure term Δn negligible compared to the ion-inertia term in (14.0.2). Up to a simple rescaling, these solutions have the universal form (Budneva, Zakharov, and Synakh 1975, Zakharov and Shur 1981)

$$\psi(r,t) = \frac{1}{(t_* - t)} u\left(\frac{r}{(t_* - t)^{2/3}} \right) e^{i(t_* - t)^{-1/3}}, \tag{14.1.1}$$

$$n(r,t) = \frac{1}{(t_* - t)^{4/3}} v\left(\frac{r}{(t_* - t)^{2/3}} \right), \tag{14.1.2}$$

where $u(\eta)$ and $v(\eta)$ are isotropic scalar functions satisfying

$$\Delta u - \frac{1}{3}u - vu = 0, \tag{14.1.3}$$

$$\frac{2}{9}(2\eta^2 v_{\eta\eta} + 13\eta v_\eta + 14v) = \Delta u^2. \tag{14.1.4}$$

In two dimensions, the collapse is no longer supersonic. The pressure term Δn in the wave equation remains relevant, and a self-similar solution of (14.0.1)–(14.0.2) is given by

$$\psi(r,t) = \frac{1}{\mu(t_* - t)} u\left(\frac{r}{\mu(t_* - t)}\right) e^{i\left(\frac{1}{\mu^2(t_* - t)} - \frac{r^2}{4(t_* - t)}\right)}, \qquad (14.1.5)$$

$$n(r,t) = \frac{1}{(\mu(t_* - t))^2} v\left(\frac{r}{\mu(t_* - t)}\right), \qquad (14.1.6)$$

where $u(\eta)$ and $v(\eta)$ satisfy

$$\Delta u - u - vu = 0, \qquad (14.1.7)$$

$$\mu^2(\eta^2 v_{\eta\eta} + 6\eta v_\eta + 6v) - \Delta v = \Delta u^2, \qquad (14.1.8)$$

with μ^2 a free positive parameter.

There is no rigorous proof of existence of solutions for the ordinary differential equations (14.1.3)–(14.1.4) governing the self-similar profile in the supersonic three-dimensional problem. In two dimensions, in contrast, rigorous results were obtained for (14.1.7)–(14.1.8) by Glangetas and Merle (1994a). They are reviewed in Section 14.3.1.

The equations for the profile were studied numerically in three dimensions by Zakharov and Shur (1981) and in two dimensions by Bergé, Pelletier, and Pesme (1990). In both cases, two pairs of localized solutions were computed, one of them (mode I) corresponding to a monotonic profile for both u and v with $v < 0$. Note that $\eta = 0$ and $\eta = \frac{1}{\mu}$ are singular points for equations (14.1.4) and (14.1.8), respectively. In order for the solutions to be smooth, proper conditions for v have to be prescribed.

Proposition 14.1. *In three dimensions, v can be solved in terms of u in the form*

$$v = \frac{9}{4\eta^2}(u^2 - u^2(0)) - \frac{9}{8\eta^{7/2}} \int_0^\eta \xi^{\frac{1}{2}}(u^2 - u^2(0))d\xi. \qquad (14.1.9)$$

Furthermore, $v(\eta) \sim -\frac{3u(0)^2}{2\eta^2}$ as $\eta \to \infty$, and $v(0) = 3\frac{u^2(0)}{14 - 9u^2(0)}$.

It follows that the value of v at the origin is not arbitrary but depends on $u(0)$. When (14.1.3)–(14.1.4) are solved by a shooting method, only $u(0)$ is used as a shooting parameter, chosen such that $u'(0) = 0$ and u decays rapidly at infinity.

Proof. Equation (14.1.4) can be written as

$$(\eta^2 v)_{\eta\eta} + \frac{5}{2\eta}(\eta^2 v)_\eta = \frac{9}{4}\frac{1}{\eta^2}\frac{d}{d\eta}\left(\eta^2 \frac{d}{d\eta}u^2\right). \qquad (14.1.10)$$

Multiplying this equation by $\eta^{5/2}$ and integrating once, one gets

$$\eta^{\frac{5}{2}}(\eta^2 v)' = \frac{9}{4}\int_0^\eta \xi^{\frac{1}{2}}(\xi^2(u^2)')'d\xi, \qquad (14.1.11)$$

where a prime denotes differentiation. After rewriting the right-hand side in the form $\frac{9}{4}\left(\eta^{\frac{5}{2}}(u^2)' - \frac{1}{2}\int_0^\eta \xi^{\frac{3}{2}}(u^2)'d\xi\right)$, one obtains

$$\eta^2 v = \frac{9}{4}(u^2 - u^2(0)) - \frac{9}{8}\int_0^\eta \frac{d\xi}{\xi^{5/2}}\int_0^\xi \zeta^{\frac{3}{2}}(u^2)'d\zeta. \qquad (14.1.12)$$

Computing the integral as

$$\int_0^\eta \frac{d\xi}{\xi^{5/2}}\int_0^\xi \zeta^{\frac{3}{2}}(u^2)'d\xi = \frac{1}{\eta^{3/2}}\int_0^\eta \xi^{\frac{1}{2}}(u^2 - u^2(0))d\xi \qquad (14.1.13)$$

then leads to (14.1.9). Since the function u decays rapidly at infinity, the large-distance behavior of v is directly obtained. To obtain the behavior of v at the origin (where u' vanishes), one writes that in this limit, $u - u(0) \approx \frac{\eta^2}{2}u''(0)$, which implies $v(0) = \frac{27}{14}u(0)u''(0)$. Equation (14.1.7) then gives $u''(0) = \frac{u(0)}{3}(v(0) + \frac{1}{3})$, which is substituted in the expression of $v(0)$.

Proposition 14.2. *In two dimensions, v can be solved in terms of u in the form,*

$$v(\eta) = \frac{u^2(\eta) - u^2(\frac{1}{\mu})}{\mu^2\eta^2 - 1} - \frac{\mu^2}{|\mu^2\eta^2 - 1|^{3/2}}\int_{1/\mu}^\eta \frac{u^2(\xi) - u^2(\frac{1}{\mu})}{|\mu^2\xi^2 - 1|^{1/2}}\xi\,d\xi. \qquad (14.1.14)$$

It follows that $v \sim -\frac{C}{\eta^3}$ with $C = \frac{1}{\mu}\int_{1/\mu}^\infty \frac{u^2}{(\mu^2\eta^2-1)^{1/2}}\eta\,d\eta$ as $\eta \to \infty$, while at the sonic point $\eta = 1/\mu$,

$$v\left(\frac{1}{\mu}\right) = \frac{2}{3\mu}u\left(\frac{1}{\mu}\right)u'\left(\frac{1}{\mu}\right) \qquad (14.1.15)$$

with

$$v'\left(\frac{1}{\mu}\right) = \frac{4}{15\mu}u^3\left(\frac{1}{\mu}\right)u'(\frac{1}{\mu}) + \frac{2}{5\mu}u'^2\left(\frac{1}{\mu}\right) - \frac{4}{5}u\left(\frac{1}{\mu}\right)u'\left(\frac{1}{\mu}\right) + \frac{2}{5\mu}u^2\left(\frac{1}{\mu}\right). \qquad (14.1.16)$$

Since the values of v at the sonic point are determined by $u(\frac{1}{\mu})$ and $u'(\frac{1}{\mu})$, the shooting method to compute the profiles must be used from this point with shooting parameters $u(\frac{1}{\mu})$, $u'(\frac{1}{\mu})$ chosen such that the solution u satisfies $u'(0) = 0$ and rapidly decays at infinity.

Proof. Defining

$$\mathcal{A}(\eta) = (\mu^2\eta^2 - 1)v + 3\mu^2\eta v - (u^2)', \qquad (14.1.17)$$

one has

$$\frac{1}{\eta}(\eta\mathcal{A})' = \mu^2(\eta^2 v'' + 6\eta v' + 6v) - \Delta v - \Delta|u|^2. \qquad (14.1.18)$$

Since v satisfies (14.1.8), it follows that $(\eta\mathcal{A})' = 0$, and since $\eta\mathcal{A}$ vanishes at $\eta = 0$,

$$(\mu^2\eta^2 - 1)v' + 3\mu^2\eta v - (u^2)' = 0. \qquad (14.1.19)$$

When multiplying this equation by $|\mu^2\eta^2 - 1|^{1/2}$ and integrating between $1/\mu$ and η, one gets

$$v(\eta) = \frac{\text{sgn}(\mu^2\eta^2 - 1)}{|\mu^2\eta^2 - 1|^{3/2}} \int_{1/\mu}^{\eta} |\mu^2\xi^2 - 1|^{1/2}(u^2)'d\xi. \tag{14.1.20}$$

By integration by parts, one obtains

$$v(\eta) = \frac{u^2(\eta)}{\mu^2\eta^2 - 1} - \frac{\mu^2}{|\mu^2\eta^2 - 1|^{3/2}} \int_{1/\mu}^{\eta} \frac{u^2}{|\mu^2\xi^2 - 1|^{1/2}} \xi \, d\xi. \tag{14.1.21}$$

From the rapid decay of u at infinity, it follows that the large distance the behavior of $v(\eta)$ is given by the (converging) integral, as indicated in the proposition.

In order to obtain $v(\frac{1}{\mu})$, it is convenient to rewrite $v(\eta)$ in the form given in the proposition. Expanding about $\eta = 1/\mu$, a straightforward calculation then leads to the result. Furthermore, $v'(\frac{1}{\mu})$ is evaluated by writing (14.1.7)–(14.1.8) at the sonic point and using the expression of $v(\frac{1}{\mu})$.

14.1.2 Dynamical Stability

The dynamical stability of the above self-similar or asymptotically self-similar solutions for both radially symmetric and anisotropic initial conditions in two or three dimensions was studied numerically in Landman, Papanicolaou, Sulem, Sulem, and Wang (1992) using the dynamic rescaling method associated to the change of dependent and independent variables of the form

$$\psi(\mathbf{x},t) = \frac{1}{L(t)} U(\boldsymbol{\xi},\tau), \tag{14.1.22}$$

$$n(\mathbf{x},t) = \frac{1}{M(t)} V(\boldsymbol{\xi},\tau), \tag{14.1.23}$$

$$n_t(\mathbf{x},t) = \frac{1}{K(t)} W(\boldsymbol{\xi},\tau). \tag{14.1.24}$$

In (14.1.22)–(14.1.24),

$$\boldsymbol{\xi} = D^{-1}(t)(\mathbf{x} - \mathbf{x}_0), \quad \tau = \int_0^t \frac{1}{\Omega^2(s)} ds, \tag{14.1.25}$$

where $\Omega(t)$ is a positive scalar function of time. Furthermore,

$$D(t) = O^T(t)\Lambda(t) \tag{14.1.26}$$

with $O(t)$ an orthogonal matrix represented by the Euler angles and $\Lambda(t)$ a diagonal matrix whose diagonal elements are denoted by λ_i $(i = 1, \ldots, d)$, with d the space dimension. The scaling factors are prescribed by the constraints

$$x_0^i = \frac{\int x_i |n|^{2p} d\mathbf{x}}{\int |n|^{2p} d\mathbf{x}}, \qquad \frac{\int \xi_i \xi_j |V|^{2p} d\boldsymbol{\xi}}{\int |V|^{2p} d\boldsymbol{\xi}} = \delta_{ij}, \qquad (14.1.27)$$

which are different from those used for the nonlinear Schrödinger equation, but convenient for the Zakharov equations because they keep W bounded away from zero near collapse. As a consequence of this scaling, the system of equations, which is integrated numerically, is nonsingular.

Numerical simulations were performed with anisotropic initial conditions corresponding to a negative Hamiltonian, both in two and three dimensions, assuming that initially $n_0 = -|\psi_0|^2$ and $(\partial_t n)_0 = 0$. It was observed that the solutions become isotropic near collapse with the same profile as that obtained when starting with isotropic initial conditions. The description of the collapse will thus be limited to the latter situation. Collapsing solutions of the scalar Zakharov equations with anisotropic contraction rates were constructed by Bergé, Pelletier, and Pesme (1990, 1991). They were, however, not observed in the numerical simulations and are thus probably unstable. It was also noted that relaxation to isotropy is significantly slower for the scalar Zakharov equations than for the nonlinear Schrödinger equation with the same initial electric field.

In the case of isotropic solutions, $\mathbf{x}_0 = 0$, the matrix O is the identity and the scaling factors λ_i, which are all equal, are denoted by λ. Consequently, $\Omega = \lambda$, $L = \lambda^{d/2}$, $M = \lambda^2$, and $K = \lambda^{2+d/2}$. The system for the rescaled quantities then reads

$$i\left(U_\tau + a\lambda^{2-d/2}\left(\frac{d}{2}U + \xi U_\xi\right)\right) + \Delta U - VU = 0, \quad (14.1.28)$$

$$V_\tau + a\lambda^{2-d/2}(2V + \xi V_\xi) - \lambda^{2-d/2}W = 0, \quad (14.1.29)$$

$$W_\tau + a\lambda^{2-d/2}\left(\left(2 + \frac{d}{2}\right)W + \xi W_\xi\right) - \lambda^{d/2}\Delta V - \lambda^{2-d/2}\Delta|U|^2 = 0, \quad (14.1.30)$$

where

$$\lambda_\tau = -a\lambda^{3-d/2} \qquad (14.1.31)$$

and

$$a = \frac{p\int (1 - \xi^2)V^{(2p-1)}W\xi^{d-1}d\xi}{\int V^{2p}\xi^{d-1}d\xi}. \qquad (14.1.32)$$

In three dimensions, it is observed that as τ increases, $|U|$, V, and W become stationary, which indicates a self-similar collapse. In this limit, the phase of U at the origin is linear in τ. Furthermore, $\lambda(\tau) \to 0$ and $a(\tau)$ tends to a finite limit A. It follows from (14.1.31) that

$$\lambda(t) = (t_* - t)^{2/3}, \qquad (14.1.33)$$

as predicted by Zakharov (1972). When substituting

$$U(\xi, \tau) = S(\xi)e^{iC\tau}, \qquad (14.1.34)$$

in (14.1.28) and taking the limit $\tau \to \infty$, one gets

$$\Delta S - CS - VS = 0, \tag{14.1.35}$$

where S can be chosen to be real by a phase translation. In the equation for the density, the pressure term ΔV is asymptotically negligible, making the collapse supersonic. From (14.1.29) and (14.1.30) one has

$$\frac{A^2}{2}(2\xi^2 V_{\xi\xi} + 13\xi V_\xi + 14V) = \Delta S^2. \tag{14.1.36}$$

By rescaling $u = \frac{2}{3A}S$, $v = \frac{1}{3C}V$ and $\eta = \sqrt{3C}\xi$, one recovers (14.1.3)–(14.1.4), which are identical to equations (2.1)–(2.2) of Zakharov and Shur (1981) up to a simple rescaling. For various initial conditions, the numerical results show that the profiles $|U|$, V quickly become peaked at the origin and (after rescaling) converge to the profiles u, v of the mode I, discussed in the previous section. Figure 14.1 (left) demonstrates this convergence in the case of initial conditions $\psi_0(\mathbf{x}) = 6e^{-|\mathbf{x}|^2}$, $n_0(\mathbf{x}) = -|\psi_0(\mathbf{x})|^2$ and $(\partial_t n)_0(\mathbf{x}) = 0$. Other types of initial conditions are presented in Landman, Papanicolaou, Sulem, Sulem, and Wang (1992). The divergence of the integrals $\int |u|^2 d\mathbf{x}$ and $\int v\,d\mathbf{x}$ associated to the self-similar solution, indicates that the self-similar profile does not extend to infinity. The large distance behavior of the solution is discussed by Bergé (1994b).

In two dimensions, numerical integrations show that in this case also the solution displays a self-similar collapse where $a(\tau)$ approaches a finite value denoted by A. This behavior corresponds to a scaling factor $\lambda(t) \approx (t_* - t)$. In contrast to the supersonic collapse observed in three dimensions, the pressure term must now be retained in the density equation. Furthermore, the phase of the self-similar solution can be calculated exactly. Writing $U(\xi, \tau) = S(\xi)\exp(-iC\tau)\exp(-(i/4)a\lambda\xi^2)$, we get in the limit where λ

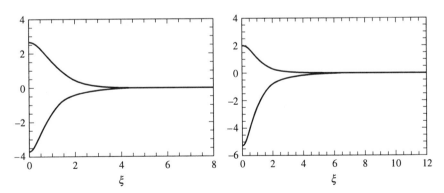

FIGURE 14.1. Behavior at large τ of $|U(\xi, \tau)|$ and $V(\xi, \tau)$ versus ξ, superimposed on the mode I self-similar profiles u and v after rescaling, in three (left) and two (right) dimensions (Landman et al. 1992).

tends to zero

$$\Delta S - CS - VS = 0, \tag{14.1.37}$$
$$A^2(\xi^2 V_{\xi\xi} + 6\xi V_\xi + 6V) - \Delta V = \Delta S^2. \tag{14.1.38}$$

After normalizing $u = \frac{1}{\sqrt{C}}S$, $v = \frac{1}{C}V$, and $\eta = \sqrt{C}\xi$, this reproduces (14.1.7)–(14.1.8). Note that this system depends on the constant $\mu^2 = \frac{A^2}{C}$. After rescaling, the profiles $|U|$, V of the initial value problem fit accurately the profiles u, v corresponding to this specific value of μ^2. This is illustrated in Fig. 14.1 (right) in the case of initial conditions $\psi_0(\mathbf{x}) = 4e^{-|\mathbf{x}|^2}$, $n_0(\mathbf{x}) = -|\psi_0(\mathbf{x})|^2$ and $(\partial_t n)_0(\mathbf{x}) = 0$.

The influence of the value of μ on the profiles of the self-similar solutions is discussed by Bergé, Pelletier, and Pesme (1990), where the collapse is defined as (moderately) subsonic if μ is smaller than unity and (moderately) supersonic if it is larger. The case $\mu = 1$ is described as sonic. Numerical simulations of the initial value problem presented in Landman, Papanicolaou, Sulem, Sulem, and Wang (1992) indicate that the constant μ depends on the initial conditions. In considering a sequence of initial conditions with an initial wave energy decreasing to $|R|^2_{L^2}$, it is observed that the computed value of μ tends to zero. In this limit, the self-similar profile becomes (strongly) subsonic and u tends to the Townes soliton R, defined, as usual, as the positive localized solution of $\Delta R - R + R^3 = 0$. This limit is, however, delicate, since as will be discussed later, all solutions of the scalar model with a critical norm $|\psi_0|^2_{L^2} = |R|^2_{L^2}$ remain smooth for all time, which is not the case in the context of the NLS equation.

14.2   Existence and Blowup Results

14.2.1   Existence Theory

We consider the initial value problem for the Zakharov system

$$i\partial_t \psi + \Delta \psi = n\psi, \tag{14.2.1}$$
$$\epsilon^2 \partial_{tt} n - \Delta n = \Delta |\psi|^2, \tag{14.2.2}$$

with $\psi(\mathbf{x}, 0) = \psi_0(\mathbf{x})$, $n(\mathbf{x}, 0) = n_0(\mathbf{x})$, and $n_t(\mathbf{x}, 0) = n_1(\mathbf{x})$. The presence of the parameter ϵ, inversely proportional to the ion acoustic speed, reflects the deviation from an adiabatic dynamics of the ions. In addition,

$$\epsilon n_t = -\nabla \cdot \mathbf{v}, \tag{14.2.3}$$

where the function $\mathbf{v}$, taken equal to a gradient, satisfies

$$\epsilon \mathbf{v}_t + \nabla(n + |\psi|^2) = 0. \tag{14.2.4}$$

We introduce the notation

$$H_k = H^k(\mathbf{R}^d) \times H^{k-1}(\mathbf{R}^d) \times H^{k-2}(\mathbf{R}^d) \tag{14.2.5}$$

and summarize the main results.

When the scalar model (14.2.1)–(14.2.2) is viewed as a simplification of the Zakharov equations for Langmuir waves, the equation for the density n results from a linearization of the hydrodynamic equations, and as a consequence $\mathbf{v}$ is irrotational. This property is explicitly used in Sulem and Sulem (1979), Added and Added (1984, 1988), Schochet and Weinstein (1986), and Ozawa and Tsutsumi (1992b). When the scalar model is considered in a most abstract setting, this hypothesis is not necessary. Results without this assumption are given by Ozawa and Tsutsumi (1992a), Kenig, Ponce, and Vega (1995), and Glangetas and Merle (1994b). The analysis is then more delicate, since the Hamiltonian conservation cannot be used.

Theorem 14.3. (Weak existence) *For initial conditions $\psi_0 \in H^1(\mathbf{R}^d)$, $n_0 \in L^2(\mathbf{R}^d)$, and $n_1 \in H^{-1}(\mathbf{R}^d)$, arbitrary in dimension 1, obeying $|\psi_0|_{L^2} < |R|_{L^2}$ in dimension 2, and satisfying $N|H| < 4/(27 \times 24^2)$ and $|\nabla\psi_0|_{L^2} \le |H|$ in dimension 3, there exists, for fixed ϵ, a global weak solution $\psi \in L^\infty(\mathbf{R}^+, H^1(\mathbf{R}^d))$, $n \in L^\infty(\mathbf{R}^+, (L^2(\mathbf{R}^d))$ to the system (14.2.1–(14.2.2).*

Theorem 14.4. (Global smooth solutions in dimension 2) *For initial conditions $(\psi_0, n_0, n_1) \in H_k$, $k \ge 2$, and $|\psi_0|_{L^2} < |R|_{L^2}$, the solution remains in H_k for all times.*

Theorem 14.5. (Local existence of strong solutions) *For initial conditions $(\psi_0, n_0, n_1) \in H_k$ ($k \ge 2$), there exists a unique solution (ψ, n, n_t) in H_k during an interval of time $[0, T)$, with either $T = \infty$, or $(\psi, n, n_t)_{H_2} \to \infty$ as $t \to T$. The time T depends only on $|(\psi_0, n_0, n_1)|_{H_2}$.*

Furthermore, during the time $[0, T)$, one has the smoothing property that for all $h(\mathbf{x}) \in C^\infty(\mathbf{R}^d)$ with compact support, $h\psi \in L^2(0, T, H^{k+1/2})$.

More recently, Bourgain and Colliander (1997) extended the methods introduced by Bourgain (1993,1994a) (see Section 3.4.3) in the context of the NLS equation in a periodic domain, to the scalar Zakharov system in $\mathbf{R}^d$, $d \le 3$. They improved the above results as follows:

Theorem 14.6. (Uniqueness of weak solutions) *The weak solution constructed in Theorem 14.3 is unique.*

Theorem 14.7. (Global smooth solutions in dimension 3) *For initial conditions $(\psi_0, n_0, n_1) \in H_k$ ($k \ge 2$), the solution remains in H_k as long as $|(\psi, n, n_t)|_{H_1}$ does not blow up. In particular, in dimension 3, it remains in H_k for all times if $|\psi_0|_{H^1}$ is small enough.*

Scattering properties (see Section 3.3) for solutions of the scalar model in dimension $d = 3$ are addressed in Ozawa and Tsutsumi (1993, 1994).

Theorem 14.8. (Subsonic limit) *Suppose in addition to the hypotheses of Theorem 14.5 that the initial conditions satisfy the compatibility conditions $(n_{0,\epsilon} + |\psi_{0,\epsilon}|^2)|_{H_{k-1}} \le C\epsilon$ and that $\psi_{0,\epsilon} \to \overline{\psi}_0$ strongly in H^k as $\epsilon \to 0$.*

Then, as $\epsilon \to 0$, $n_\epsilon + |\psi_\epsilon|^2 \to 0$ in $C^0([0, T) \times \mathbf{R}^d)$, $\nabla(n_\epsilon + |\psi_\epsilon|^2) \to 0$ in $C^0([0, T), H^{k-2})$, and $\psi_\epsilon \to \overline{\psi}$, the unique solution of the NLS equation with initial condition $\overline{\psi}_0$, in $C^1([0, T) \times \mathbf{R}^d) \cap C^1([0, T), C^2)$.

The above convergence theorem is due to Schochet and Weinstein (1986). Further results including estimate of the rate of convergence were given by Added and Added (1988) and in an optimal form by Ozawa and Tsutsumi (1992b), who proved the following theorem. Additional results are established in weighted Sobolev spaces.

Theorem 14.9. *In dimension $d \leq 3$, for initial conditions $(\psi_0, n_0, n_1) \in H_m$ independent of ϵ, the solution satisfies for small enough ϵ,*

$$\sup_{0 \leq t \leq T} |n_\epsilon(t) + |\psi_\epsilon(t)|^2 - Q_\epsilon^{(1)}(t) - Q_\epsilon^{(2)}(t)|_{H^m} \leq M\epsilon \qquad (14.2.6)$$

with

$$Q_\epsilon^{(1)}(t) = \cos\left(\frac{t}{\epsilon}(-\Delta)^{\frac{1}{2}}\right)(n_0 + |\psi_0|^2) \qquad (14.2.7)$$

$$Q_\epsilon^{(2)}(t) = \epsilon(-\Delta)^{-\frac{1}{2}} \sin\left(\frac{t}{\epsilon}(-\Delta)^{\frac{1}{2}}\right)(n_1 + 2\Im\psi_0^*\Delta\psi_0). \qquad (14.2.8)$$

Furthermore,

$$\sup_{0 \leq t \leq T} |\psi_\epsilon(t) - \psi(t)|_{H^k} \leq M\epsilon, \qquad (14.2.9)$$

or, if the compatibility condition $n_0 + |\psi_0|^2 = 0$ is satisfied,

$$\sup_{0 \leq t \leq T} |\psi_\epsilon(t) - \psi(t)|_{H^k} \leq M\epsilon^2. \qquad (14.2.10)$$

In relation to the initial value problem, it is of interest to find the space with minimal regularity in which the problem is locally well-posed, beyond the space H_1 in which the Hamiltonian is defined. As for the NLS equation, it is possible to get local existence of solutions in "larger" spaces. For this purpose, Ginibre, Tsutsumi, and Velo (1997) extended to the Zakharov system, the notion of criticality for the initial value problem H^k, discussed in Section 3.2 in the context of the NLS equation. It turns out that (14.0.1) and (14.0.2) of the Zakharov system do not display the same scaling invariance. These authors define the criticality by considering the scaling

$$\psi \to \psi_\lambda = \lambda^{3/2}\psi(\lambda \mathbf{x}, \lambda^2 t), \quad n \to n_\lambda = \lambda^2 n(\lambda \mathbf{x}, \lambda^2 t) \qquad (14.2.11)$$

that would leave the Zakharov system invariant in the absence of the term Δn. It is of interest to notice that this is the relevant scaling to study the blowing up solutions of the Zakharov system in dimension 3 (see Section 14.1). The critical values for the initial problem in $H^k \times H^l \times H^{l-1}$ are $k = \frac{d}{2} - \frac{3}{2}$ and $l = \frac{d}{2} - 2$. Note that $k - l = \frac{1}{2}$ while for the more classical setting in H_1, one has $k - l = 1$. They proved

Theorem 14.10. *In dimension 1, the Zakharov system is locally well posed in $H^k \times H^l \times H^{l-1}$, provided that $-\frac{1}{2} \le k - l \le 1$, $2k \ge l + \frac{1}{2} \ge 0$.*

Theorem 14.11. *In dimension $d \ge 2$, the Zakharov system is locally well posed in $H^k \times H^l \times H^{l-1}$ with the conditions*

$$
\begin{aligned}
l \le k \le l + 1 && \text{for all dimensions } d, \\
l > d/2 - 2,\ 2k - (l+1) > d/2 - 2 && \text{for } d \ge 4, \\
l \ge 0, \qquad 2k - (l+1) \ge 0 && \text{for } d = 2 \text{ or } 3. \qquad (14.2.12)
\end{aligned}
$$

For example, in dimension 1, the problem is locally well-posed in $L^2 \times H^{-1/2} \times H^{-3/2}$. In dimension 2 or 3, it is locally well-posed in $H^{1/2} \times L^2 \times H^{-1}$. In dimension $d \ge 4$, the whole range of subcritical values $k > \frac{d}{2} - \frac{3}{2}$ and $l > \frac{d}{2} - 2$ is covered by the theorem.

This analysis was extended to the vector Zakharov system with a coupling to an induced magnetic field by Tzvetkov (1998).

Periodic Boundary Conditions

Using a Fourier analysis approach similar to that introduced for the NLS equation, Bourgain (1994c) proved local existence for the scalar model in a one-dimensional periodic domain T in fractional Sobolev spaces of order $s < 1$. He also proved that the system is globally well-posed in $H_2(T)$ and also on a set of data in $\cup_{s < \frac{1}{2}} H_s$ carrying the Gibbs measure that is invariant under the flow.

Remarks.

(i) A formal expansion leading to corrections to the subsonic limit and resulting in a perturbed NLS equation is presented by Gibbons (1978) in one space dimension.

(ii) It is noticeable that although the paraxial approximation for the (elliptic) Helmholtz equation with nonlinear refractive index and the (scalar) nonlinear wave equation lead to the same NLS equation for the envelope dynamics, these two approximations display different features. In the former case, the blow up of the NLS solutions is due to a breakdown of the asymptotics and is arrested when a weak nonparaxiality is retained (see Section 9.4.1). Note that the initial value problem is then changed into a boundary value problem. In contrast, solutions of the nonlinear wave equation can blow up even in dimensions that are subcritical for the NLS equation (see Strauss 1989 for a review). The NLS limit is addressed in this context by Bergé and Colin (1995) (see also the review by Bergé 1998), where it is shown that the convergence of the solution of the wave equation is strong during the existence time but that the behavior of its time derivative is more complex if its initial value differs from that prescribed by the NLS equation. The discussion is extended to the case where the density obeys (14.2.2) by Bergé, Bidégaray, and Colin (1996). It is in particular shown that this limit is not compatible with the supersonic regime corresponding to $\Delta n \ll \epsilon^2 \partial_{tt} n$.

14.2.2 Blowup Results

Throughout this section we suppose that there exists a smooth solution $(\psi, n, \mathbf{v})$ to the system (14.2.1) in the sense that in the interval $[0, t_0]$,

$$(\psi, n, \mathbf{v}) \in C([0, t_0], H_2), \qquad (14.2.13)$$

where H_2 is defined in (14.2.5). In particular, this solution satisfies the conservation of the wave energy (or mass) and of the Hamiltonian. We report in this section a result of Merle (1996a) on possible blowup for solutions of the system (14.2.1). In this context, the usual variance $\int |\mathbf{x}|^2 |\psi|^2 d\mathbf{x}$, is replaced by the quantity

$$\mathcal{W}(t) = \frac{1}{4} \int |\mathbf{x}|^2 |\psi|^2 d\mathbf{x} + \epsilon \int_0^t \int (\mathbf{x} \cdot \mathbf{v}) n \, d\mathbf{x} \, dt, \qquad (14.2.14)$$

which is well-defined for functions in the space

$$\Sigma' = \{(\psi, n, \mathbf{v}) \in H_1, \int \Big(|\mathbf{x}|^2 |\psi|^2 + |\mathbf{x}|(|n|^2 + |\mathbf{v}|^2) \Big) d\mathbf{x} < \infty\}. \quad (14.2.15)$$

The function $\mathcal{W}(t)$ naturally comes into play by writing the continuity equations for the density of wave energy and linear momentum, obtained as a consequence of the Noether theorem. One has

$$\partial_t |\psi|^2 + \nabla \cdot \mathbf{J} = 0 \qquad (14.2.16)$$

with

$$\mathbf{J} = \frac{i}{2}(\psi \nabla \psi^* - \psi^* \nabla \psi) \qquad (14.2.17)$$

and

$$\partial_t (J_j + 2\epsilon n v_j) = \partial_i T_{ij}, \qquad (14.2.18)$$

where the tensor $\mathbf{T} = (T_{ij})$ is given by

$$T_{ij} = -2\Re(\partial_i \psi^* \partial_j \psi) + 2\Re(\psi^* \partial_{ij} \psi) - 2\delta_{ij} n |\psi|^2 - \delta_{ij} |n|^2 - 2v_i v_j + \delta_{ij} |\mathbf{v}|^2. \qquad (14.2.19)$$

It follows that

$$\frac{\partial^2}{\partial t^2} \Big(|\psi|^2 - 2\epsilon \int_0^t \nabla \cdot (n\mathbf{v}) dt \Big) = -\partial_{ij} T_{ij} \qquad (14.2.20)$$

and thus,

$$\frac{d^2 \mathcal{W}}{dt^2}(t) = -\frac{1}{2} \int \operatorname{tr} \mathbf{T} \, d\mathbf{x}$$
$$= dH - (d-2)|\nabla \psi|_{L^2}^2 - (d-1)|\mathbf{v}|_{L^2}^2, \qquad (14.2.21)$$

which is the equivalent of the variance identity for the NLS equation. It implies that in dimension $d \geq 2$, one has $\frac{d^2 W}{dt^2} < 0$ for $H < 0$. Nevertheless, one cannot conclude at this step about the existence of blowup solutions since W does not have a fixed sign (as shown by Merle 1996b, it tends to $-\infty$ in two dimensions, as the singularity is approached). The problem is overcome by replacing the weight $|\mathbf{x}|^2$ in the variance by a smooth function $p(\mathbf{x})$ that behaves like $|\mathbf{x}|^2$ near the origin and like $|\mathbf{x}|$ at infinity, and by working with the time derivative $-y = \frac{d\mathcal{U}}{dt}$ of

$$\mathcal{U}(t) = \frac{1}{2} \int p(x)|\psi|^2 \, d\mathbf{x} + \epsilon \int_0^t \int (\nabla p \cdot \mathbf{v}) \, n \, d\mathbf{x} \, dt, \qquad (14.2.22)$$

as in the case of solutions of the NLS equation with infinite variance. Indeed,

$$y = -\Im \int ((\nabla p \cdot \nabla \psi)\psi^* - \epsilon(\nabla p \cdot \mathbf{v})n) \, d\mathbf{x} \qquad (14.2.23)$$

obeys

$$|y| \leq C(|\psi_0|_{L^2}^2 + |\nabla \psi|_{L^2}^2 + \epsilon|\mathbf{v}|_{L^2}^2 + \epsilon|n|_{L^2}^2). \qquad (14.2.24)$$

The existence of a blowup thus reduces to the divergence of $|y|$. Nevertheless, as detailed in Proposition 14.12, the use of a function p which is not purely quadratic induces additional contributions in the estimate of $-\frac{dy}{dt} = \frac{d^2 \mathcal{U}}{dt^2}$ obtained by multiplying (14.2.20) by $p(\mathbf{x})$ and integrating in space, in the form

$$\frac{d^2 \mathcal{U}}{dt^2}(t) = \frac{1}{2} \int \partial_{ij} p T_{ij} d\mathbf{x}. \qquad (14.2.25)$$

As in the case of solutions of the NLS equation with an infinite variance, one considers in Proposition 14.13 a sequence of weight functions $p_m = m^2 \widetilde{p}(r/m)$ obtained by rescaling a suitable function $\widetilde{p}$, and monitors the effect of the additional terms by a convenient choice of the parameter m. Under the assumption of negative Hamiltonian and radial symmetry, the norm $|(\psi, n, \mathbf{v})|_{H_1}$ is then shown in Theorem 14.14 to blow up in a finite or infinite time in the sense that either $|(\psi, n, \mathbf{v})|_{H_1} \to \infty$ as $t \to t_*$ with t_* finite, or (ψ, n, v) exists for all time and $|(\psi, n, \mathbf{v})|_{H_1} \to \infty$ as $t \to \infty$.

Proposition 14.12.   *Suppose that $(\psi_0, n_0, \mathbf{v}_0) \in \Sigma'$ and that $(\psi, n, \mathbf{v})$ is a smooth solution of (14.2.1). Suppose in addition that $\big| |\mathbf{x}|^{1/2}\psi \big|_{L^\infty}$ is uniformly bounded on $[0, t_0]$. Then $\forall t \in [0, t_0]$, $(\psi(t), n(t), \mathbf{v}(t)) \in \Sigma'$. Furthermore, if p is a smooth weight function such that*

$$|p(\mathbf{x})| \leq c(1 + |\mathbf{x}|^2), \quad |\nabla p| \leq c(1 + |\mathbf{x}|), \quad |\Delta p| + |\Delta^2 p| \leq c, \quad (14.2.26)$$

then, $\forall t > 0$, the function

$$\mathcal{U}(t) = \frac{1}{2} \int p(x)|\psi|^2 \, d\mathbf{x} + \epsilon \int_0^t \int (\nabla p \cdot \mathbf{v}) \, n \, d\mathbf{x} \, dt \qquad (14.2.27)$$

satisfies

$$\frac{d^2\mathcal{U}}{dt^2}(t) = \int \Delta p\left(n|\psi|^2 + \frac{1}{2}|n|^2\right) dx + 2\Re \int \partial_{ij} p\, \partial_i \psi^* \partial_j \psi\, dx$$

$$+ \int \left(\partial_{ij} p\, v_i v_j - \frac{1}{2}\Delta p\, |\mathbf{v}|^2\right) dx - \frac{1}{2}\int \Delta^2 p\, |\psi|^2\, dx, \quad (14.2.28)$$

and in the case of radially symmetric functions,

$$\frac{d^2\mathcal{U}}{dt^2}(t) = \int \Delta p\left(n|\psi|^2 + \frac{1}{2}|n|^2\right) dx + 2\int \partial_{rr} p|\partial_r \psi|^2 dx$$

$$+ \frac{1}{2}\int \left(\partial_{rr} p - \frac{d-1}{r}\partial_r p\right)|\mathbf{v}|^2 dx - \frac{1}{2}\int \Delta^2 p|\psi|^2 dx. \quad (14.2.29)$$

The assumption of radial symmetry allows one to use the Strauss lemma (Lemma 5.6), in order to controls the sup-norm of a function away from the origin in terms of its H^1-norm, in the form $r^{d-1}|f(r)|^2 \leq |f|_{L^2}\left|\frac{\partial f}{\partial r}\right|_{L^2}$.

Specifying the function $\tilde{p} = \int_0^r h(s)ds$ where $h \in C^3$ is a radially symmetric function defined by

$$h(r) = \begin{cases} r & \text{for} \quad 0 \leq r \leq 1 \\ r - (r-1)^4 & \text{for} \quad r \geq 1 \quad \text{near} \quad 1 \\ \dfrac{3}{2} & \text{for} \quad r \geq 3 \end{cases} \quad (14.2.30)$$

and $h'(r) < 1$ for $r > 1$, and using the scaled functions

$$p_m = m^2 \tilde{p}\left(\frac{r}{m}\right) \quad \text{and} \quad h_m = mh\left(\frac{r}{m}\right), \quad (14.2.31)$$

one writes

$$y_m(t) = -\Im \int \left((\nabla p_m \cdot \nabla \psi)\psi^* - \epsilon(\nabla p_m \cdot \mathbf{v})\, n\right) dx \quad (14.2.32)$$

and establishes the following results.

Proposition 14.13. *Let $(\psi, n, \mathbf{v})$ be a smooth, radially symmetric solution of (14.2.1) for all time. Then for every t and every m, one has*
(i) in dimension $d = 3$

$$\frac{dy_m}{dt}(t) \geq -dH + \frac{1}{4}|\nabla\psi|_{L^2}^2 + c_2|\mathbf{v}|_{L^2}^2 - c_1\left(\frac{1}{m^2} + \frac{1}{m^4}\right), \quad (14.2.33)$$

(ii) in dimension $d = 2$

$$\frac{dy_m}{dt}(t) \geq -dH + c_2|\mathbf{v}|_{L^2}^2 - c_1\left(\frac{1}{m} + \frac{1}{m^2}\right) + 2\left(1 - \frac{c_1}{m}\right)\int (1 - \partial_{rr} p_m)|\nabla\psi|^2 dx,$$

$$(14.2.34)$$

where c_1 is a constant depending only on the initial condition $|\psi_0|_{L^2}$ and $c_2 > 0$.

Proof. From (14.2.29), one has

$$\frac{dy_m}{dt}(t) = -dH + \int (d - \Delta p_m)\Big(n|\psi|^2 + \frac{1}{2}|n|^2\Big)d\mathbf{x}$$

$$+2\int (1 - \partial_{rr}p_m)\ |\partial_r\psi|^2 d\mathbf{x} + (d-2)|\partial_r\psi|^2_{L^2}$$

$$+\int \Big(\frac{1}{2}(d - \Delta p_m) + \frac{d-1}{r}\partial_r p_m\Big)|\mathbf{v}|^2 d\mathbf{x} + \frac{1}{2}\int \Delta^2 p_m\ |\psi|^2 d\mathbf{x}.$$

$$(14.2.35)$$

Now, the conditions prescribed on p imply that $-\Delta p_m + d = 0$ if $r \le m$, and $-\Delta p_m + d \ge 0$ otherwise, together with the inequality $(-\Delta p_m + d)^2 \le (1 - \partial_{rr}p_m)$. Since

$$\left|\int \Delta^2 p_m |\psi|^2 d\mathbf{x}\right| \le \frac{C}{m^2}\int |\psi|^2 d\mathbf{x} = \frac{c_1}{m^2} \qquad (14.2.36)$$

and $\frac{1}{2}(d - \Delta p_m) + \frac{d-1}{r}\partial_r p_m \ge c_2 > 0$, the only term to be estimated is $\int (d - \Delta p_m)n|\psi|^2 d\mathbf{x}$. We have

$$\int (d - \Delta p_m)n|\psi|^2 d\mathbf{x} \ge -\frac{1}{2}\int (d - \Delta p_m)(|n|^2 + |\psi|^4)\,d\mathbf{x}. \qquad (14.2.37)$$

Thus,

$$\frac{dy_m}{dt}(t) \ge -dH - \frac{c}{m^2} + c_2|\mathbf{v}|^2_{L^2} + 2\int (1 - \partial_{rr}p_m)\ |\partial_r\ \psi|^2 d\mathbf{x}$$

$$+(d-2)|\partial_r\ \psi|^2_{L^2} - \frac{1}{2}\int (d - \Delta p_m)|\psi|^4 d\mathbf{x}, \qquad (14.2.38)$$

where the last integral on the right-hand side of (14.2.38) is taken on $r > m$. It is then convenient to consider the dimensions $d = 2$ and $d = 3$ separately.
(i) *Dimension $d = 3$.* Since $0 \le d - \Delta p_m \le c$ and $1 - \partial_{rr}p_m \ge 0$, the inequality (14.2.38) can be rewritten as

$$\frac{dy_m}{dt}(t) \ge -dH - \frac{c}{m^2} + c_2|\mathbf{v}|^2_{L^2} + |\partial_r\psi|^2_{L^2} - c|\psi_0|^2_{L^2}|\psi|^2_{L^\infty(r\ge m)}.$$

$$\ge -dH - \frac{c}{m^2} + c_2|\mathbf{v}|^2_{L^2} + |\partial_r\psi|^2_{L^2} - \frac{c_1}{m^2}|\partial_r\psi|_{L^2}, \qquad (14.2.39)$$

and thus (14.2.33).
(ii) *Dimension $d = 2$.* One has

$$\left|\int (d - \Delta p_m)|\psi|^4 d\mathbf{x}\right| \le c|\psi_0|^2_{L^2}\Big|(1 - \partial_{rr}p_m)^{1/4}|\psi|\Big|^2_{L^\infty(r\ge m)}. \qquad (14.2.40)$$

In the last inequality, for $r > m$,

$$\left|(1 - \partial_{rr}p_m)^{1/2}|\psi(r)|^2\right| = \left|\int_r^\infty \partial_\rho\Big((1 - \partial_{\rho\rho}p_m)^{1/2}|\psi|^2\Big)\frac{1}{\rho}(\rho d\rho)\right|$$

$$\le \frac{1}{m}\int_{|\mathbf{x}|>r}\left|\nabla\Big((1 - \partial_{\rho\rho}p_m)^{1/2}|\psi|^2\Big)\right|d\mathbf{x}$$

$$\leq \frac{1}{m} \int_{|\mathbf{x}|>r} |\nabla(1-\partial_{\rho\rho}p_m)^{1/2}||\psi|^2 dx$$

$$+\frac{2}{m} \int_{|\mathbf{x}|>r} (1-\partial_{\rho\rho}p_m)^{1/2}|\nabla\psi||\psi| dx$$

$$\leq \frac{1}{m}\left(\frac{1}{m}|\psi_0|^2_{L^2} + \int (1-\partial_{rr}p_m)|\nabla\psi|^2 dx + |\psi_0|^2_{L^2}\right),$$

(14.2.41)

and (14.2.34) follows.

Theorem 14.14. *Suppose that $H < 0$ and that the solution $(u, n, \mathbf{v})$ is radially symmetric. Then, either $|\psi|_{H^1} + |n|_{L^2} + |\mathbf{v}|_{L^2} \to \infty$ as $t \to t_*$ with t_* finite, or $(\psi, n, \mathbf{v})$ exists for all time and $|\psi|_{H^1} + |n|_{L^2} + |\mathbf{v}|_{L^2} \to \infty$ as $t \to \infty$.*

Proof.   Assume that the solution (ψ, n, v) does not blow up in a finite time. It is thus defined for all time. In order to prove that as $t \to \infty$ $|\nabla\psi(t)|^2_{L^2} + |n(t)|^2_{L^2} + |\mathbf{v}(t)|^2_{L^2} \to \infty$, using that for all t

$$y_m(t) \leq c|h_m|_{L^\infty}\left(|\psi_0|^2_{L^2} + |\nabla\psi|^2_{L^2} + |\mathbf{v}|^2_{L^2} + |n|^2_{L^2}\right),$$ (14.2.42)

it is sufficient to show that as $t \to \infty$, $y_m(t) \to \infty$ for a suitable constant m. Since $H < 0$, the right-hand sides of (14.2.33) (in 3 dimensions) and of (14.2.34) (in 2 dimensions) are larger than $\frac{d}{2}|H|$ for m sufficiently large. Thus, by time integration,

$$y_m(t) - y_m(0) \geq \frac{d}{2}|H|t,$$ (14.2.43)

which tends to infinity with t.

14.3   Further Analysis in Two Dimensions

14.3.1   Existence of Self-Similar Blowup Solutions

For the scalar system (14.2.1)–(14.2.2) in two dimensions, self-similar solutions that blow up in a finite time read

$$\psi(r,t) = \frac{1}{\mu(t_*-t)}u\left(\frac{r}{\mu(t_*-t)}\right)e^{i\left(\theta+\frac{1}{\mu^2(t_*-t)}-\frac{r^2}{4(t_*-t)}\right)},$$ (14.3.1)

$$n(r,t) = \frac{1}{(\mu(t_*-t))^2}v\left(\frac{r}{\mu(t_*-t)}\right),$$ (14.3.2)

under the condition that (u, v) satisfies the system of ordinary differential equations in the radial variable

$$\Delta u - u - vu = 0,$$ (14.3.3)

$$\lambda^2(\eta^2 v_{\eta\eta} + 6\eta v_\eta + 6v) - \Delta v = \Delta u^2$$ (14.3.4)

with $\lambda = \epsilon\mu$. In this section, following Glangetas and Merle (1994a), we discuss the existence of radially symmetric solutions of (14.3.3)–(14.3.4) and then review a few properties of these solutions. As usual, R denotes the (radial) positive solution of $\Delta R - R + R^3 = 0$.

Theorem 14.15. (Existence and uniqueness for small λ)   *There exists $\lambda^+ > 0$ such that $\forall\lambda$ with $0 < \lambda < \lambda^+$, there is a solution $(u_\lambda, v_\lambda) \in H^1 \times L^2$ of (14.3.3)–(14.3.4) with $u_\lambda > 0$.*

When $\lambda \to 0$, (u_λ, v_λ) tends to $(R, -R^2)$ in $H^1 \times L^2$. Furthermore, for all $c > |R|_{L^2}$, there exists $\lambda_c > 0$, such that for all λ with $0 < \lambda < \lambda_c$ there is a unique solution (u_λ, v_λ) with $u_\lambda > 0$ and $|u_\lambda|_{L^2} < c$.

Theorem 14.16. (Existence of a branch of solutions)   *There exists a branch of solutions $(\lambda; (u_\lambda, v_\lambda))$ with $u_\lambda > 0$ that is unbounded in $\mathbf{R}^+ \times (H^1 \times L^2)$, in the sense that either there is a solution $(u_\lambda, v_\lambda) \in H^1 \times L^2$, $u_\lambda > 0$, for all λ, or there exists $\lambda_* > 0$ finite and a sequence $(u_{\lambda_n}, v_{\lambda_n})$ of solutions, $u_{\lambda_n} > 0$ with $\lambda_n \to \lambda_* < +\infty$ and $|(u_{\lambda_n}, v_{\lambda_n})|_{H^1 \times L^2} \to +\infty$ as $n \to \infty$.*

Theorem 14.17. (Regularity and large-distance behavior)   *If (u_λ, v_λ) is a solution of (14.3.3)–(14.3.4) in $H^1 \times L^2$ (in the sense of distributions), then it is a classical solution, and u_λ and v_λ are C^∞. In addition, there exists $\delta > 0$ and $c_k > 0$ such that the derivatives of order k obey*

$$|u_\lambda^{(k)}(\eta)| \leq c_k e^{-\delta\eta}, \qquad |v_\lambda^{(k)}(\eta)| \leq \frac{c_k}{1+\eta^{k+3}}. \tag{14.3.5}$$

Proposition 14.18. *For every solution (u_λ, v_λ) of (14.3.3)–(14.3.4) with $\lambda > 0$,*

$$\int u_\lambda^2 \eta\, d\eta > \int R^2 \eta\, d\eta. \tag{14.3.6}$$

This proposition implies that for any self-similar solution (ψ, n) defined in (14.3.1)–(14.3.2), one necessarily has $|\psi_{L^2} > |R|_{L^2}$.

Proof. One first establishes the Pohozaev-type identities

$$\int |\nabla u_\lambda|^2 \eta\, d\eta + \int u_\lambda^2 \eta\, d\eta = -\int v_\lambda u_\lambda^2 \eta\, d\eta, \tag{14.3.7}$$

$$\int u_\lambda^2 \eta\, d\eta = \frac{1}{2}\int (\lambda^2\eta^2 + 1)v_\lambda^2 \eta\, d\eta. \tag{14.3.8}$$

The former equality results from multiplication of (14.3.3) by u_λ and integration in space. The latter is obtained by multiplying the same equation by $\eta\partial_\lambda u_\lambda$ and using (14.1.19). Combining these two identities, one gets

$$\int \Big(|\nabla u_\lambda|^2 + v_\lambda u_\lambda^2 + \frac{1}{2}(\lambda^2\eta^2 + 1)v_\lambda^2\Big)\eta\, d\eta = 0 \tag{14.3.9}$$

or

$$\int \Big(|\nabla u_\lambda|^2 - \frac{1}{2}u_\lambda^4 + \frac{1}{2}(v_\lambda + u_\lambda^2)^2 + \frac{1}{2}\lambda^2\eta^2 v_\lambda^2\Big)\eta\, d\eta = 0. \tag{14.3.10}$$

It follows that

$$\int \left(|\nabla u_\lambda|^2 - \frac{1}{2} u_\lambda^4\right) \eta \, d\eta < 0. \tag{14.3.11}$$

Using the Sobolev inequality with the optimal constant

$$\int u_\lambda^4 \, \eta d\eta \le \frac{2}{\int R^2 \, \eta \, d\eta} \int u_\lambda^2 \, \eta d\eta \int |\nabla u_\lambda|^2 \, \eta \, d\eta, \tag{14.3.12}$$

one then has

$$\left(1 - \frac{\int u_\lambda^2 \, \eta \, d\eta}{\int R^2 \, \eta \, d\eta}\right) \int |\nabla u_\lambda|^2 \, \eta \, d\eta < 0. \tag{14.3.13}$$

Theorem 14.19. *For any $M > |R|_{L^2}^2$, there exists λ_M such that there is a solution $(u_{\lambda_M}, v_{\lambda_M})$ of (14.3.3)–(14.3.4) with $|u_{\lambda_M}|_{L^2}^2 = M$.*

Proof. The proof consists in considering the connected component $\mathcal{C}$ of solutions $(\lambda, u_\lambda, v_\lambda)$ in $\mathbf{R}^+ \times H^1 \times L^2$ of (14.3.3)–(14.3.4) containing $(0, R, -R^2)$. As shown in Glangetas and Merle (1994a), this component is unbounded in $\mathbf{R}^+ \times H^1 \times L^2$. There thus exists a sequence (λ_k, u_k, v_k) obeying the following alternative: either there is λ^{**} finite such that $\lambda_k \to \lambda^{**}$ and $|u_k|_{H^1} + |v_k|_{L^2} \to \infty$, or $\lambda_k \to \infty$. One shows that in both cases, $|u_k|_{L^2} \to \infty$ as $k \to \infty$.

One then considers the self-similar solutions $(\psi_\lambda, n_\lambda)$ of the evolution equations, constructed from $(\lambda, u_\lambda, v_\lambda) \in \mathcal{C}^* = \mathcal{C} \backslash \{0, R, -R^2\}$. Since $\mathcal{C}^*$ is connected and the application $(\lambda, u_\lambda, v_\lambda) \mapsto |u_\lambda|_{L^2}$ is continuous, the set $I = \{|u_\lambda|_{L^2}, (\lambda, u_\lambda, v_\lambda) \in \mathcal{C}^*\}$ is an interval of $\mathbf{R}^+$ included in $(|R|_{L^2}, \infty)$, since $|u_\lambda|_{L^2} > |R|_{L^2}$ for all $\lambda > 0$. In fact, because of the existence of a sequence $(\lambda_k, u_k, v_k) \in \mathcal{C}^*$ for which $|u_k|_{L^2} \to |R|_{L^2}$ and of another one such that $|u_k|_{L^2} \to \infty$, one concludes that $I = (|R|_{L^2}, \infty)$, which proves the theorem.

14.3.2 Mass Concentration Properties

As $t \to t_*$, the self-similar solutions $(\psi_\lambda, n_\lambda)$ constructed from the profiles (u_λ, v_λ), concentrate all their mass at the singularity in the sense that

$$|\psi_\lambda(r, t)|^2 \rightharpoonup |u_\lambda|_{L^2}^2 \delta_r, \tag{14.3.14}$$

$$|n_\lambda(r, t)| \rightharpoonup |v_\lambda|_{L^1} \delta_r, \tag{14.3.15}$$

where δ_r stands for the Dirac distribution at the origin.

Concerning general blowup solutions, a first result given in Glangetas and Merle (1994b) states that if an initially smooth initial condition develops a finite-time singularity, all the H_m norms blow up at the same time, and one can just consider blowing-up solutions in H_1. In this context, the following theorem provides a uniform lower bound for the concentration of the "mass" $|\psi|_{L^2}$.

Theorem 14.20.
(i) *Suppose that (ψ, n) is a radially symmetric solution of the scalar model that blows up in a finite time in the sense that*

$$|\psi|_{H^1} + |n|_{L^2} + |n_t|_{H^{-1}} \to \infty \qquad (14.3.16)$$

as $t \to t_*$. Then there exists a constant m that depends on the initial conditions such that for all $A > 0$, $\liminf_{t \to t_*} |\psi|_{L^2(B(0,A))} \geq |R|_{L^2}$ and $\liminf_{t \to t_*} |n|_{L^1(B(0,A))} \geq m$, where $B(0, A)$ denotes the ball of radius A centered at the origin.
(ii) *In the non-radially-symmetric case, there exists a function $t \to \mathbf{x}(t)$ such that for all $A > 0$,*

$$\lim_{t \to t_*} \inf |\psi|_{L^2(B(\mathbf{x}(t),A))} \geq |R|_{L^2} \qquad (14.3.17)$$

and

$$\lim_{t \to t_*} \inf |n|_{L^1(B(\mathbf{x}(t),A))} \geq m. \qquad (14.3.18)$$

The above results concerning the mass concentration of self-similar solutions correspond to the physical picture that an isolated caviton is approximated by a self-similar solution whose dimension tends to zero as $t \to t_*$, with the entire energy trapped in the caviton becoming localized at the singularity, a situation corresponding to a strong wave collapse (Malkin and Khudik 1989). In this context, we note that numerical simulations (in one space dimension and a quartic ponderomotive force), including weak dissipative terms modeling Landau damping and ion damping show a total burnout of the wave energy trapped in the cavity (D'yachenko, Zakharov, Rubenchik, Sagdeev, and Shvets 1988, Newell, Rand, and Russel 1988). This prediction is supported by two-dimensional particles in cell simulations of the set of Vlasov equations for the electrons and the ions, which indicate that during the collapse, the electric field energy density increases by about one order of magnitude and that after this, the oscillation energy is almost completely dissipated and transferred to fast electrons (Anisimov, Berezovskiĭ, Ivanov, Petrov, Rubenchik, and Zakharov 1982). This contrasts with the case where the density is slaved to the wave amplitude, which is then governed by the (damped) NLS equation. As discussed in Section 9.1.1, the burnout is then only partial.

14.3.3   A Sharp Existence Result

Theorem 14.21. (Glangetas and Merle 1994b) *For initial conditions (ψ, n) in $H^1 \times L^2$ satisfying $|\psi_0|_{L^2} \leq |R|_{L^2}$, the solution of the scalar model exists for all time.*

Proof. Consider the Hamiltonian

$$H = \int \left(|\nabla \psi|^2 + n|\psi|^2 + \frac{1}{2}|\mathbf{v}|^2 + \frac{1}{2}n^2 \right) dx \qquad (14.3.19)$$

that is rewritten

$$H = \int \left(|\nabla \psi|^2 - \frac{1}{2}|\psi|^4 + \frac{1}{2}(n + |\psi|^2)^2 + \frac{1}{2}|\mathbf{v}|^2 \right) d\mathbf{x}. \qquad (14.3.20)$$

The case $|\psi_0|_{L^2} < |R|_{L^2}$ is straightforward when the Sobolev inequality (14.3.12) is used.

When $|\psi_0|_{L^2} = |R|_{L^2}$, one proceeds by contradiction. Assume that there is $T_0 > 0$ such that

$$|\nabla \psi(t)|_{L^2} + |n(t)|_{L^2} + |\mathbf{v}(t)|_{L^2} \to \infty. \qquad (14.3.21)$$

From the Sobolev inequatlity with the optimal constant and the hypothesis $|\psi_0|_{L^2} = |R|_{L^2}$, one has

$$\int \left(|\nabla \psi|^2 - \frac{1}{2}|\psi|^4 \right) d\mathbf{x} \geq 0. \qquad (14.3.22)$$

The conservation of the Hamiltonian then implies that for $t < T_0$,

$$0 \leq \int \left(|\nabla \psi|^2 - \frac{1}{2}|\psi|^4 \right) d\mathbf{x} \leq H, \qquad (14.3.23)$$

$$\int (n + |\psi|^2)^2 d\mathbf{x} \leq H, \qquad (14.3.24)$$

$$\int |\mathbf{v}|^2 d\mathbf{x} \leq H. \qquad (14.3.25)$$

In addition, one has

$$|\psi(t)|_{H^{-1}} \leq C. \qquad (14.3.26)$$

This estimate is obtained by writing $|\psi|^2 = (n + |\psi|^2) - n$. The first term in the right-hand side, which is bounded in L^2, is also bounded in H^{-1}. For the second one, one writes

$$|n(t)|_{H^{-1}} \leq |n(0)|_{H^{-1}} + \int_0^t |n_t(s)|_{H^{-1}} ds$$

$$\leq |n(0)|_{H^{-1}} + \frac{1}{\epsilon} \int_0^t |\nabla \cdot \mathbf{v}(s)|_{H^{-1}} ds$$

$$\leq |n(0)|_{H^{-1}} + \frac{1}{\epsilon} \int_0^t |\mathbf{v}(s)|_{L^2} ds \leq \frac{C}{\epsilon}. \qquad (14.3.27)$$

From the singularity assumption and the above upper bounds, it follows that $|\nabla \psi(t)|_{L^2} \to \infty$ as $t \to T_0$. From the concentration property, there exists $\mathbf{x}(t)$ such that $|\psi(\mathbf{x} + \mathbf{x}(t), t)|^2 \rightharpoonup |R|_{L^2}^2 \delta_{\mathbf{x}(t)}$. Since $|\psi(\mathbf{x} + \mathbf{x}(t), t)|^2$ is bounded in H^{-1}, it follows that the two-dimensional δ distribution is in H^{-1}, which leads to a contradiction.

Notice that the estimate on $|n(t)|_{H^{-1}}$ is lost in the limit $\epsilon \to 0$. The result indeed differs from the case of the NLS equation, for which there exist blow-up solutions with the minimal L^2-norm.

14.3.4 Instability of Standing Wave Solutions

There exist standing waves for the scalar model, of the form

$$\psi(r,t) = e^{i\omega t}\phi(r), \quad n(r,t) = -|\phi(r)|^2, \tag{14.3.28}$$

where ϕ is a radial solution of

$$\Delta\phi - \omega\phi + \phi^3 = 0. \tag{14.3.29}$$

The question is whether the solutions (14.3.28) are orbitally stable in the space where the initial value problem is solved.

Theorem 14.22. (Glangetas and Merle 1994b) *Let (ψ, n) be a standing-wave solution of the scalar model (14.2.1–14.2.2) of the form*

$$\psi(r,t) = e^{i\omega_0 t}\phi(r), \quad n(r,t) = -\phi^2(r) \tag{14.3.30}$$

with θ_0, $\omega_0 > 0$, constructed from the ground state

$$\phi(r) = e^{i\theta_0}\omega_0^{1/2}R(\omega_0^{1/2}r). \tag{14.3.31}$$

This solution is orbitally unstable in the sense that there exists a sequence $(\psi_{0\mu}, n_{0\mu}, n_{1\mu}) \to (\phi, -|\phi|^2, 0)$ in H_k, $k \geq 1$ as $\mu \to 0$, such that the solution (ψ_μ, n_μ) corresponding to these initial conditions blows up in a finite time in H_1.

By a simple rescaling, one can restrict oneself to the case $\omega_0 = 1$ and $\theta_0 = 0$. The proof is then based on the existence of the sequence of blowup self-similar solutions

$$\psi_\mu = \frac{1}{1-\mu t}e^{i\left(\frac{\mu|\mathbf{X}|^2}{4(\mu t-1)}+\frac{t}{1-\mu t}\right)}u_\mu\left(\frac{|\mathbf{x}|}{1-\mu t}\right), \tag{14.3.32}$$

$$n_\mu = \frac{1}{(1-\mu t)^2}v_\mu\left(\frac{|\mathbf{x}|}{1-\mu t}\right) \tag{14.3.33}$$

obtained by choosing $t_* = -\theta = \mu^{-1}$ in (14.3.1)–(14.3.2), the profiles obeying (14.3.3)–(14.3.4) with $\lambda = \epsilon\mu$. The theorem follows from the convergence in H^k $\forall k > 1$, of the initial values $(\psi_\mu(0), n_\mu(0), \partial_t n_\mu(0))$ of $(\psi_\mu, n_\mu, \partial_t n_\mu)$ to the initial conditions $(R, -R^2, 0)$, of the standing-wave solution as $\mu \to 0$. This proof requires several steps, which are detailed in the original paper.

Standing waves whose profile is an excited state are also shown to be unstable under an additional nondegeneracy condition on the linearized operator.

14.3.5 An Optimal Lower Bound for the Blowup Rate

Theorem 14.23. (Merle 1996b) *Suppose that the solution (ψ, n, n_t) blows up in a finite time T_* in H_1. There exist constants $c_1 > 0$ and $c_2 > 0$*

depending only on $|\psi_0|_{L^2}$ such that for t close to t_*,

$$|\nabla \psi(t)|_{L^2} \geq \frac{c_1}{t_* - t},$$ (14.3.34)

$$|n(t)|_{L^2} \geq \frac{c_2}{t_* - t}.$$ (14.3.35)

The constants c_1 and c_2 scale like $(|\psi_0|_{L^2}^2 - |R|_{L^2}^2)^{-1/2}$ as the initial L^2-norm $|\psi_0|_{L^2}$ tends to $|R|_{L^2}$.

This result is optimal in the sense that the blowup solutions constructed in Section 14.3.1 satisfy

$$|\nabla \psi(t)|_{L^2} = \frac{1}{\mu(t_* - t)}|\nabla u|_{L^2},$$ (14.3.36)

$$|n(t)|_{L^2} = \frac{1}{\mu(t_* - t)}|v|_{L^2}.$$ (14.3.37)

Proof. The proof of the above theorem is in the same spirit (although more delicate) as that of Theorem 5.9 for the NLS equation. A scaling of the time variable associated to the wave equation rather than to the Schrödinger operator is used to get the optimal results. One defines

$$\widetilde{\psi}(s) = \frac{1}{\lambda(t)}\psi\left(\frac{\mathbf{x}}{\lambda(t)}, t + \frac{s}{\lambda(t)}\right),$$ (14.3.38)

$$\widetilde{n}(s) = \frac{1}{\lambda^2(t)}n\left(\frac{\mathbf{x}}{\lambda(t)}, t + \frac{s}{\lambda(t)}\right),$$ (14.3.39)

$$\widetilde{\mathbf{v}}(s) = \frac{1}{\lambda^2(t)}\mathbf{v}\left(\frac{\mathbf{x}}{\lambda(t)}, t + \frac{s}{\lambda(t)}\right),$$ (14.3.40)

where t is viewed as a parameter and

$$\lambda(t) = \int \left(|\nabla \psi|^2 + \frac{1}{2}n^2 + \frac{1}{2}|\mathbf{v}|^2\right)dx.$$ (14.3.41)

Equivalently, at $s = 0$,

$$\int \left(|\nabla\widetilde{\psi}(0)|^2 + \frac{1}{2}\widetilde{n}(0)^2 + \frac{1}{2}|\widetilde{\mathbf{v}}(0)|^2\right)dx = 1.$$ (14.3.42)

One also has

$$\int |\widetilde{\psi}(0)|^2 dx = \int |\psi_0|^2 dx.$$ (14.3.43)

Furthermore, $\lambda(t) \to \infty$ as $t \to t_*$.

The rescaled functions satisfy

$$i\lambda\partial_s\widetilde{\psi} + \Delta\widetilde{\psi} = \widetilde{n}\widetilde{\psi},$$ (14.3.44)

$$\epsilon\partial_s\widetilde{n} + \nabla \cdot \widetilde{\mathbf{v}} = 0,$$ (14.3.45)

$$\epsilon\partial_s\widetilde{\mathbf{v}} + \nabla\widetilde{n} = -\nabla|\widetilde{\psi}|^2.$$ (14.3.46)

The associated Hamiltonian $\widetilde{H}$ is given by $\widetilde{H} = \frac{H}{\lambda^2(t)}$, where H is the (conserved) Hamiltonian of the primitive scalar model.

In contrast with the NLS equation, the scaling factor $\lambda(t)$ appears in the above equations, because the carrier envelope and the density wave evolve on different time scales. Furthermore, only the combination (14.3.42) of the norms that characterizes the initial value problem (14.3.44)–(14.3.46) is fixed to unity, each of them being dependent on t. The first step consists in establishing bounds for these norms. Lower bounds are obtained by noting that $\widetilde{H} = 1 + \int \widetilde{n}(0)|\widetilde{\psi}(0)|^2 dx$, and $\widetilde{H}$ decreases as $t \to t_*$. These estimates are, however, not sufficient to obtain the behavior of $\widetilde{\psi}(0)$ and $\widetilde{n}(0)$ as $t \to t_*$. The problem, which requires a refined analysis, is overcome by using compactness properties of $\widetilde{\psi}(0)$ to derive compactness for $\widetilde{n}(0)$. This leads to establish the following proposition.

Proposition 14.24.   *There are $\delta > 0$, c_1 and c_2 such that $\forall t \in [t_* - \delta, t_*)$, $0 < c_1 < |\nabla\widetilde{\psi}(0)|_{L^2} \leq c_2$, $0 < c_1 < |\widetilde{n}(0)|_{L^2} \leq c_2$ and $0 \leq |\nabla\widetilde{v}(0)|_{L^2} \leq c_2$, where c_1 and c_2 depend only on $|\psi_0|_{L^2}$.*

The next step is the estimate of $\widetilde{\psi}(s)$, $\widetilde{n}(s)$, $\widetilde{v}(s)$, as $t \to t_*$. From (14.3.42) and the assumption

$$\lim_{s \to (t_* - t)\lambda(t)} \int \left(|\nabla\widetilde{\psi}(s)|^2 + \frac{1}{2}\widetilde{n}(s)^2 + \frac{1}{2}|\widetilde{v}(s)|^2 \right) dx = \infty \qquad (14.3.47)$$

it follows that for any $A > 1$ there exists, by continuity, $\theta(t) > 0$ such that

$$\forall s \in [0, \theta(t)], \quad \int \left(|\nabla\widetilde{\psi}(s)|^2 + \frac{1}{2}\widetilde{n}(s)^2 + \frac{1}{2}|\widetilde{v}(s)|^2 \right) dx \leq A, \qquad (14.3.48)$$

and

$$\int \left(|\nabla\widetilde{\psi}(\theta(t))|^2 + \frac{1}{2}\widetilde{n}(\theta(t))^2 + \frac{1}{2}|\widetilde{v}(\theta(t))|^2 \right) dx = A. \qquad (14.3.49)$$

One then has to prove that there exists θ_0 independent of t such that for t close to t_*, $\theta(t) \geq \theta_0$. This will imply that $\lambda(t) \geq \frac{\theta_0}{t_* - t}$, which concludes the proof.

In order to establish the existence of θ_0 that only depends on $|\psi_0|_{L^2}$, and also to capture the behavior of the constants as $|\psi_0|$ approaches $|R|_{L^2}$, one proceeds in two steps. Using Proposition 14.24 and (14.3.48), one shows that $\forall s \in [0, \theta(t)]$,

$$|\nabla\widetilde{\psi}(s)|_{L^2} \leq Ac_2, \qquad (14.3.50)$$
$$|\widetilde{n}(s)|_{L^2} \leq Ac_2, \qquad (14.3.51)$$
$$|\nabla\widetilde{v}(s)|_{L^2} \leq Ac_2. \qquad (14.3.52)$$

The last inequality can be made more precise in the form

$$|\nabla\widetilde{v}(s)|_{L^2} \leq Ac_2(|\psi_0|_{L^2}^2 - |R|_{L^2}^2)^{1/2}. \qquad (14.3.53)$$

To prove this last point, one rewrites

$$\widetilde{H} = \int \{ |\nabla \widetilde{\psi}|^2 - \frac{1}{2}|\widetilde{\psi}|^4 + \frac{1}{2}(\widetilde{n} + |\widetilde{\psi}|^2)^2 + \frac{1}{2}|\widetilde{\mathbf{v}}|^2 \} dx = \frac{H}{\lambda^2}. \tag{14.3.54}$$

Therefore, using

$$\left(1 - \frac{|\psi_0|_{L^2}^2}{|R|_{L^2}^2} \right) |\nabla \widetilde{\psi}|_{L^2}^2 \leq |\nabla \widetilde{\psi}|_{L^2}^2 - \frac{1}{2}|\widetilde{\psi}|_{L^4}^4, \tag{14.3.55}$$

one gets

$$|\widetilde{\mathbf{v}}|_{L^2}^2 \leq \frac{2H}{\lambda^2} + 2\frac{c_2}{|R|_{L^2}^2}(|\psi_0|_{L^2}^2 - |R|_{L^2}^2), \tag{14.3.56}$$

which leads to the result for t close enough to t_*.

One then establishes compactness on $\widetilde{\psi}(\theta(t))$ and $\widetilde{n}(\theta(t))$ to find a lower bound to their weak limit as $t \to t_*$ in H^1 and L^2, respectively. This plays a central role in proving that there exists a constant $c > 0$ such that $\liminf_{t \to t_*} \int_0^{\theta(t)} |\widetilde{\mathbf{v}}(s)|_{L^2} ds \geq c$. From the estimate

$$\int_0^{\theta(t)} |\widetilde{\mathbf{v}}(s)|_{L^2} ds \leq \int_0^{\theta(t)} Ac_2(|\psi_0|_{L^2}^2 - |R|_{L^2}^2)^{1/2}) ds$$
$$\leq Ac_2(|\psi_0|_{L^2}^2 - |R|_{L^2}^2)^{1/2})\theta(t), \tag{14.3.57}$$

it follows that

$$\liminf_{t \to t_*} \theta(t) \geq \frac{c}{Ac_2(|\psi_0|_{L^2}^2 - |R|_{L^2}^2)^{1/2}} > 0. \tag{14.3.58}$$

15
Progressive Waves in Plasmas

15.1 Interaction with a Nonmagnetic Medium

15.1.1 Laser Beams in Plasmas

Density fluctuations are a main source of nonlinearity in laser plasmas, a subject of great importance in the context of inertial confinement fusion. The dynamical equations describing the interaction of the laser beam with plasma low-frequency fluctuations driven by ponderomotive forces can be derived using a multiple-scale expansion. In fact, due to their genericity, the dynamical equations can, to a large extent, be predicted from general arguments (Zakharov, Musher, and Rubenchik 1985). We assume for the sake of simplicity that the wave is linearly polarized (scalar field) and that the plasma, assumed non magnetized, is isotropic. We also assume that the index of refraction of the medium is sensitive to the low-frequency variations of the density. For a small-amplitude quasi-monochromatic carrying wave $\epsilon\psi(\mathbf{X}, T)e^{i(kx-\omega t)}$ with $\mathbf{X} = \epsilon x$ and $T = \epsilon t$, and low-frequency density fluctuations $\epsilon^2\rho(\mathbf{X}, T)$, the linear dispersion relation $\omega = \omega(k)$ is affected in the form (see Section 1.1.1)

$$(\omega + \epsilon i\partial_T - \Omega(k - \epsilon i\partial_X, -\epsilon i\nabla_\perp, \epsilon^2\rho))\psi = 0, \qquad (15.1.1)$$

with $\Omega(k, 0, 0) = \omega(k)$. The ϵ-expansion of the above equation leads to the paraxial wave equation

$$i(\partial_T\psi + v_g\partial_X\psi) + \epsilon\left(\frac{\omega''}{2}\partial_{XX}\psi + \frac{v_g}{2k}\Delta_\perp\psi + q\rho\psi\right) = 0. \qquad (15.1.2)$$

At the order of the expansion, the (rescaled) low-frequency density variations ρ and hydrodynamic potential φ obey the equations of the linearized hydrodynamics (conservation of mass and momentum)

$$\partial_T \rho = -\rho_0 \Delta \varphi \tag{15.1.3}$$

$$\partial_T \varphi = -c^2 \frac{\rho}{\rho_0} - q|\psi|^2, \tag{15.1.4}$$

where c denotes the sound velocity, ρ_0 the uniform density in the absence of modulation, and where the ponderomotive force originates from the interaction of oscillating contributions and results in a wave pressure effect. More elaborated models are discussed by Schmitt (1988). Note that the coefficient q is the same in eqs. (15.1.2) and (15.1.4) (this is always possible by a rescaling of ψ), in order to preserve the Hamiltonian character of the system (see below). From eqs. (15.1.3)–(15.1.4), it is seen that the density obeys a wave equation with a stirring originating from the ponderomotive force:

$$(\partial_{TT} - c^2 \Delta)\rho = q\rho_0 \Delta |\psi|^2, \tag{15.1.5}$$

as in the Zakharov systems for Langmuir oscillations.

An application arises in the context of hot spots (local maxima in the intensity) that are likely to initiate self-focusing in laser plasmas. When the transit time of the laser light along a hot spot is short compared to the hydrodynamic time scales, the time derivative is negligible in the equation for the envelope E of the electric field. On an extended longitudinal scale $\xi = \epsilon X$, the effect of the group velocity dispersion is negligible, leading to the nondimensional system

$$i\partial_\xi E + \Delta_\perp E = \rho E, \tag{15.1.6}$$

$$(\partial_{TT} - \Delta_\perp)\rho = \Delta_\perp |E|^2. \tag{15.1.7}$$

This problem was studied in details by Rose and Dubois (1993). Numerical simulations suggest the existence of a singularity when the wave amplitude is larger than critical, with formation of density holes where the light is trapped. This results in the formation of "laser filaments" in the plasma whose signature was observed experimentally (Willi, Afshar-rad, Coe, and Giulietti 1990, Labaune, Baton, Jalinaud, Baldis, and Pesme 1992). Laser-plasma filamentation is considered as possibly the most dangerous laser-plasma instability for inertial confinement fusion.

15.1.2 Hamiltonian Formalism

In other instances, the frequency shift of the carrier due to the interaction with the medium depends not only on the density variations but also on the medium velocity v_x in the direction of propagation (Doppler shift), leading to an additional contribution rv_x to the potential in the envelope equation. An example provided by the Alfvén wave filamentation in magnetohydrodynamics is analyzed in detail in the next section, using a systematic multiple-scale expansion. An Hamiltonian approach to this coupling is considered by Zakharov and Rubenchik (1972) (see also Zakharov 1974, Zakharov, and Kuznetsov 1984, 1997, Zakharov and Schulman 1991). It is easily obtained by writing a priori the Hamiltonian of the system in the form

$$H = H_w + H_s + H_i, \qquad (15.1.8)$$

where H_w is the Hamiltonian of the wave packet, H_s that of the sound waves, and H_i is associated to the interaction between them. In the weakly nonlinear regime, the Hamiltonian of a quasi-monochromatic plane wave is written

$$H_w = \int \omega(\mathbf{k}) a_{\mathbf{k}}^* a_{\mathbf{k}} d\mathbf{k}, \qquad (15.1.9)$$

where

$$a_{\mathbf{k}} = \psi_{\boldsymbol{\kappa}} e^{i(k_0 x - \omega(k_0)t)}, \qquad (15.1.10)$$

$\boldsymbol{\kappa} = \mathbf{k} - \mathbf{k}_0$, and $\mathbf{k}_0 = (k_0, 0, 0)$, with $\psi_{\boldsymbol{\kappa}}$ defined as the Fourier transform of the wave-packet amplitude

$$\psi(\mathbf{x}) = \frac{1}{(2\pi)^{3/2}} \int \psi_{\boldsymbol{\kappa}} e^{i\boldsymbol{\kappa}\cdot\mathbf{x}} d\boldsymbol{\kappa}. \qquad (15.1.11)$$

In an isotropic medium, the frequency is expanded as

$$\omega(\mathbf{k}) = \omega(k_0) + \mathbf{v}_g \cdot \boldsymbol{\kappa} + \frac{1}{2}\omega''(k_0)\kappa_x^2 + \frac{v_g^2}{2k_0}(\kappa_y^2 + \kappa_z^2) + p|\psi|^2 + \cdots, \quad (15.1.12)$$

where we have also included the nonlinear frequency shift arising from the possible self-interaction of the wave in the weakly nonlinear regime. It follows that

$$H_w = \omega(k_0) \int |\psi|^2 d\mathbf{x} + \frac{iv_g}{2} \int \left(\psi \frac{\partial\psi^*}{\partial x} - \psi^* \frac{\partial\psi}{\partial x}\right) d\mathbf{x}$$
$$+ \int \left(\frac{\omega''(k_0)}{2}\left|\frac{\partial\psi}{\partial x}\right|^2 + \frac{v_g}{2k_0}|\nabla_\perp \psi|^2 + \frac{1}{2}p|\psi|^4\right) d\mathbf{x} + \cdots, \quad (15.1.13)$$

where the first two integrals are also conserved quantities (mass or energy) and the longitudinal component of the momentum. On the other hand, in terms of the canonical variables, which in the absence of magnetic field are the density variation ρ from the equilibrium ρ_0 and the hydrodynamic scalar potential ϕ (with $v_x = \frac{\partial\phi}{\partial x}$), the Hamiltonian associated to the acoustic waves is

$$H_s = \frac{1}{2} \int \left(\frac{c^2\rho^2}{\rho_0} + \rho|\nabla\phi|^2\right) d\mathbf{x}. \qquad (15.1.14)$$

Finally, the interaction potential is written

$$H_i = \int \delta\omega|\psi|^2 d\mathbf{x}, \tag{15.1.15}$$

where $\delta\omega = rv_x + q\rho$. The Hamilton equations

$$i\psi_t = \frac{\delta H}{\delta\psi^*}, \quad \rho_t = \frac{\delta H}{\delta\phi}, \quad \phi_t = -\frac{\delta H}{\delta\rho} \tag{15.1.16}$$

then are

$$i\left(\frac{\partial\psi}{\partial T} + v_g\frac{\partial\psi}{\partial X}\right) + \epsilon\left(\frac{\omega''}{2}\frac{\partial^2\psi}{\partial X^2} + \frac{v_g}{2k}\Delta_\perp\psi - (p|\psi|^2 + q\rho + r\phi_X)\psi\right) = 0 \tag{15.1.17}$$

with

$$\partial_T\rho = -\rho_0\Delta\phi - r\partial_X|\psi|^2, \tag{15.1.18}$$

$$\partial_T\phi = -c^2\frac{\rho}{\rho_0} - q|\psi|^2, \tag{15.1.19}$$

where the equations are written in the rescaled variables. We note that in a situation where the field ρ reduces to the long-wave density variations, the coefficient r should be small, since the interactions between the density and velocity fluctuations are subdominant in the density equation. The situation is different in the case of the gravity waves at the surface of an incompressible fluid, discussed in Section 11.1 or in the context of Alfvén wave filamentation (see Section 15.2.2).

Existence during a finite time of classical solutions for (15.1.17)–(15.1.19) was recently established by Ponce and Saut (1998) in a class of Sobolev spaces. They proved that if the initial condition (ψ_0, ρ_0, ϕ_0) belongs to $H^s(\mathbf{R}^d) \times H^{s-1/2}(\mathbf{R}^d) \times H^{s+1/2}(\mathbf{R}^d)$ with $s > d/2$, there exists a time $T > 0$ and a unique solution $\psi \in C([0,T), H^s(\mathbf{R}^d))$, $\rho \in C([0,T), H^{s-1/2}(\mathbf{R}^d))$, $\varphi \in C([0,T), H^{s+1/2}(\mathbf{R}^d))$. The idea is to solve the inhomogeneous wave equations satisfied by ρ and ϕ in terms of the carrying wave amplitude. The equation for the carrier then reduces to a generalized nonlinear Schrödinger equation with an integrodifferential potential. The order of derivatives of ψ appearing in the nonlinearity is roughly of order one (see, e.g., Kenig, Ponce and Vega 1995) since $(\partial_{TT} - \Delta)^{-1}\partial_{XX}$ is basically an operator of order one. The proof that generalizes the result of Kenig, Ponce, and Vega (1997) for a potential depending ψ and $\nabla\psi$ (see Section 3.4.1) then uses the structure of the nonlinearity combined with estimates involving the local smoothing effect associated to the linear Schrödinger operator (see Section 3.1.2), together with special properties of the solutions of the wave equation, and concludes with a fixed point argument.

When writing (15.1.17)–(15.1.19) in a frame moving at the group velocity of the wave-packet by defining $\xi = X - v_gT$ and $\tau = \epsilon T$, and taking the adiabatic limit for the acoustic waves by neglecting the $O(\epsilon)$ time derivative, one gets the system

$$i\frac{\partial \psi}{\partial \tau} + \frac{\omega''}{2}\frac{\partial^2 \psi}{\partial \xi^2} + \frac{v_g}{2k}\Delta_\perp \psi = (p|\psi|^2 + q\rho + r\varphi_\xi)\psi \tag{15.1.20}$$

$$(c^2 - v_g^2)\partial_{\xi\xi}\rho + c^2\Delta_\perp \rho = -(v_g r + \rho_0 q)\partial_{\xi\xi}|\psi|^2 - \rho q \Delta_\perp |\psi|^2, \tag{15.1.21}$$

introduced by Zakharov & Schulman (1991) and which generalizes the Davey-Stewartson equations to higher dimensions (see also Chapter 12 for references on rigorous results).

Various regimes are, in fact, described by (15.1.17)–(15.1.19), according to the values of the parameters. Their characteristics are analyzed in Passot, Sulem, and Sulem (1996) in the case of localized solutions considered by the existence theorem. The hot-spot dynamics briefly mentioned in Section 15.1.1 and the Alfvén wave filamentation discussed Section 15.2.2 can be viewed as two typical regimes described by this system.

An extension of (15.1.17)–(15.1.19) was proposed by Relke and Rubenchik (1988) to the case of MHD waves where the velocity field induced by the modulation is not potential, using the Hamiltonian description of MHD flows (Zakharov and Kuznetsov 1971). The resulting equations are, however, not confirmed by the systematic multiple-scale expansion presented in Section 15.2.1, in the case of Alfvén waves propagating along an ambient magnetic field.

15.2 Alfvén Wave Filamentation

An example where the Doppler shift is relevant concerns the Alfvén waves propagating along an ambient magnetic field in dispersive magnetohydrodynamics (MHD). A systematic derivation of the modulation equations from the MHD equations is presented in Champeaux, Passot, and Sulem (1997). We review it in detail in the next section, because it provides an example of relatively elaborated modulation analysis leading to original dynamics for the low-frequency magnetosonic waves that develop fronts and not cavities as in the situations where the low-frequency density variations are governed by (15.1.7).

15.2.1 Modulation Equations

When the Hall effect associated to the inertia of the ions is retained in Ohm's law, the magnetohydrodynamic (MHD) equations become, using non-dimensional variables where the Alfvén wave velocity $c_A = \sqrt{\frac{B_0}{4\pi\rho_0}}$ constructed with the magnitude of the ambient magnetic field B_0 and the equilibrium density ρ_0, is taken as unity,

$$\partial_t \rho + \nabla \cdot (\rho \mathbf{u}) = 0 \tag{15.2.1}$$

$$\rho(\partial_t \mathbf{u} + \mathbf{u}.\nabla \mathbf{u}) = -\frac{\beta}{\gamma}\nabla\rho^\gamma + (\nabla \times \mathbf{b}) \times \mathbf{b} \tag{15.2.2}$$

$$\partial_t \mathbf{b} - \nabla \times (\mathbf{u} \times \mathbf{b}) = -\frac{1}{R_i}\nabla \times (\frac{1}{\rho}(\nabla \times \mathbf{b}) \times \mathbf{b}) \tag{15.2.3}$$

$$\nabla \cdot \mathbf{b} = 0, \tag{15.2.4}$$

where the constants R_i and γ respectively denote the nondimensional gyromagnetic frequency of the ions and the polytropic gas constant. The parameter β is defined as $\beta = c_A^2/c_s^2$ where c_s is the sound speed.

Consider a quasi-monochromatic Alfvén wave train with an amplitude of magnitude $\mu \ll 1$, propagating along an ambient magnetic field pointing in the x-direction and whose modulus is taken as unity. The modulation is described in terms of the slow variables $X = \mu x$, $Y = \mu y$, $Z = \mu z$, $T = \mu t$. The field components are expanded as

$$
\begin{aligned}
b_y &= \mu b_{y_1} + \mu^2 b_{y_2} + \cdots, & u_y &= \mu u_{y_1} + \mu^2 u_{y_2} + \cdots, \\
b_z &= \mu b_{z_1} + \mu^2 b_{z_2} + \cdots, & u_z &= \mu u_{z_1} + \mu^2 u_{z_2} + \cdots, \\
b_x &= 1 + \mu^2 b_{x_2} + \mu^3 b_{x_3} + \cdots, & u_x &= \mu^2 u_{x_2} + \mu^3 u_{x_3} + \cdots, \\
\rho &= 1 + \mu^2 \rho_2 + \mu^3 \rho_3 + \cdots,
\end{aligned}
$$

where the different magnitudes of the transverse and longitudinal components select the Alfvén wave eigenmode.

At leading order, one gets the linear equations for strictly parallel propagation,

$$\partial_t u_{y_1} - \partial_x b_{y_1} = 0, \qquad \partial_t b_{y_1} - \partial_x u_{y_1} - \frac{1}{R_i}\partial_{xx} b_{z_1} = 0, \tag{15.2.5}$$

$$\partial_t u_{z_1} - \partial_x b_{z_1} = 0, \qquad \partial_t b_{z_1} - \partial_x u_{z_1} + \frac{1}{R_i}\partial_{xx} b_{y_1} = 0. \tag{15.2.6}$$

The presence of dispersion implies that harmonic solutions of the form

$$b_{y_1} = B_{y_1} e^{i(kx-\omega t)} + \text{c.c.}, \qquad b_{z_1} = B_{z_1} e^{i(kx-\omega t)} + \text{c.c.} \tag{15.2.7}$$

satisfy $B_{y_1} = i\zeta B_{z_1}$ with $\zeta = -1$ or $\zeta = 1$ for right- or left-hand side circularly polarized waves $b_y + i\zeta b_z \sim e^{i(kx-\omega t)}$, respectively. The dispersion relation reads

$$\left(\frac{\omega}{k}\right)^2 = 1 - \frac{\zeta\omega}{R_i}. \tag{15.2.8}$$

Writing

$$b_1 = b_{y_1} + i\zeta b_{z_1} = B_1 e^{i(kx-\omega t)}, \qquad u_1 = u_{y_1} + i\zeta u_{z_1} = U_1 e^{i(kx-\omega t)}, \tag{15.2.9}$$

we obtain that the transverse velocity is to leading order proportional to the transverse magnetic field:

$$U_1 = -\frac{k}{\omega}B_1. \tag{15.2.10}$$

At order μ^2, the equations for the density and for the longitudinal components of the velocity and magnetic field read (writing $\partial_\perp = \partial_y + i\zeta\partial_z$)

$$\partial_t\rho_2 + \partial_x u_{x_2} = -\Re[\partial_\perp^* u_1] \tag{15.2.11}$$

$$\partial_t u_{x_2} + \beta\partial_x\rho_2 = 0 \tag{15.2.12}$$

$$\partial_t b_{x_2} = -\Re[\partial_\perp^* u_1] - \frac{\zeta}{R_i}\Im[\partial_x(\partial_\perp^* b_1)] \tag{15.2.13}$$

$$\partial_x b_{x_2} = -\Re[\partial_\perp^* b_1]. \tag{15.2.14}$$

They are solved as

$$\rho_2 = \tilde{\rho}_2 e^{i(kx-\omega t)} + \text{c.c.} + \bar{\rho}_2 \tag{15.2.15}$$

$$u_{x_2} = \tilde{u}_{x_2} e^{i(kx-\omega t)} + \text{c.c.} + \bar{u}_{x_2} \tag{15.2.16}$$

$$b_{x_2} = \tilde{b}_{x_2} e^{i(kx-\omega t)} + \text{c.c.} + \bar{b}_{x_2}, \tag{15.2.17}$$

where overbars refer to non-oscillating contributions that appear to be driven at the next order of the expansion. The amplitudes of the oscillating parts are given by

$$\tilde{u}_{x_2} = \frac{\beta k}{\omega}\tilde{\rho}_2, \quad \tilde{\rho}_2 = -\frac{ik}{2(\beta k^2 - \omega^2)}\partial_\perp^* B_1, \quad \tilde{b}_{x_2} = \frac{i}{2k}\partial_\perp^* B_1. \tag{15.2.18}$$

The divergence for $\omega/k = \beta^{1/2}$ of the coefficient entering the definition of $\tilde{\rho}_1$ corresponds to the coincidence of the phase velocities of the Alfvén and magneto-sonic waves. The neighborhood of this resonance is not amenable to the above envelope description: The small-scale magnetosonic waves (which are not dispersive) are no longer slaved to the Alfvén wave, as reflected by the formation of shocks in this regime (see Hada 1993, Passot & Sulem 1995, Gazol, Passot & Sulem 1999 for a discussion of the long wavelength dynamics). As stressed by Spangler (1989) and Medvedev and Diamond (1995), kinetic effects should, in fact, be retained near this resonance.

At order $O(\mu^2)$, the equations for the transverse fields read

$$\partial_t u_2 - \partial_x b_2 = -\partial_T u_1 + \partial_X b_1, \tag{15.2.19}$$

$$\partial_t b_2 - \partial_x u_2 + \zeta\frac{i}{R_i}\partial_{xx}b_2 = -\partial_T b_1 + \partial_X u_1 - \zeta\frac{2i}{R_i}\partial_{Xx}b_1. \tag{15.2.20}$$

The solvability condition is

$$k(-\partial_T U_1 + \partial_X B_1) - \omega(-\partial_T B_1 + \partial_X U_1 + \zeta\frac{2k}{R_i}\partial_X B_1) = 0, \tag{15.2.21}$$

or in terms of the Alfvén group velocity, $v_g = \omega' = \frac{2\omega^3}{k(k^2+\omega^2)}$,

$$\partial_T B_1 + v_g \partial_X B_1 = 0. \tag{15.2.22}$$

Equations (15.2.19)–(15.2.20) are then solved in the form

$$\begin{pmatrix} b_2 \\ u_2 \end{pmatrix} = \left[\begin{pmatrix} 1 \\ \frac{-k}{\omega} \end{pmatrix} B_2 + \begin{pmatrix} 0 \\ \frac{i}{\omega}(\partial_X + \frac{k}{\omega}\partial_T)B_1 \end{pmatrix} \right] e^{i(kx-\omega t)},$$

$$\tag{15.2.23}$$

where the B_2-term refers to a null-space element that in the present approach appears as a correction to the carrier amplitude.

At order $O(\mu^3)$, we have

$$\partial_t u_3 - \partial_x b_3 = -\partial_T u_2 + \partial_X b_2 - \partial_\perp(\beta \rho_2 + b_{x_2})$$
$$-\rho_2 \partial_t u_1 - u_{x_2}\partial_x u_1 + b_{x_1}\partial_x b_1$$
$$+\frac{\zeta}{2}(\partial_\perp^* b_1 - \partial_\perp b_1^*)b_1 - \frac{1}{2}(u_1^*\partial_\perp + u_1\partial_\perp^*)u_1 \tag{15.2.24}$$

$$\partial_t b_3 - \partial_x u_3 + \frac{i\zeta}{R_i}\partial_{xx}b_3 = -\partial_T b_2 + \partial_X u_2 - \frac{i\zeta}{R_i}\partial_{XX}b_1$$
$$+\partial_x(u_1 b_{x_2} - u_{x_2}b_1) + \frac{i\zeta}{2}\partial_\perp(u_1 b_1^* - u_1^* b_1)$$
$$+\frac{i\zeta}{R_i}\partial_x(\rho_2\partial_x b_1 - b_{x_2}\partial_x b_1) + \frac{i\zeta}{R_i}\partial_x\partial_\perp\left(b_{x_2} - \frac{|b_1|^2}{2}\right)$$
$$-\frac{i\zeta}{R_i}\partial_x\left(\frac{b_1}{2}(\partial_\perp^* b_1 - \partial_\perp b_1^*) + 2\partial_X b_2\right), \tag{15.2.25}$$

$$\partial_t \rho_3 + \partial_x u_{x_3} = -\partial_T \rho_2 - \partial_X u_{x_2} - \Re[\partial_\perp^* u_2], \tag{15.2.26}$$

$$\partial_t u_{x_3} + \beta\partial_x \rho_3 = -\partial_T u_{x_2} - \partial_X(\beta\rho_2 + \frac{|b_1|^2}{2}) - \partial_x\Re[b_2^* b_1]), \tag{15.2.27}$$

$$\partial_x b_{x_3} = -\partial_X b_{x_2} - \Re[\partial_\perp^* b_2], \tag{15.2.28}$$

$$\partial_t b_{x_3} = -\partial_T b_{x_2} - \Re[\partial_\perp^* u_2] - \frac{\zeta}{R_i}\Im[\partial_X\partial_\perp^* b_1 + \partial_x\partial_\perp^* b_2]. \tag{15.2.29}$$

The solvability condition that eliminates the resonant oscillating modes in the right-hand side of (15.2.24) and (15.2.25) reads

$$i(\partial_T B_2 + v_g\partial_X B_2) - \frac{\omega}{k^2+\omega^2}\left(-\frac{\omega^2}{k^2}\partial_{XX} - \frac{k^2}{\omega^2}\partial_{TT} - \frac{2k}{\omega}\partial_{XT}\right)B_1$$
$$+ \alpha\Delta_\perp B_1 - kv_g(\frac{1}{v_g}\bar{u}_{x_2} + \frac{k^2}{2\omega^2}\bar{b}_{x_2} - \frac{1}{2}\delta_2)B_1 = 0, \tag{15.2.30}$$

where $\delta_2 = \bar{\rho}_2 - \bar{b}_{x_2}$ and

$$\alpha = \frac{k\omega}{2(k^2+\omega^2)}\left(\frac{\omega^2}{k^3} - \frac{\beta k}{\beta k^2 - \omega^2}\right). \tag{15.2.31}$$

Combining (15.2.22) and (15.2.30), we obtain for $B = B_1 + \mu B_2$ the equation

$$i(\partial_T B + v_g \partial_X B) - \mu \frac{\omega}{k^2 + \omega^2} \Big(-\frac{\omega^2}{k^2} \partial_{XX} - \frac{k^2}{\omega^2} \partial_{TT} - \frac{2k}{\omega} \partial_{XT} \Big) B_1$$

$$+ \mu \alpha \Delta_\perp B + \mu k v_g \Big(-\frac{1}{v_g} \bar{u}_{x_2} - \frac{k^2}{2\omega^2} \bar{b}_{x_2} + \frac{1}{2} \delta_2 \Big) B = 0, \quad (15.2.32)$$

where the dispersive term can, in fact, be rewritten as a second derivative either in X (absolute dynamics) or in T (convective dynamics). This equation is also given by Hasegawa (1970).

Elimination of the nonoscillating secular terms in (15.2.24) and (15.2.29) leads to

$$\partial_T \bar{u}_2 - \partial_X \bar{b}_2 + \partial_\perp (\beta \delta_2 + (\beta+1)\bar{b}_{x_2} + \frac{k^2}{2\omega^2}|B_1|^2) = 0, \quad (15.2.33)$$

$$\partial_T \bar{b}_2 - \partial_X \bar{u}_2 = 0, \quad (15.2.34)$$

$$\partial_T \delta_2 + \partial_X \bar{u}_{x_2} = 0, \quad (15.2.35)$$

$$\partial_T \bar{u}_{x_2} + \beta \partial_X (\delta_2 + \bar{b}_{x_2}) + \partial_X \frac{|B_1|^2}{2} = 0, \quad (15.2.36)$$

$$\partial_X \bar{b}_{x_2} + \Re[(\partial_\perp^* \bar{b}_2] = 0, \quad (15.2.37)$$

$$\partial_T \bar{b}_{x_2} + \Re[\partial_\perp^* \bar{u}_2] = 0. \quad (15.2.38)$$

Combining (15.2.37), (15.2.38), and (15.2.33), we obtain

$$\partial_{TT} \bar{b}_{x_2} - \partial_{XX} \bar{b}_{x_2} - \Delta_\perp \Big(\beta \delta_2 + (\beta+1)\bar{b}_{x_2} + \frac{k^2}{2\omega^2}|B_1|^2 \Big) = 0. \quad (15.2.39)$$

Equations (15.2.32), (15.2.35), (15.2.36) and (15.2.39) constitute a closed system from which reference to $\bar{b}_1$ and $\bar{u}_1$ has been eliminated. However, (15.2.34)–(15.2.36) display a degeneracy when the modulation is purely transverse, and higher-order terms are to be retained in a spectral "angular boundary layer" associated to quasi-transverse perturbations. A similar effects occurs when the modulation is time-independent. One is thus led to consider at order μ^4 the equations

$$\partial_t \rho_4 + \partial_x u_{x_4} = -\partial_\tau \rho_3 - \partial_x(\rho_2 u_{x_2}) - \partial_X u_{x_3} - \Re[\partial_\perp^* (u_3 + \rho_2 u_1)], \quad (15.2.40)$$

$$\partial_t u_{x_4} + \beta \partial_x \rho_4 = -\partial_\tau u_{x_3} - \rho_1 \partial_t u_{x_2} - \beta \partial_X \rho_3$$

$$-\frac{1}{2} \partial_x [u_{x_2}^2 + \beta(\gamma-1)\rho_2^2 + 2|b_2|^2]$$

$$-\Re[(u_1^* \partial_\perp - b_1^* \partial_\perp) b_{x_2} + \partial_x(b_1 b_3^*) + \partial_X(b_1 b_2^*)], \quad (15.2.41)$$

$$\partial_t b_{x_2} = -\partial_\tau b_{x_3} + \Re[\partial_\perp^* (u_{x_2} b_1) - \partial_\perp^* (u_1 b_{x_2}) - \partial_\perp^* u_3]$$

$$-\frac{\zeta}{R_i} \Im[\partial_X \partial_\perp^* b_2 + \partial_x \partial_\perp^* b_3 + \partial_\perp^* ((b_{x_2} - \rho_2)\partial_x b_1)]$$

$$-\frac{\zeta}{R_i} \Re[\partial_\perp^* b_1 + b_1 \partial_\perp^*] \Im[\partial_\perp^* b_1]. \quad (15.2.42)$$

The solvability conditions corresponding to the elimination of the nonoscillating secular terms are

$$\partial_T \bar{\rho}_3 + \partial_X \bar{u}_{x_3} = -\Re[\partial_\perp^* \bar{u}_3 + \partial_\perp^*(\tilde{\rho}_2^* U_1)], \tag{15.2.43}$$

$$\partial_T \bar{u}_{x_3} + \beta \partial_X \bar{\rho}_3 = -\omega \Im[\tilde{\rho}_1^* \tilde{u}_{x_1}]$$
$$- \Re[U_1^* \partial_\perp \tilde{u}_{x_1} - B_1^* \partial_\perp \tilde{b}_{x_2} + \partial_X(B_1 B_2^*)], \tag{15.2.44}$$

$$\partial_T \bar{b}_{x_3} = -\Re[\partial_\perp^* \bar{u}_3 + \partial_\perp^*(\tilde{b}_{x_2}^* U_1 - \tilde{u}_{x_2}^* B_1) + \frac{\zeta k}{R_i}\partial_\perp^*((\tilde{b}_{x_2}^* - \tilde{\rho}*_{x_2})B_1)]$$
$$+ \frac{\zeta}{R_i}\Im[\frac{1}{2}B_1 \Delta_\perp B_1^* - \partial_X \partial_\perp^* \bar{b}_2]. \tag{15.2.45}$$

Using (15.2.8), (15.2.10), and (15.2.18), the equation satisfied by $\delta_3 = \bar{\rho}_3 - \bar{b}_{x_3}$ reduces to

$$\partial_T \delta_3 + \partial_X \bar{u}_{x_3} = \frac{\zeta}{R_i}\partial_X \Im[(\partial_\perp^* \bar{b}_1], \tag{15.2.46}$$

while (15.2.44) becomes

$$\partial_T \bar{u}_{x_3} + \beta \partial_X(\delta_3 + \bar{b}_{x_3}) = -2\lambda \Im[B_1 \Delta_\perp B_1^*] - \partial_X \Re[B_1^* B_2], \tag{15.2.47}$$

with $\lambda = \frac{1}{4}(\frac{\beta k^3}{\omega^2(\beta k^2 - \omega^2)} - \frac{1}{k}) = -\frac{\alpha}{v_g}$.

Combining (15.2.35) with (15.2.46) and also (15.2.36) with (15.2.47) for the mean longitudinal fields, we get that $\delta = \delta_1 + \mu\delta_3$, $\bar{u}_x = \bar{u}_{x_1} + \mu\bar{u}_{x_3}$, and $\bar{b}_x = \bar{b}_{x_1} + \mu\bar{b}_{x_3}$ obey

$$\partial_T \delta + \partial_X \bar{u}_x = 0, \tag{15.2.48}$$

$$\partial_T \bar{u}_x + \beta \partial_X(\delta + \bar{b}_x) + \partial_X \frac{|B|^2}{2} = -2\lambda\mu\Im[B\Delta_\perp B^*], \tag{15.2.49}$$

where the term $\frac{\mu}{2R_i}\Im[\partial_X(\partial_\perp^* \bar{b})]$ has been neglected in (15.2.48) because it is always smaller than $\partial_X \bar{u}_x$. In contrast, in (15.2.49) the ordering can be broken in the case of quasi-transverse or quasi-stationary modulations, for which the corrective term $\frac{2\alpha\mu}{v_g}\Im[B^*\Delta_\perp B]$ is no longer subdominant. Furthermore, $\alpha\mu\Im[B\Delta_\perp B^*]$ is usefully replaced by $(\partial_T + v_g\partial_X)|B|^2 + \mu\mathcal{D}$ where $\mathcal{D}$ denotes a contribution originating from the dispersion which is negligible in the present limit. The presence of the corrective term thus leads to a ponderomotive force in the form given by Shukla, Feix, and Stenflo (1988). Also, due to the absence of degeneracy, it is enough to retain for $\bar{b}_x$ the leading order and drop the subscript indices in eq. (15.2.39). The resulting system is given by

$$i(\partial_T B + v_g\partial_X B) + \mu\alpha\Delta_\perp B + \frac{\mu\omega}{k^2 + \omega^2}\left(\frac{\omega^2}{k^2}\partial_{XX} + \frac{k^2}{\omega^2}\partial_{TT} + \frac{2k}{\omega}\partial_{XT}\right)B$$

$$- \mu k v_g\left(\frac{1}{v_g}\bar{u}_x + \frac{k^2}{2\omega^2}\bar{b}_x - \frac{1}{2}\delta\right)B = 0, \tag{15.2.50}$$

together with

$$\partial_T \delta + \partial_X \bar{u}_x = 0, \tag{15.2.51}$$

$$\partial_T \bar{u}_x + \beta \partial_X (\delta + \bar{b}_x) = \left(\frac{1}{v_g}\partial_T + \frac{1}{2}\partial_X\right)|B|^2, \tag{15.2.52}$$

$$\partial_{TT}\bar{b}_x - \partial_{XX}\bar{b}_x = \Delta_\perp\left(\beta\delta + (\beta+1)\bar{b}_x + \frac{k^2}{2\omega^2}|B|^2\right). \tag{15.2.53}$$

When instead of using a second-order equation in time for $\bar{b}_x$, the equation for the mean transverse velocity is retained, the above system can be written in the Hamiltonian form similar to that given by Relke and Rubenchik (1988), but for slightly different variables. The source term in the density equation does not originate from the interactions between the velocity and density fluctuations (which are subdominant), but from the presence of a time-derivative contribution in the ponderomotive force, whose existence is suggested by several authors, starting from the primitive equations (see, e.g., Washimi and Karpman 1976, Karpman and Shagalov 1982, Kogiso, Ueda, and Yajima 1982, Ovenden, Statham, and ter Haar 1983). This leads to a redefinition of the velocity that is affected by the carrying wave momentum and to the presence of a self-interaction of the carrier in the potential of the Schrödinger equation.

The system (15.2.50)–(15.2.53) can be viewed as an initial value problem either in time (absolute dynamics) or relatively to the propagation coordinate (convective dynamics). Using that to leading order $\partial_T B + v_g \partial_X B = 0$ and that $\omega'' = \frac{2\omega}{k^2+\omega^2}\left(\frac{\omega^2}{k^2} + \frac{k^2}{\omega^2}v_g^2 - \frac{2k}{\omega}v_g\right)$, the longitudinal and temporal dispersive terms arising in (15.2.50) can be written in the usual form $\frac{\omega''}{2}\partial_{XX}B$ for the absolute regime or $\frac{\omega''}{2v_g^2}\partial_{TT}B$ for the convective one.

The last step consists in eliminating the transport of the wave packet at the group velocity. For this purpose, in the absolute regime, the equations are written in a reference frame moving with the Alfvén-wave group velocity, by defining $\xi = X - v_g T$. A slower time scale $\tau = \mu T$ typical of the envelope dynamics, is then introduced. In the convective regime, the origin of time is taken to be space-dependent by defining a delayed time $\tau = T - \frac{1}{v_g}X$, and a longer scale $\xi = \mu X$ is used. It follows that magnetosonic waves evolve on a time scale (absolute regime) or a length scale (convective regime) shorter by a factor μ than the Alfvén wave envelope. The problem then simplifies in two limit cases.

When the modulation is purely longitudinal, the limit $\mu \to 0$ is easily taken formally. The fields b_x, u_x, and δ become slaved to the Alfvén wave modulation, and the system then reduces to the one-dimensional nonlinear Schrödinger equation. The validity conditions of this adiabatic approximation is discussed in Champeaux, Laveder, Passot and Sulem (1999) where comparisons with direct numerical simulations of the dispersive MHD equations in one space dimension are presented.

For a purely transverse absolute modulation or for a time-independent convective modulation, the magnetosonic waves disappear, leading to a two-dimensional nonlinear Schrödinger equation that in the former case reads

$$i\partial_\tau B + \alpha \Delta_\perp B - k v_g \left(\frac{1}{v_g^2} - \frac{k^4}{4(\beta+1)\omega^4} \right) |B|^2 B = 0, \qquad (15.2.54)$$

and in the latter one

$$i\partial_\xi B + \frac{\alpha}{v_g} \Delta_\perp B + \frac{k}{4} \left(\frac{1}{\beta} + \frac{(k^2+\omega^2)^2}{\omega^4} \right) |B|^2 B = 0. \qquad (15.2.55)$$

In this context, the self-focusing singularity is usually referred to as Alfvén wave filamentation. This has recently been considered as a possible heating mechanism in the solar corona and the warm ionized phase of the interstellar medium (Champeaux, Gazol, Passot, and Sulem 1997).

15.2.2   Influence of the Magnetosonic Waves

The magnetosonic waves become relevant in considering deviations from the idealized cases of purely transverse (absolute) or stationary (convective) modulations. It is then convenient to perform additional rescalings.

Absolute Regime

One writes $\tilde{\xi}' = \mu\eta^{-1}\xi$ with $\mu\eta^{-1} \ll 1$. Equation (15.2.53) then reduces to

$$\bar{b}_x = -\frac{\beta}{\beta+1}\delta - \frac{k^2}{2(\beta+1)\omega^2}|B|^2. \qquad (15.2.56)$$

Furthermore, the longitudinal and temporal dispersions are negligible in eq. (15.2.50). Writing the ponderomotive force in the form $\frac{\eta}{v_g}\partial_\tau|B|^2$, one is led to define $v_x = \bar{u}_x - \frac{1}{v_g}|B|^2$. It is then convenient to introduce the nondimensional variables

$$\tilde{\xi} = |k|\xi', \quad \tilde{\tau} = |k||v_g|\tau, \quad \tilde{Y} = \sqrt{\frac{|kv_g|}{|\alpha|}}Y, \quad \tilde{Z} = \sqrt{\frac{|kv_g|}{|\alpha|}}Z, (15.2.57)$$

$$|B|^2 = \frac{|\tilde{B}|^2}{Q_1}, \quad v_x = -\frac{A_1}{|v_g|Q_1}\tilde{v}_x, \quad \delta = -\frac{A_1}{Q_1 v_g^2}\tilde{\delta}, \qquad (15.2.58)$$

where $A_1 = \frac{1}{2}\left(1 + \frac{\beta k^2}{(\beta+1)\omega^2}\right)$ and $Q_1 = \frac{1}{v_g^2}\left(1 - \frac{k^2\omega^2}{(\beta+1)(k^2+\omega^2)^2}\right)$. After dropping the tildes, one gets

$$i\partial_\tau B + \sigma_1 \Delta_\perp B = \sigma_2 \left[W_1(\delta + D_1 v_x) + |B|^2 \right] B, \qquad (15.2.59)$$

$$\eta\partial_\tau \delta - \sigma_3 \partial_\xi \delta = -\partial_\xi v_x - D_1 \partial_\xi |B|^2, \qquad (15.2.60)$$

$$\eta\partial_\tau v_x - \sigma_3 \partial_\xi v_x = -\frac{1}{M_1^2}\partial_\xi \delta - \partial_\xi |B|^2, \qquad (15.2.61)$$

where $M_1 = |v_g| \sqrt{\frac{\beta+1}{\beta}}$, $D_1 = -\frac{\sigma_3}{A_1}$, and $W_1 = \frac{A_1^2}{v_g^2 Q_1}$. Furthermore, $\sigma_1 = \text{sign}(\alpha)$, $\sigma_2 = \text{sign}(kv_g)$ and $\sigma_3 = \text{sign}(v_g)$.

Convective Regime

One defines $\tau' = \mu\eta^{-1}\tau$ with $\mu\eta^{-1} \ll 1$. The ponderomotive force is written as $\eta\partial_\xi |B|^2$, and one defines $d = \frac{\delta}{\beta+1} - \frac{1}{2}(\frac{1}{\beta} + \frac{k^2}{(\beta+1)\omega^2})|B|^2$. One now rescales ξ, τ', Y, and Z as in the absolute regime, and

$$|B|^2 = \frac{|\tilde{B}|^2}{Q_2}, \quad \bar{u}_x = \frac{A_2 |v_g|}{Q_2} \tilde{u}_x, \quad d = \frac{A_2 v_g^2}{\beta Q_2} \tilde{d}, \qquad (15.2.62)$$

with $A_2 = \frac{\beta+1}{2\beta}\left(1 + \frac{\beta k^2}{(\beta+1)\omega^2}\right)$ and $Q_2 = \frac{1}{4}\left(\frac{1}{\beta} + (\frac{k^2+\omega^2}{\omega^2})^2\right)$. After dropping the tildes, one obtains

$$i\partial_\xi B + \sigma_1\sigma_3\Delta_\perp B = -\sigma_2\sigma_3\left[W_2(d + D_2 u_x) + |B|^2\right]B, \quad (15.2.63)$$

$$\eta\partial_\xi d - \sigma_3\partial_\tau d = -\partial_\tau u_x - D_2\partial_\tau|B|^2, \qquad (15.2.64)$$

$$\eta\partial_\xi u_x - \sigma_3\partial_\tau u_x = -\frac{1}{M_2^2}\partial_\tau d - \partial_\tau|B|^2, \qquad (15.2.65)$$

where $M_2 = \frac{1}{|v_g|}\sqrt{\frac{\beta}{\beta+1}}$, $D_2 = -\frac{\sigma_3}{|v_g|^2 A_2}$, and $W_2 = \frac{|v_g|^2 A_2^2}{Q_2}$.

The purely transverse absolute or steady convective regimes correspond to the limit $\eta \to \infty$. The opposite limit $\eta \to 0$ associated to an adiabatic dynamics of the magnetosonic waves leads in both absolute and convective regimes to a nonlinear Schrödinger equation with a potential

$$V = \frac{k}{4}\left(\frac{k^4}{(\beta+1)\omega^4} - \frac{\beta+1}{(\beta+1)v_g^2 - \beta}(1 - \frac{\beta k^2}{(\beta+1)\omega^2})^2\right)|B|^2,$$

which vanishes in the long wavelength limit $k \to 0$. For finite values of η, (15.2.59)–(15.2.61) and (15.2.63)–(15.2.65) can be viewed as a special case of a system of equations derived by Zakharov and Rubenchik (1972) (see also Zakharov and Schulman 1991) for the interaction of a high-frequency wave with low-frequency acoustic waves. In the present case, however, the acoustic-wave equations include only longitudinal derivatives, because the transverse velocity is, to leading order, slaved to the transverse magnetic field (see (15.2.10)).

A Unified Formulation

The equations governing the absolute and the convective dynamics can be rewritten in the unified form

$$i\partial_t B + \sigma\Delta_\perp B = \sigma'[W(d + Du) + |B|^2]B, \qquad (15.2.66)$$

$$\eta\partial_t d - \sigma_3\partial_x d = -\partial_x u - D\partial_x|B|^2, \qquad (15.2.67)$$

$$\eta\partial_t u - \sigma_3\partial_x u = -\frac{1}{M^2}\partial_x d - \partial_x|B|^2, \qquad (15.2.68)$$

where the variable t refers either to time or to the propagation coordinate (in the latter case x refers to time). The coefficients in (15.2.66)–(15.2.68) are specified by identification with the absolute or the convective systems (15.2.59)–(15.2.61) or (15.2.63)–(15.2.65), respectively.

The system (15.2.66)–(15.2.68) for the magnetosonic waves is diagonalized by introducing the Riemann invariants ($j = 1, 2$)

$$Z_j = \frac{W}{2}\left(1 + (-1)^j \frac{D}{M}\right)\left(d + (-1)^j M u\right) \tag{15.2.69}$$

which obey

$$\eta \partial_t Z_j + \Lambda_j \partial_x Z_j = (-1)^{j+1} \frac{W(M + (-1)^j D)^2}{2M} \partial_x |B|^2 \tag{15.2.70}$$

with $\Lambda_j = -\sigma_3 + (-1)^j \frac{1}{M}$. The equation for the Alfvén-wave amplitude then becomes

$$i\partial_t B = -\sigma \Delta_\perp B + \sigma'(|B|^2 + Z_1 + Z_2)B. \tag{15.2.71}$$

Note that a resonance associated to the vanishing of the eigenvalue Λ_2 is a priori possible but turns out to be physically irrelevant (Passot, Sulem, and Sulem 1994). Integrating along the characteristics $x_j(a, t) = \lambda_j t + a$, with $\lambda_j = \frac{\Lambda_j}{\eta}$, one gets

$$Z_j(x_j(a, t), \mathbf{r}, t) - Z_j(a, \mathbf{r}, 0) =$$
$$(-1)^{j+1} \frac{WM}{2\eta}\left(1 + (-1)^j \frac{D}{M}\right)^2 \int_0^t |B|_x^2(x_j(a, \tau), \mathbf{r}, \tau)d\tau, \tag{15.2.72}$$

where $\mathbf{r}$ denotes the transverse vector coordinate. At early time, the magnetosonic waves Z_1 and Z_2 are negligible in (15.2.71), where the variable x then reduces to a parameter. Assuming $\sigma\sigma' < 0$, the carrying wave first focuses in the transverse planes for which the initial L^2-norm $|B(x, ., 0)|_{L^2(\mathbf{R}^2)}$ exceeds the critical value for collapse. Since the rate of focusing in these planes depends on $|B(x, ., 0)|_{L^2(\mathbf{R}^2)}$, strong gradients develop along the x-direction, producing a significant ponderomotive force. The latter drives magnetosonic waves which, in turn can react on the focusing process. The analysis of these dynamics presented in Passot, Sulem, and Sulem (1994, 1996) is based on the following ansatz. Assuming that the Alfvén wave focuses mostly independently in the different planes transverse to the propagation, one writes

$$|B(x, \mathbf{r}, t)|^2 \approx G(t_*(x) - t, \mathbf{r}), \tag{15.2.73}$$

where the main dependence in the x variable arises through the dependence of the singularity time. Differentiating with respect to x, one has

$$|B|_x^2 \approx -t'_*(x)G'_1 \tag{15.2.74}$$

where G_1' is the partial derivative of G with respect to the first variable. Along the characteristics, one has

$$|B|_x^2(x_j(t), \mathbf{r}, t) \approx \frac{t_*'(x_j(t))}{\lambda_j t_*'(x_j(t)) - 1} \frac{d}{dt} |B(x_j(t), \mathbf{r}, t)|^2. \quad (15.2.75)$$

Substituting in eq. (15.2.72) and dropping the initial contributions which are negligible near collapse, one gets

$$Z_j(x_j(t), \mathbf{r}, t) \approx \frac{A_j}{2\eta} \int_0^t \frac{t_*'(x_j(t))}{\lambda_j t_*'(x_j(t)) - 1} \frac{d}{dt} |B(x_j(t), \mathbf{r}, t)|^2 dt \quad (15.2.76)$$

with

$$A_j = (-1)^{j+1} \frac{MW}{2} \left(1 + (-1)^j \frac{D}{M}\right)^2. \quad (15.2.77)$$

Two regimes can then be distinguished according to the speed of the magnetosonic wave and the distance $t_* - t$ from the singularity time: At fixed $t_* - t$, for rapidly propagating magnetosonic waves ($\eta \ll 1$), the adiabatic regime is obtained in the form

$$Z_j(\mathbf{r}, x, t) \approx (-1)^{j+1} \frac{MW}{2\Lambda_j} \left(1 + (-1)^j \frac{D}{M}\right)^2 |B|^2(x, \mathbf{r}, t). \quad (15.2.78)$$

Differently, at fixed velocity of the magnetosonic waves, in the limit $t_* - t \to 0$, the dominant variation insides the integral comes from the wave amplitude, which to leading order near the singularity leads to

$$Z_j(\mathbf{r}, x, t) \approx (-1)^j \frac{WM}{2\Lambda_j} \left(1 + (-1)^j \frac{D}{M}\right)^2 t_*'(x) |B|^2(x, \mathbf{r}, t). \quad (15.2.79)$$

Equation (15.2.79) indicates that close enough to the singularity, the adiabatic approximation becomes invalid, and a transition takes place, where the Z_j's become antisymmetric with respect to the center of the collapsing pulse, the function $t_*'(z)$ being itself antisymmetric. If the collapse is not arrested during this adiabatic phase, the developing antisymmetric profiles of Z_1 and Z_2 will affect the potential in such a way that the focus will be shifted in the x-direction on the side where the waves reinforce the potential. The resulting dynamics deserve a numerical investigation.

Numerical integrations of (15.2.66)–(15.2.68) were performed in Champeaux, Passot, & Sulem (1997) for axisymmetric solutions using the anisotropic dynamic rescaling method introduced in Landman, Papanicolaou, Sulem, Sulem, and Wang (1991) and reviewed in Chapter 6. Results of the simulations are presented in the convective case for a fixed initial (or incident) wave packet $B(x, \mathbf{r}, 0) = 4e^{-(x^2+r^2)}$ with $d(x, \mathbf{r}, 0) = u(x, \mathbf{r}, 0) = 0$ (Fig. 15.1). Reducing the parameter η corresponds in the primitive variables to increasing the length or the duration of the initial pulse. Typically, for $\eta = 0.1$, $k/R_i = 0.1$, and $\beta = 0.5$ (and left-hand side polarization), the filamentation is rapidly inhibited under the effect of the developing

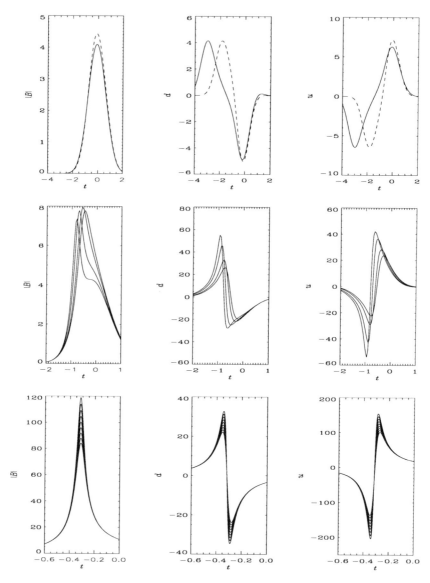

FIGURE 15.1. *Upper:* Arrest of convective filamentation with adiabatic dynamics of the magnetosonic waves, for $\eta = 0.1$, $k/R_i = 0.1$, $\beta = 0.5$, and left-hand size polarization. Plots at the center of the beam at the points $\xi = 0.0734$ (dashed line) and $\xi = 0.119$ (solid line), respectively behind and beyond the filamentation arrest. *Middle:* Arrest of the filamentation with formation of strong acoustic fronts for $\eta = 0.3$, $k/R_i = 1$, $\beta = 0.1$, and left-hand side polarization. Plots at the center of the beam at the point $\xi = 0.139$ behind the filamentation arrest and at the points $\xi = 0.149$, $\xi = 0.166$, $\xi = 0.177$ beyond it. *Lower:* Finite distance wave collapse for $\eta = 1$, $k/R_i = 0.65$, $\beta = 0.5$, and right-hand side polarization. Plots at the center of the beam $r = 0$ and various distances from $\xi = 0.156$ to $\xi = 0.16$ (Champeaux et al. 1997).

magnetosonic waves. Along the symmetry axis $\mathbf{r} = 0$, these waves first display an antisymmetric profile in the time variable whose left part is, for small η, rapidly advected away from the Alfvén wave packet. The other part stays at the location of the pulse and reaches its adiabatic limit producing a caviton that, in the small k limit, cancels out the nonlinearity in the equation for the Alfvén-wave envelope. For $\eta = 0.3$ with $k/R_i = 1$, $\beta = 0.1$ (and left-hand side polarization), the focusing of the Alfvén wave packet is arrested. In this case, however, the magnetosonic waves develop at the location of the Alfvén pulse a sharp antisymmetric acoustic front whose strength increases with η as a consequence of the amplification of the potential in the nonlinear Schrödinger equation. This leads to strong hydrodynamic effects not included in the envelope formalism, which retains only a linear description of the hydrodynamic phenomena, driven by the ponderomotive force. Finally, for $\eta = 1$ with $k/R_i = 0.65$, $\beta = 0.5$ (and right-hand side polarization), the filamentation is not arrested. A sharp antisymmetric magnetosonic front develops, and the Alfvén wave collapses in a finite time (or on a finite distance).

15.3   Weakly Dispersive Alfvén Waves

15.3.1   Parallel Propagation

In the situation where the carrying wavelength is large compared to the gyromagnetic radius, the dispersion becomes small at the scale of the carrying wave ($R_i \to \infty$). This limit can be formally taken in the amplitude equations (15.2.54) and (15.2.55) (or more generally in those including the coupling to the magnetosonic waves). The resulting equations can also be derived by a modulation analysis performed on a generalization (see Passot and Sulem 1993, 1995) of the derivative nonlinear Schrödinger equation mentioned in Section 3.4. This system describes the long-wave dynamics of Alfvén waves and their coupling to longitudinal magnetosonic waves. As argued in Champeaux, Passot, and Sulem (1998), the above regime requires that the wave amplitude be reduced according to the dispersion in order that the constraint of circular polarization remain enforced. More precisely, denoting by λ the wavelength of the carrier and by k its wavenumber, this situation corresponds to a regime where for a dispersion scale l_d with $\frac{l_d}{\lambda} \equiv \frac{k}{2R_i}$ of order ϵ, and a longitudinal modulation scale L_M such that $\frac{\lambda}{L_M}$ of order μ^2 (where ϵ and μ are small parameters), the relative amplitude of the wave $\frac{\delta B}{B_0}$ is of order $\epsilon^{1/2}\mu$.

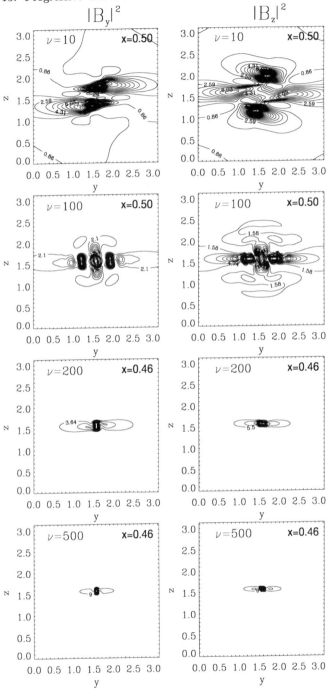

FIGURE 15.2. Influence of the dispersion, as measured by the parameter ν on the contours of $|B_y|^2$ and $|B_z|^2$ in the subdomain $[0,\pi] \times [0,\pi]$ (Champeaux et al. 1998).

If the dispersion is weaker (or the wave amplitude larger), the circular polarization is no longer preserved by the dynamics [1]. In this case, the left- and right-hand side polarized waves interact, and in an asymptotics where l_d/λ and λ/L_M are of order μ^2, while $\delta B/B_0$ is of order μ, the vector amplitude $\mathbf{B}$ of the wave (transverse to the propagation direction) obeys a vector NLS equation that in the simple case of a time-independent (convective) modulation reads

$$i\partial_\xi \mathbf{B} + \sigma\nu k\mathbf{B} + i\nu k(\mathbf{B}\times\mathbf{e}_1) + \frac{1}{2k(1-\beta)}\nabla(\nabla\cdot\mathbf{B})$$

$$+ \frac{(-8\beta^2 + 5\beta + 2)k}{4\beta(1-\beta)}|\mathbf{B}|^2\mathbf{B} + \frac{k}{4(1-\beta)}\mathbf{B}\times(\mathbf{B}\times\mathbf{B}^*) = 0. \quad (15.3.1)$$

Here $\mathbf{e}_1$ is the unit vector along the propagation axis and $\nu = \frac{k^2}{2R_i}(\frac{\delta B}{B_0})^{-2}$. The anisotropic diffraction term originates from the gradient of the total pressure whose fluctuations are proportional to the divergence of the vector amplitude. In the limit $\nu \to \infty$, corresponding to the former asymptotics, the initially circular polarization of the wave is preserved, and (15.3.1) reduces to the scalar two-dimensional NLS equation. In this case, when $\beta < 1$, wave packets with a sufficient mass collapse in localized foci, with a blowup of the amplitude. In contrast, when the parameter ν is moderate, numerical simulations show that the wave amplitude remains moderate, but stronger magnetic field gradients develop on elongated structures. Whether a blowup of the gradients occurs at a finite ξ remains an open question. In this context, an interesting question concerns the quantitative estimate of the dissipation in the presence of a weak linear damping of Landau type. Numerical simulations show that while in the strongly dispersive system the dissipation takes place in very localized spots, it is much more homogeneously distributed in the present case, resulting in a more efficient heating of the plasma (Champeaux, Passot, and Sulem 1999).

The transition between weakly and strongly dispersive regimes is illustrated in Fig. 15.2. For large but finite value of the parameter ν (typically a few hundred), the collapse proceeds with a roughly circular polarization until a critical transverse scale $l_\perp$ is reached. This scale is given by $\frac{l_\perp}{\lambda} \approx (\frac{l_d}{\lambda})^{-1/2}(\frac{\delta B}{B_0})^{-1}$. Then the wave ceases to be circularly polarized. Its amplitude saturates, and the further evolution is that observed for a moderate value of ν.

---

[1] Circular polarization can also be broken when the coupling to large-scale two-dimensional dynamics in the planes transverse to the magnetic field is retained (Gazol, Passot, and Sulem 1999).

15.3.2 Oblique Propagation

The envelope dynamics are significantly different when the direction of propagation of the Alfvén wave makes a finite angle θ with the ambient magnetic field. In the weak dispersion limit, such an Alfvén-wave train is linearly polarized in the z-direction perpendicular to the plane defined by the ambient magnetic field and the direction of propagation. It turns out that in this regime all the diffraction terms scale like $1/R_i^2$. Furthermore, the nonlinear terms vanish identically in the case of a longitudinal modulation, but not in the multidimensional problem. In the simple example of absolute modulation where the perturbation only depends on the transverse z-coordinate, one has (Gazol, Passot, and Sulem 1998)

$$i\partial_\tau B_z + \alpha\partial_{ZZ}B_z + k\cos\theta\,\frac{(3+4\beta)}{2(1+\beta)}|B_z|^2B_z = 0 \qquad (15.3.2)$$

with

$$\alpha = \frac{k\cos\theta}{R_i^2\sin^4\theta}(\cos^4\theta - \beta\cos 2\theta). \qquad (15.3.3)$$

The limit $R_i \to \infty$ corresponds to the "semiclassical" limit (Bronski and McLaughlin 1994, JinnameJin, Levermore, and McLaughlin 1994, Miller and Kamvissis 1998). For $\beta < 1$, (15.3.2) is of the focusing type. The most modulationally unstable mode appears at a scale of order $1/R_i$ which, if larger than the wavelength of the carrying wave and smaller than the scale of the modulation, introduces an intermediate scale in the problem. Numerical simulations presented by Bronski & McLaughlin (1994) show that in this case, a wave packet starts to focus, leading to a finite-amplitude solitonic profile on which oscillations at the intermediate scale are superimposed. The phase gradient obeys, at large scale, an (inviscid) Burgers equation, corresponding to the geometrical optics approximation for the linear NLS equation. After a shock has formed, the amplitude displays a well-delimited region of short-wavelength, finite-amplitude oscillations (referred to as a "sea of solitons"), which extends as time elapses, while in Fourier space the solution develops a broad energy spectrum, precursor of a fully-turbulent dynamics.

Synopsis

This monograph is concerned with the multi-dimensional envelope dynamics of a weakly nonlinear dispersive wave packet described by the NLS equation and some of its generalizations resulting from additional couplings. Special attention is paid to the phenomenon of wave collapse, which provides an efficient mechanism for energy transfer from large to small scales.

The cubic NLS equation in two dimensions appears as the simplest model for the self-focusing of a stationary laser beam in a Kerr medium. This dimension is critical in the sense that in subcritical dimensions, the problem is globally well-posed and the standing waves associated to the ground state are orbitally stable, while in supercritical dimensions, a self-similar blowup can occur at localized foci. At critical dimension, the self-similarity is weakly broken. The profile of blowup solutions adiabatically approaches the ground state that is marginally unstable, while the scaling factor differs from that of self-similar collapse by a double logarithmic correction. In this special case, a finite amount of the wave energy is absorbed in each focus, and eventually dissipated under the action of damping processes which become relevant as the singularity is approached. For this reason, critical collapse is said to be strong.

Near collapse, other initially subdominant effects such as higher-order nonlinearities, time dispersion, or small paraxiality breaking become relevant. An interesting question concerns their capability to arrest collapse. It is nevertheless unclear whether the long-time dynamics remain amenable to an envelope formalism.

The Davey–Stewartson system provides a canonical description for wave modulation in two space dimensions when a mean field is driven by the modulation. Various regimes are possible, which in the example of surface water waves depend on the relative importance of gravity and surface tension.

Coupling to mean fields also arises in plasma physics where low-frequency acoustic waves driven by the carryer modulation can affect the collapse dynamics. In the case of Langmuir waves, density cavitons are formed, where the electric field gets trapped, and collapse takes place. On the other hand, in the context of the filamentation instability of Alfvén waves propagating along an ambient magnetic field, sharp magnetosonic fronts can develop, leading to strong hydrodynamical effects not captured by the envelope formalism.

In spite of the progress made during the last decades in understanding the phenomenon of wave collapse described by envelope equations, several problems remain open. This formalism assumes in particular an idealized situation of quasi-monochromatic wave propagating in a homogeneous medium. It provides a good description of the propagation of a laser beam in a dielectric, but could nevertheless be too restrictive in other contexts like plasma waves whose spectra are not necessarily sharply peaked, and where modulational instabilities may be competing with other effects.

References

Ablowitz, M.J. and Clarkson, P.A., 1991 *Solitons, nonlinear evolution equations and inverse scattering*, London Mathematical Society Lecture Note Series **149**, Cambridge University Press.

Ablowitz, M., Kaup, D.J., Newell, A.C., and Segur, H., 1974 *The inverse scattering transform — Fourier analysis for nonlinear problems*, Stud. Appl. Math. **53**, 249–315.

Ablowitz, M.J., Manakov, S.V., and Schultz, C.L., 1990 *On the boundary conditions of the Davey–Stewartson equation*, Phys. Lett. **148 A**, 50–52.

Ablowitz, M.J. and Segur, H., 1979 *On the evolution of packets of water waves*, J. Fluid Mech. **92**, 691–715.

Ablowitz, M. and Segur, H., 1981 Solitons and the inverse scattering transform, SIAM Studies in Applied Mathematics (Philadelphia).

Abramovitz, M. and Stegun, I.A., 1965 *Handbook of mathematical functions*, Dover (New York).

Added, H. and Added, S., 1984 *Existence globale de solutions fortes pour les équations de la turbulence de Langmuir en dimension deux*, C.R. Acad. Sc. Paris A **299**, 551–554.

Added, H. and Added, S., 1988 *Equation of Langmuir turbulence and nonlinear Schrödinger equation: Smoothness and approximation*, J. Funct. Anal., **79**, 183–210.

Aichison, J.S., Weiner, A.M., Silberberg, Y., Oliver, M.K., Jackel, J.L., Leaird, D.E., Vogel, E.M., and Smith, P.W.E., 1990 *Observation of spatial optical solitons in a nonlinear glass waveguide*, Opt. Lett. **15**, 471–473.

Akhmediev, N.N., Korneev, V.I., and Nabiev, R.F., 1992 *Modulational instability of the ground state of the nonlinear wave equation: optical machine gun*, Opt. Lett. **17**, 393–395.

Akhmediev, N. and Sote-Crespo, J.M., 1993 *Generation of a train of three-dimensional solitons in a self-focusing medium*, Phys. Rev. A **47**, 1358–1364.

Akrivis, G.D. Dougalis, V.A., Karakashian, O.A., and McKinney, W.R., 1993 *Galerkin finite element methods for the nonlinear Schrödinger equation*, in "Advances Computers Math. Appl.", (E. Lipitakis ed.), World Scientific, 85–106.

Akrivis, G.D., Dougalis, V.A., and Karakashian, O.A., 1993 *On fully discrete Galerkin methods of second-order temporal accuracy for the nonlinear Schrödinger equation*, Numer. Math. **59**, 31–53.

Akrivis, G.D., Dougalis, V.A., Karakashian, O.A., and McKinney, W.R., 1997 *Numerical approximation of blow-up of radially symmetric solutions of the nonlinear Schrödinger equation*, submitted to SIAM J. Scient. Comput.

Akylas, T.R. and Kung, T.J., 1990 *On nonlinear wave envelopes of permanent form near a caustic*, J. Fluid Mech. **214**, 489–502.

Allen, J.E., 1998 *The early history of solitons (solitary waves)*, Physica Scripta **57**, 436–441.

Anderson, D.L.T., 1971 *Stability of time dependent particlelike solutions in nonlinear field theories. II*, J. Math. Phys. **11**, 945–952.

Anderson, D.L.T. and Derrick, G.H., 1970 *Stability of time dependent particlelike solutions in nonlinear field theories. I*, J. Math. Phys. **11** , 1336–1346.

Anisimov, S.I., Berezovskiĭ, M.A., Ivanov, M.F., Petrov, I.V., Rubenchick, A.M., and Zakharov, V.E., 1982 *Computer simulation of the Langmuir collapse*, Phys. Lett. A **92**, 32–34.

Arkadiev, V.A., Pogrebkov, A.K., and Polivanov, M.C., 1989 *Inverse scattering transform and soliton solution for Davey–Stewartson II equation*, Physica **D 36**, 189–197.

Assalauov, Zh. A. and Zakharov, V.E., 1985 *Modulational instability and collapse of plasma waves in a magnetic field*, Sov. J. Plasma Phys. **11**, 762–766.

Bailly, F., Clouet, J.F., and Fouque, J.P., 1996 *Parabolic and gaussian white noise approximation for wave propagation in random media*, SIAM J. Appl. Math. **56**, 1445–1470.

Balabane, M., Lochak, P., and Sulem, C. (eds.) 1989, *Integrable systems and Applications*, Lecture Notes in Physics, **342**, Springer-Verlag.

Barashenkov, I.V., Gocheva, A.D., Makhankov, V.G., and Puzynin, I.V., 1989 *Stability of the solution-like "bubbles"*, Physica D **34**, 240–254.

Barashenkov, I.V. and Padova, E.Y., 1993 *Stability and evolution of the quiescent and travelling solitonic bubbles*, Physica D **69**, 114–134.

Bardos, C., 1980 *Apparition éventuelle de singularités dans des problèmes d'évolution non linéaires* , Séminaire Bourbaki, Lecture Notes in Math. **842**, 215–224. Springer-Verlag.

Barthelemy, A., Maneuf, S., and Froehly, C., 1985 *Propagation soliton et auto-confinement de faisceaux laser par nonlinearité optique de Kerr*, Opt. Comm. **55**, 201–206.

Bekiranov, D., Ogawa, T., and Ponce, G., 1996 *On the well-posedness of Benney's interaction equation of short and long waves*, Adv. Diff. Eq. **1**, 919–937.

Bel'kov, S.A. and Tsytovich, S.A., 1979 *Modulation excitation of magnetic fields*, Sov. Phys. JETP **49**, 656–661.

Bender, C.M. and Orszag, S.A., 1978 *Advanced mathematical methods for scientists and engineers*, McGraw-Hill.

Benjamin, T.B., 1967a *Instability of periodic wavetrains in nonlinear dispersive systems*, Proc. R. Soc. London A **299**, 59–75.

Benjamin, T.B., 1967b *Internal waves of permanent form in fluids of great depth*, J. Fluid Mech. **29**, 559–592.

Benjamin, T.B. and Feir, J.F., 1967 *The disintegration of wave trains on deep water. Part 1. Theory*, J. Fluid Mech. **27**, 417–430.

Benney, D.J., 1977 *A general theory for interactions between short and long waves*, Stud. Appl. Math. **56**, 81–94.

Benney, D.J. and Newell, A.C., 1967 *The propagation of nonlinear wave envelopes*, J. Math. Phys. **46**, 133–139.

Benney, D.J. and Roskes, G.J., 1969 *Wave instabilities*, Stud. Appl. Math. **48**, 377–385.

Berestycki, H. and Cazenave, T., 1981 *Instabilité des états stationnaires dans les équations de Schrödinger et Klein-Gordon non linéaires*, C.R. Acad. Sci. Paris **293**, Série I, 489–492.

Berestycki, H., Gallouët, T., and Kavian, O., 1983 *Equations de champs scalaires euclidiens non linéaires dans le plan*, C. R. Acad. Sci. Paris **297**, Série I, 307–310.

Berestycki, H. and Lions, P.L., 1980 *Existence of stationary states in nonlinear scalar field equations*, 269–292, in "Bifurcation phenomena in mathematical physics and related topics" (C. Bardos and D. Bessis, eds.), Nato Adv. Study Institutes Series, Ser. C, Reidel.

Berestycki, H. and Lions, P.L., 1983 *Nonlinear scalar field equations, I Existence of a ground state*, Arch. Rat. Mech. Anal. **82**, 313–345; *II Existence of infinitely many solutions*, Arch. Rat. Mech. Anal. **82**, 347–369.

Berestycki, H., Lions, P.L., and Peletier, L.A., 1981 *An ODE approach to the existence of positive solutions for semilinear problems in* $\mathbf{R}^N$, Indiana Univ. Math. J. **30**, 141–157.

Bergé, L., 1994a *Dynamics stability analysis of strong/weak wave collapses*, J. Math. Phys. **35**, 5765–5780.

Bergé, L., 1994b *Non-self-similar inertial regimes of the scalar supersonic Langmuir collapse*, Physica D **72**, 87–94.

Bergé, L., 1997 *Self-focusing dynamics of nonlinear waves in media with parabolic-type inhomogeneities*, Phys. Plasmas **4**, 1227–1237.

Bergé, L., 1998 *Wave collapse in physics: principles and applications to light and plasma waves*, Phys. Reports **303**, 259–370.

Bergé, L., de Bouard, A., and Saut, J.C., 1995 *Collapse of Chern–Simons-gauged matter fields*, Phys, Rev. Lett. **74**, 3907–3911; *Blowing up time-dependent solutions of the planar, Chern-Simons gauged nonlinear Schrödinger equation*, Nonlinearity **8**, 235–253.

Bergé, L., Bidégaray, B., and Colin, T., 1996 *A perturbative analysis of the time envelope approximation in strong Langmuir turbulence*, Physica D **95**, 351–379.

Bergé, L. and Colin, T., 1995 *A singular perturbation problem for an envelope equation in plasma physics*, Physica D **84**, 437–459.

Bergé, L., Dousseau, P., Pelletier, G., and Pesme, D., 1990 *Scalar wave collapse at critical dimension*, Phys. Rev. A **42**, 4952–4961.

Bergé, L., Kuznetsov, E., and Rasmussen, J.J., 1996 *Defocusing regimes of nonlinear waves in media with negative dispersion*, Phys. Rev. E **53**, R1340–1343.

Bergé, L., Pelletier, G., and Pesme, D., 1990 *Langmuir wave collapse with anisotropic contraction rates*, Phys. Rev. A **42**, 4962–4971.

Bergé, L., Pelletier, G., and Pesme, D., 1991 *Collapsing solutions of the Zakharov equations with anisotropic contraction rates*, Physica D **52**, 59–62.

Bergé, L. and Pesme, D., 1992 *Time dependent solutions of wave collapse*, Phys. Lett. A **166**, 116–122.

Bergé, L. and Pesme, D., 1993a *Bounded spatial extension of the self-similar collapsing solution of the nonlinear Schrödinger equation*, Physica Scripta **43**, 323–327.

Bergé, L. and Pesme, D., 1993b *Non-self-similar collapsing solutions of the nonlinear Schrödinger equation at the critical dimension*, Phys. Rev E **48**, 684–687.

Bergé, L. and Rasmussen, J.J., 1996 *Multisplitting and collapse of self-focusing anisotropic beams in normal/anomalous dispersive media*, Phys. Plasmas **3**, 824–843; *Pulse splitting of self-focusing beam in normally dispersive media*, Phys. Rev. A. **53**, 4476–4480.

Bergé, L., Rasmussen, J.J., Kuznetsov, E.A., Shapiro, E.G., and Turitsyn, S.K., 1996 *Self-focusing of chirped optical pulses in media with normal dispersion*, J. Opt. Soc. Am. B **13**, 1879–1891.

Bergé, L., Schmidt, M.R., Rasmussen, J.J., Christiansen, P.L., and Rasmussen, K.Ø., 1997 *Amalgamation of interacting light beamlets in Kerr-type media*, J. Opt. Soc. Am. B **14**, 2550–2562.

Bergh, J. and Löfström, J., 1976 *Interpolation Spaces, an Introduction*, Springer–Verlag.

Berkshire, F.H. and Gibbon, J.D., 1983 *Collapse in the n-dimensional nonlinear Schrödinger equation - A parallel with Sundman's results in the N-body problem*, Stud. Appl. Math. **69**, 229–262.

Besse, C., 1997 *Relaxation schemes for nonlinear Schrödinger equations and Davey–Stewartson systems*, Rapport interne 9731, Mathématiques Appliquées de Bordeaux.

Besse, C. and Bruneau, C.H., 1997 *Numerical study of elliptic–hyperbolic Davey–Stewartson system: dromions simulations and blow up*, Math. Models Methods Appl. Sci., in press.

Bethuel, F., Brezis, H., and Hélein, F., 1994 *Ginzburg–Landau vortices*, Birkhäuser.

Bethuel, F. and Saut, J.C., 1998 *Travelling waves for the Gross–Piatevskii equation I*, preprint.

Bidégaray, B., 1995 *Invaraint measures for some partial differential equations* Physica D **82**, 340–364.

Bluman, G.W. and Kumei, S., 1989 *Symmetries and differential equations*, Applied Mathematical Sciences Series **81**, Springer-Verlag.

Boiti, M., Leon, J.J-P., Martina, L., and Pempinelli, F., 1988 *Scattering of localized solitons in the plane*, Phys. Lett. A **132**, 432–439.

de Bouard, A., Hayashi, N., and Saut, J.C., 1997 *Global existence of small solutions to a relativistic nonlinear Schrödinger equation*, Comm. Math. Phys. **189**, 73–105.

Bourgain, J., 1993 *Fourier transform restriction phenomena for certain lattice subsets and applications to nonlinear evolution equations*, Geom. and Funct. Anal. **3**, Part I, 107-156; Part II, 209-262; *Exponential sums and nonlinear Schrödinger equations*, Geom. and Funct. Anal. **3**, 157–178.

Bourgain, J., 1994a *Periodic nonlinear Schrödinger equations and invariant measures*, Comm. Math. Phys. **166**, 1–26.

Bourgain, J., 1994b *Construction of quasi-periodic solutions of Hamiltonian perturbations of linear equations and applications to nonlinear PDE*, International Math. Research Notices **11**, 475–497.

Bourgain, J., 1994c *On the Cauchy and invariant measure problem for the periodic Zakharov system*, Duke J. Math. **76**, 176–202.

Bourgain, J., 1996a *Gibbs measures and quasi-periodic solutions for nonlinear Hamiltonian partial differential equations*, "The Gelfand Math Seminar 1993-1995", (I.M. Gelfand, J. Lepowsky, and M. Smirnov, eds.), Birkhaüser.

Bourgain, J., 1996b *Invariant measures for the 2D-defocusing nonlinear Schrödinger equation*, Comm. Math.Phys. **176**, 421–445.

Bourgain, J., 1997a *On growth in time of Sobolev norms of smooth solutions of nonlinear Schrödinger equation in $\mathbf{R}^d$*, J. Anal. Math. **72**, 299–310.

Bourgain, J., 1997b *Invariant measures for the Gross–Piatevskii equation*, J. Math. Pures Appl. **76**, 649–702.

Bourgain, J., 1998a *Refinements of Stritchartz's inequality and application to 2D-NLS with critical nonlinearity*, Internat. Math. Res. Notices **5**, 253–383.

Bourgain, J., 1998b *Quasi-periodic solutions of Hamiltonian perturbations of 2D nonlinear Schrödinger equation*, preprint IHES. Annals of Math., **148**, 363–439.

Bourgain, J. and Colliander, J., 1996 *On wellposedness of the Zakharov system*, Int. Math. Res. Notices **11**, 515–546.

Brenner, P., Thoméen, V., and Wahlbin, L.B., 1975 *Besov spaces and applications to difference methods for initial value problems*, Lecture Notes in Mathematics, **434**, Springer-Verlag.

Brezis, H. and Gallouet, T., 1980 *Nonlinear Schrödinger evolution equation*, Nonlinear Analysis, Theory Methods Appl. **4**, 677–681.

Brezis, H., Merle, F., and Rivière, T., 1994 *Quantization effects for $\Delta u = u(1 - |u|^2)$ in $\mathbf{R}^2$*, Arch. Rat. Mech. Anal. **126**, 35–58.

Bronski, J.C. and McLaughlin, D.W., 1994 *Semi-classical behavior in the NLS equation: optical shocks — focusing instabilities*, "Singular limits of dispersive waves", (N.M. Ercolani, I.R. Gabitov, C.D. Levermore and D. Serre, eds.) pp. 21–38, NATO ASI Series, B **320**, Plenum Press (New York).

Brydges, D.C. and Slade, G., 1996 *Statistical mechanics of the 2-dimensional focusing nonlinear Schrödinger equation*, Comm. Math. Phys. **182**, 485–504.

Budd, C.J., Chen, S., and Russell, R.D., 1999 *New self-similar solutions of the nonlinear Schrödinger equation with moving mesh computations*, preprint.

Budd, C.J., Huang, W., and Russell, R.D. 1996 *Moving mesh methods for problems with blow up*, SIAM J. Sci. Comput. **17**, 305–327.

Budneva, O.B., Zakharov, V.E., and Synakh, V.S., 1975, *Certain models for wave collapse*, Sov. J. Plasma Phys. **1**, 335–338.

Buslaev, V.S. and Perel'man, G.S., 1993 *Scattering for the nonlinear Schrödinger equation : states close to a soliton*, St. Petersburg Math. J. **4**, 1111–1142.

Buslaev, V.S. and Perel'man, G.S., 1995 *On the stability of solitary waves for nonlinear Schrödinger equations*, Amer. Math. Soc. Transl. **164**, 75–98.

Campillo, A.J., Shapiro, S.L., and Suydam, B.R., 1973 *Periodic breakup of optical beams due to self-focusing*, Appl. Phys. Lett. **11**, 628–629.

Campillo, A.J., Shapiro, S.L., and Suydam, B.R., 1974 *Relationship of self-focusing to spatial instability modes*, Appl. Phys. Lett. **24**, 178–180.

Cao, X.D., Agrawal, G.P., and McKinstrie, C.J., 1994 *Self-focusing of chirped optical pulses in nonlinear dispersive media*, Phys. Rev. A **49**, 4085–4092.

Cazenave, T., 1979 *Equations de Schrödinger non linéaires en dimension deux*, Proc. Roy. Soc. Edinburg **84 A**, 327–346.

Cazenave, T., 1989 *An introduction to nonlinear Schrödinger equation*, Textos de Métodos Matemáticos **26**, Instituto de Matemática, UFRJ, Rio de Janeiro.

Cazenave, T., 1994 *Blow up and scattering in the nonlinear Schrödinger equation*, Textos de Métodos Matemáticos **30**, Instituto de Matemática, UFRJ, Rio de Janeiro.

Cazenave, T. and Lions, P.L., 1982 *Orbital stability of standing waves for some nonlinear Schrödinger equations*, Comm. Math. Phys. **85**, 549–561.

Cazenave, T. and Weissler, F., 1988 *The Cauchy problem for the Nonlinear Schrödinger equation in H^1*, Manuscripta Math. **61**, 477–494.

Cazenave T. and Weissler, F., 1989 *Some remarks on the nonlinear Schrödinger equation in the critical case*, Nonlinear Semigroups, Partial Differential Equations and Attractors, T. Gill and M.W. Zachary eds., 18–29, Lecture Notes in Mathematics, **394**, Springer-Verlag.

Cazenave T. and Weissler, F., 1990 *The Cauchy problem for the critical nonlinear Schrödinger equation in H^s*, Nonlinear Anal.. Theory, Methods and Appl., **14**, 807–836.

Cazenave T. and Weissler, F., 1992 *Rapidly decaying solutions of the nonlinear Schrödinger equation*, Comm. Math. Phys., **147**, 75–100.

Cazenave T. and Weissler, F., 1998a *Asymptotically self-similar global solutions of the nonlinear Schrödinger and heat equations*, Math. Z. **228**, 83–120.

Cazenave T. and Weissler, F., 1998b *More self-similar solutions of the nonlinear Schrödinger equation*, Nonlinear Diff. Equ. Appl. **5**, 355–365.

Chadam, J.M. and Glassey, R.T., 1975 *Global existence of solutions to the Cauchy problem for time-dependent Hartree equations*, J. Math. Phys. **16**, 1122–1130.

Champeaux, S., Gazol, A., Passot, T., and Sulem, P. L., 1997 *Plasma heating by Alfvén wave filamentation : a relevant mechanism in the solar corona and the interstellar medium*, Astrophys. J. **486**, 477–483.

Champeaux, S., Laveder, D. , Passot, T., and Sulem, P. L., 1999 *Remarks on the parallel propagation of dispersive Alfvén waves*, preprint.

Champeaux, S., Passot, T., and Sulem, P.L., 1997 *Alfvén wave filamentation*, J. Plasma Phys. **58**, 665–690.

Champeaux, S., Passot, T., and Sulem, P.L., 1998 *Transverse collapse of Alfvén wave-trains with small dispersion*, Phys. Plasmas **5**, 100–110.

Champeaux, S., Passot, T., and Sulem, P.L., 1999 *Dissipation of weakly dispersive Alfvén waves*, Phys. Plasmas, **6**, 413–416.

Chernev, P. and Petrov, V., 1992 *Self-focusing of light pulses in the presence of normal group-velocity dispersion*, Optics Lett. **17**, 172–174.

Cheung, P.Y. and Wong, A.Y., 1985 *Nonlinear evolution of electron-beam-plasma interactions*, Phys. Fluids **28**, 1538–1548.

Chiao, R.Y., Garmire, E., and Townes, C.H., 1964 *Self-trapping of optical beams*, Phys. Rev. Lett. **13**, 479–482.

Chihara, H., 1996 *The initial value problem for semilinear Schrödinger equations*, Publ. RIMS, Kyoto Univ., **32**, 445–471.

Christodoulou, D., 1986 *Global solutions of nonlinear hyperbolic equations for small initial data*, Comm. Pure Appl. Math. **39**, 267–282.

Cipolatti, R., 1992 *On the existence of standing waves for the Davey–Stewartson system*, Comm. Part. Diff. Eq. **17**, 967–988.

Cipolatti, R., 1993 *On the instability of ground states for a Davey–Stewartson system*, Ann. Inst. H. Poincaré, Phys. Théor. **58**, 85–104.

Coffman, C.V., 1972 *Uniqueness of positive solutions of $\Delta u - u + u^3 = 0$ and a variational characterization of other solutions*, Arch. Rat. Mech. Anal. **46**, 81–95.

Coifman, R.R. and Meyer, Y., 1985 *Nonlinear harmonic analysis and analytic dependence*, Proc. Symp. Pure Math. **43**, 71–78.

Cohen, B.I., Lasinski, B.F, Langdon, A.B., and Cummings, J.C., 1991 *Dynamics of ponderomotive self-focusing plasmas*, Phys, Fluids B **3**, 766–775.

Coleman, S., Glazer, V., and Martin, A., 1978 *Action minima among solutions to a class of Euclidean scalar field equations*, Comm. Math. Phys. **58**, 211–221.

Colin, T., 1993a *On the Cauchy problem for a nonlocal nonlinear Schrödinger equation occurring in plasma physics*, Diff. Integ. Eq. **6**, 1431–1450.

Colin, T., 1993b *On the standing wave solutions to a nonlocal nonlinear Schrödinger equation occuring in plasma physics*, Physica D **64**, 215–236.

Colin, T. and Weinstein, M.I., 1996 *On the ground states of vector nonlinear Schrödinger equation*, Ann. Inst. H. Poincaré, Phys. Théor. **65**, 57–79.

Collet, P. and Eckmann, J.P., 1990 *The time dependent amplitude equation for the Swift–Hohenberg problem*, Comm. Math. Phys., **132**, 139–153.

Colliander J.E. and Jerrard, R., 1998 *Vortex dynamics for the Ginzburg–Landau Schrödinger equation*, Internat. Math. Res. Notices **7**, 333–358.

Constantin, P. and Saut, J.C., 1988 *Local smoothing properties of dispersive equations*, J. Amer. Math. Soc. **1**, 413–446.

Cooper, F. Lucheroni, C., and Shepard, H., 1992 *Variational method for studying self-focusing in a class of nonlinear Schrödinger equation*, Phys. Lett. A **170**, 184–188.

Coste, C., 1998 *Nonlinear Schrödinger equation and superfluid hydrodynamics*, Eur. Phys. J. B **1**, 245–253.

Craig, W., 1998 Problèmes de petits diviseurs en équations aux dérivées partielles, Société mathématique de France, Panaromas et Synthèses.

Craig, W. and Groves, M., 1994 *Hamiltonian long-wave approximations to the water-wave problem*, Wave Motion **19**, 367–389.

Craig, W., Kappeler, T., and Strauss, W., 1995 *Microlocal dispersive smoothing for the Schrödinger equation*, Comm. Pure Appl. Math. **48** , 769–860.

Craig, W., Schanz, U., and Sulem, C., 1997 *The modulational regime of three-dimensional water wave and the Davey–Stewartson system*, Ann. Inst. H. Poincaré, Analyse non linéaire, **14**, 615–667.

Craig, W., and Sulem, C., 1993 *Numerical simulation of gravity waves*, J. Comp. Phys. **108**, 73–83.

Craig, W., Sulem, C., and Sulem, P.L., 1992 *Nonlinear modulation of gravity waves : a rigorous approach*, Nonlinearity **108**, 497–522.

Craig, W. and Wayne, C.E., 1994 *Periodic solutions of Nonlinear Schrödinger equations and Nash–Moser method*, in "Hamiltonian mechanics (Toruń 1993)" (J. Semanis, ed.), 103–122, NATO Adv. Sci. Inst. Series B Physics **331**, Plenum Press NY.

Craik, A.D.D., 1985 *Wave interaction and fluid flows*, Cambridge Monographs on Mechanics and Applied Mathematics, Cambridge University Press.

Creswick R.J. and Morrisson, H.L., 1980 *On the dynamics of quantum vortices*, Phys. Lett. **76 A**, 267–268.

Davey, A. and Stewartson, K., 1974 *On three-dimensional packets of surface waves*, Proc. Roy. Soc. Lond. A **338** (1974), 101–110.

Davies, R.E. and Acrivos, A. 1967 *Solitary internal waves in deep water*, J. Fluid Mech. **29**, 593–607.

Dawes, E.L. and Marburger, J.H., 1969 *Computer studies in self-focusing*, Phys. Rev. **179**, 862–868.

Degtyarev, L.M., Zakharov, V.E., and Rudakov, L.I., 1975 *Two examples of Langmuir wave collapse*, Sov. Phys. JETP, **41**, 57–61.

Degtyarev, L.M., Zakharov, V.E., and Rudakov, L.I., 1976 *Dynamics of Langmuir collapse*, Sov. J. Plasma Phys. **2**, 240–246.

Denavit, J., Pereira, N.R., and Sudan, R.N., 1994 *Two-dimensional stability of Langmuir solitons*, Phys. Rev. Lett. **33**, 1435–1438.

Diddams, S.A., Eaton, H.K. Zozulya, A.A., and Clement, T.S., 1998 *Amplitude and phase measurements of femtosecond pulse splitting in nonlinear dispersive media*, Optics Lett. **23**, 379–381.

Djordjevic V.D. and Redekopp, L.G., 1977 *On two-dimensional packets of capillary–gravity waves*, J. Fluid Mech. **79**, 703–714.

Dodd, R.K., Eilbeck, J.C., Gibbon, J.D., and Morris, H.C., 1982 *Solitons and nonlinear wave equations*, Academic Press.

Donnelly, R.J., 1991 *Quantized Vortices in Helium II*, Cambridge Univ. Press.

Drazin, P.G. and Johnson, R.S., 1989 *Solitons: an introduction*, Cambridge Text in Applied Mathematics, Cambridge University Press.

Dyachenko, S., Newell, A.C., Pushkarev, A., and Zakharov, V.E., 1992 *Optical turbulence: weak turbulence, condensates and collapsing filaments in the nonlinear Schrödinger equation*, Physica D **57**, 96–160.

Dyachenko, S., Pushkarev A.M, Rubenchik, A.M., Sagdeev, R.Z., Shvets, V.F., and Zakharov, V.E., 1991 *Computer simulation of Langmuir collapse*, Physica D **52**, 78–102.

D'yachenko, A.I., Zakharov, V.E., Rubenchik, A.M., Sagdeev, R.Z., and Shvets, V.F., 1988 *Numerical simulation of two-dimensional Langmuir collapse*, Sov. Phys. JETP **67**, 513–518.

Dyshko, A.L., Lugovoĭ, V.N., and Prokhorov, A.M., 1972 Multifocus structure of a light beam in a nonlinear medium, Sov. Phys. JETP **34**, 1235–1241.

Eckhaus, W. and Van Harten, A., 1981 *The inverse scattering transformation and the theory of solitons: an introduction*, Mathematical Studies **50**, North-Holland (Amsterdam).

Edmundson, D.E., 1997 *Unstable higher modes in a three-dimensional nonlinear Schrödinger equation*, Phys. Rev. E **55**, 7636–7644.

Ercolani, N. and Montgomery, R., 1993 *On the fluid approximation to a nonlinear Schrödinger equation*, Phys. Lett. A **180**, 402–408.

Esteban, M.J., 1980 *Existence d'une infinité d'ondes solitaires pour des équations des champs non linéaires*, Ann. Fac. Sci. Toulouse, Math. **2**, 181–191.

Evans, L.C., 1998 *Partial differential equations*, Graduate Studies in Mathematics, **19**, Amer. Math. Soc. (Providence).

Feit, M.D. and Fleck, J.A., 1988 *Beam nonparaxiality, filament formation, and beam break up in the self-focusing of optical beams.*, J. Opt. Soc. Am. B **5**, 633–640.

Fermi, E., Pasta, J. and Ulam, S., 1955 *Studies of nonlinear problems*, Document LA-1940 (May 1955), reproduced in "Fermi, Enrico, Collected papers", vol. 2 pp. 977–988, University of Chicago Press 1965, and in "Stanislaw Ulam: Sets, Numbers and Universes," W.A. Beyer, J. Mycielski and C.G. Rota, eds., Mathematicians of our Time series, vol. 9, pp. 491–501, MIT Press, 1974.

Fibich, G., 1995 *Time dispersive effects in ultrashort laser-tissue interactions*, Adv. in Heat Mass Transfer in Biotechnology HTD 322 and BTD 32: 27–31.

Fibich, G., 1996a *Femtosecond laser-tissue interactions*, Ophthalmic Technologies VI, Proc. SPIE 2673, 93–101.

Fibich, G., 1996b *An adiabatic law for self-focusing of optical beams*, Optic Lett. **21**, 1735–1737; erratum **22**, 194 (1997).

Fibich, G., 1996c *Small beam nonparaxiality arrests self-focusing of optical beams*, Phys. Rev. Lett. **76**, 4356–4359.

Fibich, G., Malkin, V.M. and Papanicolaou, G.C., 1995 *Beam self-focusing in the presence of a small normal time dispersion*, Phys. Rev. A **52**, 4218–4228.

Fibich, G. and Papanicolaou, G.C., 1997a *Self-focusing in the presence of small time dispersion and nonparaxiality*, Optics Lett. **22**, 1379–1381.

Fibich, G. and Papanicolaou, G.C., 1997b *Self-focusing in the perturbed and unperturbed nonlinear Schrödinger equation in critical dimension*, submitted to SIAM J. Appl. Math. (Computational and Applied Mathematics Report 97-21, UCLA, 1997, available at http://www.math.tau.ac.il/~fibich)

Fibich, G. and Papanicolaou, G.C., 1998 *A modulation method for self-focusing in the perturbed critical nonlinear Schrödinger equation*, Phys. Lett. A, **239**, 167–173.

Fokas, A.S. and Santini, P.M., 1990 *Dromions and a boundary value problem for the Davey–Stewartson I equation*, Physica D **44**, 99–130.

Fokas, A.S. and Sung, L.Y., 1992 *On the solvability of the n-wave, Davey–Stewartson and Kadomtsev–Petviashvili equations*, Inverse Problems, **8**, 673–708.

Fraiman, G.M., 1979 *On energetically forbidden supersonic collapse regime of Langmuir oscillations*, JETP Lett. **30**, 525–528.

Fraiman, G.M., 1985 *Asymptotic stability of manifold of self-similar solitions in self-focusing*, Sov. Phys. JETP **61**, 228–233.

Freeman, N.C. and Davey, A., 1975 *On the evolution of packets of long surface waves*, Proc. R. Soc. Lond. A **344**, 427–433.

Frisch, T., Pomeau, Y., and Rica, S., 1992 *Transition to dissipation in a model of superflow*, Phys. Rev. Lett. **69**, 1644–1647.

Gaididei, Yu. B. and Christiansen, P.L., 1998 *Spatiotemporal collapse in a nonlinear waveguide with a randomly fluctuating refractive index*, Optics Lett. **23**, 1090–1092.

Gaididei, Yu. B., Rasmussen, K. Ø, and Christiansen, P.L., 1995 *Nonlinear exci-
tations in thwo-dimensional molecular structures with impureties*, Phys. Rev.
E **52**, 2951–2962.

Galusinski, C., 1998 *A singular perturbation problem in a system of nonlinear
Schrödinger equations occurring in a Langmuir turbulence*, Rapport Interne
98011, Mathématiques Appliquées de Bordeaux.

Garmire, E., Chiao, R.Y., and Townes, C.H., 1966 *Dynamics and characteristics
of the self-trapping of intense light beams*, Phys. Rev. Lett. **16**, 347–349.

Gazol, A., Passot, T., and Sulem, P.L., 1998 *Nonlinear dynamics of obliquely
propagating Alfvén waves*, J. Plasma Phys., **60**, 95–109.

Gazol, A., Passot, T., and Sulem, P.L., 1999 *Nonlinear Alfvén waves and
their coupling to reduced magnetohydrodynamics for compressible fluids*, Phys.
Plasmas, to appear July 1999.

Gelfand, I.M. and Fomin, S.V., 1963 *Calculus of variations*, Prentice-Hall Inc.
(New Jersey).

Ghidaglia, J.M. and Saut, J.C., 1990 *On the initial value problem for the Davey–
Stewartson systems*, Nonlinearity **3**, 475–506.

Ghidaglia, J.M. and Saut, J.C., 1992 *On the Zakharov–Schulman equations*, in
"Nonlinear dispersive waves", (L. Debnath, ed.), pp. 83–97, World Scientific.

Ghidaglia, J.M. and Saut, J.C., 1993 *Nonelliptic nonlinear Schödinger equations*,
J. Nonlinear Sci. **3**, 169–195.

Ghidaglia, J.M. and Saut, J.C., 1996 *Non existence of travelling wave solutions
to non elliptic nonlinear Schrödinger equations*, J. Nonlinear Sci. **6**, 139–145.

Gibbons, J., 1978 *Behaviour of slow Langmuir solitons*, Phys. Lett. **67 A**, 22–24.

Gibbons, J., Thornhill, S.G., Wardrop, M.J., and ter Haar, D., 1977 *On the theory
of Langmuir solitons*, J. Plasma Phys. **17**, 153–170.

Gidas, B., Ni, W.M., and Nirenberg, L., 1981 *Symmetry of positive solutions of
nonlinear elliptic equations*, Math. Anal. and Appl., (L. Nachbin ed.), Adv.
Math. Supp. Studies, **7A**, 369–402, Academic Press.

Gilson, C.R. and Nimmo, J.J.C., 1991 *A direct method for dromion solutions of
the Davey–Stewartson equations and their asymptotic properties*, Proc. R. Soc.
Lond. A **435**, 339–357.

Ginibre, J., 1996 *Le problème de Cauchy pour des EDP semi-linéaires
périodiques en variables d'espace [d'après Bourgain)]*, Séminaire Bourbaki 796
(1994-1995), Astérisque **237**, 163–187, Soc. Math. de France.

Ginibre, J., 1998 *Introduction aux équations de Schrödinger non linéaires*, Cours
de DEA 1994-1995, Paris Onze Edition.

Ginibre, J., Ozawa,T., and Velo, G., 1994 *On the existence of the wave operators
for a class of nonlinear Schrödinger equation*, Ann. Inst. H. Poincaré, Phys.
Théor. **60**, 211–239.

Ginibre, J., Tsutsumi, Y., and Velo, G., 1997 On the Cauchy problem for the
Zakharov system, J. Funct. Anal. **151**, 384–436.

Ginibre, J. and Velo, G., 1978 *On a class of Schrödinger equations. III: Special
theories in dimensions 1, 2 and 3*, Ann. Inst. H. Poincaré, Sec. A **28**, 287–316.

Ginibre, J. and Velo, G., 1979a *On a class of Schrödinger equations. I: The
Cauchy problem, a general case*, J. Funct.Anal. **32**, 1–32.

Ginibre, J. and Velo, G., 1979b *On a class of Schrödinger equations. II: Scattering
theory, General case*, J. Funct.Anal. **32**, 33–71.

Ginibre, J. and Velo, G., 1980 *On a class of non linear Schrödinger equations with non local interaction*, Math. Z. **170**, 109–136.

Ginibre, J. and Velo, G., 1985a *The global Cauchy problem for the nonlinear Schrödinger equation revisited*, Ann. Inst. H. Poincaré, Analyse Non Linéaire **2**, 309–327.

Ginibre, J. and Velo, G., 1985b *Scattering theory in the energy space for a class of nonlinear Schrödinger equations*, J. Math. Pures Appl., **64**, 363–401; *Time decay of finite energy solutions of the nonlinear Klein-Gordon and Schrödinger equations*, Ann. Inst. H. Poincaré, Phys. Théor., **43**, 399–442.

Ginibre, J. and Velo, G., 1996 *The Cauchy problem in local spaces for the complex Ginzburg-Landau equation, I Compactness methods*, Physica D **95**, 191–228.

Ginibre, J. and Velo, G., 1997a *Localized estimates and Cauchy problem for the logarithmic complex Ginzburg-Landau equation*, J. Math. Phys. **38**, 2475–2482.

Ginibre, J. and Velo, G., 1997b *The Cauchy problem in local spaces for the complex Ginzburg-Landau equation, II Contraction methods*, Comm. Math. Phys. (in press).

Ginzburg, V.L. and Piatevskii, L.P., 1958 *On the theory of superfluidity*, Sov. Phys. JETP **34**, 858–861.

Glangetas, L. and Merle, F., 1994a *Existence of self-similar blow-up solutions for Zakharov equation in dimension two, Part I*, Comm. Math. Phys. **160**, 173–215.

Glangetas, L. and Merle, F., 1994b *Concentration properties of blow-up solutions and instability results for Zakharov equation in dimension two, Part II*, Comm. Math. Phys. **160**, 349–389.

Glassey, R.T., 1977a *Asymptotic behavior of solutions to certain Nonlinear Schrödinger-Hartree equations*, Comm. Math. Phys. **53**, 9–18.

Glassey, R.T., 1977b *On the blowing-up of solutions to the Cauchy problem for the nonlinear Schrödinger equation*, J. Math. Phys., **18**, 1794–1797.

Goldman, M.V., 1983 *Progress and problems in the theory of type III solar radio emission*, Solar Phys. **89**, 403–442.

Goldman, M.V., 1984 *Strong turbulence of plasma waves*, Rev. Mod. Phys. **56**, 709–735.

Goldman, M.V. and Nicholson, D.R., 1978 *Virial theory of direct Langmuir collapse*, Phys. Rev. Lett. **41**, 406–410.

Goldman, M.V., Rypdal, K., and Hafizi, B., 1980 *Dimensionality and dissipation in Langmuir collapse* Phys. Fluids **23**, 945–955.

Grikurov, V.E., 1997 *Perturbation of unstable solitons for generalized NLS with saturating nonlinearity*, International Seminar "Day on Diffraction-97", 170–179, available at http://mph.phys.spbu.ru/~grikurov/publlist.html

Grillakis, M., 1988 *Linearized instability for Nonlinear Schrödinger and Klein-Gordon equations*, Comm. Pure Appl. Math **41**, 747–774.

Grillakis, M., 1990 *Existence of nodal solutions of semilinear equations in $\mathbf{R}^n$*, J. Diff. Eq. **85**, 367–400.

Grillakis, M., Shatah, J., and Strauss, W., 1987 *Stability theory of solitary waves in the presence of symmetry, Part I*, J. Funct. Anal. **74**, 160–197.

Grillakis, M., Shatah, J., and Strauss, W., 1990 *Stability theory of solitary waves in the presence of symmetry, Part II*, J. Funct. Anal. **94**, 308–348.

Grenier, E., 1995 *Limite semi-classique de l'équation de Schrödinger en temps petit*, C. R. Acad. Sci. Paris **320**, Série I, 691–694.

Gross, E.P., 1963 *Hydrodynamics of a superfluid condensate*, J. Math. Phys. **4**, 195–207.

Gurnett, D.A., Maggs, J.E., Gallagher, D.L. Kurth, W.S, and Scarf, F.L., 1981 *Parametric interaction and spatial collapse of beam-driven Langmuir waves in the solar wind*, J. Geophys. Research **86**, 8833–8841.

Gustafson, S., 1997a *Stability of vortex solutions of the Ginzburg–Landau heat equation*, in *Partial differential equations and their applications*, (P. Greiner, V. Ivrii, L. Seco, and C. Sulem, eds.), CRM Proc. Lect. Notes, **12**, 159–165.

Gustafson, S., 1997b *Symmetric solutions of the Ginzburg–Landau equation in all dimensions*, Int. Math. Res. Notices **16**, 807–816.

Gustafson, S. and Sigal, I.M., 1998 *The stability of magnetic vortices*, preprint.

Guzmán-Goméz, M., 1994 *Asymptotic behaviour of the Davey–Stewartson system*, C.R. Math. Rep. Acad. Sci. Canada **16**, 91–96.

ter Haar, D., 1979 *Solitons*, Instituut voor Theoretische Fysica der Rijksuniversiteit Utrecht.

ter Haar, D. and Tsytovich, V.N., 1981 *Modulation instabilities in astrophysics*, Phys. Reports **73**, 175–236.

Hada, T., 1993 *Evolution of large amplitude Alfvén waves in the solar wind with $\beta \sim 1$*, Geophys. Res. Lett. **20**, 2415–2418.

Hadžievski, Lj.R. and Škonić, M.M. 1993 *the spectral signature of a Langmuir soliton instability*, Phys. Fluids B **5**, 2076–2079.

Hadžievski, Lj.R., Škonić, M.M., Rubenchick, A.M., Shapiro, E.G., and Turitsin, S.K., 1990 *Langmuir soliton and collapse in a weak magnetic field*, Phys. Rev. A **42**, 3561–3570.

Hafizi, B., 1981 *Nonlinear evolution equations, recurrence and stochasticity*, Phys. Fluids **24**, 1791–1798.

Hasegawa, A. and Tappert, F., 1973 *Transmission of stationary nonlinear optical pulses in dispersive dielectric fibers I. Anomalous dispersion*, Appl. Phys. Lett. **23**, 142–144.

Hasimoto, H. and Ono, H., 1972 *Nonlinear modulation of gravity waves*, J. Phys. Soc. Japan **33**, 805–811.

Hayashi, N., 1996 *Local existence in time of small solutions to the Davey–Stewartson systems*, Annales Inst. H. Poincaré, Phys. Théor. **65**, 313–366.

Hayashi, N., 1997 *Local existence in time of solutions to the elliptic-hyperbolic Davey–Stewartson system without smallness condition on the data*, J. Anal. Math. **LXXIII**, 133–164.

Hayashi, N. and Hirata, H., 1996 *Global existence and asymptotic behaviour in time of small solutions to the elliptic-hyperbolic Davey–Stewartson system*, Nonlinearity **9**, 1387–1409.

Hayashi, N. and Hirata, H., 1997 *Local existence in time of small solutions to the elliptic-hyperbolic Davey–Stewartson system in the usual Sobolev space*, Proc. Edinburg Math. Soc. **40**, 563–581.

Hayashi, N. and Kaikina E.I., 1998 *Local existence of solutions to the Cauchy problem for nonlinear Schrödinger equations*, preprint.

Hayashi, N. and Kato, K., 1997 *Analyticity in time and smoothing effect of solutions to nonlinear Schrödinger equations*, Comm. Math. Phys. **184**, 273–300.

Hayashi, N. and Naumkin, P.I., 1998a *On the Davey–Stewartson and the Ishimori systems*, preprint.

Hayashi, N. and Naumkin, P.I., 1998b *Asymptotics for large time solutions to the nonlinear Schrödinger and Hartree equations*, Amer. J. Math. **120**, 369-389; *Remarks on scattering theory and large time asymptotics of solutions to Hartree type equations with a long range potential*, SUT J. Math. **34**, 13-24.

Hayashi, N., Naumkin, P.I., and Ozawa, T., 1998 *Scattering theory for Hartree equation*, SIAM J. Math. Anal., to appear.

Hayashi, N. and Ozawa, T., 1987 *Time decay of solutions to the Cauchy problem for time-dependent Schrödinger-Hartree equations*, Comm. Math. Phys. **110**, 467-478.

Hayashi, N. and Ozawa, T., 1992 *On the derivative nonlinear Schödinger equation*, Physica D **55**, 14-36.

Hayashi, N. and Ozawa, T., 1994a *Remarks on nonlinear Schrödinger equations in one space dimension*, Diff. Int. Eq. **2**, 453-461.

Hayashi, N. and Ozawa, T., 1994b *Finite energy solutions of nonlinear Schrödinger equations of derivative type*, SIAM J. Math. Anal. **25**, 1488-1503.

Hayashi, N. and Ozawa, T., 1995 *Global, small radially symmetric solutions to nonlinear Schrödinger equations and a gauge transformation*, Diff. Int. Eq. **8**, 1061-1072.

Hayashi, N. and Saitoh, S., 1990 Analyticity and global existence of small solutions to some nonlinear Schrödinger equations, Comm. Math. Phys. **129**, 27-49; *Analyticity and smoothing effect for the Schrödinger equation*, Ann. Inst. H. Poincaré, Phys. Théor. **52**, 163-173.

Hayashi, N. and Saut, J.C., 1995 *Global existence of small solutions to the Davey-Stewartson and the Ishimori systems*, Diff. Int. Eq. **8**, 1657-1675.

Hayashi, N. and Saut, J.C., 1996 *Global existence of small solutions to the Ishimori system without exponential decay on the data*, Diff. Int. Eq. **9**, 1183-1195.

Hayashi, N. and Tsutsumi, Y., 1987 *Scattering theory for Hartree type equations*, Ann. Inst. H. Poincaré, Phys. Théor., **46**, 187-313.

Hasegawa, A., (1970) *Stimulated modulational instabilities of plasma waves*, Phys. Rev. A **1**, 1746-1750.

Hierarinta, J. and Hirota, R., 1990 *Multidromion solutions to the Davey-Stewartson equation*, Physics Lett. A **145**, 237-244.

Hogan, S.J., 1985 *The fourth-order evolution equation for deep-water gravity-capillary waves*, Proc. R. Soc. Lond. A **402**, 359-372.

Huang, W. and Russell, R.D. 1996 *A moving collocation method for time dependent differential equations*, Appl. Numer. Math. **20**, 101-116.

Infeld, E., 1981 *Quantitative theory of the Fermi-Pasta-Ulam Recurrence in the nonlinear Schrödinger equation*, Phys. Rev. Lett. **47**, 717-718.

Infeld, E. and Rowlands, G. 1990 *Nonlinear waves, solitons and chaos*, Cambridge University Press.

Ishimori, Y. 1984 *Multi-vortex solutions of a two-dimensional nonlinear wave equation*, Prog. Theor. Phys. **72**, 33-37.

Janssen, P.A.E.M., 1981 *Modulational instability and the Fermi-Pasta-Ulam recurrence*, Phys. Fluids **24**, 23-26.

Janssen, P.A.E.M. and Rasmussen, J.J., 1983 *Nonlinear evolution of the transverse instability of plane-envelope solitons*, Phys. Fluids **26**, 1279-1287.

Jensen, A. and Kato, T., 1979 *Spectral properties of Schrödinger operators and time-decay of the wave functions*, Duke Math. J. **46**, 583-611.

Jensen, A., 1984 *Spectral properties of Schrödinger operators and time-decay of the wave functions. Results un* $L^2(\mathbf{R}^4)$ J. Math. Anal. Appl. **101** 397–422.

Jin, S., Levermore, C.D., and McLaughlin, D.W., 1994 *The behavior of solutions of the NLS equation in the semiclassical limit,* Singular limits of dispersive waves, NATO ASI, (N.M. Ercolani, C.D. Gabitov, C.D. Levermore and D. Serre, eds.), 235–256, Series B : Physics **320**, Plenum Press.

Johnson, R.S., 1977 *On the modulation of water waves in the neighbourhood of* $kh \approx 1.363$, Proc. Roy. Soc. A **357**, 131–141.

Johnson, R.S., 1997 *A modern introduction to the mathematical theory of water waves,* Cambridge Texts in Applied MAthematics, Cambridge University Press.

Johnson, R. and Pan, X., 1993 *On an elliptic equation related to the blow-up phenomenon in the nonlinear Schrödinger equation,* Proc. Roy. Sc. Edinburgh **123A**, 763–782.

Johnston, T.W., Vidal, F., and Fréchette, D., 1997 *Laser-plasma filamentation and the spatially periodic nonlinear Schrödinger equation,* Phys. Plasmas **4**, 1582–1588.

Jones, C.K.R.T., 1988 *An instability Mechanism for radially symmetric standing waves of a nonlinear Schrödinger equation,* J. Diff. Eq. **71**, 34–62.

Jones, C.K.R.T., Küpper, T., and Plakties, H., 1990 *A shooting argument with oscillation for semilinear elliptic radially symmetric equations,* Proc. Roy. Soc. Edinburgh **108A**, 165–180.

Jones, C.A., Putterman, S.J., and Roberts, P.H., 1986 *Motions in a Bose condensate : V. Stability of solitary wave solutions of non-linear Schrödinger equation in two and three dimensions,* J. Phys. A **19**, 2991–3011.

Jones, C.A., and Roberts, P.H., 1982 *Motion in a Bose condensates: IV Axisymmetric solitary waves,* J. Phys. A **15**, 2599–2619.

Josserand, C. and Rica, S., 1997 *Coalescence and droplets in the subcritical nonlinear Schrödinger equation,* Phys. Rev. Lett. **78**, 1215–1218.

Josserand, C. and Pomeau, Y., 1995 *Generation of vortices in a model od superfluid* ^{4}*He by the Kadomtsev-Petviashvili instability,* Europhys. Lett. **30**, 43–48.

Josserand, C., Pomeau, Y., and Rica, S., 1995 *Cavitation versus vortex nucleation in a superfluid model,* Phys. Rev. Lett. **75**, 3150–3153.

Journé, J.L., Soffer, A., and Sogge, C.D., 1991 *Decay estimates for Schrödinger operators,* Comm. Pure Appl. Math **44**, 573–604.

Kadomtsev, B.B., 1979 *Phénomènes collectifs dans les plasmas,* MIR (Moscow), translated from Russian (original edition 1976).

Kadomtsev, B.B. and Karpman V.I., 1971 *Nonlinear waves,* Sov. Phys. Uspekhi **14**, 40–60.

Kadomtsev, B.B. and Petviashvili, V.I., 1970 *On the stability of solitary waves in weakly dispersive media,* Sov. Phys. Dokl. **15**, 539–541.

Karakashian, O.A., Akrivis, G.D., and Dougalis, V.A., 1993, *On optimal order error estimates for the nonlinear Schrödinger equation,* SIAM J. Numer. Anal. **30**, 377–400.

Karpman, V.I., 1975 *Non-linear waves in dispersive media,* International series of monographs in natural philosophy, **71**, Pergamon Press.

Karpman, V.I. and Shagalov A.G., 1982 *The ponderomotive force of a high-frequency electromagnetic field in a cold magnetized plasma*, J. Plasma Phys. **27**, 215–224.

Kato, T., 1987 *On Nonlinear Schrödinger equations*, Ann. Inst. H. Poincaré, Phys. Théor. **46**, 113–129.

Kato, T., 1989 *Nonlinear Schrödinger equations* in " Schrödinger operators", (H. Holden and A. Jensen, eds.), Lecture Notes in Physics **345**, Springer-Verlag.

Kaup, D.J. and Newell, A.C., 1978 An exact solution for the derivative nonlinear Schrödinger equation, J. Math. Phys. **19**, 798–801.

Kavian, O., 1987 *A remark on the blowing-up of solutions to the Cauchy problem for nonlinear Schrödinger equation*, Trans. Amer. Math. Soc. **299**, 193–203.

Kawahara, T., 1975 *Nonlinear self-modulation of capillary–gravity waves on liquid layer*, J. Phys. Soc. Japan **38**, 265–270.

Kelley, P.L., 1965 *Self-focusing of optical beams*, Phys. Rev. Lett. **15**, 1005–1008.

Keller, C., 1983 *Stable and unstable manifolds for the nonlinear wave equation with dissipation*, J. Diff. Eq. **50**, 330–347.

Kenig, C., Ponce, G., and Vega, L., 1993 *Small solutions to nonlinear Schrödinger equations*, Ann. Inst. H. Poincaré, Anal. Non Linéaire **10**, 255–288.

Kenig, C., Ponce, G., and Vega, L., 1995 *On the Zakharov and Zakharov–Shulman systems*, J. Funct. Anal. **127**, 204–234.

Kenig, C., Ponce, G., and Vega, L., 1997 *Smoothing effects and local existence theory for the generalized nonlinear Schrödinger equations*, preprint.

Kenig, C., Ponce, G., and Vega, L., 1998 *On the initial value problem for the Ishimorei system*, preprint.

Kirrmann, P., Schneider, G., and Mielke, A., 1992 *The valididy of modulation equations for extended systems with cubic nonlinearities*, Proc. Royal Soc. Edinburgh **122A**, 85–91.

Kivshar, Y.S. and Luther-Davies, B., 1998 *Dark optical solitons: physics and applications*, Phys. Reports **298**, 81–197.

Kivshar, Y.S. and Pelinovsky, D.E., 1999 *Self-focusing and transverse instabilities of solitary waves*, submitted to Phys. Reports.

Klainerman, S., 1986 *The null condition and global existence to nonlinear wave equations*, Lectures in Appl. Math. **23**, 293–326.

Klainerman, S. and Ponce, G., 1983 *Global, small amplitude solutions to nonlinear evolutions equations*, Comm. Pure Appl. Math. **36**, 133–141.

Klyatskin, V.I., 1971 *The effects of the turbulent atmosphere on wave propagation*, translated for NOAA by Israel Program for Scientific Translations (Jerusalem), Available from U.S. Department of Commerce, NTIS, Springfield, Va. 22151.

Klyatskin, V.I. and Tatarskii, V.I., 1970 *The parabolic equation approximation for propagation of waves in a medium with random inhomogeneities*, Sov. Phys. JETP **31**, 335–339.

Kogiso, K., Ueda, S., and Yajima, N., 1982 *Generalized pressure tensor, stress tensor and ponderomotive force in collisionless plasma*, J. Phys. Soc. Japan **51**, 269–279.

Kono, M., Škorić, M.M., and ter Harr, D., 1980 *The kinetic theory of magnetic-field generation in a Langmuir plasma*, Phys. Lett. A **77**, 27–29.

Kono, M., Škorić, M.M., and ter Harr, D., 1981 *Spontaneous excitation of magnetic fields and collapse dynamics in a Langmuir plasma*, J. Plasma Phys. **26**, 123–146.

Konno, K. and Suzuki, H., 1979 *Self-focusing of laser beam in nonlinear media*, Phys. Scripta. **20**, 382–386.

Koppel, N. and Landman, M., 1995 *Spatial structure of the focusing singulariry of the nonlinear Schrödinger equation: a geometrical analysis*, SIAM J. Appl. Math. **55**, 1297–1323.

Kosmatov, N.E., Petrov, I.V., Shvets V.F., and Zakharov, V.E., 1988 *Large amplitude simulation of wave collapses in nonlinear Schrödinger equations*, Preprint No 1365, Space Research Institute (Moscow).

Kosmatov, N.E., Shvets V.F., and Zakharov, V.E., 1991 *Computer simulation of wave collapses in the nonlinear Schrödinger equations*, Physica D **52**, 16–35.

Krasnosel'skikh, V.V. and Sotnikov, V.I., 1977 *Plasma-wave collapse in a magnetic field*, Sov. Plasma Phys. **3**, 491–495.

Kuksin, S., 1993 *Nearly integrable infinite-dimensional Hamiltonian systems*, Lecture Notes in Mathematics, **1556**, Springer-Verlag.

Kuznetsov, E.A., 1974 *The collapse of electromagnetic waves in plasmas*, Sov. Phys. JETP **39**, 1003–1007.

Kuznetsov, E. A., 1996 *Wave collapse in plasmas and fluids*, CHAOS **6**, 381–390.

Kuznetsov, E.A. and Rasmussen, J.J., 1995 *Self-focusing instability of two-dimensional solitons and vortices*, JETP Lett. **62**, 105–112; *Instability of two-dimensional solitons and vortices in defocusing media*, Phys. Rev. E **51**, 4479–4484.

Kuznetsov, E.A., Rasmussen, J.J., Rypdal, K., and Turitsyn, S.K., 1995 *Sharper criteria for the wave collapse*, Physica D **87**, 273–284.

Kuznetsov, E.A., Rubenchik, A.M., and Zakharov, V.E., 1986 *Soliton stability in plasmas and hydrodynamics*, Phys. Reports **142**, 103–165.

Kuznetsov, E.A. and Škonić, M.M., 1988 Hierarchy of collapse regimes for upper-hybrid and lower hybrid waves, Phys. Rev. A **38**, 1422–1426.

Kuznetsov, E.A. and Turitsyn, S.K., 1985 *Talanov transformations in self-focusing problems and instability of stationary waveguides*, Phys. Lett. **112 A**, 273–275.

Kuznetsov, E.A. and Turitsyn, S.K., 1988 *Instability and collapse of solitons in medis with a defocusing nonlinearity*, Sov. phys. JETP **67**, 1583–1588.

Kuznetsov, E.A. and Turitsyn, S.K., 1990 *Quasiclassical Langmuir wave collapse in a magnetic field*, Sov. J. Plasma Phys. **16**, 524–527.

Kuznetsov, E.A. and Zakharov, V.E., 1998 *Nonlinear coherent phenomena in continuous media*, in " Nonlinear science at the dawn of the 21 st century", P. Christiansen and M. Soerensen, eds. Springer–Verlag (in press).

Kwong, M.K., 1989 *Uniqueness of positive solutions of $\Delta u - u + u^p = 0$ in $\mathbf{R}^n$*, Arch. Rat. Mech. Anal. **65**, 243–266.

Labaune, C., Baton, S., Jalinaud, T., Baldis, H.A., and Pesme, D., 1992 *Filamentation in long scale length plasmas: Experimental evidence and effect of laser spatial incoherence*, Phys. Fluids B **4**, 2224–2231.

Laedke, E.W., Blaha, R., Spatschek, K.H., and Kuznetsov, E.A., 1992 On the stability of collapse in the critical case, J. Math. Phys. **33**, 967–973.

Laedke, E.W. and Spatschek, K.H., 1978 *Nonlinear stability of envelope solitons*, Phys. Rev. Lett. **41**, 1798–1801.

Laedke, E.W. and Spatschek, K.H., 1979 *Exact stability criteria for finite amplitude solitons*, Phys. Rev. Lett. **42**, 1534–1537.

Laedke, E.W. and Spatschek, K.H., 1984 *Stability properties of multidimensional finite-amplitude solitons.* Phys. Rev. A **30**, 3297–3288.

Laedke, E.W. and Spatschek, K.H., 1985 *Variational principles in soliton physics*, in "Differential geometry, calculus of variations and their applications", 335–357. Lectures Notes in Pure and Appl. Math. **100**, (J.M. Rassias and T.M. Rassias, eds.), M. Decker Inc.

Laedke, E.W., Spatschek, K.H., and Stenflo, L., 1983 *Evolution theorem for a class of perturbed envelope solition solutions* J. Math. Phys. **24**, 2764–2769.

Landman, M.J., LeMesurier, B.J, Papanicolaou, G.C., Sulem, C., and Sulem, P.L., 1989 *Singular solutions of the cubic Schrödinger equation*, "Integrable systems and Application", (M. Balabane, P. Lochak, and C. Sulem, eds.), 207–217, Lecture Notes in Physics, **342**, Springer-Verlag.

Landman, M.J., Papanicolaou, G.C., Sulem, C., and Sulem, P.L., 1988 *Rate of blowup for solutions of the Nonlinear Schrödinger equation at critical dimension*, Phys. Rev. A **38**, 3837–3843.

Landman, M.J., Papanicolaou, G.C., Sulem, C., Sulem, P.L., and Wang, X.P., 1991 *Stability of isotropic singularities for the nonlinear Schrödinger equation*, Physica D **47** , 393–415.

Landman, M.J., Papanicolaou, G.C., Sulem, C., Sulem, P.L., and Wang, X.P., 1992 *Stability of isotropic self-similar dynamics for scalar-wave collapse*, Phys. Rev. A **46**, 7869–7876.

Lange, C.G. and Newell, A.C., 1971 *The post-buckling problem for thin elastic shells*, SIAM J. Appl. Math. **21**, 605–629.

Laurey, C., 1995 The Cauchy problem foe a generalized Zakharov system, Diff. Int. Eq. **8**, 105–130.

Lebowitz, J., Rose, H., and Speer, E., 1988 *Statistical mechanics of the nonlinear Schrödinger equations* J. Stat. Phys. **50**, 657–687.

Lebowitz, J., Rose, H., and Speer, E., 1989 *Statistical mechanics of the nonlinear Schrödinger equations. II Mean field approximation* J. Stat. Phys. **54**, 17–56.

Lee, J.H., 1989 *Global solvability of the derivative nonlinear Schrödinger equation*, Trans. Amer. Math. Soc. **314**, 107–118.

LeMesurier, B.J., Papanicolaou, G.C., Sulem. C., and Sulem, P.L., 1987 *The focusing singularity of non-linear Schrödinger equation.* Directions in Partial Differential Equations. (M.G. Crandall, P.H. Rabinovitz, R.E. Tuner, eds.), 159–201, Academic Press.

LeMesurier, B.J., Papanicolaou, G.C., Sulem, C., and Sulem, P.L., 1988a *Focusing and multifocusing solutions of the nonlinear Schrödinger equation*, Physica D **31**, 78–102.

LeMesurier, B.J., Papanicolaou, G., Sulem, C., and Sulem, P.L., 1988b *Local structure of the self-focusing singularity of the nonlinear Schrödinger equation*, Physica D **32**, 210–226.

Leontovich, M.A., 1944 Izv. Akad. Nauk SSSR, Ser. Fiz. **8**, 16. (in Russian)

Li, L.H., 1993 *Langmuir turbulence equations with the self-generated magnetic field*, Phys. Fluids B **5**, 350–356.

Lieb, E.H., 1977 *Existence and uniqueness of the minimizing solution of Choquard's nonlinear equation*, Studies in Appl. Math. **57**, 93–105.

Lieb, E.H. and Loss, M., 1997 *Analysis*, Graduate Studies in Mathematics **14**, Amer. Math. Soc. (Providence).

Lin, J.E. and Strauss, W., 1978 *Decay and scattering of solutions of a nonlinear Schrödinger equation*, J. Funct. Anal. **30**, 245–263.

Lin, F.H. and Xin, J.X., 1997 *On the incompressible fluid limit and the vortex motion law of the nonlinear Schrödinger equation*, preprint.

Linares, F. and Ponce, G., 1993 *On the Davey–Stewartson systems*, Ann. Inst. H. Poincaré, Analyse Non Linéaire **10** , 525–548.

Lions, P.L., 1984 *The concentration-compactness principle in the calculus of variations. The locally compact case, Parts I and II* , Ann. Inst. H. Poincaré, Analyse Non Linéaire **1**, 109–145; 223-283.

Litvak, A.G., Petrova, T.A., Sergeev, A.M., and Yunakovskii, A.D., 1983 *A nonlinear wave effect in plasmas*, Sov. J. Plasma Phys. **9**, 287–290.

Lochak, P., 1990 *The blow-up of the supercritical nonlinear Schrödinger equation and a class of spectral problems*, in Nonlinear World, IVth international workshop on nonlinear and turbulent processes in physics (Kiev, 9–22 oct. 1989) (V.G. Bar'yarkhtar, V.M. Chernousenko, N.S. Erokhin, A.G. Silenko, and Z.E. Zakharov, eds.), 196–211, World Scientific.

Lund, F., 1991 *Defect dynamics for the nonlinear Schrödinger equation derived from a variational principle*, Phys. Lett. A **159**, 245–251.

Lushnikov, P.M., 1995 *Dynamic criterion for collapse*, JETP Lett. **62**, 461–467.

Luther, G.G., Newell A.C., and Moloney, J.V., 1994 *The effect of normal dispersion on collapse events*, Physica D **74**, 59–73.

Ma, Y.C., 1978 *The complete solution of the long-wave short-wave resonance equations*, Stud. Appl. Math. **59**, 201–221.

Ma, Y.C. and Ablowitz, M.J., 1981 *The periodic cubic Schrödinger equation*, Stud. Appl. Math. **65**, 113–158.

Maillotte, H., Monneret, J., Barthelemy, A., and Froehly, C. 1994 *Laser beam self-splitting into solitons by optical Kerr nonlinearity*, Opt. Comm. **109**, 265–271.

Malkin, V.M., 1993 *On the analytical theory for stationary self-focusing of radiation*, Physica D **64**, 251–266.

Malkin, V.M., 1991 *Post-collaptical effects in strong Langmuir turbulence*, Physica D **52**, 103–115.

Malkin, V.M., 1997 *Singularity formation for nonlinear Schrödinger equation and self-focusing of laser beams*, in "Partial Differential Equations and their applications", P.C. Greiner, V. Ivrii, L.A. Seco, and C. Sulem, eds., CRM Proceedings and Lecture Notes **12**, 183–198. Amer. Math. Soc.

Malkin, V.M. and Khudik, V.N., 1989 *Point spectrum in the problem of stability of self-similar scalar collapse*, Sov. Phys. JETP **68**, 947–954.

Malkin, V.M., Khudik, V.N., and Fedoruk, V.P. 1990 *Self-similar regimes of supersonic Langmuir collapse*, Sov. Phys. JETP **70**, 1023–1030.

Malkin, V.M. and Shapiro E.G., 1990 *Singular wave collapse*, Sov. Phys. JETP **70**, 102–107.

Martel, Y., 1997 *Blow-up for the nonlinear Schrödinger equation in nonisotropic spaces*, Nonlinear Analysis, Theory, Method, and Appl., **28**, 1903–1908.

Martin, D.U. and Yuen, H.C., 1980 *Quasi recurring energy leakage in the two-dimensional nonlinear Schrödinger equation*, Phys. Fluids **23**, 881–883.

Matsuba, K. and Nozaki, K., 1997 *Derivation of amplitude equations by the renormalisation group method*, Phys. Rev. E **56**, R4926–R9227.

McGoldrick, L.F., 1972 *On the rippling of small waves : a harmonic nonlinear nearly resonant interaction*, J. Fluid Mech. **52**, 725–751.

McKean, H.P., 1995 *Statistical mechanics of nonlinear wave equations (4): cubic Schrödinger*, Comm. Math. Phys. **168**, 479–491.

McKinstrie, C.J. and Russell, D.A. 1988 *Nonlinear focusing of coupled waves*, Phys. Rev. Lett. **61**, 2929–2932.

McLaughlin, D.W., Papanicolaou, G.C., Sulem, C., and Sulem, P.L., 1986 *The focusing singularity of the cubic Schrödinger equation*, Phys. Rev. A **34**, 1200–1210.

McLeod, K. and Serrin, J., 1987 *Nonlinear Schrödinger equation. Uniqueness of positive solutions of $\Delta u + f(u) = 0$ in $\mathbf{R}^n$*, Arch. Rat. Mech. Anal. **99**, 115–145.

Medvedev, M.V. and Diamond, P.H., 1995 *Fluid models for kinetic effects on coherent nonlinear Alfvén waves. I. Fundamental theory*, Phys. Plasmas **3**, 863–873.

Merle, F., 1989 *Limit of the solution of a nonlinear Schrödinger equation at blow-up time*, J. Funct. Anal, **84** 201–214.

Merle, F., 1990 *Construction of solutions with exactly k blow-up points for the Schrödinger equation with critical nonlinearity*, Comm. Math. Phys. **129**, 223–240.

Merle, F., 1992a *On the uniqueness and continuation properties blow-up time of self-similar solutions on nonlinear Schrödinger equation with critical exponent and critical mass*, Comm. Pure Appl. Math. **45**, 203–254.

Merle, F., 1992b *Limit behavior of saturated approximations of nonlinear Schrödinger equation*, Comm. Math. Phys. **149**, 377–414.

Merle, F., 1993 *Determination of blow-up solutions with minimal mass for nonlinear Schrödinger equation with critical power*, Duke Math. J. **69**, 427–454.

Merle, F., 1996a *Lower bounds for the blow-up rate of solutions of the Zakharov equation in dimension two*, Comm. Pure Appl. Math. **49**, 765–794.

Merle, F., 1996b *Non-existence of maximal blow-up solutions of equations $iu_t = -\Delta u - k(\mathbf{x})|u|^{4/N}u$ in $\mathbf{R}^N$*, Ann. Inst. H. Poincaré, Phys. Théor. **64**, 33–85; *Asymptotics for minimal L^2-minimal blow-up solutions of critical nonlinear Schrödinger equation*, Ann. Inst. H. Poincaré, Analyse Non Linéaire **13**, 553–565.

Merle, F., 1996c *Blow-up results of virial type for Zakharov equations*, Comm. Math. Phys. **175**, 433–455.

Merle, F. and Vega, L., 1998 *Compactness at blow-up time for L^2 solutions of the critical nonlinear Schrödinger equation in 2D*, Internat. Math. Res. Notices **8**, 399–425.

Merle, F. and Tsutsumi, Y., 1990 *L^2-concentration of blow-up solutions for the nonlinear Schrödinger equation with critical power nonlinearity*, J. Diff. Eq. **84**, 205–214.

Miller, P.D.and Kamvissis, S., 1998 *On the semiclassical limit of the focusing nonlinear Schrödinger equation*, Phys. Lett. A **247**, 75–86.

Mio, K., Ogino, T., Minami, K., and Takeda, S., 1976 *Modified nonlinear Schrödinger equation for Alfvén waves propagating along the magnetic field in cold plasmas*, J. Phys. Soc. Japan, **41**, 265–271.

Mjølhus, E., 1976 *On the modulational instability of hydromagnetic waves parallel to the magnetic field*, J. Plasma Phys. **16**, 321–334.

Mjølhus, E. and Hada T., 1997 *Soliton theory of quasi-parallel MHD waves*, in "Nonlinear waves and chaos in space plasmas", pp. 121–169, T. Hada and H. Matsumoto, eds., Terra Scientific Publishing Company (Tokyo).

Mollenauer, L.F., Stolen, R.H., and Gordon, J.P., 1980 *Experimental observation of picosecond pulse narrowing and solitons in optical fibers*, Phys. Rev. Lett. **45**, 1095–1098.

Moyua, A., Vargas, A., and Vega, L., 1998 *Restriction theorems and maximal operators related to oscillatory integrals in* $\mathbf{R}^3$, to appear in Duke Math. J.

Morawetz, C. and Strauss, W., 1972 *Decay and scattering properties of solutions of a nonlinear relativistic wave equation*, Comm. Pure Appl. Math **25**, 1–31.

Musher, S.L. and Sturman, B.I., 1975 *Collapse of plane waves near the lower hybrid resonance*, JETP Lett. **22**, 265–267.

Myra, J.R. and Liu, C.S., 1980 *Self-modulation of ion Berstein waves*, Phys. Fluids **23**, 2258–2264.

Myrzakulov, R., Vijayalakshmi, S,, Syzdykova, R.N., and Lakshmanan, N., 1998 *On the simplest(2+1) dimensional integrable spin systems and their equivalent nonlinear Schrödinger equation*, J. Math. Phys. **39**, 2122–2140.

Nakamura, M. and Ozawa T., 1998 *Nonlinear Schrödinger equations in the Sobolev space of critical order*, J. Funct. Anal. **155**, 364–380.

Nawa, H., 1990 *"Mass concentration" phenomenon for the nonlinear Schroödinger equation with the critical power nonlinearity II*, Koda Math. J. 333–348.

Nawa, H., 1992 *Mass concentration phenomenon for the nonlinear Schrödinger equation with critical power nonlinearity*, Funkc. Ekv. **35** , 1–18.

Nawa, H., 1993 *Asymptotic profiles of blow-up solutions of the nonlinear Schrödinger equation* in Singularities in Fluids, Plasmas and Optics, (R. Caflisch and G. Papanicolaou, eds.), NATO ASI series, vol. 404, Kluwer Academic Publishers 221–253.

Nawa, H., 1994 *Asymptotic profiles of blow-up solutions of the nonlinear Schrödinger equation with critical power nonlinearity*, J. Math. Soc. Japan **46**, 557–585.

Nawa, H., 1998a *Asymptotic and limiting profiles of blow-up solutions of the nonlinear Schrödinger equation with critical power*, Comm. Pure Appl. Math., in press.

Nawa, H., 1998b *Two points blow up in solutions of the nonlinear Schrödinger equation with quartic potential on* R, J. Stat. Phys., **91**, 439–458.

Nawa, H. and Ozawa, T., 1992 *Nonlinear scattering with nonlocal interaction*, Comm. Math. Phys. **146**, 259–275.

Nawa, H. and Tsutsumi, M., 1998 *On blow-up of the pseudo-conformally invariant nonlinear Schrödinger equation II*, Comm. Pure Appl. Math. **51**, 373–383.

Neu, J., 1990 *Vortices in the complex scalr fields*, Physica D, **43**, 385–406.

Newell, A.C., 1974 *Envelope equations*, Lectures in Applied Mathematics **15**, 157–163.

Newell, A.C., 1985 *Solitons in Mathematics and Physics*, CBMS-NSF Regional Conference Series in Applied Mathematics **48**, SIAM (Philadelphia, Pennsylvania).

Newell, A.C. and Moloney, J.V., 1992 *Nonlinear Optics*, Addison-Wesley.

Newell, A.C., Rand, D.A., and Russel D., 1988 *Turbulent dissipation rates and the random occurrence of turbulent events*, Phys. Lett. A **132**, 112–123; *Turbulent* *transport and the random occurrence of coherent events*, Physica D **33**, 281–303.

Newman, D.L., Goldman, M.V., and Ergun, R.E., 1994 *Langmuir turbulence in moderately magnetized space plasmas*, Phys. Plasmas **1**, 1691–1699.

Newman, D.L., Robinson, P.A., and Goldman, M.V., 1989 *Field structure of collapsing wave packets in 3D strong Langmuir turbulence*, Phys. Rev. Lett. **62**, 2132–2134.

Newman, D.L., Robinson, P.A., and Goldman, M.V., 1989 *Angular-momentum structure of 3D collapsing wave packets*, Physica D **52**, 49–54.

Nicholson, D.R., 1983 *Introduction to plasma theory*, Wiley (reprint edition 1992 by Krieger).

Nicholson, D.R., Payne, G.L., Sheerin, J.P., and Mei-Mei Shen, 1991 *Does Langmuir collapse produce wave breaking?*, Physica D **52**, 73–77.

Nishinari, K., Abe, K., and Satsuma, J., 1994 *Multidimensional behavior of an electrostatic ion wave in a magnetized plasma*, Phys. Plasmas **1**, 2559–2565.

Nore, C., Abid, M., and Brachet, M., 1994 *Simulation numérique d'écoulements cisaillés tridimensionnels à l'aide de l'équation de Schrödinger non linéaire*, C.R. Acad. Sci. Paris **319**, Série II, 733–735.

Nore, C., Brachet, M., and Fauve, S., 1993 *Numerical study of hydrodynamics using the nonlinear Schrödinger equation*, Physica D **65**, 154–162.

Novikov, S., Manakov, S.V., Pitaevskii, L.P., and Zakharov, V.E., 1984 *Theory of solitons. The inbverse scattering method*, Contemporary Soviet Mathematics, Consultants Bureau (New York, London).

Ogawa, T. and Tsutsumi, Y., 1990 *Blow-up of solutions for the nonlinear Schrödinger equation with quartic potential and periodic boundary condition*, in " Functional analytic methods for partial differential equations", Lecture Notes in Math. **1450**, 256–251, (H. Fujita, T. Ikebe, and S.T. Kutoda, eds.), Springer-Verlag.

Ogawa, T. and Tsutsumi, Y., 1991a *Blow-up of H^1 solution for the one-dimensional nonlinear Schrödinger equation with critical power nonlinearity*, Proc. Amer. Math. Soc. **111**, 487–496.

Ogawa, T. and Tsutsumi, Y., 1991b *Blow-up of H^1 solution for the nonlinear Schrödinger equation*, J. Diff. Eq. **92**, 317–330.

Ohta, M., 1995 *Stability and instability of standing waves for the generalized Davey–Stewartson system*, Diff. Int. Eq., **8**, 1775–1788; *Instability of standing waves for the generalized Davey–Stewartson system*, Ann. Inst. H. Poincaré, Phys. Théor. **62**, 69–80; *Blow-up solutions and strong instability of standing waves for the generalized Davey–Stewartson system in $\mathbf{R}^2$*, Ann. Inst. H. Poincaré, Phys. Théor. **63**, 111–117.

Ono, H., 1975 *Algebraic solitary waves in stratified fluids*, J. phys. Soc. Japan **30**, 1082–1091.

Onsager, L., 1949 *Statistical hydrodynamics*, Nuevo Cimento, Suppl. al vol. VI, Serie IX (2), 279–287.

Ovchinnikov, Y.N. and Sigal, I.M., 1997 *Ginzburg–Landau equation I. Static vortices* in *Partial Differential equations and their applications*, (P. Greiner, V. Ivrii, L. Seco, and C. Sulem eds.), CRM Proc. Lect. Notes, **12**, 199–220.

Ovchinnikov, Y.N. and Sigal, I.M., 1998a *Ginzburg–Landau equation III. Vortex dynamics*, Nonlinearity **11**, 1277-1294 ; *Long-time behaviour of Ginzburg–Landau vortices*, Nonlinearity, **11**, 1295–1309.

Ovchinnikov, Y.N. and Sigal, I.M., 1998b *On the Ginzburg–Landau and related equations*, Séminaire de l'Ecole Polytechnique (Equations aux dérivées partielles), Exposé XXI, 1997–1998.

Ovenden, C.R., Statham, G., and ter Haar, D., 1983 *Strong turbulence of a magnetized plasma. I. The generalized Zakharov equations. II. The ponderomotive force*, Plasma Phys. **25**, 665–679; 681–698.

Ozawa, T., 1992 *Exact blow-up solutions to the Cauchy problem for the Davey–Stewartson systems*, Proc. R. Soc. Lond. A **436**, 345–349.

Ozawa, T., 1996 *On the nonlinear Schrödinger equation of derivative type*, Indiana Univ. Math. J. **45**, 137–163.

Ozawa, T. and Tsutsumi, Y., 1992a *Existence and Smoothing effect of solutions for the Zakharov equations*, Publications of the RIMS **28**, 329–361.

Ozawa, T. and Tsutsumi, Y., 1992b *The nonlinear Schrödinger limit and the initial layer of the Zakharov equations*, Diff. Int. Equ. **5** 721–745.

Ozawa, T. and Tsutsumi, Y., 1993/94 *Global existence and asymptotic behavior of solutions for the Zakharov equations in three space dimensions* Adv. Math. Sci. Appl. Gakkōtosho, Tokyo **3**, 301–334.

Papanicolaou, C.G., McLaughlin, D., and Weinstein, M. 1982 *Focusing singularity for the nonlinear Schroedinger equation*, Lecture Notes in Num. Appl. Anal. **5**, 253–257.

Papanicolaou, G.C., Sulem, C., Sulem, P.L., and Wang, X.P., 1991 *Singular solutions of the Zakharov equations for Langmuir turbulence*, Phys. Fluids B **3**, 969–980.

Papanicolaou, G.C., Sulem, C., Sulem, P.L., and Wang, X.P., 1994 *The focusing singularity of the Davey–Stewartson equations for gravity–capillary surface waves*, Physica D **72**, 61–86.

Passot, T., Sulem, C., and Sulem, P.L., 1994 *Effect of longitudinal modulational on Alfvén wave filamentation*, Phys. Rev. E **50**, 1427–1436.

Passot, T., Sulem, C., and Sulem, P.L., 1996 *Generation of acoustic fronts by focusing wave-packets*, Physica D **94**, 168–187.

Passot, T. and Sulem, P.L., 1993 *Multidimensional modulation of Alfvén waves*, Phys. Rev. E **48**, 2966–2974.

Passot, T. and Sulem, P.L., 1995 *Nonlinear dynamics of dispersive Alfvén waves*, "Small-scale structures in three-dimensional hydrodynamic and magnetohydrodynamic turbulence", (M. Meneguzzi, A. Pouquet, P.L. Sulem, eds.), pp. 405–410, Lecture Notes in Physics, **462**, Springer-Verlag.

Pecher, H., 1997 *Solutions of semi-linear Schrödinger equations*, Ann. Inst. H. Poincaré, Phys. Théor. **67**, 259–296.

Pelinovsky, D., 1995 *Intermediate nonlinear Schrödinger equation for internal waves in a fluid of finite depth*, Phys. Lett. A **197**, 401–406.

Pelinovsky, D., 1998 *Radiative effects to the adiabatic dynamics of envelope-wave solitons*, Physica D, **119**, 301-313.

Pelinovsky, D. and Grimshaw, R., 1995 *A spectral transform for the intermediate nonlinear Schrödinger equation*, J. Math. Phys. **36**, 4203–4219.

Pelinovsky, D. and Grimshaw, R., 1996 *Nonlocal models for envelope waves in a stratified fluid*, Studies Appl. Math. **97**, 369–391.

Pelinovsky, D. and Grimshaw, R.H.J., 1997 *Asymptotic methods in soliton stability theory*, in "Nonlinear instability analysis" (L. Debnath and S.R.

Choudhury), Advances in Fluid Mechanics **12**, 245–312, Computational Mechanics Publications (Southampton, Boston).

Pelletier, G., 1987 *The anisotropic collapse of Langmuir waves*, Physica **27 D**, 187–200.

Pelletier, G., Sol, H., and Asseo, E., 1988 *Magnetized Langmuir wave packets excited by a strong beam-plasma interaction*, Phys. Rev. A **38**, 2552–2563.

Pereira, N.R., Sen, A., and Bers, A., 1978 *Nonlinear development of lower hybrid cones*, Phys. Fluids **21**, 117–120.

Pereira N.R., Sudan, R.N., and Denavit, J., 1977 *Numerical study of two-dimensional generation and collapse of Langmuir solitons*, Phys. Fluids **20**, 936–945.

Pierce, R. and Wayne, C.E., 1995 *On the validity of mean-field amplitude equations for counterpropagating wavetrains*, Nonlinearity **8**, 769–780.

Pillet, C.A. and Wayne, C.E., 1998 *Invariant manifolds for a class of dispersive, Hamiltonian, partial differential equations*, J. Diff. Eq., in press.

Pitaevskii, L.P., 1961 *Vortex lines in an imparfect Bose gas*, Sov. Phys. JETP **13**, 451–454.

Pohožaev, S.I., 1965 *Eingenfunctions of the equation* $\Delta u + \lambda f(u) = 0$, Sov. Math. Doklady **165**, 1408–1411.

Ponce, G. and Saut, J.C., 1998 in preparation. Oral presentation *On the Benney-Roskes system* at the IUTAM Symposium "Three-dimensional aspects of air-sea interaction", May 17-21 1998, Nice, France.

Powell, J.A., Moloney, A.C., and Albanese, R., 1993 *Beam collapse as an explanation for anomalous ocular domage*, J. Opt. Soc. Am. B **10**, 1230–1241.

Rasmussen, J.J. and Rypdal, K., 1986 *Blow-up in nonlinear Schrödinger equation I*, Phys. Scripta **33**, 481–497.

Ranka, J.K., Schirmer, R.W., and Gaeta, A.L., 1996 *Observation of pulse splitting in nonlinear dispersive media*, Phys. Rev. Lett. **77**, 3783–3786.

Rauch, J., 1978 *Local decay of scattering solutions to Schrödinger equations*, Comm. Math. Phys. **61**, 149–168.

Relke, I.V. and Rubenchik, A.M., 1988 *The interaction of high-frequency and low-frequency waves in magnetized plasmas*, J. Plasma Phys. **39**, 369–384.

Remoissenet, M., 1996 *Waves called solitons: Concepts and Experiments*, Springer-Verlag.

Ren, W. and Wang, X.P., 1999 *An iterative grid redistribution method for singular problems in multiple dimensions*, submitted to Comm. Pure Appl. Math.

Robinson, P.A., 1997 *Nonlinear wave collapse and strong turbulence* Rev. Mod. Phys. **69**, 507–573.

Robinson, P.A. and Newman, D.L., 1990 *Two-component model of strong Langmuir turbulence: Scalings, spectra and statistics of Langmuir waves*, Phys. Fluids B **2**, 2999–3016.

Robinson, P.A., Newman, D.L., and Goldman, M.V., 1988 *Three-dimensional strong Langmuir turbulence and wave collapse*, Phys. Rev. Lett. **61**, 702–705.

Rolland, P. and Tagare, S.G., 1982 *Filamentation and collapse of Langmuir waves in a weak magnetic field*, J. Plasma Phys. **28**, 19–36.

Rose, H.A. and Dubois, D.F., 1993 *Initial development of ponderomotive filaments in plasma from intense hot spots produced by a random phase plate*, Phys. Fluids B, **5**, 3337–3356.

Rose, H.A., Dubois, D.F., Russel, D., and Bezzerides, B., 1991 *Experimental signature of localization in Langmuir wave turbulence*, Physica D **52**, 116–128.

Rose H.A. and Weinstein, M.I., 1988 *On the bound states of the nonlinear Schrödinger equation with a linear potential*, Physica D **30**, 207–218.

Rothenberg, J.E., 1992 *Pulse splitting during self focusing in normally dispersive media*, Optics Lett. **17**, 583–585.

Rowlands, G., 1980 *Time recurrent behaviour in the nonlinear Schrödinger equation*, J. Phys. A **13**, 2395–2399.

Rowland, H.L., Lyon, J.G., and Papadopoulos, K., 1981 *Strong Langmuir turbulence in one and two dimensions*, Phys. Rev. Lett. **46**, 346–349.

Rubenchik, A.M., Sagdeev, R.Z., and Zakharov, V.E., 1985 *Collapse versus cavitons*, Comments Plasma Phys. Controlled Fusion, **9**, 183–206.

Rudakov, L.I. and Tsytovich, V.N., 1978 *Strong Langmuir turbulence*, Phys. Reports **40**, 1–73.

Rypdal, K. and Rasmussen, J.J., 1986 *Blow-up in nonlinear Schrödinger equations II*, Phys. Scripta **33**, 498–504.

Rypdal, K. and Rasmussen, J.J., 1989 *Stability of solitary structures in the nonlinear Schrödinger equation*, Phys. Scripta **40**, 192–201.

Rypdal, K., Rasmussen, J.J., and Thomsen, K., 1985 *Singularity structure of wave collapse*, Physica D **16**, 339–359.

Schmidt, G., 1975 *Stability of enevelope solitons*, Phys. Rev. Lett. **34**, 724–726.

Schmitt, A.J., 1988 *The effect of optical smoothing techniques on filamentation in laser plasmas*, Phys. Fluids **31**, 3079–3101.

Schochet, S. and Weinstein, M., 1986 *The nonlinear Schrödinger limit of the Zakharov equations governing Langmuir turbulence*, Comm. Math. Phys., **106**, 569–580.

Schultz, C.L. and Ablowitz, M.J., 1989 *Action-angle variables and trace formula for D-bar limit case of Davey–Stewartson I*, Phys. Lett. A, **135**, 433–437.

Scott, A.C., Chu, F.T., and McLaughlin, D.W., 1973 *The soliton: a new concept in applied science*, Proc. IEEE **61**, 1443–1483.

Segev, M. and Stegeman, G., 1998 *Self-trapping of optical beams: spatial solitons*, Physics Today, August 1998, 42–48.

Segur, H., 1978 *Solitons as Approximate descriptions of physical phenomena*, Rocky Mountain J. Math. **8**, 15–24.

Sigov, S.Yu. and Zakharov, V.E., 1979 *Strong turbulence and its computer simulation*, J. Phys. (Paris) C7, **40**, 63–79.

Shapiro, V.D., Shevshenko, V.I., Solov'vev, G.I., Kalimin, V.P., Bingham, R., Sagdeev, R.Z., Ashour-Abdalla, M., Dawson, J., and Su, J.J., 1993 *Wave collapse at the lower-hybrid resonance*, Phys. Fluids B **5**, 3148–3162.

Shatah, J., 1982 *Global existence of small solutions to nonlinear evolution equations*, J. Diff. Eq. **46**, 409–425.

Shatah, J., 1983 *Stable standing waves of nonlinear Klein-Gordon equations*, Comm. Math. Phys. **91**, 313–317.

Shatah, J., 1985a *Normal forms and quadratic nonlinear Klein-Gordon equations*, Comm. Pure Appl. Math. **38**, 685–696.

Shatah, J., 1985b *Unstable ground state of nonlinear Klein-Gordon equations*, Trans. Amer. Math. Soc. **290**, 701–710.

Shatah, J. and Strauss, W., 1985 *Instability of nonlinear bound states*, Comm. Math. Phys. **100**, 173–190.

Shen, Y.R., 1976 *Recent Advances in nonlinear optics*, Rev. Mod. Phys. **48**, 1–32.

Shukla, P.K., Feix, G., and Stenflo, L., 1988 *Nonlinearity coupled electromagnetic ion-cyclotron and magnetosonic waves in astrophysical plasmas*, Astrophys. Space Sci. **147**, 383–388.

Shul'man, E.I., 1983 *On the integrability of equations of Davey–Stewartson type*, Theor. and Math. Phys. **56**, 720–724.

Shvets, V.F., 1990 *On superstrong wave collapse in lower dimensions*, Phys. Lett. A **150**, 74–78.

Shvets, V.F., Kosmatov, N.E., and LeMesurier, B.J., 1993 *On collapsing solutions of the nonlinear Schrödinger equation in supercritical case*, in " Singularities in fluids, plasmas and optics," R. Caflisch and G.C. Papanicolaou, eds.), Kluwer Acad. Publ., 317–321.

Shvets, V.F. and Zakharov, V.E., 1990 *Computer simulation of wave collapses and wave turbulence*, in "Nonlinear World, IVth international workshop on nonlinear and turbulent processes in physics" (Kiev, 1989), eds. V.G. Bar'yarkhtar, V.M. Chernousenko, N.S. Erokhin, A.G. Silenko, and Z.E. Zakharov, 671–692, World Scientific.

Silberberg, Y., 1990 *Collapse of optical pulses*, Optics Lett. **15**, 1282–1284.

Sjölin, P., 1987 *Regularity of solutions of Schrödinger equations*, Duke Math. J. **55**, 699–715.

Skryabin, D.V. and Firth, W.J., 1998 *Dynamics of self-trapped beams with phase dislocation in saturable Kerr and quadratic nonlinear media*, Phys. Rev. E **58**, 3916–3930.

Smirnov, A.I. and Fraiman, G.M., 1991 *The interaction representation in the self-focusing theory*, Physica D **52**, 2–15.

Soffer, A. and Weinstein, M., 1990 *Multichannel nonlinear scattering for nonintegrable equations*, Comm. Math. Phys. **133**, 119–146.

Soffer, A. and Weinstein, M., 1992 *Multichannel nonlinear scattering, II. The case of anisotropic potentials and data*, J. Diff. Equ. **98**, 376–390.

Soljačić, M., Sears, S., and Segev, M., 1998 *Self-trapping of "necklace" beams in self-focusing Kerr media*, Phys. Rev. Lett. **81**, 4851–4854.

Soto-Crespo, J.M., Weight, E.M., and Akhamediev, N.N. 1991 *Stability of the higher-bound states in a saturable self-focusing medium*, Phys. Rev. A **44**, 636–644.

Soto-Crespo, J.M., Weight, E.M., and Akhamediev, N.N., 1992 *Recurrence and azimuthal-symmetry breaking of a cylindrical Gaussian beam in a saturable self-focusing medium*, Phys. Rev. A **45**, 3168–3175.

Spangler, S., 1989 *Kinetic effects on Alfvén wave nonlinearity: The modified nonlinear wave equation*, Phys. Fluids B **2**, 407–418.

Spiegel, E.A., 1980 *Fluid dynamical form of the linear and nonlinear Schrödinger equation*, Physica **1D**, 236–240.

Staffilani, G., 1997 *Quadratic forms for a 2D semilinear Schrödinger equation*, Duke Math. J. **86**, 79–108; *On the growth of high Sobolev norms of solutions for KdV and Schrödinger equations*, Duke Math. J. **86**, 109–142.

Stein, E.M., 1970 *Singular integrals and differentiability properties of functions*, Princeton Univ. Press.

Strauss, W., 1974 *Nonlinear scattering theory* in "Scattering theory in Mathematical Physics", (J.A. La Vita and J.P Marchard, eds.), Reidel, Dordrecht, 53–78.

Strauss, W., 1977 *Existence of solitary waves in higher dimensions*, Comm. Math. Phys. **55**, 149–162.

Strauss, W., 1981 *Nonlinear scattering theory at low energy*, J. Funct. Anal. **41**, 110–133; *Nonlinear scattering theory at low energy: sequel*, J. Funct. Anal. **43**, 281–293.

Strauss, W., 1989 *Nonlinear wave equations*, Regional Conference Series in Mathematics, **73**, Amer. Math. Soc.

Strichartz, R.S., 1977 *Restriction of Fourier transform to quadradic surfaces and decay of solutions of wave equations*, Duke Math. J. **44**, 705–714.

Stubbe, J., 1989 *Linear stability theory of solitary waves arising from Hamiltonian systems with symmetry*, Portugaliae Math. **46**, 17–32 ; in "Integrable systems and Applications", (M. Balabane, P. Lochak, and C. Sulem, eds.), 329–336, Lecture Notes in Physics, **342**, Springer-Verlag.

Stubbe, J., 1991 *Global solutions and stable ground states of nonlinear Schrödinger equations*, Physica D **48**, 259–272.

Stubbe, J. and Vázquez, L., 1989 *A nonlinear Schrödinger equation with magnetic field effect: solitary waves with internal rotation*, Portugaliae Mathematica **46**, 493-499; *Nonlinear Schödinger equations with magnetic field effect: existence and stability of solitary waves*, Integrable Systems and Applications, (M. Balabane, P. Lochak, and C. Sulem eds.), pp. 336–342, Lecture Notes in Physics **342**, Springer-Verlag.

Sturman, V.I., 1976 *Plasma turbulence neqr the lower-hybrid resonance*, Sov. Phys. JETP **44**, 322–329.

Sulem, C. and Sulem, P.L., 1979 *Quelques resultats de régularité pour les équations de la turbulence de Langmuir*, C.R. Acad. Sc. Paris **289**, A 173–176.

Sulem, C. and Sulem, P.L., 1997 *Focusing nonlinear Schrödinger equation and wave-packet collapse*, Nonlinear Analysis, Theory, Methods, and Applications **30**, 833–844.

Sulem, P.L., Sulem, C., and Bardos, C., 1986 *On the continuous limit for a system of classical spins*, Comm. Math. Phys. **107**, 431–454.

Sulem, P.L., Sulem, C., and Frisch, H., 1983 *Tracing complex singularities with spectral method*, J. Comp. Phys. **50**, 138–161.

Sulem, P.L., Sulem, C., and Patera, A., 1984 *Numerical simulation of singular solutions of the two-dimensional cubic Schrödinger equation*, Comm. Pure Appl. Math. **37**, 755–778.

Sung, L.Y., 1994 *An inverse scattering transform for the Davey–Stewartson II equations, I, II, III*, J. Math. Anal. Appl. **183**, 121-154, 289–325, 477-494.

Sung, L.Y., 1995 *Long-time decay of the solutions of the Davey–Stewartson II equations*, J. Nonlinear Science **5**, 433–452.

Tai, K. Hasegawa, A., and Tomita, A., 1986 *Observation of modulational instability in optical fibers*, Phys. Rev. Lett. **56**, 135–138.

Tajima, T., Goldman, M.V., Leboeuf, J.N., and Dawson, J.M., 1981 *Break-up and reconstitution of Langmuir wave packets*, Phys. Fluids **24**, 182–183.

Tanaka, M., 1982 Nonlinear self-modulation problem of Benjamin–Ono equation, J. Phys. Soc. Japan **51**, 2686–2692.

Talanov, V.I., 1964, Izv. VUZ'ov, Radiofizika **7**, 564 (in Russian).

Talanov, V.I., 1965 *Self-focusing of wave beams in nonlinear media*, Sov. Phys. JETP Lett. **2**, 138–141.

Talanov, V.I., 1966 *Self modelling wave beams in a nonlinear dielectric* , Izvestia VUZ. Radiofizica **9**, 410–412 ; English translation in Radiophys. and Quantum Electronics **9** (1967), 260–261.

Talanov, V.I., 1970 *Focusing of light in cubic media*, JETP Lett. **11**, 199-201.

Thornhill, S.G. and ter Haar, D., 1978 *Langmuir turbulence and modulational instability*, Phys. Reports **43**, 43–99.

Thyragaraja, A., 1979 *Recurrent motions in certain continuum dynamical systems*, Phys. Fluids **22**, 2093–2096.

Tikhonenko, V., Christou, J., and Luther-Davies, B. 1996 *Three-dimensional bright spatial soliton collision and fusion in a saturable nonlinear medium*, Phys. Rev. Lett. **76**, 2698–2701.

Tourigny, Y. and Sanz-Serna, J.M., 1992 *The numerical study of blowup with application to a nonlinear Schrödinger equation*, J. Comp. Phys. **102**, 407–416.

Tsutsumi, M., 1978 *Nonexistence and instability of solutions of nonlinear Schrödinger equations*, Unpublished.

Tsutsumi, M., 1984 *Nonexistence of global solutions to the Cauchy problem for the damped nonlinear Schrödinger equations*, SIAM J. Math. Anal. **15**, 357–366.

Tsutsumi, Y., 1985 *Scattering problem for nonlinear Schrödinger equations*, Ann. Inst. H. Poincaré, Phys. Théor. **43**, 321–347.

Tsutsumi, Y., 1987 L^2 *solutions for the nonlinear Schrödinger equation and nonlinear groups*, Funkcial. Ekvac. **30**, 115–125.

Tsutsumi, Y., 1990 *Rate of L^2-concentration of blow-up solutions for the non-linear Schrödinger equation with critical power*, Nonlinear Analysis, Theory, Methods. and Appl. **15**, 719–724.

Tsutsumi Y. and Yajima, K., 1984   *The asymptotic behavior of nonlinear Schrödinger equations*, Bull. Amer. Math. Soc. **11**, 186–188.

Turitsyn, S.K., 1993 *Nonstable solitons and sharp criteria for wave collapse*, Phys. Rev. E **47**, R13–R16.

Tzvetkov, N., 1998 *Low regularity solutions for a generalized Zakharov system*, Diff. Int. Eq., in press.

Vakhitov, N.G. and Kolokolov, A.A., 1973 *Stationary solutions of the wave equation in a medium with nonlinearity saturation*, Radiophys. and Quantum Electronics **16**, 783–789.

Vega, L., 1987 *Schrödinger equations: pointwise convergence to the initial data*, Proc. Amer. Math. Soc. **102**, 874–878.

Velo, G., 1996 *Mathematical aspects of the nonlinear Schrodinger equation*, in "Nonlinear Klein-Gordon and Schrödinger systems: Theory and Applications", 39–67, (L. Vázquez, L. Strit, and V.M. Pérez-García, eds.), Word Scientific.

Vidal, F. and Johnston, T.W., 1997 *Asymptotic behavior of radially symmetric self-focused beam*, Phys. Rev. E **55**, 3571–3579.

Vlasov, S.N., Petrishchev, V.A., and Talanov, V.I., 1971, *Averaged description of wave beams in linear and nonlinear media (the method of moments)*, Izv. Vys. Uchebn. Zaved. Radiofiz. **14**, 1353 [Radiophys. Quantum Electron. **14**, 1062–1070 (1974)].

Vlasov, S.N., Piskunova, L.V., and Talanov, V.I., 1978 *Structure of the field near a singularity arising from self-focusing in a cubically nonlinear medium*, Sov. Phys. JETP **48**, 808–812.

Vlasov, S.N., Piskunova, L.V., and Talanov, V.I., 1989 *Three-dimensional wave collapse in the nonlinear Schrödinger equation model*, Sov. Phys. JETP **68**, 1125–1128.

Vlasov, S.N. and Talanov, V.I., 1997 *Wave self-focusing*, Publications of the Institute of Applied Physics, Nizhy Novogorov, Russian Academy of Sciences.

Wang, X.P. 1990 *On singular solutions of the nonlinear and Zakharov equations*, Ph. D. thesis, New York University.

Wardrop, M.J. and ter Haar, D., 1979 *The stability of three-dimensional planar Langmuir solitons*, Phys. Scripta **20**, 493–501.

Washimi, H. and Karpman, V.I., 1976 *The ponderomotive force of a high-frequency electromagnetic field in a dispersive medium*, Sov. Phys. JETP **44**, 528–530.

Washimi, H. and Tanuiti, T., 1966 *Propagation of ion-acoustic solitary waves of small amplitude*, Phys. Rev. Lett. **17**, 996–998.

Weinstein, M.I., 1983 *Nonlinear Schrödinger equations and sharp interpolation estimates*, Comm. Math. Phys. **87**, 567–576.

Weinstein, M.I., 1985 *Modulational stability of ground states of nonlinear Schrödinger equations*, SIAM J. Math. Anal., **16** 472–491.

Weinstein, M.I., 1986a *Lyapunov stability of ground states of nonlinear dispersive evolution equations*, Comm. Pure Appl. Math. **39**, 51–68.

Weinstein, M.I., 1986b *On the structure and formation of singularities in solutions to nonlinear dispersive evolution equations*, Comm. Part. Diff. Eq., **11**, 545–565.

Weinstein, M.I., 1987 *Solitary waves of nonlinear dispersive evolution equations with critical powe nonlinearities*, J. Diff. Eq. **69**, 192–203.

Weinstein, M.I., 1989 *The nonlinear Schrödinger equation — Singularity formation, Stability and Dispersion*, in "The connection between Infinite Dimensional and Finite Dimensional Dynamical Systems" (B. Nicolaenco, C. Foias, and R. Temam, eds.), Contemporary Mathematics. **99**, 213–232. Amer, Math. Soc. (Providence).

Weinstein, M.I. and Xin J.X., 1996 *Dynamic stability of the vortex solutions of the Ginzburg–Landau and nonlinear Schrödinger equations*, Comm. Math. Phys. **180**, 389–428.

Weissman, M.A., 1979 *Nonlinear wave packet in the Kelvin–Helmholtz instability*, Phil. Trans. Roy. Soc. Lond. A **290**, 639–685.

Whitham, G.B., 1974 *Linear and nonlinear waves*, Wiley-Interscience.

Willi, O., Afshar-rad, T., Coe, S., and Giulietti, A., 1990 *Study of instabilities in long scale-length plasmas with and without laser-beam-smoothing techniques*, Phys. Fluids B **2**, 1318–1324.

Wong, A.Y. and Cheung, P.Y., 1984 *Three-dimensional self-collapse of Langmuir waves*, Phys. Rev. Lett. **52**, 1222–1225.

Wood, D., 1984 *The self-focusing singularity in the nonlinear Schrödinger equation*, Stud. Appl. Math. **71**, 103–115.

Xin, J.X., 1998 *Modeling light bullets with the two-dimensional Sine-Gordon equation*, preprint.

Yajima, K., 1987 *Existence of solutions for Schrödinger evolution equations*, Comm. Math. Phys. **110**, 415–426.

Yajima, N., 1974 *Stability of envelope soliton*, Prog. Theor. Phys. **52**, 1066–1067.

Yajima, N. and Oikawa, M., 1976 *Formation and interaction of sonic-Langmuir solitons*, Prog. Theor. Phys. **56**, 1719–1739.

Yarmchuk, E.J., Gordon, J.V., and Packard, R.E., 1979 *Observation of stationary vortex arrays in rotating superfluid helium*, Phys. Rev. Lett. **43**, 214–217.

Yuen, H.C. and Ferguson, W.E., 1978a *Benjamin-Feir instability and recurrence in the nonlinear Schrödinger equation*, Phys. Fluids **21**, 1275–1278.

Yuen, H.C. and Ferguson, W.E., 1978b *Fermi–Pasta–Ulam recurrence in the two-space dimensional nonlinear Schrödinger equation*, Phys. Fluids **21**, 2116–2118.

Yuen, H.C. and Lake, B.M., 1975 *Nonlinear deep water waves: theory and experiment*, Phys. Fluids **18**, 956–960.

Yuen, H.C. and Lake, B.M., 1980 *Instabilities of waves on deep water*, Ann. Rev. Fluid. Mech. 303–334.

Zakahrov, V.E., 1968a *Instability of self-focusing of light*, Sov. Phys. JETP **26**, 994–998.

Zakharov, V.E., 1968b *Stability of periodic waves of finite amplitude on the surface of a deep fluid*, J. Appl. Mech. Tech. Phys. **9**, 190–194.

Zakharov, V.E., 1972 *Collapse of Langmuir waves*, Sov. Phys. JETP **35**, 908–914.

Zakharov, V.E., 1974 *The Hamiltonian formalism for waves in nonlinear media having dispersion*, Radiofizica **17**, 431–453 ; English translation in Radiophys. Quantum Electronics **17**, 326–343 (1975).

Zakharov, V.E., 1984 *Collapse and self-focusing of Langmuir waves*, Handbook of plasma physics, (M.N. Rosenbluth and R.Z. Sagdeev, eds.), vol. 2, (A.A. Galeev and R.N. Sudan, eds.), 81–121, Elsevier.

Zakharov, V.E., 1998 *Weakly nonlinear waves on the surface of an ideal finite depth fluid*, Amer. Math. Soc. Transl. **182**, 167–197.

Zakharov, V.E., and Kosmatov, N.E., and Shvets, V.F., 1989 *Ultrastrong wave collapse*, JETP Lett. **49**, 492–495.

Zakharov, V.E. and Kuznetsov, E.A., 1971 *Variational principle and canonical variables in magnetohydrodynamics*, Sov. Phys. Dokl. **15**, 913–914.

Zakharov, V.E. and Kuznetsov, E.A., 1984 *Hamiltonian formalism for systems of hydrodynamic type*, Sov. Sci. Rev., Section C: Math. Phys. Rev. **4** (S.P. Novikov, ed.), 167–220, Harwood Academic Publishers.

Zakharov V.E. and Kuznetsov, E.A., 1986a *Multi-scale expansion in the theory of systems integrable by the inverse scattering transform*, Physica **18 D** , 455–463.

Zakharov, V.E. and Kuznetsov, E.A., 1986b *Quasi-classical theory of three-dimensional wave collapse*, Sov. Phys. JETP **64**, 773–780.

Zakharov, V.E. and Kuznetsov, E.A., 1997 *Hamiltonian formalism for nonlinear waves*, Uspekhi Fizicheskikh Nauk,**167**, 1137–1168.

Zakharov, V.E., Litvak, A.G., Rakova, E.I., Sergeev, A.M, and Shvets, V.F., 1988 *Structural stability of wave collapse in media with a local instability*, Sov. Phys. JETP **67**, 925–927.

Zakharov, V.E., L'vov V.S., and Falkovich, C., 1992 *Kolmogorov spectra of turbulence I. Wave turbulence*, Series in nonlinear dynamics, Springer-Verlag.

Zakharov, V.E., Mastryukov A.F., and Synakh, V.S., 1975 *Dynamics of plasma-wave collapse in a hot plasma*, Sov. J. Plasma Phys. **1**, 339–343.

Zakharov, V.E., Musher, S.L., and Rubenchik, A.M., 1985 *Hamiltonian approach to the description of non-linear plasma phenomena*, Phys. Reports **129**, 285–366.

Zakharov, V.E. and Rubenchik, A.M., 1972 *Nonlinear interaction between high and low frequency waves.* Prikl. Mat. Techn. Fiz. **5**, 84–98 (in Russian).

Zakharov, V.E. and Rubenchik, A.M., 1974 *Instability of waveguides and solitons in nonlinear media,* Sov. Phys. JETP **38**, 494–500.

Zakharov, V.E. and Schulman, E.I., 1991 *Integrability of nonlinear systems and perturbation theory,* in "What is integrability ?", (V.E. Zakharov, ed.), 185–250, Springer Series on Nonlinear Dynamics, Springer-Verlag.

Zakharov, V.E. and Shabat, A.B., 1972 *Exact theory of two-dimensional self-focusing and one-dimensional self-modulation of waves in nonlinear media,* Soviet Phys. JETP **34**, 62–69.

Zakharov, V.E. and Shabat, A.B., 1974 *A scheme for integrating the nonlinear equations of numerical physics by the method of the inverse scattering problem I,* Funct. Anal. Appl. **8**, 226–235.

Zakharov V.E. and Shur, L.N., 1981 *Self-similar regimes of wave collapse,* Sov. Phys. JETP **54**, 1064–1070.

Zakharov, V.E. and Shvets, V.F., 1988 *Nature of wave collapse in the critical case,* JETP Lett. **47**, 275–278.

Zakharov, V.E. and Synakh, V.S., 1976 *The nature of the self-focusing singularity,* Sov. Phys. JETP **41**, 465–468.

Zharova, N.A., Litvak, A.G., Petrova, T.A., Sageev, A.M., and Yunakovskii, A.D., 1986 *Multiple fractionation of wave structures in a nonlinear medium,* JETP Lett. **44**, 13–17.

Zhidkov, P.E., 1991 *On an invariant measure for the nonlinear Schrödinger equations,* Sov. Math. Dokl. **43**, 431–434.

Zhidkov, P.E., 1995 *On invariant measures for some one-dimensional dynamical systems,* Ann. Inst. H. Poincaré, Phys. Théor. **62**, 267–287.

Name Index

Subject Index

Action, 27, 28, 30, 76, 86, 226
adaptive
  Galerkin finite elements, 116, 123
  mesh, 116
admissible
  pair, 45, 46, 48
  solution, 142, 147, 150
asymptotic
  completeness, 59, 60, 63
  self-similarity, 58, 59
  states, 59

Benjamin–Ono equation, v, 201
blowup, vi, viii, 35, 38, 93
  critical, viii
  explicit solution, 36, 56
  in k points, 106, 108
  rate, 103–105, 108, 115
  self-similar, viii
Bose condensate, vii, 25, 66
bound states, 71, 77, 91
Boussinesq equation, 196
breathers, 240

Canonical variables, 29, 289
Chern-Simons gauged planar NLS
  equation, 240

collapse, vii–ix, 14, 20, 38, 93, 164
  arrest of, vii, viii, 303
  critical, ix, 29, 115, 123, 141
  post, 245
  self-similar, 115, 116, 133, 269
  sonic, 270
  strong, viii, 107, 137, 161, 281
  super-strong, 139
  supersonic, 269
  transverse, ix
  weak, 137
collective coordinates, 29
conservation laws, 27, 30, 225
critical
  dimension, viii, 35, 39, 52, 55, 74,
    75, 116, 141
  mass, 56, 104–106
  point, 76–78, 87, 226
criticality, 52
  at the level of H^k, 52, 272
  at the level of L^2, 35

Davey–Stewartson system, viii,
    194, 213, 221
  hyperbolic–elliptic, 228
  blowup, 232, 234
  boundary conditions, 215, 221

Applied Mathematical Sciences

(continued from page ii)

(continued on next page)

Applied Mathematical Sciences

(continued from previous page)